Microorganisms for Sustainability

Volume 28

Series editor
Naveen Kumar Arora, Environmental Microbiology, School for Environmental, Babasaheb Bhimrao Ambedkar University, Lucknow, Uttar Pradesh, India

Microorganisms have been in existence since the origin of life on earth and can survive the most extreme habitats or conditions on earth. Microorganisms are involved in regulating biogeochemical cycles, maintaining plant and animal health, and sustaining the global food chain. Moreover, they play crucial roles in addressing the challenges of climate change and achieving the targets of Sustainable Development Goals (SDGs).

This multidisciplinary book series captures the role of microbes towards building a sustainable world, while encompassing cutting-edge technologies and current needs across various fields such as agriculture sustainability, bioremediation, restoration of degraded habitats and wastelands, and food security. Additionally, this series explores microbial applications in industries, and building their utilization in clean and green energy solutions. Furthermore, themes like microbial secondary metabolites, extremophilic microbes and modern omics, including next generation sequencing and metagenomics, are also covered in this series.

With contributions from researchers across the globe, this series addresses the important call of 'One Planet-One Health-One Future'. It comprises a collection of diverse volumes that provides insights for scientists, young researchers, educators and decision-makers in the government, private sector, and non-governmental organizations, empowering their efforts to achieve the global goals.

The series invites, evaluates, and accepts book proposals to ensure a diverse, inclusive, and evolving program. The final decision regarding acceptance rests with the series editor.

Peer-review

This book series follows a stratified review process. Proposals for individual volumes are reviewed by the series editor and then the editorial board members. On a case-to-case basis, external reviewers are also invited for further evaluation of the book proposal. Review of the chapters is the responsibility of the volume editor(s). A manuscript submission platform has been recently made available to the authors, volume editors and the series editor.

Meenu Saraf • Dweipayan Goswami
Dinesh Kumar Maheshwari
Editors

Climate Change and Soil Microorganisms for Environmental Sustainability

Springer

Editors
Meenu Saraf
Department of Microbiology and Biotechnology
University School of Sciences, Gujarat University
Navrangpura, Ahmedabad, Gujarat, India

Dweipayan Goswami
Department of Microbiology and Biotechnology
University School of Sciences, Gujarat University
Navrangpura, Ahmedabad, Gujarat, India

Dinesh Kumar Maheshwari
Department of Botany and Microbiology
Gurukul Kangri University
Haridwar, Uttarakhand, India

ISSN 2512-1901 ISSN 2512-1898 (electronic)
Microorganisms for Sustainability
ISBN 978-981-96-3424-8 ISBN 978-981-96-3425-5 (eBook)
https://doi.org/10.1007/978-981-96-3425-5

This Springer imprint is published by the registered company Springer Nature Singapore Pte Ltd.
The registered company address is: 152 Beach Road, #21-01/04 Gateway East, Singapore 189721, Singapore

Preface

Soil is a vibrant, dynamic, and communicative web of diverse microorganisms that underpin the vitality and sustainability of our planet. It hosts an extensive network of soil bacteria and fungi that interact symbiotically with plants and is responsible for a multitude of processes essential for ecosystem health and resilience. This intricate world of plant-microbe interactions holds vital implications for addressing the global challenges of climate change, food security, and environmental conservation. The climate is one of the key elements impacting several cycles connected to soil and plant systems, as well as plant production, soil quality, and environmental quality. Due to heightened human activity, the amount of carbon dioxide is rising in the atmosphere. Changing climatic conditions (such as temperature, CO_2, and precipitation) influence plant nutrition in a range of ways, comprising mineralization, decomposition, leaching, and losing nutrients in the soil. Soil carbon sequestration plays an essential function—not only in climate change mitigation but also in plant nutrient accessibility and soil fertility. As a result, there is a significant interest globally in soil carbon capture from atmospheric CO_2 and sequestration in the soil via plants. Boosting agricultural yield is necessary for food security, all over the world. Several unanswered problems remain regarding climate change and its impacts on plant nutrition and global food output, which will be elucidated over time. Aligning with the United Nations Development Programme's (UNDP) Sustainable Development Goals (SDGs), the book acknowledges soil microbes' significant contribution to environmental health and their potential to address critical global issues.

Climate Change and Soil Microorganisms for Environmental Sustainability invites readers into the awe-inspiring world of plant-microbe interplay within the soil, an area of research gaining rapid attention due to its significant role in various environmental aspects. These tiny powerhouses are critical to our planet's health, from nurturing soil fertility to influencing our climate system. The authors elegantly unfold the complexities of microbial communication, their synergistic interactions with plants, and their indispensable role in environmental sustainability. The result is a unique tapestry of knowledge that bridges microbiology, botany, ecology, and environmental sciences. We have tried to decipher the interconnectedness within the microbial world and its association with plants. The authors delve into the intriguing phenomenon of quorum sensing, a bacterial communication method that facilitates collective decision-making. This revelation underscores soil microbes' pivotal role in global sustainability, elucidating a tangible link between the microscopic and macroscopic realms of life on our planet.

Facing the pressing global issue of climate change, the book accentuates the critical role of plant-microbe interactions. We have focused on the often-underestimated impact of these symbiotic relationships on climate regulation, tracing how these interactions influence the carbon cycle and greenhouse gases balance. The implications of global warming on these symbiotic relationships are also explored, with possible far-reaching consequences for Earth's climate system. The role of microbial metabolic activities, particularly those involved in plant-microbe symbiosis, influences nutrient cycling and availability, consequently affecting plant growth and shaping terrestrial ecosystems, which are justified in many chapters of the book. This enlightening read pushes the boundaries of traditional understanding and invites readers to appreciate the complex, unseen world of these microscopic allies. It reinforces the message that our environmental sustainability and climate change efforts must acknowledge and harness the critical role of plant-microbe symbiosis in shaping and sustaining the world we inhabit.

This book will be useful to all knowledge seekers in the area of life sciences and can be read by students, researchers, teachers, and those involved in multidisciplinary subjects.

We wish to acknowledge all the subject experts, who were instrumental and cooperative for making invaluable contributions to make this book a success. Thanks are due to Karan, Priyanka, and Rohit who generously assisted in compilation and completion of this book. We also wish to acknowledge our family members who were always an invisible support. Lastly, we gratefully acknowledge help of Sanchi and Nandhini for their valuable support in making our dream project come true.

Ahmedabad, India — Meenu Saraf
Ahmedabad, India — Dweipayan Goswami
Haridwar, India — Dinesh Kumar Maheshwari

Contents

Editors and Contibutors

About the Editors

Meenu Saraf is the Dean, Faculty of Sciences and Director of University School of Sciences. She is a Professor at the Department of Microbiology and Biotechnology, Gujarat University, India. She obtained her PhD in microbiology from Barkatullah University and has been working at Gujarat University since 1992. Plant-microbe interactions, biofertilizers, biocontrol, biofortification, biofuel production, and nitrogen cycle in agriculture are the main areas of her research. She has been honored with numerous awards, fellowships, and funding from government agencies. Her research work is depicted through her strong publication record.

Dweipayan Goswami is an Assistant Professor at the Department of Microbiology and Biotechnology, Gujarat University, India. He has a PhD in biotechnology from Charotar University of Science and Technology, Gujarat, India. He has published 41+ journal articles, 5+ book chapters, 4 conference proceedings, and 8 reviews in the field of biotechnology. His research interests include plant-microbe interactions, biofertilizers, biocontrol, biofortification, biofuel production, nitrogen cycle, antimicrobial resistance, and metagenomics.

Dinesh Kumar Maheshwari is a well-known botanist and microbiologist, who has done important work on nitrogen fixing bacteria, halophiles, endophytes, and sustainable agriculture. He was the Vice-Chancellor of Gurukul Kangri Vishwavidyalaya, Haridwar, and awardee of MHRD-UGC-BSR Eminent Faculty Fellowship. He has written many books and research articles on Microbiology and related subjects. Prof. Maheshwari got his PhD in botany from Agra University in 1977 and became a Professor in the Department of Botany and Microbiology at Gurukul Kangri University in 1990. Dr. Maheshwari's research interests are the isolation, characterization, and application of useful microorganisms for crop productivity and protection, bioremediation of saline soils, biofertilizers, biopesticides, and biotechnology.

Contributors

Gupta Aishani Department of Psychology, Indraprastha Women's College, University of Delhi, New Delhi, India

Bhavdip Anavadiya Department of Microbiology and Biotechnology, University School of Sciences, Gujarat University, Ahmedabad, Gujarat, India

Sugandha Asthana Department of Biotechnology, Birla Institute of Technology & Science, Pilani, Dubai Campus, Dubai International Academic City, Dubai, UAE

Nitin Baliyan Research and Development, Xenesis, Absolute Biologicals, Gurugram, Haryana, India

Department of Botany and Microbiology, Gurukula Kangri University, Haridwar, Uttarakhand, India

Raksha Bawankar Department of Biotechnology, Silver Oak Institute of Science, Ahmedabad, Gujarat, India

Meera Bharat Vaishnav Anatek Services Pvt Ltd, Mumbai, Maharashtra, India

Kinjal Bhatt Department of Biochemistry – Biotechnology, St. Xavier's College, Ahmedabad, Gujarat, India

Shivangi Bhatt Department of Microbiology and Biotechnology, University School of Science, Gujarat University, Ahmedabad, Gujarat, India

Somnath Chakrobortty Department of Forestry, Mizoram University, Mizoram, India

Rajdeep Chanda Department of Forestry, School of Natural Resource and Management, Mizoram University, Aizawl, Mizoram, India

Balu Ananda Chopade Department of Biotechnology, AKS University, Satna, MP, India

Kamlesh Choure Department of Biotechnology, AKS University, Satna, MP, India

Milan Dabhi Department of Microbiology and Biotechnology, University School of Sciences, Gujarat University, Ahmedabad, Gujarat, India

Khoisnam Vramari Devi Department of Forestry, School of Natural Resource and Management, Mizoram University, Aizawl, Mizoram, India

Ningthoujam Ranjana Devi Department of Forestry, School of Natural Resource and Management, Mizoram University, Aizawl, Mizoram, India

Pinakin Dhandhukia Department of Microbiology, School of Science and Technology, Vanita Vishram Women's University, Surat, Gujarat, India

Sandhya Dhiman Department of Oncology, Wayne State University, Detroit, MI, USA

Avni Divatia Department of Biochemistry – Biotechnology, St. Xavier's College, Ahmedabad, Gujarat, India

Shraddha Gang Department of Health and life Sciences, School of Sciences, Coventry University, Coventry, UK

Sanghdeep Gautam Institute of Management Studies Ghaziabad University Course Campus, Ghaziabad, Uttar Pradesh, India

Bernard R. Glick Department of Biology, University of Waterloo, Waterloo, ON, Canada

Dweipayan Goswami Department of Microbiology and Biotechnology, University School of Science, Gujarat University, Ahmedabad, Gujarat, India

Vandana Gupta Ram Lal Anand College University of Delhi, Delhi, India

M. Tofazzal Islam Institute of Biotechnology and Genetic Engineering (IBGE), Bangabandhu Sheikh Mujibur Rahman Agricultural University, Gazipur, Bangladesh

Chaitanya Kumar Jha Microbiology Department, Gujarat Arts and Science College, Ahmedabad, India

Salome John Department of Microbiology, Ram Lal Anand College, University of Delhi, New Delhi, India

Nishra Joshi Microbiology Department, Gujarat Arts and Science College, Ahmedabad, India

Rushikesh Joshi Department of Biochemistry and Forensic Sciences, School of Sciences, Gujarat University, Ahmedabad, Gujarat, India

Dinesh Kumar Department of Biological Sciences, Birla Institute of Technology and Science (BITS) Pilani, Hyderabad Campus, Telangana, India

Emilynruwaka Lamare Department of Forestry, School of Natural Resource and Management, Mizoram University, Aizawl, Mizoram, India

Dinesh Kumar Maheshwari Department of Botany and Microbiology, Gurukula Kangri University, Haridwar, Uttarakhand, India

Sudeshna Menon Department of Biochemistry – Biotechnology, St. Xavier's College, Ahmedabad, Gujarat, India

Rakeshkumar R. Panchal Department of Microbiology and Biotechnology, Gujarat University, Ahmedabad, Gujarat, India

Jaimin Pandya Department of Microbiology and Biotechnology, Gujarat University, Ahmedabad, Gujarat, India

Dimple Pardhi Department of Microbiology and Biotechnology, School of Sciences, Gujarat University, Ahmedabad, Gujarat, India

Aanal Patani Department of Microbiology and Biotechnology, Gujarat University, Ahmedabad, Gujarat, India

Drashti Patel Department of Microbiology and Biotechnology, School of Sciences, Gujarat University, Ahmedabad, Gujarat, India

Rohit Patel Department of Microbiology & Biotechnology, University School of Sciences, Gujarat University, Ahmedabad, Gujarat, India

Karan Prajapati Department of Microbiology and Biotechnology, University School of Sciences, Gujarat University, Ahmedabad, Gujarat, India

Kamal Prasad Research and Development, Xenesis, Absolute Biologicals, Gurugram, Haryana, India

Piyush Kant Rai Department of Biotechnology, AKS University, Satna, MP, India

Kiransinh N. Rajput Department of Microbiology and Biotechnology, Gujarat University, Ahmedabad, Gujarat, India

Khushbu Rathod Department of Biotechnology, P D Patel Institute of Applied Sciences, Charotar University of Science and Technology, Gujarat, India

Tejalba Rathod Department of Microbiology and Biotechnology, School of Sciences, Gujarat University, Ahmedabad, Gujarat, India

Vikram H. Raval Department of Microbiology and Biotechnology, Gujarat University, Ahmedabad, Gujarat, India

Janki Ruparelia Microbiology Department, Gujarat Arts and Science College, Ahmedabad, India

Lovelesh Singh Saini Department of Life Sciences, Parul Institute of Applied Sciences, Parul University, Vadodara, Gujarat, India

Patra Sandeep Department of Microbiology, Ram Lal Anand College, University of Delhi, New Delhi, India

Meenu Saraf Department of Microbiology and Biotechnology, University School of Sciences, Gujarat University, Navrangpura, Ahmedabad, Gujarat, India

Sheetal Sharma Construction Testing Solution, Leicester, UK

Ngangbam Somen Singh Department of Forestry, Mizoram University, Mizoram, India

Salam Suresh Singh Department of Forestry, School of Natural Resource and Management, Mizoram University, Aizawl, Mizoram, India

Anjali Solanki Department of Microbiology and Biotechnology, School of Sciences, Gujarat University, Ahmedabad, Gujarat, India

Neeru Sood Department of Biotechnology, Birla Institute of Technology & Science, Pilani, Dubai Campus, Dubai International Academic City, Dubai, UAE

Janki N. Thakker Department of Biotechnology, P D Patel Institute of Applied Sciences, Charotar University of Science and Technology, Gujarat, India

Jahanvi Thakor Department of Microbiology and Biotechnology, Gujarat University, Ahmedabad, Gujarat, India

P. K. Toushif Department of Forestry, Mizoram University, Mizoram, India

Shri Kant Tripathi Department of Forestry, School of Natural Resource and Management, Mizoram University, Aizawl, Mizoram, India

Ayushi Tyagi Department of Forestry, Mizoram University, Mizoram, India

Keshav Kumar Upadhyay Department of Forestry, School of Natural Resource and Management, Mizoram University, Aizawl, Mizoram, India

Sebastian Vadakan Department of Biochemistry – Biotechnology, St. Xavier's College, Ahmedabad, Gujarat, India

Apoorva Vaghela Department of Microbiology and Biotechnology, Gujarat University, Ahmedabad, Gujarat, India

Himanshu Vasava Department of Microbiology and Biotechnology, Gujarat University, Ahmedabad, Gujarat, India

Nidhi Verma Department of Microbiology, Ram Lal Anand College, University of Delhi, New Delhi, India

Charu Vyas Department of Biotechnology, Faculty of Life Science and Technology, AKS University, Satna, MP, India

Ashwini A. Waoo Department of Biotechnology, Faculty of Life Science and Technology, AKS University, Satna, MP, India

Ashish Warghane School of Applied Sciences and Technology (SAST), Gujarat Technological University, Ahmedabad, Gujarat, India

Part I

Foundations of Soil Microbial Communities

A Comprehensive Introduction to Soil Microbial Communities and Microbiomes

1

Piyush Kant Rai, Kamlesh Choure, and Balu Ananda Chopade

Abstract

Background: The soil microbiome consisting of bacteria, fungi, and archaea is essential for maintaining life and balancing ecosystems. Studying this microbiome is necessary in the aspect of threats such as weather change and soil degradation.

Scope: Culture-independent methods, for example, 16S rRNA gene study, metagenomics, and metatranscriptomics, supplement traditional methods to provide detailed insights into microbial diversity and function. High-throughput sequencing, metatranscriptomics, and metaproteomics contribute to advancements in studying microbiome and microbial communities.

Conclusion: Advancements in computer technology and bioinformatics tools like DADA2 and specialized pipelines are essential for processing biological data. Understanding soil microbiomes aids in developing sustainable practices for agriculture and forest ecosystems amid environmental challenges.

Keywords

Soil microbiome · Biofertilizers · Microbial diversity · Plant microbiome · Metagenomics · Bioinformatics

1.1 Introduction

In the complex web of Earth's ecosystems, the balance required for our ecosystem is maintained by the significant soil microbiome. These microscopic organisms help to facilitate plant growth and the cycling of essential elements, contributing to the

P. K. Rai · K. Choure · B. A. Chopade (✉)
Department of Biotechnology, AKS University, Satna, Madhya Pradesh, India

M. Saraf et al. (eds.), *Climate Change and Soil Microorganisms for Environmental Sustainability*, Microorganisms for Sustainability 28,
https://doi.org/10.1007/978-981-96-3425-5_1

robustness and vitality of various habitats, from the forest's canopy to the rhizosphere. However, this balance is threatened by continuously changing climate patterns, soil degradation, and unsustainable land management practices, jeopardizing the essential functions performed by the soil microbial community. For example, in "forest ecosystems," microorganisms thrive in habitats such as foliage, wood, bark, and the rhizosphere (Sahu et al., 2017; Khangura et al., 2023). Owing to the upsurge in population globally, the demand for food has increased, leading to increased use of cost-effective commercial fertilizers and pesticides in conventional agriculture. Unfortunately, this reliance on these chemicals has resulted in the loss of biodiversity, negative impact on air and water, and decomposition of soil and human health. However, managing all the integrated nutrient systems is crucial to counteract these challenges and foster a healthy environment, with microbial inoculants playing a vital role.

Importantly, microbial inoculants also known as "biofertilizers" are inhabited by living microorganisms that occupy the rhizosphere (Roberts & Mattoo, 2018). All these microorganisms enhance plant growth by transforming soil elements into soluble forms, including phosphate solubilization and nitrogen fixation. Some of these beneficial microbial inoculants (plant-growth-promoting rhizobacteria) include N_2 fixers, phosphate and sulfur, and zinc solubilizers. Such soil "microbiotas" contribute toward improving microflora, soil vitality, and overall vegetative growth as well; they remain essential for agriculture and play a pivotal role in expanding forest ecosystems, which cover over 40 million km^2, approximately 30% of the world's terrestrial area (Shahwar et al., 2023; Kumar et al., 2022). In recent years, ecological research, in the spotlight, has historically centered on plant succession, leaving the intricate contributions of microbial communities in these processes overshadowed and underappreciated. As ecosystems advance through stages of gradualness, diversity burgeons, and nutrients become intricately interwoven, influencing the buildup of organic detritus in soil and the closure of mineral cycles.

The complex interplay between plant roots and the microbiome involves various factors, including host genotypes, developmental stages, temperature, soil pH, seasonal variations, and rhizosphere deposits. The intricate relationship between microbial diversity and biomass in various ecosystems is usually assessed by soil carbon levels (Ciccazzo et al., 2016; Li et al., 2023). The stress gradient hypothesis suggests that in regions with low carbon levels, like arid soils, positive species associations, including facilitation and specialization, take precedence over competition, which is the primary factor in areas with higher carbon content. This hypothesis is supported by studies showing that in arid environments, plant species tend to benefit from interactions with other species rather than competing for limited resources. These positive species associations help to increase overall biodiversity and ecosystem resilience in these challenging environments. Some of the classical microbiological techniques are limited in capturing the complexity of soil microorganisms, leaving around 80–90% of remaining uncultured microorganisms (Hammarlund & Harcombe, 2019). Microbiome research nowadays has shifted its trend toward cultivation-independent techniques, which provide a broader perspective. This approach reveals that a given soil is a mosaic of life from all three domains of the biosphere—bacteria, eukarya, and archaea. Despite the challenges in exploring and managing soil microbiomes, the prospect of unraveling their secrets holds great relevance to harnessing ecosystem services and unlocking the prospect of soil microbiomes' capability under diverse conditions. Exploring the microbial realm offers a promising opportunity to understand the earth's hidden stewards better (Bay et al., 2020). Soil microbial diversity and microbiome research, including in silico and isolation techniques, are the focus of this chapter.

1.2 The "Soil Microbial Communities" and "Microbiomes in Soil"

In order to maintain biodiversity and control the flow of materials, above-ground ecosystem services provided by soil microbial communities are crucial (Suman et al., 2022). Soil microbial communities play a vital role in maintaining biodiversity and regulating the flow of materials, making above-ground ecosystem services vital. Microbes have a substantial impact on the nutrient content of plant nutriment as they mineralize degradable organic matter. A healthy civilization is directly related to healthy soils (Kumar et al., 2022). These microorganisms significantly impact vegetation growth and well-being. The rhizosphere microbial is an active group of microbes associated with plants and soil, including fungi, archaea, and bacteria, which can have both advantageous and detrimental effects on plant growth and crop yield. Approximately, the positive effects include enhancing plant growth and nutrient uptake, while negative effects involve causing plant diseases and reducing crop yield. Some factors such as pH, depth, salinity, rainfall, organic matter, temperature, carbon dioxide concentration, and UV radiation as well as porosity, oxygen, and others. There are many examples of plant growth-promoting (PGP) microbes in the soil, among some bacteria such as *Bacillus*, *Pseudomonas*, *Rhizobacteria*, and *Azotobacter* can enhance plant development by colonizing the plant root, producing phytohormones by increasing nutrient availability in soil, and improving resistance against pathogens. Some of the major fungi kingdom members, like *Aspergillus*, *Penicillium*, *Fusarium*, and *Trichoderma*, can also promote plant growth by increasing soil nutrient content, producing plant hormones, and improving plant resistance to different stressors. Additionally, some microbes like cyanobacteria can harness atmospheric nitrogen, transforming it into organic forms and making it usable to plants. An exciting new approach to improving crop yields and reducing environmental impact is the use of bacteria (Suman et al., 2022). These microorganisms can augment plant development and protect plants from pathogens and abiotic stresses by colonizing the plant root, producing phytohormones, accumulating the accessibility of soil nutrients, and improving plant resistance against biological and physical challenges. Distinct microbial assemblages in the soil can suppress soil-borne diseases through nutrient competition, antagonism, hyperparasitism, and inducing plant resistance, influencing microbial function, diversity, and activity in the rhizospheric soil, acting as the initial line of protection against pathogens. Furthermore, during pathogen invasion, beneficial microorganisms may be recruited to the plant rhizosphere, which can boost disease resistance and promote disease management. Therefore, soil microbial communities, especially in the rhizosphere, significantly promote plant health and disease resistance. Soil microbiome research has significant implications for sustainable agriculture by leading to the development of more efficient and environmentally friendly agricultural practices. These practices include biofertilization, biocontrol agents, plant growth promotion, abiotic stress tolerance, and carbon sequestration. To realize the full potential of soil microbiome research for regenerative farming, it is vital to develop a thoughtful of the intricate interplay among "soil microbiota, plants, and the environment, enabling the optimization of agricultural practices and the development of more sustainable and resilient agricultural systems".

1.2.1 Role in Agriculture

Now farmers can use "soil microbiome" research to improve soil health by implementing practices that promote beneficial microbial communities in the soil, such as reducing tillage, crop rotation, cover cropping, organic amendments, biofertilizers, and biocontrol agents (Fig. 1.1). Incorporating soil microbiome research into sustainable agriculture policies can yield several potential benefits: increased crop yields, reduced environmental impact, improved soil health, resilience, carbon sequestration, and reduced costs. Incorporating soil microbiome research into

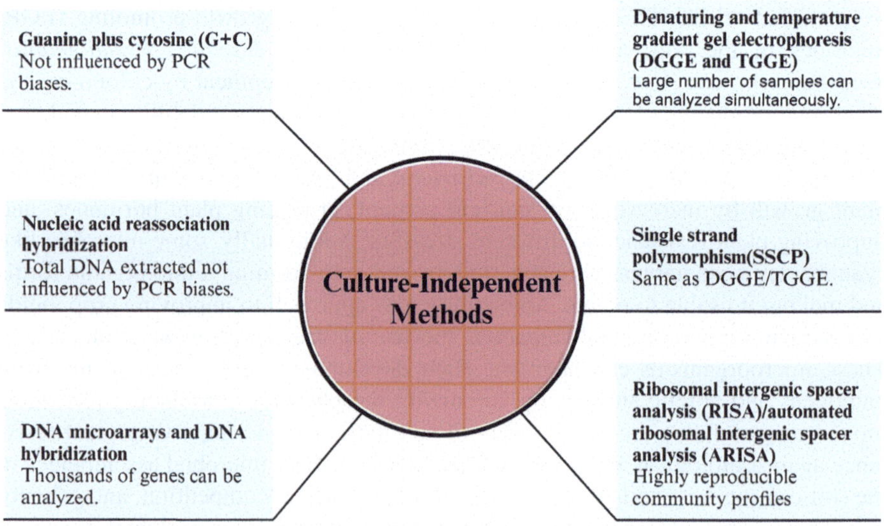

Fig. 1.1 Culture-independent methods

sustainable agriculture policies fosters more efficient and environmentally friendly practices, enhancing soil health, increasing crop yields, and reducing environmental impact.

To measure the impact of "soil microbiome" research on crops, farmers can employ various approaches. Moreover, monitoring crop growth and yield allows for a comparison between fields, and implementing soil microbiomes with focused management practices but those that are not should be determining the effectiveness of these methods in improving crop performance. Additionally, analyzing soil microbial communities using molecular techniques such as next-generation sequencing helps identify specific microbial communities contributing to enhanced crop performance (Wydro, 2022). Furthermore, assessing nutrient accessibility in the soil, monitoring plant health, and evaluating soil structure and aggregation provide insights into the effectiveness of soil microbiome-focused management practices.

1.2.2 Role in Greenhouse Gas Emission

Another aspect is analyzing greenhouse gas emissions to determine if such practices reduce emissions and improve carbon sequestration (Udara Willhelm Abeydeera et al., 2019). Monitoring soil microbial biomass can indicate whether these practices will increase overall microbial abundance. Additionally, evaluating the impact of specific microbial consortia and their contributions to crop performance helps in implementing targeted management practices for their enhancement. It is crucial to evaluate these practices' environmental and economic impacts to inform policy and research recommendations supporting sustainable agriculture.

Incorporating soil microbiome research into sustainable agriculture practices offers numerous benefits with some potential risks. These potential risks include unintended consequences, limited understanding, variability, economic and technical challenges, knowledge gaps, regulatory and policy challenges, social and cultural barriers, and the potential for misuse. To minimize these risks, investing in research and development shall build a strong evidence base by developing supportive policies and regulations while engaging with farmers and stakeholders with essential steps. This ensures an understanding of both the benefits and risks of soil microbiome-focused management practices, enabling effective implementation.

1.2.3 Role in Sustainable Agriculture

Incorporating "soil microbiome" research into sustainable agriculture policies can lead to numerous benefits. These benefits include understanding the contribution of soil microbes to nutrient cycling, soil composition, and pest and disease control. This can help develop practices promoting soil vitality and reducing the need for chemical inputs. Ultimately enhancing soil quality and reducing environmental impact. However, managing soil microbiota can enhance nutrient availability and promote plant development, leading to increased crop yields. Additionally, promoting beneficial microbial communities in the soil can improve crop resilience to environmental stressors such as drought and heat, contributing to enhanced soil biodiversity and carbon sequestration.

Some examples of sustainable agriculture policies that incorporate "soil microbiome" research include the "European Union's Green Deal," which aims to reduce chemical inputs and promote practices enhancing soil health. Similarly, the "United Nations' Sustainable Development Goals" emphasize "sustainable agriculture and soil health". The U.S. Department of "Agriculture's National Institute of Food and Agriculture" supports investigations on "microbiomes," while the "Global Research Alliance" on Agricultural Greenhouse Gases promotes research on soil microbiomes and their role in carbon sequestration. The collaboration between researchers, farmers, and industry stakeholders is a crucial step in developing innovative solutions supporting global food security and environmental sustainability.

Thus, soil microbiome research can help to address climate change in agriculture by enhancing carbon sequestration, resilience to environmental stress, and informing climate-friendly soil interventions. Moreover, it also aids in understanding the impact of climate change on soil microbes, essential for predicting broader implications on agricultural ecosystems.

These incorporations in soil microbiome research into sustainable agriculture policies can yield economic benefits such as reduced costs, increased crop yields, and enhanced soil fertility. This leads to improved market demand for sustainable products and overall environmental sustainability by highlighting the importance of integrating soil microbiome research into agricultural practices and policies for a more sustainable and resilient agricultural system.

All these soil microbiomes vary across different biomes owing to a range of aspects, including climate, soil form, and vegetation. The research has shown that a few abundant species dominate soil bacterial communities, and their abundance is linked to unique species whose ecological functions and biogeography are still mainly a mystery. In soil bacterial community analysis, findings reliably demonstrate that a minority of bacterial species are widespread across numerous biomes, while the majority are considered rare, with low relative richness and restricted distribution. The dominant role of soil moisture has been emphasized in shaping bacterial uniqueness, with drier soils suppressing the richness of bacterial species. Additionally, soil's aqueous-phase content and connectivity have been found to strongly influence bacterial variety for each soil kind and rainwater. Furthermore, long-term soil management practices have been shown to influence the rhizosphere microbial community, indicating that dissimilar soil administration practices can change the soil microbiome. Therefore, the variation in soil microbiomes across different biomes is influenced by a mixture of environmental aspects and soil management practices.

1.2.4 Biomes in Microbiomes

Understanding the variation in soil microbiomes across different biomes is essential for developing eco-friendly agricultural practices that promote soil health and ecosystem resilience. It is important to note that soil microbiomes play a crucial role in soil carbon sequestration by removing carbon dioxide from the air and storing it in underground carbon reservoirs. Because of this, the rates of biochemical or physical maintenance of organic carbon supplies are greatly controlled. Ultimately, these microorganisms facilitate the creation of durable soil aggregates, shielding soil carbon reservoirs from degradation and erosion. Simultaneously, microorganisms transform soil organic carbon into soluble or gaseous forms of carbon via a process called "Mineralization." This may possibly entail the dynamics of soil carbon, which are greatly impacted by soil microbes. However, it is likely to expedite the progression of soil carbon sequestration by altering the soil microbiome to promote certain groups of organisms. There is a possibility of improving the efficiency of microbes, which can eventually result in increased carbon storage in soils. However, it is imperative to conduct more research to comprehend the ways to accomplish this. Therefore, understanding the role of soil microbiomes in soil carbon sequestration is essential for developing sustainable agricultural practices that promote soil health and ecosystem resilience.

1.2.5 The Factors Influencing Soil Microbiomes

They may vary across different soil types due to a range of factors, including soil texture, pH, nutrient availability, and organic matter content. Understanding these factors through research has exposed that soil microbial groups exhibit distinct configurations and functional capabilities in different soil types. Additionally, bacterial diversity has been found to positively correlate with clay content, possibly due to the higher water-holding capacity of clay soils, which supports a more diverse microbial community. Furthermore, soil pH has been shown to influence microbial community composition, with acid soils supporting different microbial communities than alkaline soils. Soil type also influences the richness and diversity of specific microbial groups, such as arbuscular mycorrhizal fungi (AMF), which are more profuse in sandy soils than in clay soils. Furthermore, soil administration practices, such as crop rotation, fertilization, and tillage, can significantly alter soil microbiomes, leading to differences in microbial communities across soil types. Therefore, understanding the variation in soil microbiomes across different soil types is essential for developing sustainable agricultural practices that promote soil health and ecosystem resilience.

1.2.6 Impact on Soil Erosion

Soil microbiomes have significant effects on soil erosion. Research has shown that erosion can cause a decrease in soil microbial diversity, network complexity, and multifunctionality, negatively impacting soil health and ecosystem resilience. The erosion has been initiated to increase the relative abundances of certain bacterial families involved in nitrogen cycling, such as *Acetobacteraceae* and *Beijerinckiaceae*, which can alter the soil microbiome's characteristics and functionality. Additionally, erosion has been shown to reduce the network complexity of soil microbiomes, leading to fewer microbial taxa and associations among microbial taxa. This indicates the variations in the soil microbiome which are strongly associated with erosion-induced changes in soil multifunctionality.

Overall, soil microbiomes play a central part in sustaining soil vigor and mitigating the negative impacts of erosion on soil functionality and ecosystem services (Suman et al., 2022). Understanding the interaction and belongings of soil microbiomes on soil erosion is essential for emerging sustainable land organization practices and mitigating the detrimental effects of erosion on soil well-being and agricultural efficiency (Gowtham et al., 2022). These soil microbiomes play a critical part in soil nutrient availability. Soil microbes are complicated in nutrient recycling through the decomposition of organic matter and ecosystem functioning (Moses et al., 2013). These communities and microbiomes in soil are essential for providing ecosystem services that occur above the earth, such as maintaining biodiversity and controlling material exchange (Suman et al., 2022). Furthermore, the makeup of a specific microbiome is determined by a multitude of elements, such as environmental conditions, soil physical qualities, nutrient availability, and plant type. A variety of aspects contribute to the composition of a specific microbiome, including environmental circumstances, physical characteristics of the soil, nutrient availability, and the type of plant (Kumar et al., 2022).

1.3 Approaches to Investigating the Diversity of Microorganisms in Soil

Soil microbiological diversity encompasses the range of microorganisms found in the soil, such as archaea, bacteria, fungi, and other microbes. Such organisms play a vital role in a range of soil activities, including nitrogen cycling, breakdown of organic debris, and maintaining plant health. Soil microbial diversity can be understood via numerous approaches to gain insights into their makeup and function (Bay et al., 2020). Discovering a variety of soil microbes has been a focal point in various studies, each contributing unique insights into the methodologies and challenges associated with unraveling the intricacies of microbial populations in soil (Dubey et al., 2020). Several methods have been proposed by the researcher over the years, which have been discussed in the following sections.

1.3.1 Culture-Specific Methods

Culture-specific methods rely on the tillage of microorganisms in the laboratory. This approach involves the isolation and growth of microbes on specific media, allowing for the identification and characterization of individual microbial species. Culture-dependent methods have been used for decades and provide valuable information about microorganisms' biochemical, morphological, and physiological characteristics (Bonnet et al., 2019; Rafiq et al., 2023).

1.3.2 Plate Count Techniques

The traditional method to assess microbial diversity in soil is by using discriminative plating and immediate assessment of viable counts (Fig. 1.1). Although these cost-effective methods provide insights into the population's active, heterotrophic segment, they have limitations (Lee et al., 2021). For instance, it can be challenging to dislodge bacteria or spores from soil residues or biofilms, and the choice of growth medium and conditions can also impact the results. The existing techniques cannot cultivate a broad spectrum of bacterial and fungal species. The plate count method is a traditional approach to measuring microbial diversity and involves cultivating viable and culturable microbes on solid nutrient media (Davey, 2011). This technique is primarily employed to investigate morphological attributes, including the configuration, dimensions, color, and production of pigments within colonies (Poddar et al., 2021). However, it is crucial to acknowledge and address the inherent limitations of plate count techniques to ensure a more accurate and comprehensive understanding of microbial variety in soil ecosystems.

1.3.3 Culture-Independent Methods

The soil is a composite and dynamic ecosystem that houses many microbes, such as archaea, bacteria, and fungi. These tiny organisms show essential parts in soil health by engaging in critical developments such as organic matter decomposition, nutrient cycling, and facilitating plant growth. Traditional methods, such as culture-dependent approaches, fall short of comprehending microbial diversity, crucial for effective soil resource management and productivity enhancement. This limitation arises from the non-cultivability of many soil microorganisms, making it challenging to estimate their abundance and diversity accurately (Creamer et al., 2022). To overcome these challenges, culture-independent techniques have emerged, leading to a revolution in the exploration of microbial populations in soil. One of these methods involves analyzing the 16S rRNA gene, which is universally present in archaea and bacteria (Fukuda et al., 2016). By amplifying and sequencing this gene from soil samples, researchers can identify microbial taxa, discover new groups, and gain insights into their ecological roles. Another culture-independent method is metagenomics, which entails sequencing all DNA in a sample, offering a comprehensive view of microbial genomes and their functional roles in the soil ecosystem. Metatranscriptomics complements these efforts by sequencing RNA transcripts, revealing actively expressed genes in microbial communities (Handelsman, 2004; Sabale et al., 2019). Meta-proteomics identifies proteins microbial communities produce, deepening our perception of active metabolic pathways. Stable isotope probing (SIP) involves adding isotopically labeled substrates to soil samples, tracking them through the soil food web, and identifying microorganisms actively involved in substrate consumption. These culture-independent methods have transformed our conception of soil microbial communities, unveiling hidden ecological intricacies, metabolic pathways, and symbiotic interactions. As we delve deeper into these innovative techniques, our knowledge of soil microbiomes expands, empowering us to preserve soil health and optimize productivity (Jehmlich et al., 2016). Studies that are culture independent, such as nucleic acid analyses, are crucial for research. This can include the study of entire genomes or specific genes like 16S rRNA. Various procedures serve for both partial and whole community analyses. One such technique is the PCR-based method, which uses total RNA/DNA extracted from ecological samples for partial community analysis, enabling the characterization of microorganisms. Integrated community study involves non-PCR techniques like G+C content estimation, whole-genome sequencing, and hybridization, providing a comprehensive view of genetic diversity. Random amplified polymorphic DNA (RAPD) is a rapid process for genetic diversity analysis that uses short primers to amplify multiple locations on the microbial genome, producing characteristic amplicons (Clarridge, 2004). Amplified fragment length polymorphism (AFLP) screens many genomic regions simultaneously, rapidly assessing genetic differences (Paun & Schönswetter, 2012). Denaturing/temperature gradient gel electrophoresis (DGGE/TGGE) separates PCR amplicons with different nucleotide sequences, allowing for the identification of genetic diversity (Kovac et al., 2022). Analyzing microbial communities based on fatty acid properties is probable with the usage of the biochemical method known as fatty acid methyl ester (FAME) analysis. Fatty acids, which constitute a continuous proportion of cellular biomass, allow differentiation of major taxonomic groups. This method, not reliant on culturing microorganisms, has been instrumental in studying microbes' overall composition and population deviations due to environmental factors (Lewe et al., 2021). Among the techniques mentioned above, DGGE and FAME were the most reliable soil microbial diversity analysis methods. Some other lesser techniques are shown in Figs. 1.1 and 1.2.

1.4 Integrated Metagenomics

Integrated metagenomics analysis is a comprehensive methodology investigating microbial communities within a given sample. This is done using read-based, assembly-based, and detection-based approaches. This technique helps gain a more thorough and nuanced knowledge of the microbial community's diversity, variety, functional potential, and the presence of particular genes (Nam et al., 2023). In

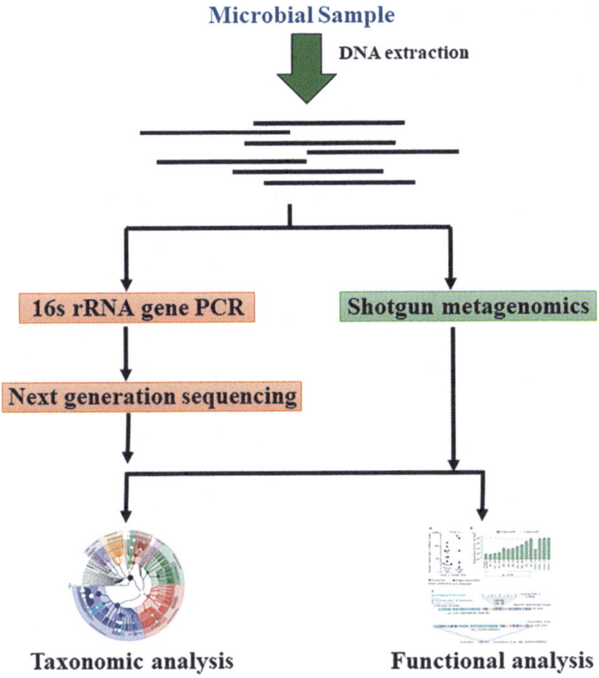

Fig. 1.2 Pipeline for shotgun metagenomics

read-based analysis, researchers look at unassembled reads to acquire quantitative information. By comparing these readings to reference databases, they may determine the quantity, distribution, and diversity of individual organisms in the sample (Navgire et al., 2022). Furthermore, read-based analysis offers insights into the functional capabilities and metabolic pathways of the microbial community. Assembly-based analysis seeks to assemble reads from one or more samples in order to rebuild the genomes of particular organisms or species found within the sample. Computational tools and algorithms are critical to this process. After genomes have been constructed, further analysis is undertaken to discover and annotate genes, anticipate their roles, and comprehend the microbial community's overall functional potential (Lapidus & Korobeynikov, 2021). Detection-based approaches are used to determine the presence of specific genes or organisms in samples. These approaches use particular molecular probes or biomarkers to identify the target genes or organisms present properly. They are widely used in medical diagnostics and environmental monitoring. Integrating these techniques entails integrating read-based insights on composition and variety with assembly-based genomic reconstructions (Califf, 2018). Researchers compare taxonomic information from reads to functional potential gained from genomic assemblies. Detection-based methods are then utilized to test and confirm the existence of critical genes or organisms discovered using read and assembly procedures (Tamames et al., 2019; Yang et al., 2021). Metagenomic sequencing analysis can be challenging due to the complex nature of the data. The method requires higher coverage than 16S rRNA gene sequencing. Additionally, the cost per sample varies based on sequencing depth, where more profound sequencing results in more data and analysis options (Durazzi et al., 2021). However, a new method called shallow shotgun sequencing combines multiple samples into a single run and uses a modified protocol to reduce costs. This technique targets a specific area of the bacterial/archaeal 16S rRNA gene to provide information about the microbial composition at the genus or species level, which is helpful for bacterial community profiling. 16S rRNA sequencing can identify bacterial or archaeal species and analyze microbial diversity within a sample (La Reau et al., 2023). By comparing the obtained sequences to a reference database, researchers can gain perceptions into the evolutionary relationships and efficient potential of the microbial communities present. 18S rRNA sequencing is similar to 16S but targets eukaryotic microorganisms, including fungi ITS sequence. Shotgun metagenomic sequencing reads all gDNA (genomicDNA) in a specimen, enabling the identification and profiling of bacteria, fungi, viruses, and other microorganisms. Research analyzing data from 16S rRNA and shotgun sequencing were used to characterize the gut microbiota taxonomically. The results showed that 16S rRNA sequencing was less accurate in taxonomy than shotgun metagenomic sequencing, recovering less than 31% of genera and less than 50% of phyla (Clarridge, 2004; Hoggard et al., 2018) (Table 1.1). However, 16S rRNA sequencing generates simpler data that nonexperts can analyze more easily, and it is a cost-effective method for microbial community characterization.

1.4.1 Soil Microbiome Study

A thorough examination of the bacteria, fungi, viruses, archaea, and other microorganisms discovered in the soil environment is the goal of soil microbiome research. The operation of the soil ecosystem as a whole, plant health, and nutrient cycling all depend on the understanding of the soil microbiome. Knowledge of the composition and activities of microbial population in soils that are vital to preserving the health and sustainability of soils requires a knowledge of the soil microbiome (Odelade & Babalola, 2019). Soil microbiomes have been studied using a variety of methodologies, mainly high-throughput sequencing techniques, molecular biology procedures, including DNA extraction, and bioinformatics tools for data processing (Wydro, 2022). Humic acids are a significant factor to consider during DNA isolation procedures as they can inhibit the process and lead to PCR reaction failure (Lim et al., 2017). Fortunately, many readymade kits are specifically designed for soil DNA extraction using the magnetic bead system or extraction on columns. To what extent the extracted DNA is suitable for further analysis and modification depends on its quality and purity (Fatima et al., 2014). Some markers, known as α-diversity and β-diversity, reveal the variety of the community inside the sample and between samples, respectively, in the investigation of the genetic entity of soil microflora. In contrast to α-diversity indices, which encompass species and phylogenetic richness/evenness/dominance estimators, β-diversity permits the evaluation of distance and

Table 1.1 Comparison between shotgun metagenomics, metagenomics, and 18S rRNA sequencing

S.no.	Feature	Shotgun metagenomics	Metagenomics	18S rRNA sequencing
1.	Identified targets	Complete microbial genomes	Total genomic DNA	Eukaryotic and nonbacterial organisms
2.	Taxonomic specificity	High	High	Taxonomic-level alignment
3.	Sample types	Suitable for diverse sample types (e.g., water soil and human microbiome)	Suitable for diverse sample types (e.g., water soil, human microbiome)	Limited to specific sample types due to primer specificity
4.	Output information	Genetic sequences	Genomic DNA sequences	Eukaryotic genome sequences
5.	Applications	Microbiome analysis and functional profiling	Comprehensive study of microbial community functions	Studying the composition and functions of fungal and protist communities
6.	Strengths	Comprehensive microbial detection	inclusive microbial detection	Phylogenetic alignment
7.	Limitations	Analytical expenses, data intricacy, and method uniformity	Cost-effectiveness, data intricacy, and method standardization	Taxonomic specificity for eukaryotes only
8.	Data analysis complexity	Proficiency in bioinformatics is necessary for data interpretation	Proficiency in bioinformatics is necessary for data interpretation	Standardized bioinformatics workflows available

similarity in the makeup of communities (Walters & Martiny, 2020). For the purpose of visualizing statistical findings, principal component analysis (PCA), statistical clustering, nonmetric multidimensional scaling (NMDS), redundancy analysis (RDA) or principal coordinate analyses (PCoA), and one-way analysis of similarity (ANOSIM) are the most popular analysis tools. Even though there have been many publications on genetic soil biology concerns in the past 10 years, there is currently no document that compiles practically all methods, from the most basic to the most sophisticated, cutting-edge approaches and tools, in one place (Paliy & Shankar, 2016; Buttigieg & Ramette, 2014; Chahouki, 2011). The impression of agricultural management on soil microbiota in Styria/Austria's vineyards and orchards is one of the significant conclusions and revelations from soil microbiome investigations. Bacterial species comprise the biggest category in terms of quantity and variety. The connection between soil and human gut microbiomes is another discovery. The gut microbiome of humans may be influenced by soil microbiomes, with soil serving as the primary predictor of the gut microbiome of baboons that live on land. Since soils are a reservoir for microorganisms, soil microbiomes can have implications for "one health." With approximately 40 soil microbiome activities, the total microbial diversity in a broad spectrum of species affects soil, plant, animal, and human health, either directly or indirectly. Using model soil consortia, the soil microbiome meta-phenome research seeks to develop a systems-level sympathetic of the phenotypic expression of the soil microbiome to changing environmental circumstances (Attwood et al., 2019; Samaddar et al., 2021).

DNA extraction, molecular techniques, 16S rRNA gene sequencing for archaea and bacteria, and data analysis using bioinformatics tools and software are all essential for studying the soil microbiome. Additionally, two more techniques can be used to advance understanding of the functional activities of microbial communities in the soil. Metatranscriptomics involves studying the total RNA to understand gene expression and active metabolic processes, while metaproteomics analyzes the entire complement of proteins that the microbial community expresses (Silva et al., 2023; Wydro, 2022).

Soil microbiome production has a noteworthy role in maintaining soil health and sustainability, which is crucial for sustainable agriculture and food production. These microbes enhance soil fertility by mediating biogeochemical cycles and improving carbon storage in soils. Based on this, this microbial community affects the nutrient content of our food. Soil microbiomes also contribute to nutrient use efficiency, control of pests and diseases, and plant growth promotion, thereby improving crop productivity and overall health. Moreover, soil microbiomes help regulate the biogeochemical cycles that affect greenhouse gas emissions, playing a key role in mitigating climate change. They also show a crucial role as a reservoir of microbiome and the overall microbial diversity in a wide range of organisms, including humans and animals. Hence, studying soil microbiomes can help us understand the interconnected health of plants, soil, and humans, promoting a holistic approach to environmental and human health. Finally, findings from a soil microbiome study contribute to understanding the ecological interactions, diversity, and functions of microbial communities in different soil environments (Suman et al., 2022).

1.4.2 Bioinformatics Tools for Microbiome Study

The development of molecular methods has led to a huge collection of experimental data. Yet, the extensive data yielded by cell-scale experiments needed advanced computational tools for meaningful analysis and interpretation. Currently, computer technology is an important part of molecular biology initiatives that are used to ascertain the order, organization, or other kind of data. Software applications can extract common elements and prepare useful predictions from these data archives (Schlick & Portillo-Ledesma, 2021). In order to achieve their systematic objectives, most initiatives start by collecting the relevant data from various online sources. The data discovered is subsequently analyzed utilizing web-based tools. Investigational research in molecular biology relies on computer databases to store its results. The data in these databases is organized in a methodical way, in accordance with the exploration software and specific resources used. Biological databases share many similarities with popular internet portals in terms of their features. They offer a wide range of search methods for different types of biological data, including sequences, structures, and molecular interaction networks (Wooley et al., 2005; Bernasconi, 2021). The investigations conducted employ algorithms that rely on the principles of biochemistry and molecular evolution. DNA sequences are the primary type of data collected for DNA examination using biological databases. There are three databases that hold all newly discovered DNA sequences: GenBank, EMBL, and DDBJ (Baxevanis, 2001). The sequence that is deposited in one of these databases is immediately transferred to the other two. Before the report is published, research effort must deposit the newly discovered sequences in one of these databases. Finding every sequence that is comparable to the recently learned sequence involves searching the sequence databases. This difficulty can be solved only by examining the sequence similarity coefficient's statistical relevance. Without a doubt, the BLAST method (now PSI-BLAST), which calculates statistical relevance as an expected value (e-value), is the most widely used database search technique. Various programs exist for conducting database searches. One such program is BLAST, which provides five different types of searches: blastp (for protein sequences), blastn (for DNA and RNA sequences), blastx, tblastn, and tblastnx. In addition, DIAMOND is an additional tool that can be employed to match protein or DNA sequence elucidations with a protein reference database (Goudey et al., 2022). Bioinformatics tools can accurately depict the evolutionary history of sequences using phylogenetic analysis, characterized as trees. Phylip and MEGA software are commonly used for these analyses (Kaari et al., 2022). There are several widely used tools for amplicon-based sequencing, including Comparative Microbiome Analysis (CoMA), QIIME2, DADA2, and Deblur. DADA2 is an R-package that is open-source and builds upon the DADA approach to enhance its performance. The DADA2 approach involves various steps such as chimera identification, sample inference, combining of assembled reads, dereplication, and screening. This algorithm is designed to infer the ASVs. QIIME2 and Mothur are powerful tools used in the field of microbiome analysis. These pipelines enable researchers to process raw sequencing data, perform control over quality, and generate meaningful visualizations and statistical analyses (Özkurt et al., 2022). However, using these solutions requires extensive knowledge and experience in the use of command-line tools and data processing based on sequences. One way to address these issues is by leveraging user blogs and online documentation available on the internet. One freely accessible popular tool is Comparative Microbiome Analysis (CoMA), a nifty pipeline intended for effortless utilization across various operating systems, including Linux, Windows, and macOS, ensuring accessibility to multiple domains of users (Hupfauf et al., 2020). With a vast array of tools and possibilities, including quality checking, clustering to operational taxonomic units (OTUs), data visualization, taxonomic assignment, data post-processing, and statistical evaluation, this technique is incredibly adaptable in its use. Studies on the soil microbiome can make use of this instrument. It is important to note that Deblur is a powerful tool that is used for calling accurate sequences from the Illumina MiSeq and HiSeq sequencing systems. This tool is particularly useful in the analysis of ecological patterns, especially those subsequent from the distinction of closely related taxa. Unlike other suboperational-taxonomic unit (sOTU) approaches, Deblur can significantly reduce processing needs while still maintaining high levels of sensitivity and specificity. One of the most significant benefits of Deblur is that it is an open-source tool that can be used for commercial purposes (Wei et al., 2021). It is released under the open-source Berkeley Software Distribution (BSD) license, which means that it is freely available to anyone who wants to use it. Researchers have tested a variety of pipelines and discovered that DADA produces more false ASVs than QIIME2-Deblur and USEARCH-UNOISE3, but it performs best in terms of selectivity and resolution. On the other hand, USEARCH-UNOISE3 offers an excellent balance between resolution and specificity. QIIME-uclust, however, tends to generate many false OTUs and may overestimate alpha diversity, indicating the need for alternative pipelines (Özkurt et al., 2022). In a recent study, researchers compared various bioinformatics workflows, including Mothur, QIIME, CoMA, and QIIME2-DADA2, using sequence of the microbial community data from three or more different soils. The results showed that all the associated tools produced comparable results and identified most genera/species in the test communities (Hupfauf et al., 2020; Schloss et al., 2009; Kuczynski et al., 2011). Overall, Deblur is a powerful tool that can be used to extract and analyze ecological patterns resulting from closely related taxonomical group. Its open-source nature and high levels of sensitivity and specificity make it an excellent choice for commercial use.

1.5 Conclusion

The study of soil microbiota and microbiomes is crucial in understanding the complex interactions that occur within soil ecosystems. Researchers have used culture-specific and culture-independent methods to better understand the diversity and function of microorganisms present in soil environments. Culture-specific methods, such as plate count methods, have traditionally been used to isolate and characterize specific microbial strains. However, such methods have limitations in capturing the full range of microbial diversity present in soil due to the challenges in cultivating most microorganisms in controlled laboratory settings. On the other hand, culture-independent methods, including integrated metagenomics, have critically prolonged our knowledge of soil microbiomes by allowing for the direct analysis of microbial DNA extracted from metagenomic samples. These techniques have uncovered a previously hidden world of microbial diversity and functionality, revealing the complexity of soil ecosystems. Moreover, soil microbial communities play critical roles in essential ecological processes, such as soil erosion, by stabilizing soil structure, enhancing nutrient cycling, and mitigating erosion through the assembly of extracellular polymeric materials and the formation of soil aggregates. Going forward, integrated approaches that combine metagenomic analyses with ecological and bioinformatics tools will be essential in advancing our understanding of soil microbiomes. By elucidating the interactions between microorganisms and their environment, we can better predict and manage soil ecosystem dynamics in the face of environmental challenges.

References

Attwood, G. T., Wakelin, S. A., Leahy, S. C., Rowe, S., Clarke, S., Chapman, D. F., Muirhead, R., & Jacobs, J. M. E. (2019). Applications of the soil, plant and rumen microbiomes in pastoral agriculture. *Frontiers in Nutrition, 6*, 107. https://doi.org/10.3389/fnut.2019.00107

Baxevanis, A. D. (2001). The molecular biology database collection: An updated compilation of biological database resources. *Nucleic Acids Research, 29*(1), 1–10.

Bay, S. K., McGeoch, M. A., Gillor, O., Wieler, N., Palmer, D. J., Baker, D. J., Chown, S. L., & Greening, C. (2020). Soil bacterial communities exhibit strong biogeographic patterns at fine taxonomic resolution. *mSystems, 5*(4), e00540–e00520. https://doi.org/10.1128/mSystems.00540-20

Bernasconi, A. (2021). Data quality-aware genomic data integration. *Computer Methods and Programs in Biomedicine Update, 1*, 100009. https://doi.org/10.1016/j.cmpbup.2021.100009

Bonnet, M., Lagier, J. C., Raoult, D., & Khelaifia, S. (2019). Bacterial culture through selective and non-selective conditions: The evolution of culture media in clinical microbiology. *New Microbes and New Infections, 34*, 100622. https://doi.org/10.1016/j.nmni.2019.100622

Buttigieg, P. L., & Ramette, A. (2014). A guide to statistical analysis in microbial ecology: A community-focused, living review of multivariate data analyses. *FEMS Microbiology Ecology, 90*(3), 543–550. https://doi.org/10.1111/1574-6941.12437

Califf, R. M. (2018). Biomarker definitions and their applications. *Experimental Biology and Medicine, 243*(3), 213–221. https://doi.org/10.1177/1535370217750088

Chahouki, M. A. Z. (2011). Multivariate analysis techniques in environmental science. *Earth and Environmental Sciences*. https://www.academia.edu/63823025/Multivariate_Analysis_Techniques_in_Environmental_Science

Ciccazzo, S., Esposito, A., Borruso, L., & Brusetti, L. (2016). Microbial communities and primary succession in high altitude mountain environments. *Annals of Microbiology, 66*(1), 43. https://doi.org/10.1007/s13213-015-1130-1

Clarridge, J. E. (2004). Impact of 16S rRNA gene sequence analysis for identification of bacteria on clinical microbiology and infectious diseases. *Clinical Microbiology Reviews, 17*(4), 840–862. https://doi.org/10.1128/CMR.17.4.840-862.2004

Creamer, R. E., Barel, J. M., Bongiorno, G., & Zwetsloot, M. J. (2022). The life of soils: Integrating the who and how of multifunctionality. *Soil Biology and Biochemistry, 166*, 108561. https://doi.org/10.1016/j.soilbio.2022.108561

Davey, H. M. (2011). Life, death, and in-between: Meanings and methods in microbiology. *Applied and Environmental Microbiology, 77*(16), 5571–5576. https://doi.org/10.1128/AEM.00744-11

Dubey, R. K., Tripathi, V., Prabha, R., Chaurasia, R., Singh, D. P., Rao, C. S., El-Keblawy, A., & Abhilash, P. C. (2020). Methods for exploring soil microbial diversity. In R. K. Dubey, V. Tripathi, R. Prabha, R. Chaurasia, D. P. Singh, C. S. Rao, A. El-Keblawy, & P. C. Abhilash (Eds.), *Unravelling the soil microbiome: Perspectives for environmental sustainability* (pp. 23–32). Springer International Publishing. https://doi.org/10.1007/978-3-030-15516-2_3

Durazzi, F., Sala, C., Castellani, G., Manfreda, G., Remondini, D., & De Cesare, A. (2021). Comparison between 16S rRNA and shotgun sequencing data for the taxonomic characterization of the gut microbiota. *Scientific Reports, 11*(1), 3030. https://doi.org/10.1038/s41598-021-82726-y

Fatima, F., Pathak, N., & Verma, S. R. (2014). An improved method for soil DNA extraction to study the microbial assortment within rhizospheric region. *Molecular Biology International, 2014*, 518960. https://doi.org/10.1155/2014/518960

Fukuda, K., Ogawa, M., Taniguchi, H., & Saito, M. (2016). Molecular approaches to studying microbial communities: Targeting the 16S Ribosomal RNA gene. *Journal of UOEH, 38*(3), 223–232. https://doi.org/10.7888/juoeh.38.223

Goudey, B., Geard, N., Verspoor, K., & Zobel, J. (2022). Propagation, detection and correction of errors using the sequence database network. *Briefings in Bioinformatics, 23*(6), bbac416. https://doi.org/10.1093/bib/bbac416

Gowtham, H. G., et al. (2022). Insight into recent progress and perspectives in improvement of antioxidant machinery upon pgpr augmentation in plants under drought stress: A review. *Antioxidants, 11*(9), 1763.

Hammarlund, S. P., & Harcombe, W. R. (2019). Refining the stress gradient hypothesis in a microbial community. *Proceedings of the National Academy of Sciences of the United States of America, 116*(32), 15760–15762. https://doi.org/10.1073/pnas.1910420116

Handelsman, J. (2004). Metagenomics: Application of genomics to uncultured microorganisms. *Microbiology and Molecular Biology Reviews, 68*(4), 669–685. https://doi.org/10.1128/MMBR.68.4.669-685.2004

Hoggard, M., Vesty, A., Wong, G., Montgomery, J. M., Fourie, C., Douglas, R. G., Biswas, K., & Taylor, M. W. (2018). Characterizing the human mycobiota: A comparison of small subunit rRNA, ITS1, ITS2, and large subunit rRNA genomic targets. *Frontiers in Microbiology, 9*, 2208. https://www.frontiersin.org/articles/10.3389/fmicb.2018.02208

Hupfauf, S., Etemadi, M., Fernández-Delgado Juárez, M., Gómez-Brandón, M., Insam, H., & Podmirseg, S. M. (2020). CoMA – An intuitive and user-friendly pipeline for amplicon-sequencing data analysis. *PLoS One, 15*(12), e0243241. https://doi.org/10.1371/journal.pone.0243241

Jehmlich, N., Vogt, C., Lünsmann, V., Richnow, H. H., & von Bergen, M. (2016). Protein-SIP in environmental studies. *Current Opinion in Biotechnology, 41*, 26–33. https://doi.org/10.1016/j.copbio.2016.04.010

Kaari, M., Joseph, J., Manikkam, R., Shamya, M., & Aruni, W. (2022). Application of bioinformatic tools for phylogenetic analysis. In D. Dharumadurai (Ed.), *Methods in actinobacteriology* (Springer protocols handbooks) (pp. 187–191). Springer US. https://doi.org/10.1007/978-1-0716-1728-1_27

Khangura, R., Ferris, D., Wagg, C., & Bowyer, J. (2023). Regenerative agriculture—A literature review on the practices and mechanisms used to improve soil health. *Sustainability, 15*(3), 2338. https://doi.org/10.3390/su15032338

Kovac, J., Rolon, M. L., Naum, M., & Lampel, K. A. (2022). DNA-based assays. In P. L. H. McSweeney & J. P. McNamara (Eds.), *Encyclopedia of dairy sciences* (3rd ed., pp. 356–362). Academic Press. https://doi.org/10.1016/B978-0-12-818766-1.00157-4

Kuczynski, J., Stombaugh, J., Walters, W. A., González, A., Caporaso, J. G., & Knight, R. (2011). Using QIIME to analyze 16S rRNA gene sequences from microbial communities. *Current Protocols in Bioinformatics*. https://doi.org/10.1002/0471250953.bi1007s36

Kumar, M., Poonam, Ahmad, S., & Singh, R. P. (2022). Plant growth promoting microbes: Diverse roles for sustainable and ecofriendly agriculture. *Energy Nexus, 7*, 100133. https://doi.org/10.1016/j.nexus.2022.100133

La Reau, A. J., Strom, N. B., Filvaroff, E., Mavrommatis, K., Ward, T. L., & Knights, D. (2023). Shallow shotgun sequencing reduces technical variation in microbiome analysis. *Scientific Reports, 13*, 7668. https://doi.org/10.1038/s41598-023-33489-1

Lapidus, A. L., & Korobeynikov, A. I. (2021). Metagenomic data assembly – The way of decoding unknown microorganisms. *Frontiers in Microbiology, 12*. https://www.frontiersin.org/articles/10.3389/fmicb.2021.613791

Lee, J., Kim, H.-S., Jo, H. Y., & Kwon, M. J. (2021). Revisiting soil bacterial counting methods: Optimal soil storage and pretreatment methods and comparison of culture-dependent and -independent methods. *PLoS One, 16*(2), e0246142. https://doi.org/10.1371/journal.pone.0246142

Lewe, N., Hermans, S., Lear, G., Kelly, L. T., Thomson-Laing, G., Weisbrod, B., Wood, S. A., Keyzers, R. A., & Deslippe, J. R. (2021). Phospholipid fatty acid (PLFA) analysis as a tool to estimate absolute abundances from compositional 16S rRNA bacterial metabarcoding data. *Journal of Microbiological Methods, 188*, 106271. https://doi.org/10.1016/j.mimet.2021.106271

Li, F., Zi, H., Sonne, C., & Li, X. (2023). Microbiome sustains forest ecosystem functions across hierarchical scales. *Eco-Environment & Health, 2*(1), 24–31. https://doi.org/10.1016/j.eehl.2023.03.001

Lim, H. J., Choi, J.-H., & Son, A. (2017). Necessity of purification during bacterial DNA extraction with environmental soils. *Environmental Health and Toxicology, 32*, e2017013. https://doi.org/10.5620/eht.e2017013

Moses, A., et al. (2013). 14 International Journal of Pure and Applied Sciences and Technology Harnessing the potentials of vesicular arbuscular mycorrhizal (vam) fungi to plant growth-a review phenotypic evaluation of heritability, agro-morphological and yield characters of sixteen amaranthus linn. *Genotypes view project drought and nitrogen toler*. https://www.researchgate.net/publication/259268139

Nam, N. N., Do, H. D. K., Loan Trinh, K. T., & Lee, N. Y. (2023). Metagenomics: An effective approach for exploring microbial diversity and functions. *Food, 12*(11), 2140. https://doi.org/10.3390/foods12112140

Navgire, G. S., Goel, N., Sawhney, G., Sharma, M., Kaushik, P., Mohanta, Y. K., Mohanta, T. K., & Al-Harrasi, A. (2022). Analysis and interpretation of metagenomics data: An approach. *Biological Procedures Online, 24*(1), 18. https://doi.org/10.1186/s12575-022-00179-7

Odelade, K. A., & Babalola, O. O. (2019). Bacteria, fungi and archaea domains in rhizospheric soil and their effects in enhancing agricultural productivity. *International Journal of Environmental Research and Public Health, 16*(20), 3873. https://doi.org/10.3390/ijerph16203873

Özkurt, E., Fritscher, J., Soranzo, N., Ng, D. Y. K., Davey, R. P., Bahram, M., & Hildebrand, F. (2022). LotuS2: An ultrafast and highly accurate tool for amplicon sequencing analysis. *Microbiome, 10*(1), 176. https://doi.org/10.1186/s40168-022-01365-1

Paliy, O., & Shankar, V. (2016). Application of multivariate statistical techniques in microbial ecology. *Molecular Ecology, 25*(5), 1032–1057. https://doi.org/10.1111/mec.13536

Paun, O., & Schönswetter, P. (2012). Amplified Fragment Length Polymorphism (AFLP)—An invaluable fingerprinting technique for genomic, transcriptomic and epigenetic studies. *Methods in Molecular Biology (Clifton, N.J.), 862*, 75–87. https://doi.org/10.1007/978-1-61779-609-8_7

Poddar, K., Padhan, B., Sarkar, D., & Sarkar, A. (2021). Purification and optimization of pink pigment produced by newly isolated bacterial strain Enterobacter sp. PWN1. *SN Applied Sciences, 3*(1), 105. https://doi.org/10.1007/s42452-021-04146-x

Rafiq, M., Hassan, N., Rehman, M., Hayat, M., Nadeem, G., Hassan, F., Iqbal, N., Ali, H., Zada, S., Kang, Y., Sajjad, W., & Jamal, M. (2023). Challenges and approaches of culturing the unculturable archaea. *Biology, 12*(12), 1499. https://doi.org/10.3390/biology12121499

Roberts, D. P., & Mattoo, A. K. (2018). Sustainable agriculture—Enhancing environmental benefits, food nutritional quality and building crop resilience to abiotic and biotic stresses. *Agriculture, 8*(1), 8. https://doi.org/10.3390/agriculture8010008

Sabale, S. N., Suryawanshi, P. P., & Krishnaraj, P. U. (2019). Soil metagenomics: Concepts and applications. In *Metagenomics—Basics, methods and applications*. IntechOpen. https://doi.org/10.5772/intechopen.88958

Sahu, N., Vasu, D., Sahu, A., Lal, N., & Singh, S. K. (2017). Strength of microbes in nutrient cycling: A key to soil health. In V. S. Meena, P. K. Mishra, J. K. Bisht, & A. Pattanayak (Eds.), *Agriculturally important microbes for sustainable agriculture: Volume I: Plant-soil-microbe nexus* (pp. 69–86). Springer. https://doi.org/10.1007/978-981-10-5589-8_4

Samaddar, S., Karp, D. S., Schmidt, R., Devarajan, N., McGarvey, J. A., Pires, A. F. A., & Scow, K. (2021). Role of soil in the regulation of human and plant pathogens: Soils' contributions to people. *Philosophical Transactions of the Royal Society B: Biological Sciences, 376*(1834), 20200179. https://doi.org/10.1098/rstb.2020.0179

Schlick, T., & Portillo-Ledesma, S. (2021). Biomolecular modeling thrives in the age of technology. *Nature Computational Science, 1*(5), Article 5. https://doi.org/10.1038/s43588-021-00060-9

Schloss, P. D., Westcott, S. L., Ryabin, T., Hall, J. R., Hartmann, M., Hollister, E. B., Lesniewski, R. A., Oakley, B. B., Parks, D. H., Robinson, C. J., Sahl, J. W., Stres, B., Thallinger, G. G., Van Horn, D. J., & Weber, C. F. (2009). Introducing mothur: Open-source, platform-independent, community-supported software for describing and comparing microbial communities. *Applied and Environmental Microbiology, 75*(23), 7537–7541. https://doi.org/10.1128/AEM.01541-09

Shahwar, D., Mushtaq, Z., Mushtaq, H., Alqarawi, A. A., Park, Y., Alshahrani, T. S., & Faizan, S. (2023). Role of microbial inoculants as bio fertilizers for improving crop productivity: A review. *Heliyon, 9*(6), e16134. https://doi.org/10.1016/j.heliyon.2023.e16134

Silva, D. P., Epstein, H. E., & Vega Thurber, R. L. (2023). Best practices for generating and analyzing 16S rRNA amplicon data to track coral microbiome dynamics. *Frontiers in Microbiology, 13*, 1007877. https://doi.org/10.3389/fmicb.2022.1007877

Suman, J., Rakshit, A., Ogireddy, S. D., Singh, S., Gupta, C., & Chandrakala, J. (2022). Microbiome as a key player in sustainable agriculture and human health. *Frontiers in Soil Science, 2*. https://www.frontiersin.org/articles/10.3389/fsoil.2022.821589

Tamames, J., Cobo-Simón, M., & Puente-Sánchez, F. (2019). Assessing the performance of different approaches for functional and taxonomic annotation of metagenomes. *BMC Genomics, 20*(1), 960. https://doi.org/10.1186/s12864-019-6289-6

Walters, K. E., & Martiny, J. B. H. (2020). Alpha-, beta-, and gamma-diversity of bacteria varies across habitats. *PLoS One, 15*(9), e0233872. https://doi.org/10.1371/journal.pone.0233872

Wei, Z.-G., Zhang, X.-D., Cao, M., Liu, F., Qian, Y., & Zhang, S.-W. (2021). Comparison of methods for picking the operational taxonomic units from amplicon sequences. *Frontiers in Microbiology, 12*. https://www.frontiersin.org/articles/10.3389/fmicb.2021.644012

Wooley, J. C., Lin, H. S., Biology, N. R. C., & National Research Council (US) Committee on Frontiers at the Interface of Computing and Biology. (2005). Computational tools. In *Catalyzing inquiry at the interface of computing and biology*. National Academies Press (US). https://www.ncbi.nlm.nih.gov/books/NBK25460/

Wydro, U. (2022). Soil microbiome study based on DNA extraction: A review. *Water, 14*(24), Article 24. https://doi.org/10.3390/w14243999

Yang, C., Chowdhury, D., Zhang, Z., Cheung, W. K., Lu, A., Bian, Z., & Zhang, L. (2021). A review of computational tools for generating metagenome-assembled genomes from metagenomic sequencing data. *Computational and Structural Biotechnology Journal, 19*, 6301–6314. https://doi.org/10.1016/j.csbj.2021.11.028

2 Decoding Quorum Sensing: The Language of Bacteria for Sustainable Development

Shivangi Bhatt, Dweipayan Goswami, and Meenu Saraf

Abstract

Bacteria engage with a mechanism called cell-to-cell communication, facilitated by the production and recognition of small molecules referred to as autoinducers, to regulate gene expression in a density-dependent manner. These autoinducer compounds are generated by a diverse array of bacterial strains, the most prevalent type being acyl homoserine lactones, commonly produced by Gram-negative bacteria. Quorum sensing is a pivotal process that integrates various ecologically and interdependently significant traits exhibited by bacteria. These encompass functions such as antibiotic production, nitrogen fixation, siderophore production, biofilm formation, enzyme secretion, the release of phytopathogenic virulence factors, and interactions between plants and microbes. The central theme of this chapter is a thorough examination of quorum sensing, providing detailed insights into the intricate communication mechanisms used by bacteria. Additionally, the chapter extensively examines versatile applications of quorum sensing across a wide range of sectors, including wastewater treatment, industrial waste degradation, advancements in crop cultivation, the management of pathogenic agents, adaptation to heavy metal exposure, and responses to challenging environmental conditions to in the pursuit of sustainable development.

Keywords

Quorum sensing (QS) · Sustainable development · Acyl homoserine lactone (AHL) · Biofilm · Wastewater treatment · Microbial fuel cell (MFC)

S. Bhatt · D. Goswami · M. Saraf (✉)
Department of Microbiology and Biotechnology, University School of Science, Gujarat University, Ahmedabad, Gujarat, India
e-mail: msaraf@gujaratuniversity.ac.in

M. Saraf et al. (eds.), *Climate Change and Soil Microorganisms for Environmental Sustainability*, Microorganisms for Sustainability 28,
https://doi.org/10.1007/978-981-96-3425-5_2

2.1 Introduction

Quorum sensing (QS) represents a fundamental process in bacterial communication, wherein bacteria engage in density-dependent interactions. This phenomenon is alternatively acclaimed as autoinduction. Within the scope of this process, acyl homoserine lactones (AHLs) predominantly serve as autoinducer molecules, particularly in the system under study (Patel et al., 2023). As bacterial populations grow, the concentration of autoinducer molecules increases, facilitating their diffusion out of and re-entry into cells. These molecules subsequently bind to specific receptors, instigating the activation of transcriptional genes.

Quorum sensing has been extensively characterized in two symbiotic marine bacteria, *Vibrio fischeri* and *Vibrio harveyi.* In these organisms, bioluminescence production is regulated by the luxCDABE structural operons. Notably, the accumulation of secreted autoinducer molecules exhibits density-dependent behavior, and it plays a pivotal role in regulating bioluminescence in both Vibrio species (Miller & Bassler, 2001).

Gram-negative bacteria predominantly enlist acyl homoserine lactones (AHL) as autoinducer molecules, a system referred to as autoinducer type I or LuxI/LuxR-based quorum sensing. In contrast, Gram-positive organisms utilize peptide-mediated quorum-sensing compounds. A distinct category, autoinducer II type quorum-sensing systems, is found in both Gram-positive and Gram-negative bacteria. This particular system, also known as interspecies quorum sensing, was initially identified in marine bacteria like *V. harveyi* and exhibits unique regulatory properties (Jaafar et al., 2022; Majdura et al., 2023). It is important to note that while N-acyl homoserine lactones have been extensively studied as autoinducer signals, Gram-negative bacteria also produce numerous other chemicals, including diketopiperazines and 4-hydroxy-2-alkylquinolines (HAQs), commonly referred to as autoinducer-3 (AI-3) (Singh et al., 2017).

Quorum-sensing signaling circuits play a significant role in governing various physiological functions in both Gram-positive and Gram-negative bacteria. These functions encompass symbiosis, pathogenicity, competence, conjugation, antibiotic production, motility, sporulation, nitrogen fixation, and biofilm formation, among others (Miller & Bassler, 2001).

Additionally, quorum-sensing molecules play a crucial role in biofilm development, contributing to the production of extracellular polymeric substances. These substances, in turn, support microbial fuel cells by acting as the primary source of solid or liquid waste for applications such as wastewater treatment, bioelectricity generation, biosensor development, bioremediation of contaminants, phytohormonal defense, soil enhancement, and plant growth promotion (Chug et al., 2021). Quorum sensing also finds diverse applications across various fields, including food production, agriculture, pigment synthesis, waste degradation, biofouling prevention, and the design of engineered quorum sensing (Choudhary & Schmidt-Dannert, 2010; In et al., 2002; Xiao et al., 2022; Zhang et al., 2019). In combination, these applications underscore the key role of bacterial cell-to-cell communication in fostering sustainable development.

Collectively, these applications highlight the critical importance of bacterial cell-to-cell communication in advancing sustainable development. This chapter primarily aims to delineate the diverse mechanisms of quorum sensing in bacteria and to elucidate the way in which quorum sensing facilitates sustainable development.

2.2 Diversity of Quorum-Sensing Molecules

Quorum sensing (QS) is a microbial communication system that modulates gene expression and coordinates the behavior of microbial communities. Bacteria synthesize and release specific signaling molecules into their environment. Once these molecules reach a critical concentration, they are detected by receptor proteins, which can then regulate the expression of target genes, either directly or indirectly, to initiate cooperative behaviors that benefit the entire population (Zhang et al., 2019).

2.2.1 Quorum-Sensing Mechanism in Gram-Negative Bacteria

The study of luminescence in the Gram-negative bacterium *Vibrio fischeri* (Miller & Bassler, 2001) was pivotal in the discovery of bacterial quorum sensing (QS). Over 25 types of Gram-negative bacteria utilize acyl homoserine lactones (AHLs), the most common autoinducers in this group. AHLs consist of a hydrophobic head, a conserved serine lactone ring, a hydrophilic tail, and a variable amide side chain, with the tail determining the specificity of the signal (Dong et al., 2001).

In a typical AHL-regulated QS system, the LuxI-type synthase produces AHLs, which bind to the LuxR-type receptor. AHLs, being membrane-permeable, diffuse out of the cell. Once their concentration in the environment reaches a threshold, they re-enter the cell and bind to the amino terminus of LuxR in the cytoplasm, forming LuxI/LuxR complexes that regulate the expression of target genes (see Fig. 2.1). These complexes also exert feedback control over the synthesis of AHLs and their receptors.

The LuxI/LuxR two-component system in *V. fischeri* serves as a model for Gram-negative bacterial QS. This system regulates luminescence, which protects the symbiotic host and helps *V. fischeri* locate suitable environments (Nealson & Hastings, 1979).

2.2.2 Signal Transduction in Gram-Positive Bacterial Quorum Sensing

In Gram-positive bacteria, quorum sensing (QS) primarily relies on oligopeptides, which cannot penetrate the cell membrane and thus require specialized transporters for secretion. The synthesis of these oligopeptides involves initial modifications and cyclization before they are released. Sensing occurs through membrane-bound two-component systems that consist of a cognate receptor and a histidine kinase

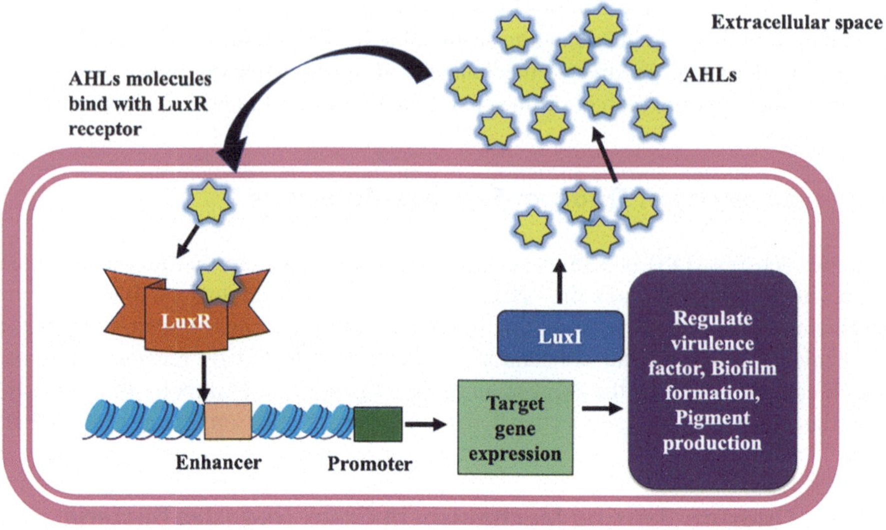

Fig. 2.1 Autoinducer I-based quorum sensing in Gram-negative bacteria

receptor. These systems transmit signals via phosphorylation, with cognate receptors acting as both cytoplasmic response regulators and transcriptional regulators (Hawver et al., 2016).

While the overall QS pathway in Gram-positive bacteria is similar to that of Gram-negative bacteria, the key difference is the reliance on autoinducing peptides (AIPs) for intercellular communication. AIPs are initially synthesized in an immature form and require a specific transport mechanism to move in and out of cells, where they undergo modifications to become functional. Throughout bacterial growth, precursor peptides are produced and modified for stability and activity. AIPs can cross the cell membrane only with the aid of ATP-binding cassette (ABC) transporters or other membrane proteins (Singh & Ray, 2014). Once the concentration of AIPs in the environment reaches a critical threshold, they bind to cell surface receptors, activating the two-component phosphokinase system (TCS) and triggering signal transduction pathways that lead to gene transcription (see Fig. 2.2).

2.2.3 Role of Quorum Sensing in Interspecies Communication Dynamics

The LuxS protein is central to the AI-2 (autoinducer-2) signaling pathway, found in both Gram-negative and Gram-positive bacteria. This system enables bacteria to detect the presence of various microorganisms in their environment, allowing them to adjust their metabolic processes accordingly. The LuxS/AI-2 quorum-sensing system was first identified in *Vibrio harveyi*, a bioluminescent bacterium that produces two signal molecules: AI-1 and AI-2.

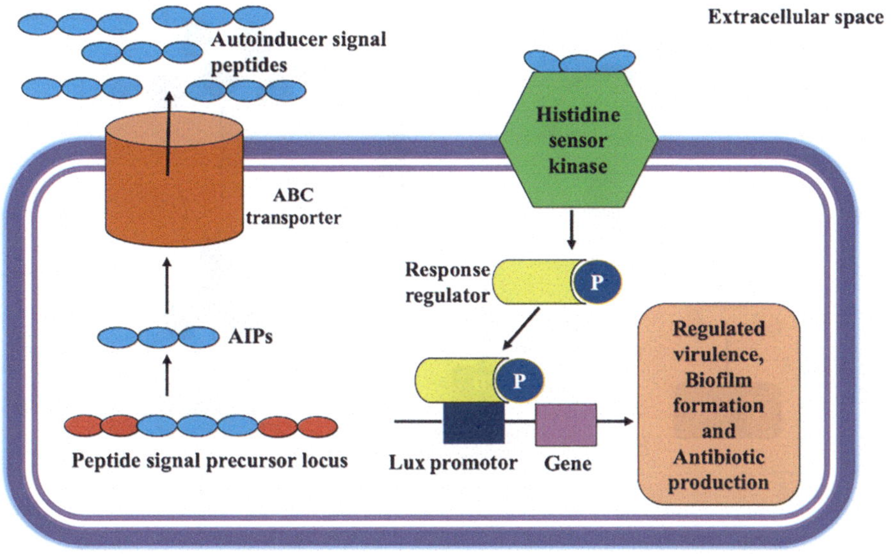

Fig. 2.2 Oligopeptide-based quorum sensing in Gram-positive bacteria

In this system, the LuxN receptor, encoded by the LuxLM gene, senses AI-1, while LuxS is responsible for generating AI-2. LuxP and LuxQ act as receptor proteins for AI-2. Communication between LuxN and LuxQ is mediated by the phosphotransferase LuxU. Once LuxU is phosphorylated, it activates specific gene expression in conjunction with the protein LuxR and the regulatory protein LuxO. These signaling interactions are depicted in Fig. 2.3 (Balan et al., 2021).

2.2.4 Diverse Quorum-Sensing Systems

Diffusible signaling factors (DSFs), categorized as cis-2-unsaturated fatty acids with varying chain lengths and double-bond configurations, are prevalent among bacteria. Intraspecies communication using DSFs primarily occurs in three genomic groups: *Xanthomonas campestris pv. campestris (Xcc)*, *Burkholderia cenocepacia (Bcc)*, and *Pseudomonas aeruginosa*. Additionally, DSF-mediated interspecies communication includes bacteria such as *B. cenocepacia*, *Stenotrophomonas maltophilia*, *Bacillus*, *Francisella novicida*, *Salmonella*, and *Bdellovibrio bacteriovorus*. There is also evidence of inter-kingdom communication between DSF-producing bacteria and Candida albicans. DSF signaling plays critical roles in various processes, including plant development, defense, the colonization of insects by *Xylella fastidiosa*, and inflammatory responses in zebrafish (He et al., 2023).

In bacterial quorum sensing, the less polar AI-3 molecule interacts with the host adrenergic signaling system. Notably, *Enterohemorrhagic Escherichia coli* (EHEC) and *Shigella* pathogenic processes involve AI-3, which was first identified in

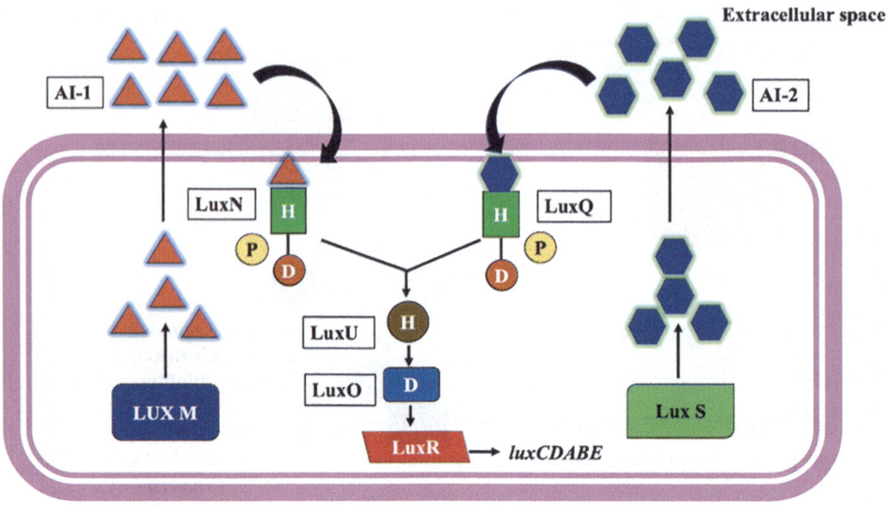

Fig. 2.3 Quorum sensing in Gram-positive and Gram-negative bacteria

infected animal intestinal tissues. Distinct QS systems, such as AHLs, Pseudomonas quinolone signals (PQS), and diketopiperazines (DKPs), regulate target gene expression in pathogenic bacteria. *Pseudomonas aeruginosa* employs not only AHLs but also PQS and DKPs as signaling molecules, with analogous genes related to DKPs and PQS found in *Pseudomonas* and *Burkholderia*.

DSF and diffusible factors (DFs), commonly isolated from agricultural pathogens and biocontrol bacteria, influence critical processes like xanthomycin production and exopolysaccharide (EPS) synthesis in *Xanthomonas campestris pv. campestris*, which are essential for plant parasitism. Additionally, *Streptomyces* species utilize butyrolactone as a signaling molecule, as demonstrated in previous studies (Lyte & Freestone, 2010) (Table 1.1).

2.3 Innovative Applications of Quorum Sensing for Sustainability

Quorum sensing is a crucial process in both Gram-positive and Gram-negative bacteria, involving the production of autoinducer molecules that regulate various bacterial activities. These bacteria are vital for sewage treatment, promoting plant growth, biocontrol, bioengineered plant production, and pigment synthesis, all of which help maintain environmental balance. However, increasing industrialization has led to heightened pollution, resulting in severe health issues and declining soil fertility. Utilizing quorum-sensing mechanisms for cell-to-cell communication offers a promising strategy to address these environmental challenges. The following sections will explore this process in detail (Fig. 2.4).

Table 1.1 Various quorum-sensing signal molecules

S. no.	Signal	Structure	Source	Signal synthesis	Targets	Functions	References
1.	N-acyl-homoserine lactones (AHLs)	A lactone ring and an acylated chain containing 4 to 18 carbons	Gram-negative bacteria, commonly produced by α,β and γ-, Proteobacteria	Controlled by LuxI/LuxR system	Member of the LuxR family, gram-negative bacteria	1. Intraspecies communication 2. Biofilm formation 3. EPS production 4. Aerobic granulation 5. Bacterial activity	Tan et al. (2014a)
2.	Autoinducer 2 (AI-2)	Furanosyl borate diester molecules	Gram-positive and Gram-negative bacteria, commonly in *Vibrio harveyi*	Two I-genes, pfs,and luxS	Gram-positive and Gram-negative bacteria	1. Interspecies interactions 2. Biofilm formation 3. Pathogenicity factors and toxins 4. Bioluminescence 5. Swarming motility	Tarighi and Taheri (2011)
3.	Autoinducing peptides (AIP)	Oligopeptides	Gram-positive bacteria	Two-component regulatory systems: Synthesized first in the cell ribosome, then processed by ATP-binding cassette	Gram-positive bacteria	1. Intercellular communication 2. Virulence and antimicrobial compounds	Sturme et al. (2002)
4.	Autoinducer 3 (AI-3)	Aromatic compound	Gastrointestinal microbial flora	luxS gene	*Escherichia coli*	1. Transcription activation of virulence genes in enterohemorrhagic *Escherichia coli*	Clarke et al. (2006) and Kim et al. (2020)

(continued)

Table 1.1 (continued)

S. no.	Signal	Structure	Source	Signal synthesis	Targets	Functions	References
5.	Pseudomonas quinolone signal (PQS)	Quinolone compounds	Gram-negative bacteria, especially *Pseudomonas*	4Q regulatory system	*Pseudomonas aeruginosa*	1. Multiple virulence factors 2. Biofilm development 3. Control C4-HSL production 4. Regulate denitrification 5. Regulate iron transport systems	Diggle et al. (2006)
6.	Diffusible signal factors (DSF) and diffusible factor	Cis-11-Methyl-2-dodecanoic acid and butyrolactone	*Xanthomonas* spp.	RI-sensory system	*Xanthomonas campestris pv. compestris*	1. Production of exoenzymes 2. The presence of DSF to bacterial dispersion, whereas the absence of DSF leads to EPS production, aggregation, and biofilm formation	Gudesblat et al. (2009) and Lee et al. (2007)
7.	3-Hydroxypalmitic acid methyl ester (3-OHPAME)	3-OH-PAME	*Ralstonia solanacearum*	S-Adenosylmethionine-dependent enzyme methyl transferase encoded by the phcB gene	*Ralstonia solanacearum* and *Ralstoniaeutropha*	1. Affect the production of AHLs, 2. Control expression of motility and siderophore synthesis	Tarighi and Taheri (2011)

8.	Diguanylate monophosphate (cyclic-di-GMP, second messenger)	Diguanylate monophosphate	Gram-negative and a few Gram-positive bacteria	Controlled by diguanylate cyclases (DGCs) and c-di-GMP phosphodiesterases (PDEs)	Gram-negative and a few Gram-positive bacteria, such as some clinically important pathogens	1. Coordinate biofilm formation and pathogenicity 2. Regulate virulence and motility factors 3. Regulate the synthesis of exopolysaccharide 4. Determine granular stability	Lee et al. (2007), Wan et al. (2013), and Zhan et al. (2024)

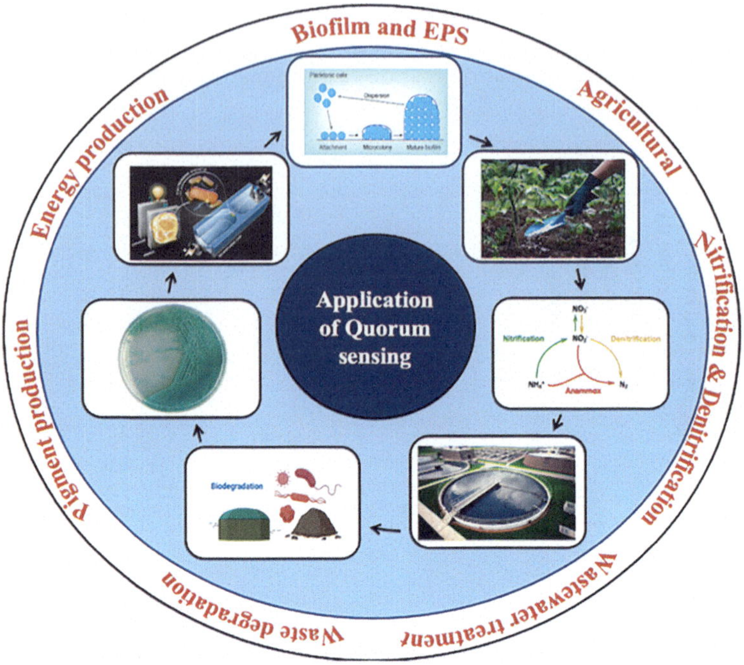

Fig. 2.4 Application of quorum sensing

2.3.1 Impact of Quorum Sensing on Agricultural Systems

The growing global population has intensified the need for sustainable food production methods that enhance food supply and quality while minimizing environmental harm. Although chemical fertilizers, manures, and pesticides contribute to issues like nitrate contamination and eutrophication, they remain integral to intensive agriculture. Addressing food demand requires solutions that increase agricultural yields, reduce spoilage, and improve quality ((Tommonaro et al., 2021).

Emerging strategies include manipulating microbial communities for plant growth promotion and disease control, and enhancing resilience against challenges like disease and drought (Abbamondi et al., 2016). Plant-associated microbiomes, integrated into plant holobionts, rely on gene expression from all partners and signal communication, likely involving quorum-sensing (QS) molecules (Zhang et al., 2017).

For example, the plant growth-promoting rhizobacterium *Stenotrophomonas rhizophila* DSM14405T utilizes a QS system and secretes diffusible signal factor (DSF), which enhances environmental adaptation and plant growth under saline-alkaline stress (Zhuang et al., 2023). Additionally, N-3-oxo-decanoyl-homoserine-lactone (3-O-C10-HL) promotes root development in mung bean seedlings ((Bai et al., 2012). Research indicates that AHL molecules can alter root morphology and phytohormonal balance, influencing root growth and trichome formation ((Schikora et al., 2016; Von Rad et al., 2008).

Quorum sensing also enhances resistance to pathogens and pests. Water-soluble AHLs can be absorbed by plants lacking lactonases, such as *Arabidopsis thaliana*, wheat, and barley, influencing enzyme activity to boost stress tolerance (Götz-Rösch et al., 2015; Sieper et al., 2014).

2.3.2 Regulation of Pigment Biosynthesis by Quorum Sensing

Quorum sensing (QS) is a key microbial communication mechanism that significantly influences pigment production, with important implications for sustainable development across multiple industries. Pigments are valuable natural compounds used in food, textiles, cosmetics, and pharmaceuticals. In *Pseudomonas aeruginosa*, QS regulates the production of pyocyanin through autoinducers like AHL and PQS, which increase with population density. The primary QS systems, LasR-LasI and RhlR-RhlI, control genes involved in pyocyanin synthesis. This versatile pigment exhibits antibacterial, antifungal, anti-inflammatory, anticancer, and antioxidant properties (Abdelaziz et al., 2023). In aquaculture, pyocyanin effectively reduces *Vibrio* spp. populations in recirculating aquaculture systems (RAS) without harming beneficial nitrifying bacteria, achieving a 100% survival rate in shrimp (*Penaeus monodon*) at a concentration of 5 mg/L (Balakrishnan et al., 2022).

Serratia marcescens is known for producing the red pigment prodigiosin, regulated by its QS system involving N-butanoyl homoserine lactone (BHL) and HHL. Disrupting these QS signals significantly reduces prodigiosin production (Fekrirad et al., 2020).

Bacterial pigments like prodigiosin have industrial potential, especially in textiles where safety regulations are less stringent than in food. Prodigiosin from marine *Serratia* species has been successfully used as a natural dye for various fabrics, demonstrating durability after washing (Darshan & Manonmani, 2015). Prodigiosin and violacein from *Serratia marcescens* and *Chromobacterium violaceum*, respectively, are effective for dyeing cotton, silk, rayon, satin, and polyester (Kumar et al., 2015; Simsek Geyik et al., 2024). These pigments also offer eco-friendliness and antimicrobial properties, enhancing their suitability for textile applications (Venil et al., 2021).

Other bacteria, such as *Vibrio* spp. and *Janthinobacterium lividum*, produce pigments regulated by QS, with applications in biotechnology, medicine, and environmental monitoring. *Vibrio* pigments can indicate the presence of these species in aquatic environments, aiding water quality monitoring (Singh et al., 2021). *Janthinobacterium lividum*'s violet pigment offers a potential eco-friendly alternative for textile dyeing (Kramar & Kostic, 2022).

Finally, *Chromobacterium violaceum* produces violacein, a purple pigment with antimicrobial and antioxidant properties, also applicable in textile dyeing (Anahas et al., 2022). The utilization of these bacterial pigments highlights the importance of understanding QS mechanisms, showcasing their potential contributions to sustainable development across various industries.

2.3.3 Quorum-Sensing Role in Wastewater Treatment

Quorum sensing (QS) plays a pivotal role in wastewater treatment processes, contributing to the optimization of various key functions. QS mechanisms enable microorganisms to engage in cell-to-cell communication, coordinating their activities based on population density. In the context of wastewater treatment, QS governs the formation of biofilms, intricate microbial communities that enhance the removal of contaminants. These biofilms, regulated by QS, are critical in the breakdown of organic matter and the biodegradation of pollutants. Additionally, QS-mediated extracellular polymeric substances (EPS) production further aids in biofilm formation, stability, and pollutant binding. Moreover, QS facilitates the efficient removal of pollutants through phytoremediation processes, where plants and associated microorganisms work synergistically to detoxify and immobilize contaminants. The application of QS in wastewater treatment not only enhances the overall efficiency of the process but also contributes to more sustainable and environmentally friendly treatment strategies. This multifaceted role of QS underscores its significance in addressing contemporary challenges in wastewater management.

QS has emerged as a potential approach to increasing the community's performance in wastewater treatment plants (WWTPs) (In et al., 2002) QS, regulates the community assembly of aerobic granules (Tan et al., 2014b), protects biofilms from unfavorable shock, improves anammox bacteria activity (Wang et al., 2021b), and improves the nitrification performance (Mellbye et al., 2016) of biological systems.

Aerobic granules, with their layered structures, are highly effective in wastewater treatment due to their significant roles in the removal of organics and inorganics (e.g., nitrogen and phosphorus) from wastewater. QS plays an important role in granule formation and stability (Li et al., 2017). Mature granules contain higher levels of signal molecules (e.g., AIs) compared to early stage or smaller granules (Liu et al., 2016; Xiong & Liu, 2012). Biofilm growth increases tenfold during aerobic granulation with higher signal molecule levels, highlighting the strong link between QS signaling and granulation (Zhang & Tay, 2015). Despite this, the exact molecular mechanisms of QS in granulation are not fully understood. AI-2 has been shown to affect aerobic granulation by enhancing EPS secretion and bacterial adhesin (Sun et al., 2016). Although floccular biomass conversion to granules appears random, it is significantly influenced by AHLs at the community level (Tan et al., 2014a). The size of acyl chains in AHLs can also impact granulation, with changes in AHL concentrations affecting granule stability and size (Halide & Eq, 2017; Jiang et al., 2013; Tan et al., 2014a). Overall, QS systems significantly influence granulation through various factors, including EPS production, microbial community composition, and AHL levels.

2.3.4 Quorum-Sensing Role in EPS Production and Biofilm Formation

Bacterial colonization is initiated through communication regulated by specific quorum-sensing (QS) systems (Dang & Lovell, 2016; Pietschke et al., 2017). Different QS systems play unique roles under varying environmental conditions. For instance, *Vibrio fischeri* utilizes two QS systems (AinS/AinR and LuxI/LuxR) for persistent colonization, with AinS/AinR governing the initial phase ((Lupp & Ruby, 2005; Miyashiro et al., 2014). In contrast, *Vibrio cholerae* has a more complex quadruple QS system (AinS/AinR, LuxI/LuxR, and LuxS/LuxPQ) that regulates surface colonization and biofilm formation (Jung et al., 2015).

Biofilm development is a dynamic process where planktonic cells transition to a stationary growth state. While the role of QS in biofilm formation is debated, it is known to influence biofilm development across various species. This process typically involves four stages: (1) Initial adsorption to a substrate, (2) formation of an extracellular polymeric substance (EPS) matrix, (3) accumulation of multilayered microbial clusters, and (4) detachment or dispersal (Coughlan et al., 2016; Wang et al., 2021a).

QS enhances bacterial growth in biofilms and modulates EPS composition, facilitating effective bioremediation of pollutants by genera such as *Pseudomonas*, *Aeromonas*, *Burkholderia*, *Xanthomonas*, and *Bacillus* (Sela et al., 2021). Bioremediation processes involving biofilm-associated microorganisms include biosorption, bioaccumulation, and biomineralization of persistent organic pollutants (POPs), often through the secretion of enzymes or entire cells as catalysts (Flemming et al., 2016). Promoting gene transfer among organisms in biofilms can enhance bioremediation efficacy (Soto-Aceves et al., 2020). Additionally, genetically engineered strains to improve pollutant bioavailability, alongside optimized physicochemical conditions for biofilm formation, can yield better outcomes (Dogsa et al., 2021).

Biofilms serve dual roles: They can be beneficial in systems like granular sludge and moving bed biofilm reactors, yet detrimental in membrane biofouling during wastewater treatment (Li et al., 2017; Meng et al., 2017). In genera such as *Aeromonas*, *Pseudomonas*, and *Xanthomonas*, the interplay between QS and EPS composition is critical for biofilm proliferation (He & Zhang, 2008; Yadav & Chandra, 2019). Microbial communities adapt and thrive in diverse environments by synthesizing extracellular polymeric substances (EPS), which form a protective matrix. EPS, composed of proteins, polysaccharides, lipids, DNA, and other high molecular weight biomolecules, provide structural support, facilitate mass transport, and adsorb organic/inorganic substances (Coughlan et al., 2016). For instance, *Klebsiella variicola* SMHMZ46 produces significant EPS that can interact with metal ions, showcasing its bioremediation potential (Sharma & Saraf, 2023).

In summary, bacterial colonization and biofilm formation are intricate processes influenced by QS, gene expression, and EPS composition, with significant implications for bioremediation and environmental adaptation. Biofilms demonstrate both beneficial and detrimental roles, underscoring their versatility within microbial ecosystems.

2.3.5 Quorum Sensing in Industrial Waste Management

Environmental pollutants from various industries, including pulp and paper mills, distilleries, tanneries, pharmaceuticals, fertilizer production, pesticides, and herbicides, pose risks to aquatic ecosystems due to their persistence and hydrophobic nature. Traditional methods for detoxifying these persistent organic pollutants (POPs), such as physical and chemical processes, can be costly and generate undesirable by-products. In contrast, biological treatment processes, facilitated by biofilm bioreactors like up-flow sludge blanket (USB), trickle-bed biofilm bioreactor (TBR), biofilm fluidized bed (BFB), rotating biological contactors (RBC), expanded granular sludge blanket (EGSB), and biofilm airlift suspension (BAS), offer a more cost-effective approach (Chandra et al., 2018; Shah et al., 2021; Tripathi et al., 2022; Yadav & Chandra, 2018).

Bioaugmentation, which involves introducing exogenous microbial strains, enhances the biodegradation and detoxification of organic pollutants in sewage treatment. This process often relies on quorum-sensing (QS) mechanisms, such as AHLs and AI-2 systems, which facilitate the colonization and enhance nicotine removal as demonstrated by (Chen et al., 2018).

Introducing functional genes, strains, or microbial consortia improves the survival and metabolic activities of indigenous communities. For instance, in bioaugmentation systems for nicotine removal, *Pseudomonas* sp. HF-1 utilizes C6-HSL, 3-oxo-C6-HSL, and 3-oxo-C8-HSL to promote EPS production, biofilm formation, swarming, and successful colonization (Balan et al., 2021). Similarly, the nicotine-degrading strain *Acinetobacter* sp. TW shows that short-chain AHLs controlled biofilm formation, while long-chain AHLs increased resistance to nicotine toxicity (Chen et al., 2016; Shi et al., 2017). Polycyclic aromatic hydrocarbons (PAHs) are hydrophobic, persistent, and often carcinogenic pollutants resistant to degradation. However, biofilms can enhance PAH solubility and biotransformation, offering a promising solution (Saxena et al., 2019). Isolates with QS-mediated biofilm formation potential can further improve PAH degradation. In the biozo process, co-metabolism of xenobiotics and PAHs within biofilms was demonstrated (Zhang et al., 2021). Upregulation of LasI and RhlI QS genes significantly enhanced PAH degradation by *Pseudomonas aeruginosa* N6P6, promoting biofilm formation and EPS production (Mangwani et al., 2015). In another case, increased AHL production in Bordetella-stimulated biofilm formation, enhancing 1, 2, 4-trichlorobenzene biodegradation (Sheng et al., 2017). *Burkholderia cepacia d*, a biofilm-forming, AHL-producing strain, displayed remediation capabilities for polychlorinated biphenyls (PCBs) and PAHs (Zhong & He, 2021). Similarly, biofilm formation in Pseudomonas species improved the degradability of crude oil (Dasgupta et al., 2013).

In situ phytoremediation, a sustainable and cost-effective approach for industrial wastewater treatment, involves using plants and rhizosphere microorganisms to promote pollutant biodegradation. Successful phytoremediation relies on plant survival, rhizophore support for an active microbial population, and microbial biofilms that enhance processes like plant biomass increase, phytoextraction, phyto stabilization, soil-to-root bioaccumulation, and root-to-shoot translocation (Shivaprasad et al., 2021). Phytoremediation also influences bacterial diversity through quorum sensing (QS) (Wu et al., 2020). Symbiotic relationships between bacteria and plants, mediated by AHL-based QS, are crucial in phytoremediation. These relationships involve chemical signal exchange, QS, and the differential expression of various genes (Miller & Gilmore, 2020). Furthermore, QS controls siderophore formation under iron-deficient conditions (Balan et al., 2021).

2.3.6 Impact of Quorum Sensing in Nitrification and Denitrification

Nitrification and denitrification involve a diverse array of microorganisms. Key genera of nitrifying bacteria include *Nitrosomonas*, *Nitrosospira*, *Nitrospira*, *Nitrobacter*, and *Nitrococcus* (Daims et al., 2016). Most nitrifiers are chemoautotrophic, making pure cultures rare, with only a few known to produce acyl homoserine lactones (AHLs) (Mellbye et al., 2015; Shen et al., 2016). *Nitrobacter winogradskyi*, a facultative chemolitautotroph, oxidizes nitrite and produces various AHLs, primarily C10:1-HSL. *Nitrobacter vulgaris* also produces C10:1-HSL, while *Nitrospira moscoviensis* generates C8-HSL. Interestingly, *Nitrobacter hamburgensis* has AHL synthetase and receptor proteins but does not produce detectable AHLs (Mellbye et al., 2015; Shen et al., 2016).

Comammox microorganisms, capable of complete nitrification, thrive in various habitats, including soils and wastewater treatment facilities. They are primary ammonia oxidizers in oligotrophic environments, particularly under low oxygen conditions. Research into comammox interactions with traditional nitrifiers is needed, but pure culture isolation remains challenging. Metagenomic analysis and enrichment cultivation are essential tools for studying these organisms (Pjevac et al., 2017; Poghosyan et al., 2019). Although the specific AHLs produced by comammox bacteria remain unknown, draft genomes of *Candidatus Nitrospira inopinata*, *Candidatus Nitrospira nitrosa*, and *Candidatus Nitrospira nitrificans* contain QS homologs, suggesting a role for QS in their environmental adaptation and interactions (Daims et al., 2015; Van Kessel et al., 2015).

Anammox bacteria, which oxidize ammonium to nitrogen gas using nitrite as an electron acceptor, blur the lines between nitrification and denitrification. Since their discovery in the 1990s, they have gained attention for their unique physiology and potential in wastewater treatment. Draft genomes of *Candidatus Kueneniastuttgartiensis* and *Candidatus Jettenia caeni* contain QS genes. *Candidatus Jettenia caeni* expresses AHL synthesis genes (jqsI-1 and jqsI-2) in *Escherichia coli*, producing various AHLs including 3-oxo-C6-HSL and C8-HSL (Tang et al., 2019).

Denitrifying bacteria, including genera like *Pseudomonas*, *Paracoccus*, *Alcaligenes*, *Bacillus*, and *Thiobacillus*, display significant biochemical and taxonomic diversity (Knowles, 1982). *Pseudomonas aeruginosa* is a prominent denitrifier with well-characterized RhlR-RhlI and LasR-LasI QS systems, producing C4-HSL and 3-oxo-C12-HSL, respectively. Aerobic denitrification pathways have also been elucidated (Ji et al., 2015). *Paracoccus denitrificans*, a model aerobic denitrifier, uses exogenous C6-HSL to regulate gene expression and N_2O production, employing the pdeI/pdeR QS system to produce C16-HSL (Zhang et al., 2018), while also utilizing AHLs of varying carbon chain lengths (Morinaga et al., 2018).

In summary, quorum sensing (QS) is prevalent among both nitrifying and denitrifying bacteria. The diversity of AHLs produced and the distinct AHL synthetases involved highlight the complexity of QS within these microbial communities.

2.3.7 Quorum Sensing and Microbial Fuel Cells

Quorum sensing (QS) is emerging as a key strategy to enhance microbial fuel cell (MFC) technology for sustainable energy production. MFCs generate electricity by decomposing organic waste using live microorganisms, transforming waste treatment processes from domestic and industrial sources into renewable energy solutions (Gajda et al., 2017; Hu et al., 2018; Yong et al., 2014).

MFCs utilize organic waste as a feedstock for various applications, including bioelectricity generation, biohydrogen production, wastewater treatment, biosensor development, and bioremediation (Butti et al., 2016). Modulating QS can enhance energy production by improving anode biofilms. For instance, bacterial QS molecules can boost lipid content and photosynthetic efficiency in *Chlorella sorokiniana*, enhancing electricity generation. Engineered QS systems also improve wastewater degradation (Zhang & Li, 2016; Zhu et al., 2020). In MFCs, extracellular electron transfer from bacteria to the anode occurs through three mechanisms: (i) Electrically conductive pili, (ii) electron mediators (either artificial or microbial), and (iii) direct transfer via outer membrane c-type cytochromes (Singh et al., 2006). In a two-chamber MFC, biofilms in the anaerobic anode chamber oxidize substrates, directing released electrons toward the electrode and generating electricity (Wang et al., 2013).

In *Pseudomonas aeruginosa*, phenazines serve as electron mediators, enhancing electron transfer to the electrode (Venkataraman et al., 2010). The transcriptional regulator LasR positively influences the pqsABCDE operon, leading to PQS production, which regulates phenazine synthesis (Sahreen et al., 2022). Overproduction of phenazines through the multi-copy plasmid pYC-rhlIR increased current output by 1.6-fold compared to wild-type strains. However, PQS can inhibit anaerobic growth, limiting electricity production (Haghighi et al., 2011; Toyofuku et al., 2010). A modified strain with defective PQS and enhanced phenazine production achieved a fivefold increase in current density (Wang et al., 2013).

Additionally, acyl-homoserine lactones (AHLs) significantly affect biofilm formation at anodes in microbial electrochemical systems. The short-chain AHL

3-oxo-C6-HSL, for instance, increased hydrogen yields by altering microbial community structures in cathodic biofilms (Cai et al., 2016). In a study with an AHL-deficient *P. aeruginosa* mutant, exogenous C4-HSL restored electricity production to wild-type levels (Yang et al., 2018). Overall, wastewater, rich in organic compounds, is an ideal substrate for MFCs, offering an eco-friendly alternative to traditional waste removal methods by leveraging microbial activity for simultaneous wastewater treatment and electricity generation.

2.4 Conclusion and Future Prospects

Quorum sensing (QS) serves as a fundamental mechanism orchestrating intra- and interspecies communication among microbial cells, exerting profound influence over critical physiological processes. These processes encompass the intricate regulation of crucial aspects such as the production of extracellular polymeric substances (EPS), initiation of siderophore biosynthesis, facilitation of genetic transformation, orchestration of cellular aggregation, modulation of colonization dynamics, promotion of granulation phenomena, optimization of nutrient removal efficiency, enhancement of pollutant solubilization capabilities, proficient degradation of a wide spectrum of biological and inorganic pollutants, and the effective management of biofouling within complex microbial consortia.

Moreover, the versatile role of quorum sensing extends its impact across various domains including agriculture, aquaculture, textile and pharmaceutical industry, industrial waste degradation, bioremediation, wastewater treatment, and energy production. These diverse applications collectively contribute to the overarching goal of sustainable development by harnessing the regulatory process of quorum sensing in microbial communities.

In conclusion, quorum sensing represents a fascinating and versatile biological mechanism with significant implications for sustainable development across various fields. By unlocking the potential of microbial communication, we can optimize biotechnological processes, enhance environmental remediation efforts, and promote sustainable practices in agriculture, bioprocessing, and beyond. Prospects focus on a deeper exploration of quorum sensing.

References

Abbamondi, G. R., Tommonaro, G., Weyens, N., Thijs, S., Sillen, W., Gkorezis, P., Iodice, C., de Melo Rangel, W., Nicolaus, B., & Vangronsveld, J. (2016). Plant growth-promoting effects of rhizospheric and endophytic bacteria associated with different tomato cultivars and new tomato hybrids. *Chemical and Biological Technologies in Agriculture, 3*(1), 1–10. https://doi.org/10.1186/s40538-015-0051-3

Abdelaziz, A. A., Kamer, A. M. A., Al-Monofy, K. B., & Al-Madboly, L. A. (2023). Pseudomonas aeruginosa's greenish-blue pigment pyocyanin: Its production and biological activities. *Microbial Cell Factories, 22*(1), 1–14. https://doi.org/10.1186/s12934-023-02122-1

Anahas, A. M. P., Kumaran, S., Kandeel, M., Muralitharan, G., Silviya, J., Adhimoolam, G. L., Panagal, M., Pugazhvendan, S. R., Suresh, G., Wilson Aruni, A., Rethinam, S., & Prasannabalaji, N. (2022). Applications of natural violet pigments from halophilic Chromobacterium violaceum PDF23 for textile dyeing with antimicrobial and antioxidant potentials. *Journal of Nanomaterials, 2022*. https://doi.org/10.1155/2022/3885396

Bai, X., Todd, C. D., Desikan, R., Yang, Y., & Hu, X. (2012). N-3-oxo-decanoyl-l-homoserine-lactone activates auxin-induced adventitious root formation via hydrogen peroxide- and nitric oxide-dependent cyclic GMP signaling in mung bean. *Plant Physiology, 158*(2), 725–736. https://doi.org/10.1104/pp.111.185769

Balakrishnan, S., Ameer, A., Pazhur Mohandas, S., Maliyekkal Sajeevan, A. K., Bhaskaran Sathyabhama, A., & Singh, B. (2022). Pyocyanin as a safe aquaculture drug for the control of vibriosis in shrimp recirculating aquaculture system (RAS). *Aquaculture International, 30*(4), 2129–2144. https://doi.org/10.1007/s10499-022-00890-y

Balan, B., Dhaulaniya, A. S., Varma, D. A., Sodhi, K. K., Kumar, M., Tiwari, M., & Singh, D. K. (2021). Microbial biofilm ecology, in silico study of quorum sensing receptor-ligand interactions and biofilm mediated bioremediation. *Archives of Microbiology, 203*(1), 13–30. https://doi.org/10.1007/s00203-020-02012-9

Butti, S. K., Velvizhi, G., Sulonen, M. L. K., Haavisto, J. M., Oguz Koroglu, E., Yusuf Cetinkaya, A., Singh, S., Arya, D., Annie Modestra, J., Vamsi Krishna, K., Verma, A., Ozkaya, B., Lakaniemi, A. M., Puhakka, J. A., & Venkata Mohan, S. (2016). Microbial electrochemical technologies with the perspective of harnessing bioenergy: Maneuvering towards upscaling. *Renewable and Sustainable Energy Reviews, 53*, 462–476. https://doi.org/10.1016/j.rser.2015.08.058

Cai, W., Zhang, Z., Ren, G., Shen, Q., Hou, Y., Ma, A., Deng, Y., Wang, A., & Liu, W. (2016). Quorum sensing alters the microbial community of electrode-respiring bacteria and hydrogen scavengers toward improving hydrogen yield in microbial electrolysis cells. *Applied Energy, 183*, 1133–1141. https://doi.org/10.1016/j.apenergy.2016.09.074

Chandra, R., Kumar, V., Tripathi, S., & Sharma, P. (2018). Heavy metal phytoextraction potential of native weeds and grasses from endocrine-disrupting chemicals rich complex distillery sludge and their histological observations during in-situ phytoremediation. *Ecological Engineering, 111*(November 2017), 143–156. https://doi.org/10.1016/j.ecoleng.2017.12.007

Chen, Y. Y., Ju, S. P., & Lee, D. J. (2016). Aerobic granulation of protein-rich granules from nitrogen-lean wastewaters. *Bioresource Technology, 218*, 469–475. https://doi.org/10.1016/j.biortech.2016.06.120

Chen, H., Li, A., Cui, D., Wang, Q., Wu, D., Cui, C., & Ma, F. (2018). N-Acyl-homoserine lactones and autoinducer-2-mediated quorum sensing during wastewater treatment. *Applied Microbiology and Biotechnology, 102*(3), 1119–1130. https://doi.org/10.1007/s00253-017-8697-3

Choudhary, S., & Schmidt-Dannert, C. (2010). Applications of quorum sensing in biotechnology. *Applied Microbiology and Biotechnology, 86*(5), 1267–1279. https://doi.org/10.1007/s00253-010-2521-7

Chug, R., Mathur, S., Kothari, S. L., Harish, & Gour, V. S. (2021). Maximizing EPS production from Pseudomonas aeruginosa and its application in Cr and Ni sequestration. *Biochemistry and Biophysics Reports, 26*(February), 100972. https://doi.org/10.1016/j.bbrep.2021.100972

Clarke, M. B., Hughes, D. T., Zhu, C., Boedeker, E. C., & Sperandio, V. (2006). The QseC sensor kinase: A bacterial adrenergic receptor. *Proceedings of the National Academy of Sciences of the United States of America, 103*(27), 10420–10425. https://doi.org/10.1073/pnas.0604343103

Coughlan, L. M., Cotter, P. D., Hill, C., & Alvarez-Ordóñez, A. (2016). New weapons to fight old enemies: Novel strategies for the (bio)control of bacterial biofilms in the food industry. *Frontiers in Microbiology, 7*(Oct), 1–21. https://doi.org/10.3389/fmicb.2016.01641

Daims, H., Lebedeva, E. V., Pjevac, P., Han, P., Herbold, C., Albertsen, M., Jehmlich, N., Palatinszky, M., Vierheilig, J., Bulaev, A., Kirkegaard, R. H., Von Bergen, M., Rattei, T., Bendinger, B., Nielsen, P. H., & Wagner, M. (2015). Complete nitrification by Nitrospira bacteria. *Nature, 528*(7583), 504–509. https://doi.org/10.1038/nature16461

Daims, H., Lücker, S., & Wagner, M. (2016). A new perspective on microbes formerly known as nitrite-oxidizing bacteria. *Trends in Microbiology, 24*(9), 699–712. https://doi.org/10.1016/j.tim.2016.05.004

Dang, H., & Lovell, C. R. (2016). Microbial surface colonization and biofilm development in marine environments. *Microbiology and Molecular Biology Reviews, 80*(1), 91–138. https://doi.org/10.1128/mmbr.00037-15

Darshan, N., & Manonmani, H. K. (2015). Prodigiosin and its potential applications. *Journal of Food Science and Technology, 52*(9), 5393–5407. https://doi.org/10.1007/s13197-015-1740-4

Dasgupta, D., Ghosh, R., & Sengupta, T. K. (2013). Biofilm-mediated enhanced crude oil degradation by newly isolated pseudomonas species. *ISRN Biotechnol, 2013*, 250749.

Diggle, S. P., Cornelis, P., Williams, P., & Cámara, M. (2006). 4-Quinolone signalling in Pseudomonas aeruginosa: Old molecules, new perspectives. *International Journal of Medical Microbiology, 296*(2–3), 83–91. https://doi.org/10.1016/j.ijmm.2006.01.038

Dogsa, I., Spacapan, M., Drago, A., & Mandic-mulec, I. (2021). Peptide signaling without feedback in signal production operates as a true quorum sensing communication system in Bacillus subtilis. *Communications Biology, 4*, 58. https://doi.org/10.1038/s42003-020-01553-5

Dong, Y. H., Wang, L. H., Xu, J. L., Zhang, H. B., Zhang, X. F., & Zhang, L. H. (2001). Quenching quorum-sensing-dependent bacterial infection by an N-acyl homoserine lactonase. *Nature, 411*(6839), 813–817. https://doi.org/10.1038/35081101

Fekrirad, Z., Gattali, B., & Kashef, N. (2020). Quorum sensing-regulated functions of serratia marcescens are reduced by Eugenol. *Iranian Journal of Microbiology, 12*(5), 451–459. https://doi.org/10.18502/ijm.v12i5.4607

Flemming, H. C., Wingender, J., Szewzyk, U., Steinberg, P., Rice, S. A., & Kjelleberg, S. (2016). Biofilms: An emergent form of bacterial life. *Nature Reviews Microbiology, 14*(9), 563–575. https://doi.org/10.1038/nrmicro.2016.94

Gajda, I., Stinchcombe, A., Greenman, J., Melhuish, C., & Ieropoulos, I. (2017). Microbial fuel cell – A novel self-powered wastewater electrolyser for electrocoagulation of heavy metals. *International Journal of Hydrogen Energy, 42*(3), 1813–1819. https://doi.org/10.1016/j.ijhydene.2016.06.161

Götz-Rösch, C., Sieper, T., Fekete, A., Schmitt-Kopplin, P., Hartmann, A., & Schröder, P. (2015). Influence of bacterial N-acyl-homoserine lactoneson growth parameters,pigments,antioxidative capacities and the xenobiotic phase II detoxification enzymesin barley and yam bean. *Frontiers in Plant Science, 6*(Apr), 1–13. https://doi.org/10.3389/fpls.2015.00205

Gudesblat, G. E., Torres, P. S., & Vojnov, A. A. (2009). Xanthomonas campestris overcomes Arabidopsis stomatal innate immunity through a DSF cell-to-cell signal-regulated virulence factor. *Plant Physiology, 149*(2), 1017–1027. https://doi.org/10.1104/pp.108.126870

Haghighi, B. J., Alizadeh, O., & Firoozabadi, A. H. (2011). The role of plant growth promoting rhizobacteria (PGPR) in sustainable agriculture. *Advances in Environmental Biology, 5*(10), 3079–3083.

Halide, O., & Eq, S. (2017). Electronic Supplementary Material (ESI) for RSC advances. *This Journal is © The Royal Society of Chemistry 2017 Supplementary Material, 250022*(111), 1–9.

Hawver, L. A., Jung, S. A., & Ng, W. L. (2016). Specificity and complexity in bacterial quorum-sensing systemsa. *FEMS Microbiology Reviews, 40*(5), 738–752. https://doi.org/10.1093/femsre/fuw014

He, Y. W., & Zhang, L. H. (2008). Quorum sensing and virulence regulation in Xanthomonas campestris. *FEMS Microbiology Reviews, 32*(5), 842–857. https://doi.org/10.1111/j.1574-6976.2008.00120.x

He, Y. W., Deng, Y., Miao, Y., Chatterjee, S., Tran, T. M., Tian, J., & Lindow, S. (2023). DSF-family quorum sensing signal-mediated intraspecies, interspecies, and inter-kingdom communication. *Trends in Microbiology, 31*(1), 36–50. https://doi.org/10.1016/j.tim.2022.07.006

Hu, J., Zhang, Q., Lee, D. J., & Ngo, H. H. (2018). Feasible use of microbial fuel cells for pollution treatment. *Renewable Energy, 129*, 824–829. https://doi.org/10.1016/j.renene.2017.02.001

In, A., Genetics, T. H. E., Bacteria, N. F., Genetik, L. F. U. R., Biologie, F. F. U. R., Bielefeld, U., Rinaudi, L. V., Giordano, W., LaSarre, B., Federle, M. J., Cubo, M. T., Economou, A., Murphy,

G., Johnston, A. W. B., Downie, J. A., Tun-Garrido, C., Bustos, P., González, V., Brom, S., et al. (2002). Aryl-homoserine lactone quorum sensing in stem-nodulating photosynthetic bradyrhizobia. *Journal of Bacteriology, 12*(1), 7183–7188. https://doi.org/10.1073/pnas.1103821108

Jaafar, F. N., Al-Bayati, M. A., Musafer, H. K., Azeez, M. A., & Raheem, Z. K. (2022). Quorum sensing and its correlation with virulence factors. *South Asian Research Journal of Pharmaceutical Sciences, 4*(3), 60–69. https://doi.org/10.36346/sarjps.2022.v04i03.003

Ji, B., Yang, K., Zhu, L., Jiang, Y., Wang, H., Zhou, J., & Zhang, H. (2015). Aerobic denitrification: A review of important advances of the last 30 years. *Biotechnology and Bioprocess Engineering, 20*(4), 643–651. https://doi.org/10.1007/s12257-015-0009-0

Jiang, W., Xia, S., Liang, J., Zhang, Z., & Hermanowicz, S. W. (2013). Effect of quorum quenching on the reactor performance, biofouling and biomass characteristics in membrane bioreactors. *Water Research, 47*(1), 187–196. https://doi.org/10.1016/j.watres.2012.09.050

Jung, S. A., Chapman, C. A., & Ng, W. L. (2015). Quadruple quorum-sensing inputs control vibrio cholerae virulence and maintain system robustness. *PLoS Pathogens, 11*(4), 1–19. https://doi.org/10.1371/journal.ppat.1004837

Kim, C. S., Gatsios, A., Cuesta, S., Lam, Y. C., Wei, Z., Chen, H., Russell, R. M., Shine, E. E., Wang, R., Wyche, T. P., Piizzi, G., Flavell, R. A., Palm, N. W., Sperandio, V., & Crawford, J. M. (2020). Characterization of Autoinducer-3 structure and biosynthesis in E. coli. *ACS Central Science, 6*(2), 197–206. https://doi.org/10.1021/acscentsci.9b01076

Knowles, R. (1982). Denitrification. *Microbiological Reviews, 46*(1), 43–70. https://doi.org/10.1128/mr.46.1.43-70.1982

Kramar, A., & Kostic, M. M. (2022). Bacterial secondary metabolites as biopigments for textile dyeing. *Text, 2*(2), 252–264. https://doi.org/10.3390/textiles2020013

Kumar, A., Vishwakarma, H. S., Singh, J., & Kumar, M. (2015). Microbial pigments: Production and their applications in various industries. *International Journal of Pharmaceutical, Chemical and Biological Sciences, 5*(1), 203–212.

Lee, V. T., Matewish, J. M., Kessler, J. L., Hyodo, M., Hayakawa, Y., & Lory, S. (2007). A cyclic-di-GMP receptor required for bacterial exopolysaccharide production. *Molecular Microbiology, 65*(6), 1474–1484. https://doi.org/10.1111/j.1365-2958.2007.05879.x

Li, Y. S., Pan, X. R., Cao, J. S., Song, X. N., Fang, F., Tong, Z. H., Li, W. W., & Yu, H. Q. (2017). Augmentation of acyl homoserine lactones-producing and -quenching bacterium into activated sludge for its granulation. *Water Research, 125*, 309–317. https://doi.org/10.1016/j.watres.2017.08.061

Liu, X., Sun, S., Ma, B., Zhang, C., Wan, C., & Lee, D. J. (2016). Understanding of aerobic granulation enhanced by starvation in the perspective of quorum sensing. *Applied Microbiology and Biotechnology, 100*(8), 3747–3755. https://doi.org/10.1007/s00253-015-7246-1

Lupp, C., & Ruby, E. G. (2005). Vibrio fischeri uses two quorum-sensing systems for the regulation of early and late colonization factors. *Journal of Bacteriology, 187*(11), 3620–3629. https://doi.org/10.1128/JB.187.11.3620-3629.2005

Lyte, M., & Freestone, P. P. E. (2010). Microbial endocrinology: Interkingdom signaling in infectious disease and health. *Microbial Endocrinology: Interkingdom Signaling in Infectious Disease and Health*, 1–316. https://doi.org/10.1007/978-1-4419-5576-0

Majdura, J., Jankiewicz, U., Gałązka, A., & Orzechowski, S. (2023). The role of quorum sensing molecules in bacterial–plant interactions. *Metabolites, 13*(1). https://doi.org/10.3390/metabo13010114

Mangwani, N., Kumari, S., & Das, S. (2015). Involvement of quorum sensing genes in biofilm development and degradation of polycyclic aromatic hydrocarbons by a marine bacterium Pseudomonas aeruginosa N6P6. *Applied Microbiology and Biotechnology, 99*(23), 10283–10297. https://doi.org/10.1007/s00253-015-6868-7

Mellbye, B. L., Bottomley, P. J., & Sayavedra-Soto, L. A. (2015). Nitrite-oxidizing bacterium Nitrobacter winogradskyi produces N-acyl-homoserine lactone autoinducers. *Applied and Environmental Microbiology, 81*(17), 5917–5926. https://doi.org/10.1128/AEM.01103-15

Mellbye, B. L., Giguere, A. T., Bottomley, P. J., & Sayavedra-Soto, L. A. (2016). Quorum quenching of nitrobacter winogradskyi suggests that quorum sensing regulates fluxes of nitrogen oxide(s) during nitrification. *MBio, 7*(5). https://doi.org/10.1128/mBio.01753-16

Meng, F., Zhang, S., Oh, Y., Zhou, Z., Shin, H. S., & Chae, S. R. (2017). Fouling in membrane bioreactors: An updated review. *Water Research, 114*, 151–180. https://doi.org/10.1016/j.watres.2017.02.006

Miller, M. B., & Bassler, B. L. (2001). Quorum sensing in bacteria. *Annual Review of Microbiology, 55*, 165–199. https://doi.org/10.1146/annurev.micro.55.1.165

Miller, C., & Gilmore, J. (2020). Detection of quorum-sensing molecules for pathogenic molecules using cell-based and cell-free biosensors. *Antibiotics, 9*(5). https://doi.org/10.3390/antibiotics9050259

Miyashiro, T., Oehlert, D., Ray, V. A., Visick, K. L., & Ruby, E. G. (2014). The putative oligosaccharide translocase SypK connects biofilm formation with quorum signaling in Vibrio fischeri. *MicrobiologyOpen, 3*(6), 836–848. https://doi.org/10.1002/mbo3.199

Morinaga, K., Yamamoto, T., Nomura, N., & Toyofuku, M. (2018). Paracoccus denitrificans can utilize various long-chain N-acyl homoserine lactones and sequester them in membrane vesicles. *Environmental Microbiology Reports, 10*(6), 651–654. https://doi.org/10.1111/1758-2229.12674

Nealson, K. H., & Hastings, J. W. (1979). Bacterial bioluminescence: Its control and ecological significance. *Microbiological Reviews, 43*(4), 496–518. https://doi.org/10.1128/mmbr.43.4.496-518.1979

Patel, R., Soni, M., Soyantar, B., Shivangi, S., Sutariya, S., & Saraf, M. (2023). A clash of quorum sensing vs quorum sensing inhibitors: An overview and risk of resistance. *Archives of Microbiology*, 1–23. https://doi.org/10.1007/s00203-023-03442-x

Pietschke, C., Treitz, C., Forêt, S., Schultze, A., Künzel, S., Tholey, A., Bosch, T. C. G., & Fraune, S. (2017). Host modification of a bacterial quorum-sensing signal induces a phenotypic switch in bacterial symbionts. *Proceedings of the National Academy of Sciences of the United States of America, 114*(40), E8488–E8497. https://doi.org/10.1073/pnas.1706879114

Pjevac, P., Schauberger, C., Poghosyan, L., Herbold, C. W., van Kessel, M. A. H. J., Daebeler, A., Steinberger, M., Jetten, M. S. M., Lücker, S., Wagner, M., & Daims, H. (2017). AmoA-targeted polymerase chain reaction primers for the specific detection and quantification of comammox Nitrospira in the environment. *Frontiers in Microbiology, 8*(Aug), 1–11. https://doi.org/10.3389/fmicb.2017.01508

Poghosyan, L., Koch, H., Lavy, A., Frank, J., van Kessel, M. A. H. J., Jetten, M. S. M., Banfield, J. F., & Lücker, S. (2019). Metagenomic recovery of two distinct comammox Nitrospira from the terrestrial subsurface. *Environmental Microbiology, 21*(10), 3627–3637. https://doi.org/10.1111/1462-2920.14691

Sahreen, S., Mukhtar, H., Imre, K., Morar, A., Herman, V., & Sharif, S. (2022). Exploring the function of quorum sensing regulated biofilms in biological wastewater treatment: A review. *International Journal of Molecular Sciences, 23*(17). https://doi.org/10.3390/ijms23179751

Saxena, P., Joshi, Y., Rawat, K., & Bisht, R. (2019). Biofilms: Architecture, resistance, quorum sensing and control mechanisms. *Indian Journal of Microbiology, 59*(1), 3–12. https://doi.org/10.1007/s12088-018-0757-6

Schikora, A., Schenk, S. T., & Hartmann, A. (2016). Beneficial effects of bacteria-plant communication based on quorum sensing molecules of the N-acyl homoserine lactone group. *Plant Molecular Biology, 90*(6), 605–612. https://doi.org/10.1007/s11103-016-0457-8

Sela, R., Hammer, B. K., & Halpern, M. (2021). Quorum-sensing signaling by chironomid egg masses' microbiota, affects haemagglutinin/protease (HAP) production by Vibrio cholerae. *Molecular Ecology, 30*(7), 1736–1746. https://doi.org/10.1111/mec.15662

Shah, M., Taylor, V. L., Bona, D., Tsao, Y., Stanley, S. Y., Pimentel-Elardo, S. M., McCallum, M., Bondy-Denomy, J., Howell, P. L., Nodwell, J. R., Davidson, A. R., Moraes, T. F., & Maxwell, K. L. (2021). A phage-encoded anti-activator inhibits quorum sensing in Pseudomonas aeruginosa. *Molecular Cell, 81*(3), 571–583.e6. https://doi.org/10.1016/j.molcel.2020.12.011

Sharma, S., & Saraf, M. (2023). Enhanced exopolysaccharide production by multi metal tolerant Klebsiella variicolaSMHMZ46 isolated from mines area and application in metal bioremediation. *International Microbiology, 26*, 1167. https://doi.org/10.1007/s10123-023-00366-w

Shen, Q., Gao, J., Liu, J., Liu, S., Liu, Z., Wang, Y., Guo, B., Zhuang, X., & Zhuang, G. (2016). A new acyl-homoserine lactone molecule generated by Nitrobacter winogradskyi. *Scientific Reports, 6*(November 2015), 1–11. https://doi.org/10.1038/srep22903

Sheng, H., Harir, M., Boughner, L. A., Jiang, X., Schmitt-Kopplin, P., Schroll, R., & Wang, F. (2017). N-acyl-homoserine lactone dynamics during biofilm formation of a 1,2,4-trichlorobenzene mineralizing community on clay. *Science of the Total Environment, 605–606*, 1031–1038. https://doi.org/10.1016/j.scitotenv.2017.06.233

Shi, Y., Huang, J., Zeng, G., Gu, Y., Chen, Y., Hu, Y., Tang, B., Zhou, J., Yang, Y., & Shi, L. (2017). Exploiting extracellular polymeric substances (EPS) controlling strategies for performance enhancement of biological wastewater treatments: An overview. *Chemosphere, 180*, 396–411. https://doi.org/10.1016/j.chemosphere.2017.04.042

Shivaprasad, D. P., Taneja, N. K., Lakra, A., & Sachdev, D. (2021). In vitro and in situ abrogation of biofilm formation in E. coli by vitamin C through ROS generation, disruption of quorum sensing and exopolysaccharide production. *Food Chemistry, 341*(July 2020), 128171. https://doi.org/10.1016/j.foodchem.2020.128171

Sieper, T., Forczek, S., Matucha, M., Krämer, P., Hartmann, A., & Schröder, P. (2014). N-acyl-homoserine lactone uptake and systemic transport in barley rest upon active parts of the plant. *New Phytologist, 201*(2), 545–555. https://doi.org/10.1111/nph.12519

Simsek Geyik, M., Efe, D., & Gormez, A. (2024). Identification of Bacteria producing red pigments and their application in the textile industry. *Arabian Journal for Science and Engineering*. https://doi.org/10.1007/s13369-024-09290-1

Singh, R., & Ray, P. (2014). Quorum sensing-mediated regulation of staphylococcal virulence and antibiotic resistance. *Future Microbiology, 9*(5), 669–681. https://doi.org/10.2217/fmb.14.31

Singh, R., Paul, D., & Jain, R. K. (2006). Biofilms: Implications in bioremediation. *Trends in Microbiology, 14*(9), 389–397. https://doi.org/10.1016/j.tim.2006.07.001

Singh, D. P., Singh, H. B., & Prabha, R. (2017). Plant-microbe interactions in agro-ecological perspectives. *Plant-Microbe Interactions in Agro-Ecological Perspectives, 1*, 1–657. https://doi.org/10.1007/978-981-10-5813-4

Singh, S., Nimse, S. B., Mathew, D. E., Dhimmar, A., Sahastrabudhe, H., Gajjar, A., Ghadge, V. A., Kumar, P., & Shinde, P. B. (2021). Microbial melanin: Recent advances in biosynthesis, extraction, characterization, and applications. *Biotechnology Advances, 53*(May), 107773. https://doi.org/10.1016/j.biotechadv.2021.107773

Soto-Aceves, M. P., Cocotl-Yañez, M., Servín-González, L., & Soberón-Chávez, G. (2020). The Rhl quorum-sensing system is at the top of the regulatory hierarchy under phosphate-limiting conditions in pseudomonas aeruginosa PAO1. *Journal of Bacteriology, 203*(5). https://doi.org/10.1128/JB.00475-20

Sturme, M. H. J., Kleerebezem, M., Nakayama, J., Akkermans, A. D. L., Vaughan, E. E., & De Vos, W. M. (2002). Sturme MHJ 2002 peptides in gram-positive bacteria. *Antonie Van Leeuwenhoek, 81*(1), 233–243.

Sun, S., Liu, X., Ma, B., Wan, C., & Lee, D. J. (2016). The role of autoinducer-2 in aerobic granulation using alternating feed loadings strategy. *Bioresource Technology, 201*, 58–64. https://doi.org/10.1016/j.biortech.2015.11.032

Tan, C. H., Koh, K. S., Xie, C., Tay, M., Zhou, Y., Williams, R., Ng, W. J., Rice, S. A., & Kjelleberg, S. (2014a). The role of quorum sensing signalling in EPS production and the assembly of a sludge community into aerobic granules. *ISME Journal, 8*(6), 1186–1197. https://doi.org/10.1038/ismej.2013.240

Tan, P. W., Tan, W. S., Yunos, N. Y. M., Mohamad, N. I., Adrian, T. G. S., Yin, W. F., & Chan, K. G. (2014b). Short chain N-Acyl homoserine lactone production in tropical marine vibrio sinaloensis strain T47. *Sensors (Switzerland), 14*(7), 12958–12967. https://doi.org/10.3390/s140712958

Tang, X., Guo, Y., Zhu, T., Tao, H., & Liu, S. (2019). Identification of quorum sensing signal AHLs synthases in Candidatus Jettenia caeni and their roles in anammox activity. *Chemosphere, 225*, 608–617. https://doi.org/10.1016/j.chemosphere.2019.02.192

Tarighi, S., & Taheri, P. (2011). Different aspects of bacterial communication signals. *World Journal of Microbiology and Biotechnology, 27*(6), 1267–1280. https://doi.org/10.1007/s11274-010-0575-4

Tommonaro, G., Abbamondi, G. R., Nicolaus, B., Poli, A., D'Angelo, C., Iodice, C., & De Prisco, R. (2021). Productivity and nutritional trait improvements of different tomatoes cultivated with effective microorganisms technology. *Agriculture (Switzerland), 11*(2), 1–10. https://doi.org/10.3390/agriculture11020112

Toyofuku, M., Nakajima-Kambe, T., Uchiyama, H., & Nomura, N. (2010). The effect of a cell-to-cell communication molecule, Pseudomonas quinolone signal (PQS), produced by p. aeruginosa on other bacterial species. *Microbes and Environments, 25*(1), 1–7. https://doi.org/10.1264/jsme2.ME09156

Tripathi, S., Chandra, R., Purchase, D., Bilal, M., & Mythili, R. (2022). Quorum sensing-a promising tool for degradation of industrial waste containing persistent organic pollutants. *Environmental Pollution, 292*, 118342.

Van Kessel, M. A. H. J., Speth, D. R., Albertsen, M., Nielsen, P. H., Op Den Camp, H. J. M., Kartal, B., Jetten, M. S. M., & Lücker, S. (2015). Complete nitrification by a single microorganism. *Nature, 528*(7583), 555–559. https://doi.org/10.1038/nature16459

Venil, C. K., Dufossé, L., Velmurugan, P., Malathi, M., & Lakshmanaperumalsamy, P. (2021). Extraction and application of pigment from Serratia marcescens SB08, an insect enteric gut bacterium, for textile dyeing. *Text, 1*(1), 21–36. https://doi.org/10.3390/textiles1010003

Venkataraman, A., Rosenbaum, M., Arends, J. B. A., Halitschke, R., & Angenent, L. T. (2010). Quorum sensing regulates electric current generation of Pseudomonas aeruginosa PA14 in bioelectrochemical systems. *Electrochemistry Communications, 12*(3), 459–462. https://doi.org/10.1016/j.elecom.2010.01.019

Von Rad, U., Klein, I., Dobrev, P. I., Kottova, J., Zazimalova, E., Fekete, A., Hartmann, A., Schmitt-Kopplin, P., & Durner, J. (2008). Response of Arabidopsis thaliana to N-hexanoyl-DL-homoserine-lactone, a bacterial quorum sensing molecule produced in the rhizosphere. *Planta, 229*(1), 73–85. https://doi.org/10.1007/s00425-008-0811-4

Wan, C., Zhang, P., Lee, D. J., Yang, X., Liu, X., Sun, S., & Pan, X. (2013). Disintegration of aerobic granules: Role of second messenger cyclic di-GMP. *Bioresource Technology, 146*, 330–335. https://doi.org/10.1016/j.biortech.2013.07.073

Wang, K., Bhandari, V., Chepustanova, S., Huber, G., O'Hara, S., O'Hern, C. S., Shattuck, M. D., & Kirby, M. (2013). Which biomarkers reveal neonatal sepsis? *PLoS One, 8*(12), 6–13. https://doi.org/10.1371/journal.pone.0082700

Wang, J., Liu, Q., Dong, D., Hu, H., Wu, B., & Ren, H. (2021a). AHLs-mediated quorum sensing threshold and its response towards initial adhesion of wastewater biofilms. *Water Research, 194*, 116925. https://doi.org/10.1016/j.watres.2021.116925

Wang, Z., Liu, X., Ni, S. Q., Zhuang, X., & Lee, T. (2021b). Nano zero-valent iron improves anammox activity by promoting the activity of quorum sensing system. *Water Research, 202*(July), 117491. https://doi.org/10.1016/j.watres.2021.117491

Wu, S., Liu, J., Liu, C., Yang, A., & Qiao, J. (2020). Quorum sensing for population-level control of bacteria and potential therapeutic applications. *Cellular and Molecular Life Sciences, 77*(7), 1319–1343. https://doi.org/10.1007/s00018-019-03326-8

Xiao, Y., Zou, H., Li, J., Song, T., Lv, W., Wang, W., Wang, Z., & Tao, S. (2022). Impact of quorum sensing signaling molecules in gram-negative bacteria on host cells: Current understanding and future perspectives. *Gut Microbes, 14*(1), 1–19. https://doi.org/10.1080/19490976.2022.2039048

Xiong, Y., & Liu, Y. (2012). Essential roles of eDNA and AI-2 in aerobic granulation in sequencing batch reactors operated at different settling times. *Applied Microbiology and Biotechnology, 93*(6), 2645–2651. https://doi.org/10.1007/s00253-011-3565-z

Yadav, S., & Chandra, R. (2018). Detection and assessment of the phytotoxicity of residual organic pollutants in sediment contaminated with pulp and paper mill effluent. *Environmental Monitoring and Assessment, 190*(10), 581. https://doi.org/10.1007/s10661-018-6947-1

Yadav, S., & Chandra, R. (2019). Biofilm-mediated bioremediation of pollutants from the environment for sustainable development. In *New and future developments in microbial biotechnology and Bioengineering: Microbial biofilms current research and future trends in microbial biofilms*. Elsevier B.V. https://doi.org/10.1016/B978-0-444-64279-0.00014-1

Yang, Y., Zhou, H., Mei, X., Liu, B., & Xing, D. (2018). Dual-edged character of quorum sensing signaling molecules in microbial extracellular electron transfer. *Frontiers in Microbiology, 9*(Jul), 1–8. https://doi.org/10.3389/fmicb.2018.01924

Yong, X. Y., Feng, J., Chen, Y. L., Shi, D. Y., Xu, Y. S., Zhou, J., Wang, S. Y., Xu, L., Yong, Y. C., Sun, Y. M., Shi, C. L., OuYang, P. K., & Zheng, T. (2014). Enhancement of bioelectricity generation by cofactor manipulation in microbial fuel cell. *Biosensors and Bioelectronics, 56*, 19–25. https://doi.org/10.1016/j.bios.2013.12.058

Zhan, X., Zhang, K., Wang, C., Fan, Q., Tang, X., Zhang, X., Wang, K., Fu, Y., & Liang, H. (2024). A c-di-GMP signaling module controls responses to iron in Pseudomonas aeruginosa. *Nature Communications, 15*(1), 1–15. https://doi.org/10.1038/s41467-024-46149-3

Zhang, W., & Li, C. (2016). Exploiting quorum sensing interfering strategies in gram-negative bacteria for the enhancement of environmental applications. *Frontiers in Microbiology, 6*(Jan), 1–15. https://doi.org/10.3389/fmicb.2015.01535

Zhang, Y., & Tay, J. H. (2015). Toxic and inhibitory effects of trichloroethylene aerobic co-metabolism on phenol-grown aerobic granules. *Journal of Hazardous Materials, 286*, 204–210. https://doi.org/10.1016/j.jhazmat.2015.01.003

Zhang, R., Vivanco, J. M., & Shen, Q. (2017). The unseen rhizosphere root–soil–microbe interactions for crop production. *Current Opinion in Microbiology, 37*, 8–14. https://doi.org/10.1016/j.mib.2017.03.008

Zhang, Y., Luo, J., Flewitt, A. J., Cai, Z., & Zhao, X. (2018). Film bulk acoustic resonators (FBARs) as biosensors: A review. *Biosensors and Bioelectronics, 116*, 1–15. https://doi.org/10.1016/j.bios.2018.05.028

Zhang, J., Feng, T., Wang, J., Wang, Y., & Zhang, X. H. (2019). The mechanisms and applications of quorum sensing (QS) and quorum quenching (QQ). *Journal of Ocean University of China, 18*(6), 1427–1442. https://doi.org/10.1007/s11802-019-4073-5

Zhang, C., Ho, S. H., Li, A., Fu, L., & Zhou, D. (2021). Co-culture of Chlorella and Scenedesmus could enhance total lipid production under bacteria quorum sensing molecule stress. *Journal of Water Process Engineering, 39*(October), 101739. https://doi.org/10.1016/j.jwpe.2020.101739

Zhong, S., & He, S. (2021). Quorum sensing inhibition or quenching in Acinetobacter baumannii: The novel therapeutic strategies for new drug development. *Frontiers in Microbiology, 12*(February). https://doi.org/10.3389/fmicb.2021.558003

Zhu, Z., Yang, Y., Fang, A., Lou, Y., Xie, G., Ren, N., & Xing, D. (2020). Quorum sensing systems regulate heterotrophic nitrification-aerobic denitrification by changing the activity of nitrogen-cycling enzymes. *Environmental Science and Ecotechnology, 2*, 100026. https://doi.org/10.1016/j.ese.2020.100026

Zhuang, X., Liu, Y., Fang, N., Bai, Z., & Gao, J. (2023). Quorum sensing improves the plant growth-promoting ability of Stenotrophomonas rhizophila under saline-alkaline stress by enhancing its environmental adaptability. *Frontiers in Microbiology, 14*(April), 1–10. https://doi.org/10.3389/fmicb.2023.1155081

Soil Actinomycetes with a Perspective of Cellulolytic Activity

3

Aanal Patani, Jaimin Pandya, Apoorva Vaghela, Jahanvi Thakor, Himanshu Vasava, Kiransinh N. Rajput, Vikram H. Raval, and Rakeshkumar R. Panchal

Abstract

This chapter explores the potential of soil actinomycetes to provide insights into the bioconversion of cellulosic waste into valuable by-products. Actinomycetes are primarily found in soil, where they represent more than 30% of the entire microbial population. To survive in this environment, actinomycetes developed complex systems for utilizing organic materials by producing enzymes to use soil components and get sustenance. The study provides a detailed characterization of cellulases producing soil actinomycetes, and the characterization of cellulase enzymes, including their optimal pH, temperature, and substrate specificity. Key findings indicate that the degradation of cellulosic waste is a critical environmental and economic challenge, necessitating efficient and eco-friendly solutions and soil actinomycetes have emerged as promising candidates for this purpose due to their robust cellulase-producing capabilities. Moreover, it discusses the scalability of using these actinomycetes at the industrial level. The chapter concludes with a discussion on the gene present in the actinomycetes responsible for the production of cellulase enzymes and positioning soil actinomycetes as a sustainable solution for industrial waste management. Future research directions include exploring metabolic pathways involved in cellulase production to enhance enzyme yields and stability.

Keywords

Soil actinomycetes · Cellulase enzymes · Cellulose degradation · Environmental biotechnology

A. Patani · J. Pandya · A. Vaghela · J. Thakor · H. Vasava · K. N. Rajput · V. H. Raval
R. R. Panchal (✉)
Department of Microbiology and Biotechnology, Gujarat University, Ahmedabad, Gujarat, India
e-mail: rrpanchal@gujaratuniversity.ac.in

M. Saraf et al. (eds.), *Climate Change and Soil Microorganisms for Environmental Sustainability*, Microorganisms for Sustainability 28,
https://doi.org/10.1007/978-981-96-3425-5_3

3.1 Introduction

A class of Gram-positive bacteria known as soil actinomycetes is well known for its filamentous growth and extensive synthesis of secondary metabolites (Adegboye & Babalola, 2016). These microbes contribute to the nutrient cycle and the decomposition of organic matter, which are important components of soil ecosystems (Condron et al., 2010). With a concentration on their contributions to the synthesis of cellulases and the breakdown of organic materials, this study aims to provide an overview of the state of knowledge on the ecological functions and biotechnological uses of soil actinomycetes. Several actinomycetes may be found in soil, but *Streptomyces* is the most well-known genus because of its high synthesis of primary and secondary metabolites. *Nocardia* and *Micromonospora* are two more notable genera contributing to soil processes (Das et al., 2021). Actinomycetes play an essential role in the breakdown of complex chemical molecules like lignin and cellulose (Kirby, 2005). Research shows that actinomycetes produce enzymes such as cellulases and ligninases, which help in the breakdown of plant matter, improving the amount of organic matter in the soil and the availability of nutrients (Prasad & Raghuwanshi, 2022).

A number of microorganisms, such as fungi and bacteria, can produce lignocellulolytic enzymes. Among bacteria, actinomycetes are an important group that has been utilized to produce cellulases. The classification and application of cellulolytic actinomycetes have been discussed in this book chapter, along with an explanation of the cellulase enzyme biosynthesis and organisms used to break down cellulose (Saini et al., 2015).

Actinomycetes are also well known for their capacity to generate a broad range of antibiotics, which have proved important for the advancement of modern medicine (Demain, 2009). The first antibiotic that effectively treated TB was streptomycin, which is produced by *Streptomyces griseus*. They continue to produce a wide variety of antibiotics, which is a valuable resource for drug development (Lewis, 2017).

Actinomycetes interact with bacteria, fungi, plants, and other living bodies present in the soil. These interactions can be antagonistic, like inhibiting the growth of pathogenic microorganisms, or synergistic, like stimulating the development of plants by nitrogen fixation (Olanrewaju & Babalola, 2018). Actinomycetes in the soil are essential to the functioning of ecosystems as well as the development of biotechnological innovations. In the future, research should concentrate on exploring the genetic and metabolic diversity of actinomycetes to achieve the increased potential of these organisms in various applications. However, there are still gaps in our knowledge of the variety and ecological significance of these bacteria, which need more study with advanced molecular methods.

3.2 Characteristics of Soil

3.2.1 Nature of Soil

Soil is the most complex biomaterial in the ecosystem, with its complex composition of organic matter, inorganic particles, minerals, and microorganisms. For the metabolic processes of microorganisms, it is thought to be a suitable medium (Biswal et al., 2020).

Furthermore, it is important to the Earth's global biogeochemical cycles, which have an impact on the ecosystem and other organisms. Majority of the organic matter in soil is derived from plants. Plant tissues consist of 60–90% moisture. At the same time, the remaining dry matter is mostly constituted of carbon (C), oxygen (O_2), and hydrogen (H), with trace amounts of nitrogen (N), phosphorus (P), potassium (K), calcium (Ca), magnesium (Mg), and sulfur (S). Despite the minimal value, these nutrients are important for microorganisms (Dayal et al., 2023; Ondrasek et al., 2019). For this reason, soil is rich in various types of nutrients and an ideal medium for actinomycetes to obtain nutrients from. The diversity of actinomycetes in the ecosystem is being affected by environmental conditions and soil types, which are described in the following sections.

3.2.2 Effect of Soil Characteristics on Diversity of Actinomycetes

3.2.2.1 Soil pH

Generally, actinomycetes prefer neutral to slightly alkaline pH (6.5 to 8.0). This range promotes the best possible growth and activity, enabling these bacteria to rule (Kontro & Yaradoddi, 2021).

In acidic soils (pH below 6), the growth of actinomycetes is typically reduced. Acidic environments can change the integrity of the cell membrane and reduce the activity of enzymes, which reduces diversity and abundance (Naz et al., 2022). Soils with a pH close to neutral tend to have a more diverse community of actinomycetes. Various ecological processes, including the decomposition of organic materials and the synthesis of antibiotics, depend on this diversity. Extremely high or extremely low pH levels act as selective pressures, allowing only a small number of specialized actinomycetes that have adapted to that environment to survive. As a result, overall, less diversity was found. *Streptomyces acidiscabies* and *Streptomyces thermoacidophilus* are species that can survive and grow in an acidic environment.

3.2.2.2 Soil Texture and Structure

Soil texture and structure refer to the relative proportions of sand, silt, and clay particles in the soil and the arrangement of soil particles into aggregates or clumps respectively (Plante et al., 2006). Sandy soils have larger particles leading to better aeration and drainage, which is favorable for the growth of actinomycetes, especially those that require high oxygen levels (Sung et al., 2017). Clayey soils have fine particles that retain more water and nutrients but may have poor aeration which can limit some actinomycetes. Other actinomycetes who adapted to low-oxygen environments may thrive here (Dar, 2009). Loamy soils are often considered the best for microbial growth, it has a balanced texture that provides good aeration, moisture retention, and nutrient availability, supporting a diverse community of actinomycetes (Tarafdar, 2022). Furthermore, well-structured soils with stable aggregates create microhabitats that protect actinomycetes from environmental stresses, such as drought or extreme temperatures. Compacted or poorly structured soils limit air and water movement, reducing the availability of oxygen and nutrients, which can hinder the growth and activity of actinomycetes (Dixon & Tilston, 2010).

3.2.2.3 Organic Matter Content

Organic matter in soil composed of decomposed plant and animal residues provides essential nutrients and energy to actinomycetes. These microorganisms also break down complex organic compounds such as cellulose, lignin, and chitin (Bhatti et al., 2017). Also, actinomycetes play a crucial role in the decomposition process, converting organic matter into simpler compounds. This activity not only provides them with energy but also releases nutrients like nitrogen, phosphorus, and sulfur back into the soil, which are essential for plant growth.

3.2.2.4 Soil Moisture and Aeration

The majority of actinomycetes are aerobic microorganisms; they require oxygen for respiration, which enables them to decompose organic materials and produce energy (Boyd, 1995). Oxygen is readily available in well-aerated soils, such as those with good structure and porosity. This encourages actinomycetes to proliferate and diversify, which in turn allows various species to grow. A diversified population of actinomycetes is supported by these types of soil properties, which are frequently seen in sandy or loamy textures. According to Grenni et al. (2018), these actinomycetes may carry out various ecological functions, such as the degradation of organic materials and the synthesis of antibiotics. This is made possible by the presence of oxygen. Furthermore, the scientists added compacted or waterlogged soils have limited oxygen availability. In such environments, the diversity of actinomycetes decreases, as only facultative anaerobes or microaerophiles adapted to low-oxygen conditions can survive. This reduction in the presence of actinomycetes can negatively impact soil health and its ability to support plant growth. Soil is not uniform, and oxygen availability can vary within microhabitats.

Actinomycetes occupy these microhabitats based on their oxygen needs. For example, *Actinomadura*, *Nocardia*, *Nonomurea*, and *Streptomyces* species may colonize the rhizosphere (root zone) where oxygen levels are higher (Khamna et al., 2008), while others may be found in deeper soil layers with less oxygen. Anaerobic actinomycetes belong to several genera, but they are less diverse than aerobic actinomycetes. Notable genera include *Actinomyces*, *Bifidobacterium*, and *Propionibacterium*. In anoxic environments like deep soil layers or aquatic sediments, anaerobic actinomycetes participate in the breakdown of organic matter. These environments may also include facultative anaerobes that can switch between aerobic and anaerobic metabolism depending on the availability of oxygen. Some anaerobic actinomycetes, such as those in the genus *Propionibacterium*, are also involved in fermentation processes (Devi et al., 2022; Selim et al., 2021).

3.2.2.5 Temperature of Soil

The survival and functional effectiveness of actinomycetes are significantly affected by temperature. Occurring in a range of biological niches, from hot composts to freezing soils, their capacity for adaptation to different temperatures is very high (Kontro & Yaradoddi, 2021).

Most soil actinomycetes prefer mesophilic temperatures, which are between 25 °C and 30 °C. Their enzymatic activity is highest in this range, facilitating processes like the breakdown of organic materials and the synthesis of secondary metabolites. Compost piles contain thermophilic actinomycetes, which can survive and remain active at temperatures as high as 50 to 55 °C (López et al., 2021). These thermophiles contribute an essential role in the composting process by degrading complex organic compounds like cellulose and lignin under high temperatures. Actinomycetes have lower metabolic rates in cold environments, which can affect their development and metabolite synthesis (Votruba & Vaněk, 2020). Although thermotolerant organisms may still survive in colder environments, their productivity will usually be less than under ideal conditions. For survival at low temperatures, a large number of actinomycetes produce spores. Because of their resistance to environmental stress like temperature changes, these spores can remain dormant until appropriate conditions arise again. (Bobek et al., 2017).

Temperature regulates the production of several proteins in actinomycetes, including polyketide synthase (PKS), non-ribosomal peptide synthetase (NRPS), heat shock proteins (HSPs)—hsp60, hsp70, and hsp100 genes; cold shock proteins (CSPs)—cspA gene; chaperone genes—dnaK and groEL genes. For instance, genes involved in secondary metabolite biosynthesis may be upregulated or downregulated depending on the environmental temperature (Blin et al., 2017).

3.2.3 Potential of Soil Actinomycetes

Actinomycetes are mostly found in soil, air, plant remains, and the silt of water bodies. These are the most abundant organisms in the soil that develop threads like filaments. They grow as hyphae-like fungi and are responsible for the characteristically "earthy" smell of freshly turned healthy soil (Bhatti et al., 2017). The actinomycetes exist in various habitats in nature and represent a ubiquitous group of microbes widely distributed in natural ecosystems around the world (George, 2012). They are primarily soil inhabitants but have been found widely distributed in a diverse range of aquatic ecosystems, including sediments obtained from deep sea even from the greatest depth Mariana Trench (Jing et al., 2022). Actinomycete population is largest in the surface layer of soils and gradually decreases with depth; individual actinomycete strains are present in all soil layers (Takahashi & Omura, 2003). In addition to bacteria, actinomycetes are abundant and extensively distributed in soil.

They are widely distributed in the soil, compost, etc. with estimated values ranging from 10^4 to 10^8 per gram of soil. These organisms are mesophilic, growing at 25–30 °C, but certain species that are frequently found in compost and manures are thermophilic, growing at 55–65 °C (e.g. *Thermoatinomycetes* and *Streptomyces*). Some may also be present in extreme environments, especially at very low temperatures ranging from −20°C to 20 °C are known as cryophilic actinomycetes (Pathania et al., 2021).

3.3 Structure of Actinomycetes in Soil

The Greek terms "atkis" (ray) and "mykes" (fungus) are the origins of the word "actinomycetes." It describes a class of organisms that share characteristics with fungi and bacteria but vary sufficiently to belong in the kingdom of bacteria (Bhatti et al., 2017). Actinomycetes are aerobic, spore-forming, Gram-positive bacteria with high levels of guanine-cytosine (57–75%) in their genomes. They are classified under the order Actinomycetales; these are a group of bacteria having branched unicellular filaments (Waksman, 1940). This filamentous structure is called hyphae, which can be divided into two types. One part that grows into the medium (soil) and absorbs nutrients is known as substrate mycelium and the other part that rises above the surface is called aerial mycelium, which is also responsible for spore production (Hopwood, 2008). Some actinomycetes form specialized structures called sporangia, where spores develop (Kalakoutskii & Agre, 1976).

They have certain deficiencies when compared to their primary competitors, bacteria and fungi. Their rate of radial growth on culture media is lower than that of fungi, and their rate of cell production is generally lower than that of bacteria. Actinomycetes reproduce either by fission or by means of special spores or conidia and are characterized by the formation of normally branching threads or rods (Bhatti et al., 2017).

Generally, hyphae are not septate; but, in certain special cases, septa may be seen. The sporulating mycelium might have a straight or spiral structure, branching or not branching at all. The spores are oval, spherical, or cylindrical in shape (Waksman, 1940). They produce initial microcolonies composed of branching system filaments that after 24–48 h fragment into diptheroids, short chain, and coccobacillary forms.

The cell wall of actinomycetes is a rigid structure that maintains the shape of the cell and prevents the bursting of the cell due to high osmotic pressure. The wall consists of a large variety of complex compounds, including peptidoglycan, teichoic and teichuronic acid, and polysaccharides (Weidenmaier & Peschel, 2008). The peptidoglycan consists of glycan (polysaccharides) chains of alternating N-acetyl-d-glucosamine (NAG) and N-acetyl-d-muramic acid (NAM) and diaminopimelic acid (DAP), which is unique in prokaryotic cell walls (Edelvio et al., 2018). Teichoic and teichuronic acid are chemically bonded to peptidoglycan. In certain species, the cell wall also has special substances called mycolic acids, which contribute to resistance to desiccation and certain antibiotics (McPhee et al., 2009).

Although the chemical composition of their cell wall has similarities to that of Gram-positive bacteria, actinomycetes are thought to be a distinct group from other common bacteria because of their well-developed morphological (hyphae) and cultural characteristics. However, unlike the true fungi, the actinomycetes have thin hyphae (0.5–1.5 mm in diameter) with genetic material coiled inside as free DNA (Hopwood, 1973). The cell wall of the hyphae is made up of a cross-linked polymer containing short chains of amino acids and long chains of amino sugars (Free, 2013). Actinomycetes have a very diverse range of cell wall compositions, which has important taxonomic significance. Four major cell wall types are distinguished in these filamentous bacteria on the basis of the three features of peptidoglycan composition and structure (Pazos & Peters, 2019). These features are (i) diaminopimelic acid isomer on tetrapeptide side chain position 3, (ii) sugar content of peptidoglycan, and (iii) the presence of glycine in interpeptide bridges (Mayer et al., 2019).

Aerial hyphae further develop into long chains of spores, giving them a distinctive "powdery" look, in the genus *Streptomyces*, which is the well-studied group of actinomycetes. Also, they often produce exopolysaccharides that form a protective matrix around the mycelium. This matrix can help in the attachment to surfaces and protection against environmental stress (Narayanasamy, 2013).

3.4 Soil Actinomycetes: Classification of Cellulase Enzyme

Cellulase is not an isolated enzyme, but rather a unique system in which many enzymes collaborate to carry out the hydrolysis of cellulose, a task that none of them can accomplish individually. Cellulose-degrading enzymes are a category of enzymes that aid in the breakdown of cellulose into glucose. Cellulose is catalyzed by several enzymes like endo-β-1,4-glucanase (EC 3.2.1.4), exo-β-1,4-glucanases (EC 3.2.1.91), and β-glucosidases (EC 3.2.1.21) (Sharma et al., 2016) (Horn et al., 2012).

Endo-β-1,4-glucanase: This consists of many components that exhibit different levels of unpredictability. One of these enzymes may be responsible for catalyzing the breakdown of crystalline cellulose. Nevertheless, its mechanism of action lacks specificity and mainly focuses on carboxymethyl cellulose (CMC), phosphoric acid-swollen cellulose, and cellodextrin. This component does not have an effect on cellobiose. The primary products consist of cellobiose and cellotriose (Lee & Fan, 1980).

Exo-β-1,4-glucanases: β-1,4-glucan glucohydrolase exists in several forms and is responsible for removing a single glucose molecule from the nonreducing end of the cellulose chain.

This enzyme has activity toward Walseth, CMC, and cellodextrin chains consisting of four to seven units, which are generated by the endo-glucanase. However, it shows limited effectiveness in breaking down insoluble cellulose. The enzyme β-1,4-glucan cellobiohydrolase (CBH) cleaves a cellobiose molecule from the nonreducing ends of the chain. Many experts now consider the CBH to be equivalent to

Table 3.1 Cellulases produced by *Actinomycetes*

Organism	Mode of production	enzyme	Reference
Microbispora bispora	Smf, SSF	Endoglucanase, β-glucosidase	Waldron et al. (1986)
Microbispora cellulosiformans	Smf	Cellulase	Gong et al. (2020)
Streptomyces flavogriseus	Smf, SSF	CMCase	Castañeda-Cisneros et al. (2020)
Streptomyces griseoaurantiacus	Smf, SSF	CMCase	
Actinomycetes HJ4(2a)	Smf, SSF	Endo-1-4, β-glucanase Exo-1-4, β-glucanase	Lopak et al. (2020)
Streptomyces sp.	Smf	CMCase	Bui (2014)
Streptomyces sp.	Smf	Endoglucanase, β-glucosidase	Das et al. (2017)
Streptomyces albogriseolus	Smf	β-glucosidase, endoglucanase, avicelase	Saini et al. (2015)
Streptomyces lividans	Smf	Cellulases	

the conventional C1 enzyme. This component has the highest affinity toward cellulose. It is unable to degrade CMC and its action on H_3PO_4-swollen cellulose is marked a very slow reaction. While it cannot effectively break down crystalline cellulose, it can hydrolyze cellulose substrates by progressively eliminating cellobiose residues from the ends of the chain. When CBH is recombined with C_x and β-glucosidase it plays a major role in hydrolysis of cotton or Avicel-like crystalline cellulose.

β-glucosidase: This catalyzes the hydrolysis of cellobiose and short-chain cello-oligosaccharides, converting them into glucose. However, it does not have any impact on cellulose. Although it quickly breaks down cellobiose and cellotriose, its rate of degradation significantly reduces as the polymerization degree increases. Unlike exo-β-1,4 glucanase, which primarily targets longer cellooligosaccharides, this is different (Lee & Fan, 1980). In Table 3.1, various species of actinomycetes have been listed with their ability to produce different classes of cellulase enzymes.

3.5 Mechanism of Cellulase Synthesis by *Streptomyces*

The suggested cellulase regulation model in *Streptomyces* sp. can be described as follows (Fig. 3.1):

1. The confined cellulolytic products produced by the hydrolysis of cellulose by "basal cellulase" are cellobiose and glucose. In addition to being a "potential" inducer, cellobiose also acts as a source of carbon.
2. By developing an adaptable transport enzyme system, cellobiose is actively transported across the cell membrane. Constitutive glucosidase, an enzyme that

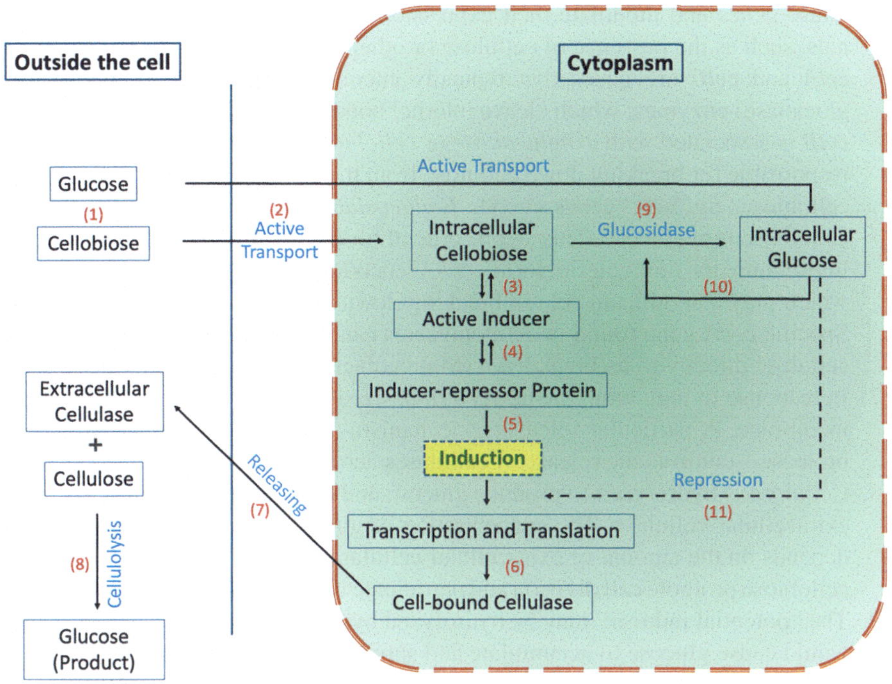

Fig. 3.1 Biosynthetic pathway of cellulase production in actinomycetes

hydrolyzes cellobiose to glucose, may exist as a membrane-bound compound. When cellobiose or other inducers like lactose or sophorose are present in the growth medium, adaptive glucosidase may also develop in response to such conditions.

3. Following cell entry, intracellular cellobiose may transform into the "active" inducer by reacting with the repressor protein to inactivate it, allowing for induction to take place. Alternatively, (8) the "potential inducer" may be hydrolyzed by intracellular glucosidase, which would cause glucose to accumulate and suppress the synthesis of cellulase.
4. The effectiveness of induction is based on how well the "active inducer" binds to the repressor protein. The *catabolite control protein A (CcpA)* is one of the key repressor proteins involved in the regulation of cellulase gene expression. CcpA mediates carbon catabolite repression (CCR), a mechanism where the presence of easily metabolizable carbon sources like glucose represses the expression of cellulase and other catabolic genes. By binding to specific DNA sequences, CcpA inhibits the transcription of these genes, ensuring that the bacteria prioritize the use of more readily available energy sources.) (Goh, 2005).
5. After the "active inducer" inactivates the repressor protein, transcription and translation proceed, leading to the induction of cellulase production. (Transcriptional regulators such as *Clp [cAMP receptor-like protein]* and *Xyr1 [xylanase regulator 1]*). These regulators bind to the promoter regions of cel-

lulase genes and modulate their expression in response to environmental signals, such as the presence of cellulose or other inducers].

6. *celA* and *celC* are genes that typically encode *endoglucanase* (endo-β-1,4-glucanase) enzymes, which cleave internal bonds within the cellulose polymer. *celB* is associated with *exoglucanase* or *cellobiohydrolase* activity; this gene is responsible for breaking down cellulose from the ends of the polymer to release cellobiose. *Bgl/bglC* genes encode *β-glucosidase* enzymes, which hydrolyze cellobiose into glucose. This step is crucial for the complete breakdown of cellulose into its monomeric form. *cbhA* encodes *cellobiohydrolase* enzymes, which play a significant role in breaking down cellulose into cellobiose units. Specific *cenA* gene found in actinomycetes can encode a *cellulase* that acts on cellulose directly to its breakdown (Main cofactor of the process is Cu^{+2}).
7. In response to induction, the newly synthesized cellulase could attached to the membrane. A particular releasing mechanism—possibly an intracellular acid protease—controls the release of cellulases across the cell membrane.
8. Cellulose is hydrolyzed to produce glucose and cellobiose by recently released extracellular cellulase. The concentration of glucose and cellobiose generated depends on the amount of extracellular cellulase released. In turn, glucose and cellobiose promote cell division and participate in both induction and repression.
9. The "potential inducer" may be hydrolyzed by intracellular glucosidase, which would cause glucose to accumulate and suppress the synthesis of cellulase.
10. Higher amount of intracellular glucose inhibits intracellular glucosidase activity.
11. Through an inhibition mechanism, the concentration of intracellular glucose affects the production of cellulase.

Intracellular glucose is the main repressor of cellulase induction process.

3.6 Cellulosic Waste Characterization

The structure of cellulose microfibrils is defined by the existence of linear glucan chains that are connected by hydrogen bonds and other noncovalent interactions (Jarvis, 2022). Cellulose decomposition and its subsequent utilization are essential for global carbon sources. The importance of cellulose as a renewable source of energy has made cellulose hydrolysis a topic of great study and industrial interest. Figure 3.2 shows the amount of cellulose waste generation during the year 2022 in India, where maximum cellulosic waste was generated by food industries and subsequently by the agricultural industries.

Except the industrial sector, particularly the paper and textile industries also generates significant amounts of cellulose waste as a byproduct of their production processes. Despite cellulose being an abundant and renewable natural polymer, the disposal of cellulose waste poses a substantial environmental challenge. Cellulose waste primarily consists of residues from wood processing, paper manufacturing,

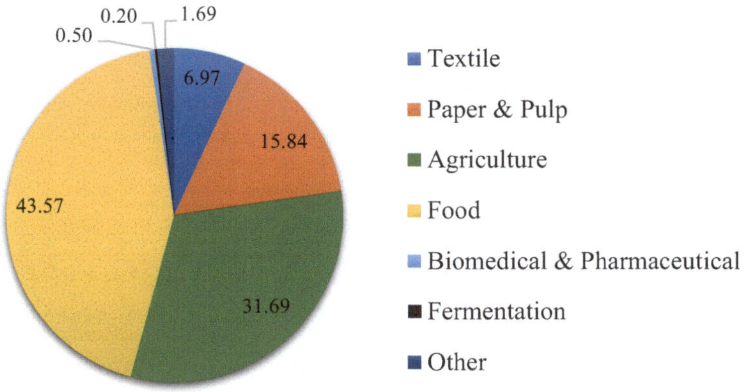

Fig. 3.2 Percentage of cellulose waste generation during the year 2022 (India)

and agricultural by-products. Table 3.2 shows the detailed chemical composition of cellulosic waste generated from different industries, with the maximum concentrations of cellulose present in waste generated from textile industries.

3.7 Cellulases Production Using Cellulosic Waste

From the different soil samples like rhizosphere soil, agricultural waste disposal sites, extreme environments and water sediments, isolation of actinomycetes has been performed. Serial dilution approach followed by the spread plate method is used for the isolation. Additionally, Pirzadah et al. (2014) employed an actinomycetes isolate agar medium, starch caseinate agar medium, and glucose asparagine agar medium for isolation. The ability of the isolated strains to produce cellulases during their growth was determined by the spot plate method. For the detection, CMC was added as a substrate in the media, and after incubation, Congo red and NaCl were used according to Castañeda-Cisneros et al. (2020) to observe the zone of hydrolysis. Strain was selected based on the maximum cellulase index calculation.

Selected species was grown on ISP2 agar plates according to Castañeda-Cisneros et al. (2020) for sporulation at 28 °C. For the spore harvest, tween 80 was added to the plate and this spore suspension was used as an inoculum for the further production of cellulases and degradation of cellulosic waste. Based on the state of the waste submerged, solid-state fermentation can be operated. Several processes regarding cellulase production using SSF and Smf are mentioned in Table 1. Detailed process of these fermentations is performed according to Ventorino et al. (2016) and Castañeda-Cisneros et al. (2020).

Table 3.2 Chemical composition of waste

Agro-industrial wastes	Chemical composition (% w/w)					References
	Cellulose (%)	Hemicellulose (%)	Lignin (%)	Others	Moisture (%)	
Rice straw	32.4 ± 1.06	28.3 ± 0.97	19.60 ± 2.02	Extractives: 11.27 ± 1.52 Ash: 8.43 ± 0.70	8.9 ± 0.2	Louis et al. (2022)
Wheat straw	35.69	29.68	18.8		2.5 ± 0.3	Riseh et al. (2024)
Sorghum	46.6	34.1	22.3		5.3 ± 0.2	
Corn stalks	32.75	31.08	10.07		1.8 ± 0.2	
Switchgrass	31.8	25	31.2		4.1 ± 0.2	
Oat straw	34.6 ± 0.5	28.0 ± 0.3	14.0 ± 0.3	Extractives: 16.0 ± 0.4		Rencoret et al. (2023)
Sugarcane bagasse	37.72	22.95	22.34			Riseh et al. (2024)
Sawdust	44	26	26	Extractives: 3 ash: 1		Rusanen et al. (2019)
Sugar beet waste	30	26.8	4.1			Ziemiński et al. (2012)
Barley straw	40.0 ± 0.2	27.2 ± 0.8	7.8 ± 0.2	Ash: 3.4	6.8 ± 0.2	Raud et al. (2021)
Cotton stalks	33–36	9–18	25 ± 2			Keshav et al. (2021)
Textile fiber wastes	Composition (% w/v)					
Cotton	90–95	3–5				Lu et al. (2023)
Linen	80–90	10–15				
Bamboo	~ 80	10–15	1–2			
Hemp	70–80	8–12	2–3			
Pulp & Paper industrial wastes	Composition (% w/v)					
Bond paper	70–85	3–5	1–2			Małachowska et al. (2020) Ververis et al. (2007)
Cardstock	70–80	4–6	1–2			
Newsprint	40–50	20–30	25–35			
Kraft paper	70–80	10–20	5–10			
Recycled paper	Variable					

Table 3.3 Application of actinomycetes with reference to cellulases

Species	Action/effect	Reference
S. ruber, *S. lividans*, and *S. rutgersensis*	Textile, detergents	Mukhtar et al. (2017)
Thermobifida sp. and *Micromonospora* sp.	Paper and pulp industry	Mukhtar et al. (2017)
S. thermocarboxydus	Agricultural biomass degradation	Castañeda-Cisneros et al. (2020)
Streptomyces sp. STCH565-A	Lignocellulosic bioconversion	Escudero-Agudelo et al. (2023)
Actinomycetes HJ4(2a)	Dietary fiber production from cassava pulp fiber	Lopak et al. (2020)
C. thermocellum	Cellulose digestion in soil	Wilson (2016)
Streptomyces plymuthica C48	Inhibit *B. cinerea* spore germination	Sharma and Salwan (2018)
Cellulomonas uda	Extraction, stabilization, and clarification of fruit and vegetable juices	Chakraborty et al. (2016)
Cellulomonas bioazotea	Enhance fermentable sugars in wine production, improve flavor (pineapple, carrots, and apple)	Annamalai et al. (2016)
Thermomonospora curvata	Extraction of olive oil	Chakraborty et al. (2016)
Thermomonospora curvata	Bio-stoning of jeans	De Souza Moreira et al. (2016)
Thermomonospora fusca, *S. lividans*	Decreased pulp viscosity; enhanced brightness	Singh et al. (2016)
Cellulomonas fimi	Reduced environmental pollution by uncontrolled dumping	Keshk (2016)

3.8 Potential Biotechnological Applications of Actinomycetes and Their Cellulases

Actinomycete-derived cellulases find extensive use in various sectors, including the production of biofuel, textiles, food, feed, and pulp and paper. Their sustainability, stability, and adaptability make them useful tools for effective and sustainable procedures. Further, the application of actinomycetes species is listed in Table 3.3.

References

Adegboye, M. F., & Babalola, O. O. (2016). Isolation and identification of potential antibiotic producing rare Actinomycetes from Rhizospheric soils. *Journal of Human Ecology, 56*(1–2), 31–41. https://doi.org/10.1080/09709274.2016.11907035

Annamalai, N., Rajeswari, M. V., & Balasubramanian, T. (2016). Thermostable and alkaline Cellulases from marine sources. In *Elsevier eBooks* (pp. 91–98). https://doi.org/10.1016/b978-0-444-63507-5.00009-5

Bhatti, A. A., Haq, S., & Bhat, R. A. (2017). Actinomycetes benefaction role in soil and plant health. *Microbial Pathogenesis, 111*, 458–467. https://doi.org/10.1016/j.micpath.2017.09.036

Biswal, T., BadJena, S. K., & Pradhan, D. (2020). Sustainable biomaterials and their applications: A short review. *Materials Today Proceedings, 30*, 274–282. https://doi.org/10.1016/j.matpr.2020.01.437

Blin, K., Kim, H. U., Medema, M. H., & Weber, T. (2017). Recent development of antiSMASH and other computational approaches to mine secondary metabolite biosynthetic gene clusters. *Briefings in Bioinformatics, 20*(4), 1103–1113. https://doi.org/10.1093/bib/bbx146

Bobek, J., Šmídová, K., & Čihák, M. (2017). A waking review: Old and novel insights into the spore germination in Streptomyces. *Frontiers in Microbiology, 8*. https://doi.org/10.3389/fmicb.2017.02205

Boyd, C. E. (1995). Soil organic matter and aerobic respiration. In *Springer eBooks* (pp. 149–193). https://doi.org/10.1007/978-1-4615-1785-6_5

Bui, H. B. (2014). Isolation of cellulolytic bacteria, including actinomycetes, from coffee exocarps in coffee-producing areas in Vietnam. *International Journal of Recycling of Organic Waste in Agriculture, 3*(1). https://doi.org/10.1007/s40093-014-0048-0

Castañeda-Cisneros, Y. E., Mercado-Flores, Y., Anducho-Reyes, M. A., Álvarez-Cervantes, J., Ponce-Lira, B., Evangelista-Martínez, Z., & Téllez-Jurado, A. (2020). Isolation and selection of Streptomyces species from semi-arid agricultural soils and their potential as producers of Xylanases and Cellulases. *Current Microbiology, 77*(11), 3460–3472. https://doi.org/10.1007/s00284-020-02160-7

Chakraborty, S., Gupta, R., Jain, K. K., Hemansi, N., Gautam, S., & Kuhad, R. C. (2016). Cellulases: Application in wine and brewery industry. In *Elsevier eBooks* (pp. 193–200). https://doi.org/10.1016/b978-0-444-63507-5.00017-4

Condron, L., Stark, C., O'Callaghan, M., Clinton, P., & Huang, Z. (2010). The role of microbial communities in the formation and decomposition of soil organic matter. In *Springer eBooks* (pp. 81–118). https://doi.org/10.1007/978-90-481-9479-7_4

Dar, G. H. (2009). *Soil microbiology and biochemistry*. New India Publishing.

Das, P., Solanki, R., & Khanna, M. (2017). Screening, quantification, and purification of Cellulases from soil Actinomycetes. In *Apple academic press eBooks* (pp. 219–247). https://doi.org/10.1201/b19978-15

Das, P., Singh, S. K., Singh, P., Zeyad, M. T., Aamir, M., & Upadhyay, R. S. (2021). Actinomycetes as biostimulants and their application in agricultural practices. In *Elsevier eBooks* (pp. 267–282). https://doi.org/10.1016/b978-0-12-822122-8.00021-2

Dayal, G., Patel, M., & Singh, H. (2023). 13. Plant Nutrient. *Basic Introduction of Soil Science*, 157.

De Souza Moreira, L. R., Lo Sciuto, D., & Filho, E. X. F. (2016). An overview of cellulose-degrading enzymes and their applications in textile industry. In *Elsevier eBooks* (pp. 165–175). https://doi.org/10.1016/b978-0-444-63507-5.00014-9

Demain, A. L. (2009). Antibiotics: Natural products essential to human health. *Medicinal Research Reviews, 29*(6), 821–842. https://doi.org/10.1002/med.20154

Devi, C. S., Keziah, S. M., Naine, S. J., & Mohanasrinivasan, V. (2022). *Actinomycetes: Microbiology to systems biology* (pp. 1–35). https://doi.org/10.1007/978-981-16-5835-8_1

Dixon, G. R., & Tilston, E. L. (2010). Soil-borne pathogens and their interactions with the soil environment. In *Springer eBooks* (pp. 197–271). https://doi.org/10.1007/978-90-481-9479-7_6

Edelvio, D. B. G., Leo, R. L. D., & De Miranda Rita, D. C. M. (2018). Actinomycetes bioactive compounds: Biological control of fungi and phytopathogenic insect. *African Journal of Biotechnology, 17*(17), 552–559. https://doi.org/10.5897/ajb2017.16323

Escudero-Agudelo, J., Martínez-Villalobos, J., Arocha-Garza, H., Galán-Wong, L. J., Avilés-Arnaut, H., & De La Torre-Zavala, S. (2023). Systematic bioprospection for cellulolytic actinomycetes in the Chihuahuan Desert: Isolation and enzymatic profiling. *PeerJ, 11*, e16119. https://doi.org/10.7717/peerj.16119

Free, S. J. (2013). Fungal cell wall organization and biosynthesis. *Advances in Genetics, 81*, 33–82. https://doi.org/10.1016/b978-0-12-407677-8.00002-6

George, N. M. (2012). Distribution and bioactive potential of soil actinomycetes from different ecological habitats. *African Journal of Microbiology Research, 6*(10). https://doi.org/10.5897/ajmr11.856

Goh, Y. J. (2005). Genetic characterization of fructooligosaccharides metabolism by Lactobacillus paracasei 1195. https://digitalcommons.unl.edu/cgi/viewcontent.cgi?article=11091&context=dissertations

Gong, X., Xiang, W., Cao, X., Yu, Y., Hao, Y., Li, L., Wang, Q., Zou, H., & Qian, C. (2020). Microbispora cellulosiformans sp. nov., a novel actinomycete with cellulase activity isolated from soil in the cold region. *Antonie Van Leeuwenhoek, 113*(12), 2053–2062. https://doi.org/10.1007/s10482-020-01477-4

Grenni, P., Ancona, V., & Caracciolo, A. B. (2018). Ecological effects of antibiotics on natural ecosystems: A review. *Microchemical Journal, 136*, 25–39. https://doi.org/10.1016/j.microc.2017.02.006

Hopwood, D. A. (1973). *Genetical and cytological studies on Actinomycetes*. University of Glasgow (United Kingdom).

Hopwood, D. A. (2008). Streptomyces in nature and medicine: The antibiotic makers. *Choice Reviews Online, 45*(07), 45–3774. https://doi.org/10.5860/choice.45-3774

Horn, S. J., Vaaje-Kolstad, G., Westereng, B., & Eijsink, V. (2012). Novel enzymes for the degradation of cellulose. *Biotechnology for Biofuels, 5*(1). https://doi.org/10.1186/1754-6834-5-45

Jarvis, M. C. (2022). Hydrogen bonding and other non-covalent interactions at the surfaces of cellulose microfibrils. *Cellulose, 30*(2), 667–687. https://doi.org/10.1007/s10570-022-04954-3

Jing, H., Xiao, X., Zhang, Y., Li, Z., Jian, H., Luo, Y., & Han, Z. (2022). Composition and ecological roles of the core microbiome along the Abyssal-Hadal transition zone sediments of the Mariana trench. *Microbiology Spectrum, 10*(3). https://doi.org/10.1128/spectrum.01988-21

Kalakoutskii, L. V., & Agre, N. S. (1976). Comparative aspects of development and differentiation in actinomycetes. *Bacteriological Reviews, 40*(2), 469–524. https://doi.org/10.1128/mmbr.40.2.469-524.1976

Keshav, P. K., Banoth, C., Kethavath, S. N., & Bhukya, B. (2021). Lignocellulosic ethanol production from cotton stalk: An overview on pretreatment, saccharification and fermentation methods for improved bioconversion process. *Biomass Conversion and Biorefinery, 13*(6), 4477–4493. https://doi.org/10.1007/s13399-021-01468-z

Keshk, S. M. (2016). Cellulase application in enzymatic hydrolysis of biomass. In *Elsevier eBooks* (pp. 185–191). https://doi.org/10.1016/b978-0-444-63507-5.00016-2

Khamna, S., Yokota, A., & Lumyong, S. (2008). Actinomycetes isolated from medicinal plant rhizosphere soils: Diversity and screening of antifungal compounds, indole-3-acetic acid and siderophore production. *World Journal of Microbiology and Biotechnology, 25*(4), 649–655. https://doi.org/10.1007/s11274-008-9933-x

Kirby, R. (2005). Actinomycetes and lignin degradation. *Advances in Applied Microbiology, 58C*, 125–168. https://doi.org/10.1016/s0065-2164(05)58004-3

Kontro, M. H., & Yaradoddi, J. S. (2021). Actinomycetes in environmental applications. In *Rhizosphere biology* (pp. 247–270). https://doi.org/10.1007/978-981-16-3353-9_14

Lee, Y. H., & Fan, L. T. (1980). Properties and mode of action of cellulase. In *Advances in biochemical engineering, biotechnology* (pp. 101–129). https://doi.org/10.1007/3-540-09955-7_9

Lewis, K. (2017). New approaches to antimicrobial discovery. *Biochemical Pharmacology, 134*, 87–98. https://doi.org/10.1016/j.bcp.2016.11.002

Lopak, S. S., Meryandini, A., & Sunarti, T. C. (2020). Dietary fibre production from cassava pulp fibre using Actinomycetes cellulase. *International Food Research Journal, 27*(4), 702–711. https://www.proquest.com/scholarly-journals/dietary-fibre-production-cassava-pulp-using/docview/2454692248/se-2

López, M. J., Jurado, M. M., López-González, J. A., Estrella-González, M. J., Martínez-Gallardo, M. R., Toribio, A., & Suárez-Estrella, F. (2021). Characterization of thermophilic Lignocellulolytic microorganisms in composting. *Frontiers in Microbiology, 12*. https://doi.org/10.3389/fmicb.2021.697480

Louis, A. C. F., Venkatachalam, S., & Gupta, S. (2022). Innovative strategy for rice straw valorization into nanocellulose and nanohemicellulose and its application. *Industrial Crops and Products, 179*, 114695. https://doi.org/10.1016/j.tifs.2022.11.020

Lu, L., Fan, W., Meng, X., Xue, L., Ge, S., Wang, C., Foong, S. Y., Tan, C. S., Sonne, C., Aghbashlo, M., Tabatabaei, M., & Lam, S. S. (2023). Current recycling strategies and high-value utilization of waste cotton. *The Science of the Total Environment, 856*, 158798. https://doi.org/10.1016/j.scitotenv.2022.158798

Małachowska, E., Dubowik, M., Boruszewski, P., Łojewska, J., & Przybysz, P. (2020). Influence of lignin content in cellulose pulp on paper durability. *Scientific Reports, 10*(1), 19998. https://doi.org/10.1038/s41598-020-77101-2

Mayer, V. M. T., Hottmann, I., Figl, R., Altmann, F., Mayer, C., & Schäffer, C. (2019). Peptidoglycan-type analysis of the N-acetylmuramic acid auxotrophic oral pathogen Tannerella forsythia and reclassification of the peptidoglycan-type of Porphyromonas gingivalis. *BMC Microbiology, 19*(1), 200. https://doi.org/10.1186/s12866-019-1575-7

McPhee, J. B., Tamber, S., Brazas, M. D., Lewenza, S., & Hancock, R. E. W. (2009). Antibiotic resistance due to reduced uptake. In *Humana Press eBooks* (pp. 97–110). https://doi.org/10.1007/978-1-59745-180-2_9

Mukhtar, S., Zaheer, A., Aiysha, D., Malik, K. A., & Mehnaz, S. (2017). Actinomycetes: A source of industrially important enzymes. *Journal of Proteomics & Bioinformatics, 10*(12). https://doi.org/10.4172/jpb.1000456

Narayanasamy, P. (2013). Detection and identification of bacterial biological control agents. In *Springer eBooks* (pp. 201–293). https://doi.org/10.1007/978-94-007-6380-7_4

Naz, M., Dai, Z., Hussain, S., Tariq, M., Danish, S., Khan, I. U., Qi, S., & Du, D. (2022). The soil pH and heavy metals revealed their impact on soil microbial community. *Journal of Environmental Management, 321*, 115770. https://doi.org/10.1016/j.jenvman.2022.115770

Olanrewaju, O. S., & Babalola, O. O. (2018). Streptomyces: Implications and interactions in plant growth promotion. *Applied Microbiology and Biotechnology, 103*(3), 1179–1188. https://doi.org/10.1007/s00253-018-09577-y

Ondrasek, G., Begić, H. B., Zovko, M., Filipović, L., Meriño-Gergichevich, C., Savić, R., & Rengel, Z. (2019). Biogeochemistry of soil organic matter in agroecosystems & environmental implications. *The Science of the Total Environment, 658*, 1559–1573. https://doi.org/10.1016/j.scitotenv.2018.12.243

Pathania, S., Solanki, P., Putatunda, C., Bhatia, R. K., & Walia, A. (2021). Adaptation to cold environment: The survival strategy of Psychrophiles. In *Springer eBooks* (pp. 87–111). https://doi.org/10.1007/978-981-16-2625-8_4

Pazos, M., & Peters, K. (2019). Peptidoglycan. *Subcell Biochem, 92*, 127–168. https://doi.org/10.1007/978-3-030-18768-2_5

Pirzadah, T., Garg, S., Singh, J., Vyas, A., Kumar, M., Gaur, N., Bala, M., Rehman, R., Varma, A., Kumar, V., & Kumar, M. (2014). Characterization of Actinomycetes and *Trichoderma spp.* for cellulase production utilizing crude substrates by response surface methodology. *Springerplus, 3*(1). https://doi.org/10.1186/2193-1801-3-622

Plante, A. F., Conant, R. T., Stewart, C. E., Paustian, K., & Six, J. (2006). Impact of soil texture on the distribution of soil organic matter in physical and chemical fractions. *Soil Science Society of America Journal, 70*(1), 287–296. https://doi.org/10.2136/sssaj2004.0363

Prasad, J. K., & Raghuwanshi, R. (2022). Mechanisms of multifarious soil microbial enzymes in plant growth promotion and environmental sustainability. In *Elsevier eBooks* (pp. 117–143). https://doi.org/10.1016/b978-0-323-90958-7.00020-0

Raud, M., Rocha-Meneses, L., Lane, D. J., Sippula, O., Shurpali, N. J., & Kikas, T. (2021). Utilization of barley straw as feedstock for the production of different energy vectors. *PRO, 9*(4), 726. https://doi.org/10.3390/pr9040726

Rencoret, J., Marques, G., Rosado, M. J., Benito, J., Barro, F., Gutiérrez, A., & Del Río, J. C. (2023). Variations in the composition and structure of the lignins of oat (Avena sativa L.) straws according to variety and planting season. *International Journal of Biological Macromolecules, 242*, 124811. https://doi.org/10.1016/j.ijbiomac.2023.124811

Riseh, R. S., Vazvani, M. G., Hassanisaadi, M., & Thakur, V. K. (2024). Agricultural wastes: A practical and potential source for the isolation and preparation of cellulose and application in agriculture and different industries. *Industrial Crops and Products, 208*, 117904.

Rusanen, A., Lappalainen, K., Kärkkäinen, J., Tuuttila, T., Mikola, M., & Lassi, U. (2019). Selective hemicellulose hydrolysis of Scots pine sawdust. *Biomass Conversion and Biorefinery, 9*, 283–291. https://doi.org/10.1007/s13399-018-0357-z

Saini, A., Aggarwal, N. K., Sharma, A., & Yadav, A. (2015). Actinomycetes: A source of Lignocellulolytic enzymes. *Enzyme Research, 2015*, 1–15. https://doi.org/10.1155/2015/279381

Selim, M. S. M., Abdelhamid, S. A., & Mohamed, S. S. (2021). Secondary metabolites and biodiversity of actinomycetes. *Journal of Genetic Engineering and Biotechnology, 19*(1), 72. https://doi.org/10.1186/s43141-021-00156-9

Sharma, V., & Salwan, R. (2018). Biocontrol potential and applications of Actinobacteria in agriculture. In *Elsevier eBooks* (pp. 93–108). https://doi.org/10.1016/b978-0-444-63994-3.00006-0

Sharma, A., Tewari, R., Rana, S. S., Soni, R., & Soni, S. K. (2016). Cellulases: Classification, methods of determination and industrial applications. *Applied Biochemistry and Biotechnology, 179*(8), 1346–1380. https://doi.org/10.1007/s12010-016-2070-3

Singh, S., Singh, V. K., Aamir, M., Dubey, M. K., Patel, J. S., Upadhyay, R. S., & Gupta, V. K. (2016). Cellulase in pulp and paper industry. In *Elsevier eBooks* (pp. 152–162). https://doi.org/10.1016/b978-0-444-63507-5.00013-7

Sung, C., Ishak, C., Abdullah, R., Othman, R., Panhwar, Q., & Aziz, M. (2017). Soil properties (physical, chemical, biological, mechanical). In *CRC Press eBooks* (pp. 103–154). https://doi.org/10.1201/9781315277189-6

Takahashi, Y., & Omura, S. (2003). Isolation of new actinomycete strains for the screening of new bioactive compounds. *The Journal of General and Applied Microbiology, 49*(3), 141–154. https://doi.org/10.2323/jgam.49.141

Tarafdar, J. C. (2022). *Role of soil biology on soil health for sustainable agricultural production* (pp. 67–81). https://doi.org/10.1007/978-981-16-8770-9_3

Ventorino, V., Ionata, E., Birolo, L., Montella, S., Marcolongo, L., De Chiaro, A., Espresso, F., Faraco, V., & Pepe, O. (2016). Lignocellulose-adapted Endo-Cellulase producing Streptomyces strains for bioconversion of cellulose-based materials. *Frontiers in Microbiology, 7*. https://doi.org/10.3389/fmicb.2016.02061

Ververis, C., Georghiou, K., Danielidis, D., Hatzinikolaou, D., Santas, P., Santas, R., & Corleti, V. (2007). Cellulose, hemicelluloses, lignin and ash content of some organic materials and their suitability for use as paper pulp supplements. *Bioresource Technology, 98*(2), 296–301. https://doi.org/10.1016/j.biortech.2006.01.007

Votruba, J., & Vaněk, Z. (2020). Physicochemical factors affecting Actinomycete growth and secondary metabolism. In *CRC Press eBooks* (pp. 263–281). https://doi.org/10.1201/9781003068600-8

Waksman, S. A. (1940). On the classification of Actinomycetes. *Journal of Bacteriology, 39*(5), 549–558. https://doi.org/10.1128/jb.39.5.549-558.1940

Waldron, C. R., Becker-Vallone, C. A., & Eveleigh, D. E. (1986). Isolation and characterization of a cellulolytic actinomycete Microbispora bispora. *Applied Microbiology and Biotechnology, 24*(6), 477–486. https://doi.org/10.1007/bf00250327

Weidenmaier, C., & Peschel, A. (2008). Teichoic acids and related cell-wall glycopolymers in gram-positive physiology and host interactions. *Nature Reviews Microbiology, 6*(4), 276–287. https://doi.org/10.1038/nrmicro1861

Wilson, D. B. (2016). Microbial diversity and Cellulase production. In *Elsevier eBooks* (pp. 43–48). https://doi.org/10.1016/b978-0-444-63507-5.00005-8

Ziemiński, K., Romanowska, I., & Kowalska, M. (2012). Enzymatic pretreatment of lignocellulosic wastes to improve biogas production. *Waste Management, 32*(6), 1131–1137.

Part II

Plant-Microbe Interactions for a Sustainable Environment

Dialogue in the Dirt: Plant–Microbe Interaction in Soil

4

Ashish Warghane, Meera Bharat Vaishnav, Lovelesh Singh Saini, Raksha Bawankar, Sanghdeep Gautam, Jigisha Thakkar, Vaibhav Bhatt, Rajeshkumar G. Chaudhari, Vandana Gupta, Balu Ananda Chopade, and Manju Shri

Abstract

Background: Soil health is critical for sustaining life, influenced by the complex interactions between soil, microorganisms, and plants; indulging in these interactions is indispensable for sustainable agriculture. This chapter explores how plant–soil–microbe dynamics affect soil ecosystem health, focusing on how microorganisms influence nutrient uptake, stress resilience, and plant development.

Scope: Plant growth-promoting microorganisms are helpful in promoting plant growth by taking part in nutrient recycling, suppressing diseases, and mitigating innumerable types of stresses. Through an extensive review of literature

A. Warghane (✉) · J. Thakkar · V. Bhatt · R. G. Chaudhari · M. Shri (✉)
School of Applied Sciences and Technology (SAST), Gujarat Technological University, Ahmedabad, Gujarat, India

M. B. Vaishnav
Anatek Services Pvt Ltd, Mumbai, Maharashtra, India

L. S. Saini
Department of Life Sciences, Parul Institute of Applied Sciences, Parul University, Vadodara, Gujarat, India

R. Bawankar
Department of Biotechnology, Silver Oak Institute of Science, Ahmedabad, Gujarat, India

S. Gautam
Institute of Management Studies Ghaziabad University Course Campus, Ghaziabad, Uttar Pradesh, India

V. Gupta
Ram Lal Anand College University of Delhi, Delhi, India

B. A. Chopade
AKS University, Satna, Madhya Pradesh, India

M. Saraf et al. (eds.), *Climate Change and Soil Microorganisms for Environmental Sustainability*, Microorganisms for Sustainability 28,
https://doi.org/10.1007/978-981-96-3425-5_4

and empirical evidence, this chapter elucidates the intricate interplay between plants and microorganisms within soil ecosystems. This chapter also underscores the role of PGPRs in biofertilization, pathogen bio-control, and abiotic stress mitigation strategies.

Conclusion: This chapter explores various omics-based studies in the area of plant–soil–microbe interactions. By understanding and leveraging the symbiotic relationship between plants and microorganisms, we can advance our efforts toward fostering resilient and productive soil ecosystems. This chapter also tries to provide a way to use these studies in the agriculture field to improve crop production and soil health.

Keywords

Soil · Plants · Microorganism · Diversity · Biofertilizers · Pathogen

4.1 Introduction

Soil is a multifaceted ecosystem, sheltering a majority of microbial population, including bacterial, fungal as well as various nematodes. Microbes residing in the soil contributes to plant health in various ways such as disease suppression, recycling of nutrients, nitrogen fixation, and helping in alleviating stresses (Berg et al., 2017; Dastogeer et al., 2020). Interactions between plants and microbes in the soil, are mediated by various chemosignals that are released by plants as well as microbes (Compant et al., 2019). Chemosignals that are secreted by plants have a prominent role in shaping the microbes within the rhizospheric and phyllospheric regions and the release of these chemosignals may vary with the environmental conditions (Compant et al., 2019; Sachdev et al., 2009). Recruitment of microbes by plants enriches the rhizosphere with specific microbes. However this is not a random process, but rather a directional and targeted process (Berg et al., 2017), which is governed by several factors such as soil conditions and geographic locations (Fierer et al., 2006; Zhou et al., 2016; Jia et al., 2020; Liu et al., 2020). The microbial diversity of the plants also depends on the plant growth stage explaining the differing microbial diversity of the organs and the stage of the same plant (Dong et al., 2019). *Acidobacteria* *Actinobacteria*, *Bacteroidetes*, *Planctomycetes*, *Proteobacteria*, and *Verrucomicrobia* are some of the dominant phyla associated with plants (Compant et al., 2019; Fierer, 2017; Rokhbakhsh-Zamin et al., 2011).

The challenging signaling between the plant roots and soil-residing microbes constitutes a fascinating interaction that profoundly impacts the health, boom, and productivity of flora. This dynamic relationship is characterized with the aid of a complex molecular dialogue taking place in the rhizosphere, the soil location surrounding plant roots (Huey et al., 2020). In this symbiotic affiliation, plants and microbes engage in sophisticated signaling processes to set up mutually useful connections, in the long run contributing to the resilience and nutrient acquisition of vegetation. Central to this interplay is the involvement of mycorrhizal fungi, which shape huge symbiotic relationships with plants (Huey et al., 2020). These fungi make their hyphae into the soil, facilitating superior absorption of water and vital nutrients, especially nitrogen and phosphorus, for the plant. In response, vegetation releases signaling molecules like strigolactones, which appear as chemical messengers that appeal to mycorrhizal fungi to establish symbiotic connections (Huey et al., 2020). In going back, the fungi produce signaling compounds along with chitin oligosaccharides, further activating the plant's defense mechanisms and nutrient uptake pathways (Huey et al., 2020). The signaling dynamics among plants and soil bacteria also play a pivotal function in the establishment of beneficial relationships. Notably, microorganisms, like rhizobia, interact in nitrogen-solving nodules on the roots of leguminous flora (Wang et al., 2018). Chemical signaling, driven via flavonoids releases from plant roots, attracts rhizobia, initiating the release of Nod factors through the bacteria. These factors, in turn, cause particular plant responses and nodule formation for nitrogen fixation. This symbiotic association complements the plant's nitrogen vitamins even providing a conducive habitat for the bacteria (Wang et al., 2018).

Beyond symbiotic relationships, plant life employs an array of signaling molecules to communicate with both beneficial and harmful microbes. Compounds along with salicylic acid (SA) and jasmonic acid (JA) are necessary for plant protection responses in opposition to pathogens (Nadarajah & Abdul Rahman, 2021). The detection of harmful microbes activates vegetation to prompt defense mechanisms, even as some microbes have evolved strategies to suppress those responses, facilitating their colonization and acquisition of nutrients (Nadarajah & Abdul Rahman, 2021; Sachdev et al., 2009). The comprehension of those intricate signaling pathways holds sizable significance in agriculture and environmental technology. Leveraging these interactions might lead to the improvement of sustainable agricultural practices, which include utilizing beneficial microbes to decorate nutrient uptake and guard flora from pathogens (Suman et al., 2022). Furthermore, unraveling the complexities of plant–microbe signaling opens new avenues for bioengineering crops with improved resistance to strain and sicknesses (Suman et al., 2022).

This chapter offers a complete exploration of the signaling dynamics among plant roots and soil microbes. By incorporating the omics approach, gain a deeper knowledge of the genetic and metabolic intricacies governing those interactions. The positive and negative factors of those relationships, along with their programs in agriculture and the past, are elucidated (Suman et al., 2022). However, recognizing and addressing the demanding situations in making use of this expertise is critical for realizing the total capability of utilizing the plant–microbe interactions for sustainable and resilient ecosystems. This chapter deals with the vigorous signaling taking place between microbes and the plant roots in the soil, and the positive and negative interactions among them. This chapter also deals with the omics approaches such as genomics and metabolomics that have been used to study the relationship between them. We also aim to provide some insights into the applications of these interactions and what are the challenges in the fields when we are applying this knowledge (Suman et al., 2022).

4.2 Microbial Diversity in Soil

Soil comprises sand, clay, and silt. Its texture affects the biophysical properties of soil and are interconnected with soil fertility. It is also connected with the porosity of soil, which regulates water flow and diffusion of gases, which directly affects microbial growth (Kowalenko & Ivarson, 1978). Soil architecture has been found to be composite three-dimensional with packed aggregates and pores (Bailey et al., 2013; Edwards & Bremner, 1967; Oades & Waters, 1991). Soil aggregates create a linkage of particles and cavities, which are responsible for the streaming of nutrients and water and can be retrieved by organisms in the soil, enhancing the way that microorganisms, plants, and the soil matrix interact (Bailey et al., 2013). For our planet to maintain its biogeochemical cycles, soil is an essential component. It contributes significantly to the regulation of plant development and the preservation of the microorganism habitat (Wilpiszeski et al., 2019).

Soil is home to macrofauna, mesofauna, microfauna, and microflora (Nannipieri et al., 2002). Soil microorganisms form inhabitants with interrelated communities that are meticulously related to soil components like minerals and carbons (Wilpiszeski et al., 2019, Dhakepalkar Chopade 1994). Soil biodiversity maintains the diverse functioning of the ecosystem and withstands food production and regulation of climate (Banerjee & van der Heijden, 2023; Byrnes et al., 2014).

Although soil microorganisms represent a microscopic fraction of soil, they play an essential role in influencing the soil's chemical and physical properties. These populations have been found to be numerous and diversified in nature as compared to other communities on Earth (Neergaard, Encyclopedia). For viable agriculture, the maintenance of diversified functional microbial populations is very essential (Beare et al., 1995; Benizri et al., 2002). Soil exhibits various ecological, social, and economic functions, which include acting as a biodiversity storehouse, purification of surface water, recycling of mineral elements, and the storage of carbon dioxide from the atmosphere (Pimentel et al., 1997). The soil ecosystem has been significantly impacted by the interruption of soil assets by a variety of human activities (Yang et al., 2017). Microorganisms inhabit various microhabitats in soil, but it is known that soil has less nutrients and energy sources than what is ideal for microbial growth in vitro (Stotzky, 1997).

4.2.1 Microbial Kingdoms in Soil

Microbial kingdoms of soil comprise soil microorganisms, soil animals, and plants (Soil Quality Institute, 2001). The five taxonomic groups of microorganisms are viruses, bacteria, fungi, protists, and algae (Prescott et al., 1996) and closely associated with soil particles (Foster, 1988).

4.2.1.1 Soil Bacteria

Soil bacteria are found to be one of the important classes of microorganisms and maintain the productive health of the soil. They are involved in the recycling of nutrients like sulfur, phosphorous, carbon, and nitrogen (Hoorman, 2016).

4.2.1.1.1 Aerobic and Anaerobic Bacteria

Oxygen level in the soil directly affects the activity of soil bacteria (Dick, 2009). Some of the bacteria favor oxygen levels in soil and are known as aerobic bacteria and functionally they are involved in the decomposition of carbon compounds. *Aerobacter* are majorly found within soil, and *Actinomycetes* is known for the good earthy smell of bacteria (Lowenfels & Lewis, 2006). There are certain bacteria that survive deep inside compact soil known as anaerobic bacteria. Some pathogenic bacteria also prefer anaerobic soil conditions (Lowenfels & Lewis, 2006). *Paenibacillus*, *Bacillus*, *Clostridium*, *Trichococcus*, *Carnobacterium*, *Pseudomonas*, *Rahnella*, and *Cellulomonas* are found to be anaerobic in nature (Kim et al., 2013). Members of *acinetobactors* are commonly found in soil (Huddedar et al., 2002; Chopade et al., 2008 USA patent).

4.2.1.1.2 Plant Growth-Promoting Rhizobacteria

Microbes and the roots of plants have a close interaction. Plants are inclined toward the root surface's microbial population (Backer et al., 2018). *Azospirillum*, *Serattia*, *Rhodococcus*, *Herbaspirillum*, *Acetobacter*, and *Paenibacillus* have displayed enhanced plant growth (Babalola, 2010; Chopade et al., 2008 USA patent). Alongside this, there are also Lactobacillus (Lamont et al., 2017), Pseudomonas (Santoyo et al., 2012; Sivasakthi et al., 2014), and Actinobacteria (Shivlata & Satyanarayana, 2017) have also shown increased plant growth. *Bacillus*, *Glomus*, and *Pseudomonas* are subjected to commercialization (Borriss, 2011).

4.2.1.2 Archaea

Archaea has been found to be most beneficial among microorganisms. They are found to be in extreme environments. The two subgroups of archaebacteria are facultative and obligate anoxybionts, respectively. Obligate anoxybionts are species that can only survive in an anaerobic environment and are rendered nonviable by it. Among these are the halophile and methanogen species (Barns et al., 1996). Although they can survive in anaerobic environments, facultative anoxybionts are found in oxygen-rich environments. *Thermoacidophiles* are their representative species (Tate III, 1995; Barns et al., 1996). According to Jung et al. (2020), bacterium promotes plant growth. Song et al. (2019) discovered that plants benefit from the nitrogen-oxidizing archaea's ability to stimulate plant development and improve resistance to external challenges (Barns et al., 1996).

4.2.1.3 Fungi

According to Giri et al. (2005), fungi dominate all soil types and are the most varied group of soil microbes. Individual hyphae make up the filamentous mycelium of fungi. The hyphae can be nonseptate or septate, and they can be uni-, bi-, or multinucleate (Hawksworth, 1991). It has been observed that fungi can colonize terrestrial plant land (and carbon deposition in geological reserves (Floudas et al., 2012).

4.3 Dialogue Delivery Between Soil Microbiota and Plants

The core knowledge about plant helpful microorganisms and their special effects on plant wellbeing and productivity may be found in the chemical dialogue between plants and helpful microorganisms. It involves the molecular architecture of rhizobium–plant symbiosis, the biological repertoire of arbuscular mycorrhizal association, and the role of endophytes in transcriptional plasticity during host colonization by endophytes (Sharma et al., 2023). The intricate interactions between the broad array of bacteria, fungi, protozoa, archaea, and viruses that makeup soil microbiota and plants have a significant impact on plant growth, health, and ecosystem functioning. For sustainable agriculture, ecosystem management, and environmental security, it is essential to comprehend the mechanisms and dynamics of these interactions (Sharma et al., 2023).

The soil area that is immediately affected by root exudates, or the rhizosphere, is home to a diverse and active microbiome. Microbial communities use root exudates, consisting of sugars, organic acids, amino acids, and secondary metabolites, as a source of energy and chemical cues (Mendes et al., 2013). The amount and make up of root exudates affect plant growth promotion, disease suppression, and nutrient cycling by forming the rhizosphere microbiome's structure and activity (Bulgarelli et al., 2013). The microbiota found in soil are vital to the cycling of nutrients because they mediate the mobilization and transformation of important elements, including sulfur, phosphorus, and nitrogen. Mycorrhizal fungi and nitrogen-fixing bacteria work in symbiotic relationships with plants to help them absorb phosphorus from the soil and nitrogen from the atmosphere, respectively. By improving nutrient uptake by plants and lowering the demand for chemical fertilizers, these interactions support sustainable farming methods (Mendes et al., 2013).

Through a variety of processes, the microbiota associated with plants, which includes beneficial bacteria and fungi, provide protection against soil-borne diseases. The development of plant systemic resistance, resource rivalry, and antibiotics are a few of the tactics beneficial bacteria use to impede the growth of pathogens and the progression of diseases (Berg et al., 2014). In agroecosystems, using soil microbiota's bio-control capability could help lessen the need for artificial pesticides and promote plant health. Some soil microbiota have characteristics that promote plant growth, such as the generation of phytohormones, the solubilization of mineral nutrients, and the improvement of plant stress tolerance (Berg et al., 2014). In a similar vein, in exchange for carbon resources, mycorrhizal fungi create symbiotic connections with plants that improve nutrient uptake and drought tolerance (Berg et al., 2014).

Microbiota in the soil and microbes and plants can communicate and coordinate more easily, thanks to symbiotic signaling mechanisms like quorum sensing.

Microbial communities can sense population densities, control gene expression, and plan group activities in response to environmental stimuli, thanks to quorum sensing (Berg et al., 2014). Plant–microbe signaling pathways influence symbiotic relationships, developmental processes, and plant immune responses. They are mediated by phytohormones, reactive oxygen species, and tiny signaling molecules (Lugtenberg & Kamilova, 2009; Pieterse et al., 2014)

4.3.1 Symbiotic Signaling Mechanisms

Symbiotic signaling mechanisms play a pivotal role in the relationship between plants and microbes. While plants are enclosed by various microbes, only a limited number have developed an intimate, endosymbiotic relationship, residing within the host cells. Root symbioses are crucial for the nutritional needs of both plants and microbes, with over 80% of terrestrial plants and arbuscular mycorrhizal fungi forming a unique partnership (Oldroyd, 2013). While most plant–microbe interactions occur on the plant's surface or between cells, this relationship takes a step further, leading to the fungus residing inside plant cells. This intimate association is initiated by the plant's recognition of fungal signals, triggering a series of events that ultimately allow the fungus to establish itself within the plant's cellular interior (Berg et al., 2014).

Nitrogen, a vital component for life, is copiously shown in a blocked-off shape within the air. Both chemical and organic forms play an imperative part in overcoming the vitality boundary required to change nitrogen into assimilable forms, such as nitrates and nitrites. Leguminous plants have developed a specialized mechanism to attract and host a group of Gram-negative bacteria called rhizobia within their roots, establishing a mutually beneficial relationship for biological nitrogen fixation (Oldroyd, 2013).

The roots of vegetable discharge flavonoids and these compounds are identified by comparing organisms within the rhizosphere. The knowledge of flavonoids creates host-specific chemotaxis and actuates the expression of rhizobia gesture qualities, which are imperative for building up an advantageous relationship with the have plant (Cooper, 2007). This enactment of gesture qualities leads to the blend of lipo-chitooligosaccharides (LCOs), commonly alluded to as nodulation variables or gesture variables (NFs). NFs are characterized by their low atomic weight, diffusible nature, semi-hydrophobic properties, thermostability, and resistance to proteases (Tikhonovich & Provorov, 2007; Oldroyd, 2013). Once discharged, gesture variables (NFs) are identified by particular plant receptors, counting gesture figure receptors (NFRs) and lysin theme receptor-like kinases (LYKs/LYRs). This event, triggered by a precise interaction, marks the initiation of the infection process.

The component nitrogen is restricted, the roots of the plants produce (iso) flavonoids within the nanomolar extend and discharge them into the rhizosphere. These flavonoids are at that point detected by consistent rhizobia (Lea et al., 2007; Liu & Murray, 2016). Separated from activating the biosynthesis of nodulation components (NFs), flavonoids also activate qualities related to nitrogen obsession, as well as proteins included within the union and adjustment of exopolysaccharides, at the side transcriptional controllers related to the Sort III emission framework (T3SS) (Kobayashi et al., 2004; Jiménez-Guerrero et al., 2017; Firmin et al., 1986).

4.3.1.1 Evolution of Signaling in Plant Symbioses

Plants are bounded by organisms, but as it were a constrained number of organisms have created a hint of endosymbiotic affiliation by dwelling interior cells. Root symbioses play a pivotal part as sources of sustenance for both plants and organisms (Genre and Russo, 2016). Intriguingly, more than 80% of all earthly plants lock in intracellular symbioses with arbuscular mycorrhizal parasites. This study explores the developmental associations among different plant root endosymbiosis, with a particular center on the advancement of signaling in four particular sorts: The antiquated arbuscular mycorrhizae, the more recent ericoid mycorrhizae, and the root nodules with nitrogen-fixing soil bacteria, including either Frankia (actinorhizal nodulation) or rhizobia (rhizobium-legume nodulation) (Genre and Russo, 2016).

In each situation, beneficial interaction begins with a signal interaction between the two accomplices, eventually driving to the facilitation of organisms' interior plant cells. This process requires modifications in the plant development and function (Geurts et al., 2016; MacLean et al., 2017). As the endosymbiosis builds up, continuous signaling between the host and organism serves to finely alter the interaction (MacLean et al., 2017)..

Recent pivotal studies and research have reshaped our understanding of the evolution of root endosymbiosis, particularly emphasizing a probable scenario of a single improvement of root nodule symbiosis in a clade of Rosids, followed by numerous occurrences of symbiosis losses (Griesmann et al., 2018; van Velzen et al., 2018). This likely involved the symbiont *Frankia* as the initial partner for actinorhizal species, with the later emergence of rhizobia in legume symbioses (van Velzen et al., 2019). Additionally, the shared utilization of orthologous nodulation genes across actinorhizal and legume species underscores the similarity in signaling pathways between legume and actinorhizal symbioses. This similarity has evolved, at least in part, from signaling mechanisms in mycorrhizal symbioses (Markmann & Parniske, 2009).

Wang et al. (2018) provide an evolutionary evaluation of various mechanisms that confer specificity, emphasizing recent insights into how immune responses are harnessed to restrict successful symbiotic interactions. Even in instances of successful infections, there is a necessity to limit these interactions to prevent an excessive drain on plant resources. The CLE protein family, particularly CLV2, has been extensively studied for its role in controlling nodule formation in legumes. Recent research by Wang et al. (2018) has revealed a broader function of CLV2 in plant–microbe interactions. They found that CLV2, which is involved in both shoot development and nodulation, also plays a key role in the regulation of mycorrhizal symbiosis in tomatoes. This finding highlights the evolutionary conservation of signaling pathways between different plant–microbe interactions, suggesting a shared mechanism for controlling symbiotic relationships (Wang et al., 2018).

Finally, the later ericoid and ectomycorrhizal symbioses, which started from parasitic ancestries unmistakable from those shaping arbuscular mycorrhizae (Wang & Qiu, 2006), are shedding light on novel ways in which organisms can flag the plant. The ericoid mycorrhizal advantageous interaction, in particular, stands out as unordinary since it can enter cells to set up an endosymbiosis; however, the signaling between organism and plant remains completely obscure. Casarrubia et al. (2018) distinguish a little, emitted protein, OmSSP1, from an ericoid mycorrhizal organism. This protein offers highlights with hydrophobin effectors created by ectomycorrhizal organisms but shows up to operate in ericoid disease or colonization of the *Vaccinium myrtillus* root, stamping one of the early atomic signals between the endosymbiont and the host in this advantageous interaction (Casarrubia et al., 2018).

4.3.2 Quorum Sensing (QS)

Quorum sensing (QS) is a cell-to-cell communication framework utilized by microbes. It includes the generation, discovery, and reaction of signaling atoms called autoinducers (AIs). As the bacterial populace develops, AI levels increment, permitting microscopic organisms to sense their populace thickness and arrange their behavior. QS directs qualities mindful for different exercises, counting bioluminescence, sporulation, competence, antimicrobial generation, biofilm arrangement, and destructiveness calculate discharge. (Novick & Geisinger, 2008; Ng & Bassler, 2009; Williams & Camara, 2009). All majority detecting (QS) frameworks, in any case of their particular components and instruments, share three center standards. To begin with, microscopic organisms deliver signaling particles called autoinducers (AIs). At low cell density, AIs scatter and their concentration remains underneath the location edge. Alternately, at high cell thickness (HCD), the collective AI generation comes about in a localized high concentration, encouraging location and activating a reaction (Kaplan & Greenberg, 1985). Second, AIs are recognized by receptors found within the cytoplasm or film. Third, in expansion to actuating the expression of qualities basic for agreeable behaviors, AI discovery leads to the actuation of AI generation (Novick et al., 1995; Seed et al., 1995).

Gram-positive bacteria use peptide-based quorum-sensing (QS) systems. They produce signaling molecules called autoinducing peptides (AIPs). After synthesis and secretion, AIPs accumulate in the extracellular environment. At high cell density, AIPs bind to specific receptors on the bacterial cell surface, triggering a QS response (Williams & Camara, 2009). Typically, this binding activates the receptor's kinase activity, leading to autophosphorylation, and subsequently, the transfer of phosphate to a matching cytoplasmic response regulator. The phosphorylated response regulator then stimulates the transcription of genes within the QS regulon. In certain instances of Gram-positive bacterial QS, AIPs may be transported back into the cell cytoplasm, where they interact with transcription factors, influencing the activity of these factors and, consequently, modulating changes in gene expression (Seed et al., 1995).

Gram-negative bacteria use small molecules, such as acylhomoserine lactones (AHLs), as signaling molecules. These molecules are produced using S-adenosylmethionine (SAM) as a building block (Wei et al., 2011). AIs are generated within the cell and can freely diffuse across both the inner and outer membranes. At high cell density (HCD), when the AI concentration is sufficiently elevated, they bind to cytoplasmic receptors, which are transcription factors. These AI-bound receptors then regulate the expression of genes within the QS regulon. In certain instances of Gram-negative bacterial QS, AIs may also be detected by two-component histidine kinase receptors, operating similarly to those described earlier for Gram-positive QS bacteria (Wei et al., 2011). Soil bacteria have long been employed in crop production due to their vital role in biogeochemical cycles. The health and fertility of plants and soil are crucial for ensuring food security and are significant indicators of the safety and sustainability of food systems. Hence, comprehending the dynamic changes and progression of root microbiomes in relation to plants, as explored in this chapter, presents novel opportunities to harness the advantages of beneficial root–microbiome interactions in order to achieve sustainable plant health (Wei et al., 2011).

4.4 Plant–Rhizospheric Microbe Interaction

4.4.1 Beneficial Rhizosphere Organisms

4.4.1.1 Nitrogen-Fixing Bacteria

Nitrogen is a constraining and significant element for plant growth. It serves as an essential component in both chlorophylls, the principal pigment necessary for photosynthesis, and amino acids, which are the principal elements in the formation of proteins. Although nitrogen is widely distributed in the Earth's atmosphere, particularly as nitrogen gas (N_2), plants are limited to utilizing decreased forms of this element (Wani et al., 2008). Plants acquire these forms of "combined" nitrogen via diverse mechanisms: (1) application of ammonia and/or nitrate fertilizer (generated via the Haber-Bosch process) or manure into the soil; (2) conversion of atmospheric nitrogen into ammonia and nitrate (NO_3) via natural processes such as lightning; (3) liberation of ammonia during the decomposition of organic substances; and (4) biologically via asymbiotic or symbiotic nitrogen fixation (Zahir et al., 2010).

4.4.1.1.1 Asymbiotic Nitrogen Fixation

It is carried out by autonomous microorganisms, particularly free-living microbes, which include *Azotobacter*, *Clostridium*, and cyanobacteria (blue-green algae) such as *Anabaena* and *Nostoc*. In free-living microorganisms, nitrogenase produces ammonium, which is then converted to glutamate via the glutamine synthetase/glutamate synthase process (Ahemad & Khan, 2011). The enzymes involved in nitrogenase activity are highly vulnerable to oxygen-induced degradation and thus require anaerobic conditions (Wang et al., 2008). Consequently, numerous bacteria halt the production of the enzyme when oxygen is present. Several nitrogen-fixing microorganisms exclusively thrive in anaerobic environments, where they respire to deplete oxygen or combine oxygen with a protein like leghemoglobin, which acts as an oxygen scavenger. Within the realm of living plants, certain bacteria, fungi, and blue-green algae (*Cyanobacteria*) are able to convert molecular nitrogen into a functional form of nitrogen, specifically NH_4. *Azotobacter vinelandii, Clostridium pasteurianum, Rhodospirillum rubrum, Chromatium, Nostoc, Anabaena, Rivularia*, and other microorganisms have the capability to perform molecular nitrogen fixation (Ahemad & Khan, 2011).

4.4.1.1.2 Symbiotic Nitrogen Fixation

It involves a mutualistic association with host plants, where plants offer an atmosphere and carbon source to bacteria in return for fixed nitrogen. The diverse instances of symbiotic biological nitrogen fixation can be classified into the following three categories: Nitrogen fixation occurs through the process of nodule formation in leguminous plants; The predominant type of symbiotic relationship leads the host plant to develop nodules, an enlarged and multicellular structures on its root or in some cases, the stem (as seen in the Sesbania plant, which has stem nodules); and the symbiont in the case of peas, clover, soybean, and alfalfa from the genera *Rhizobium*, *Bradyrhizobium*, and *Azorhizobium*.

Nodule development as a means of nitrogen fixation in nonleguminous plants: Actinomycetes from the genus *Frankia* are the symbionts in nonleguminous plants like Alnus, Casuarina, and Myrica. Nitrogen fixation through non-nodulation: Some plants participate in synergistic nitrogen fixation without the development of nodules. These associations are classified as pseudo-symbiotic or pseudo-symbiosis. One example is *Anabaena azollae*, which is a type of *Cyanobacterium* that resides in the pores found on the leaves of a water plant known as *Azolla* (Ahemad & Khan, 2011).

4.4.1.2 Mycorrhizal Fungi

Mycorrhizae, which translates to "fungus root," is a symbiotic relationship between plant roots and fungi. Albert Bernhard Frank proposed the term for this phenomenon. Mycorrhizae colonize plant roots both within the cells and outside the cells. In the kingdom of plants, mycorrhizal association is a universal phenomenon. There are several types of mycorrhizal associations. Ectomycorrhiza: From a phylogenetic context, the range of ectomycorrhizal fungi (EM fungi) is vast. In their 2002 study, Högberg and Högberg noticed that fungal partners in ectomycorrhizal (EM) associations account for 30% of the microbial biomass in forest soils (Marschner & Dell, 1994).

A mycelial sheath surrounds the cortical tissue of the host root, while a Hartig net (a network of hyphae) encloses the outer layer and/or cortical cells of the root, facilitating an extensive surface and enabling the movement of sustenance and minerals. Mycorrhizal associations aid in the soil's inorganic nutrient uptake, shield roots from pathogen intrusion, and break down organic matter. Ericoid mycorrhiza: Ericoid mycorrhiza refers to the symbiotic interaction between *Ericaceae plants* with ericoid mycorrhizal fungi (ERM). During this association, ERM fungi develop specific hyphal coils within the root cells of the plants. Ericoid mycorrhizae are a type of fungi that associates with certain plant species belonging to Ericaceae, Epacridaceae, and Empetraceae families. Some examples are rhododendrons, blueberries, and cranberries. Most *Ericaceous* species typically prosper in soil that is acidic and lacks nutrients, with a greater amount of ammonium than nitrate; thus, Ericoid mycorrhizal fungi facilitate the absorption of nitrogen and phosphorus. Orchid mycorrhiza: Orchid mycorrhiza refers to the mutual relationship between orchids and certain fungi. During seed coat rupture, "protocorm" emerges, and for the development of the plant, it is necessary for the mycorrhizal fungus to establish an infection within it. This takes up to six months. The seed will succumb if the fungus does not infect it. The fungus can survive as a saprophyte if the host is unavailable (Meding & Zasoski, 2008).

Arbuscular mycorrhizae (AM): Arbuscular mycorrhizae (AM) are a type of fungi that is obligate, symbiotic, and reproduces asexually. They possess distinct morphological and genomic features that allow them to establish themselves in plant roots and soil. Arbuscular mycorrhizal associations are predominantly found in most perennial plants, tree species of moderate and tropical areas, and most agricultural crops. Almost 100,000 plant species are thought to be engaged in a mutually beneficial relationship called arbuscular mycorrhiza symbiosis. The genera *Glomus*, *Gigaspora*, *Scutellospora*, *Acaulospora*, and *Entrophospora* are all common arbuscular mycorrhizal (AM) fungi. The term "arbuscular" pertains to two specific structures: Arbuscules, which are present in the cortical cells of many plant roots, and certain mycothalli that are formed by AM fungi (Meding & Zasoski, 2008). AM fungi found in the soil generate robust and well-protected dormant spores called extramatrical chlam ydospores. These spores are able to withstand unfavorable soil conditions and germinate if circumstances become suitable. The outgrowth succumbs if it fails to come in contact with and effectively infiltrate the host root. Upon developing an appressorium over the root surface, the hyphae enter the root and proliferate within the root cortex. Intercellular hyphae extend laterally and penetrate cortical cells, giving rise to arbuscules. The simultaneous introduction of arbuscular mycorrhizal (AM) fungus and other beneficial soil bacteria has been demonstrated to be more effective at stimulating plant development. This suggests that suitable "microbial consortia" should be established in order to inoculate different agricultural plants (Meding & Zasoski, 2008).

4.4.1.3 Plant Growth-Promoting Rhizobacteria (PGPR)

Plant growth-promoting rhizobacteria (PGPR) are described as beneficial microbes; the term was coined in 1978 by Kloepper and Schroth. Antoun and Prevost (2005) estimate that the percentage of rhizosphere microorganisms that are PGPR ranges from 2% and 5%. Plant growth-promoting nonpathogenic bacteria, or PGPR, are autonomous soil microbes that can colonize root systems and increase plant yields by one or more methods. Rhizobacteria that promote plant development can influence plant development in a variety of ways, both directly and indirectly. These microbes become adsorbed onto soil particles via a mechanism known as simple ion exchange. By nourishing the plant with a substance produced by bacteria, such as phytohormones, or by improving the plant's absorption of specific nutrients from the environment, PGPR stimulates plant development (Mujumdar et al., 2014; Maindad et al., 2014; Sachdev et al., 2009; Glick, 1995). Antibiosis, parasitism, induced systemic resistance, and the generation of compounds such as hydrogen cyanide and siderophores that reduce toxic rhizobacteria are all examples of indirect plant development processes. The genera *Acetobacter*, *Azospirillum*, *Azotobacter*, *Bacillus*, *Gluconacetobacter*, *Klebsiella*, and *Pseudomonas* are some of the genera that are classified as PGPR.

4.4.2 Deleterious Rhizosphere Organisms

4.4.2.1 Plant Pathogenic Fungi

Numerous symptoms are caused by fungus infections. Fungal plant pathogens are traditionally divided into two main groups: Necrotrophic pathogens, which destroy tissue to obtain nutrients, and biotrophic pathogens, which form close relationships with plants and are able to survive inside and utilize living tissues. Aside from these, there exists a distinct group known as hemibiotrophic pathogens, which initially function as biotrophs and subsequently transition into necrotrophs. Necrotrophs induce necrosis, which eventually leads to the death of affected plants. Phyla *Ascomycota* and *Basidiomycota* contain only fungi-based plant diseases. *Ascomycetes* include a variety of groups into which plant infections are divided, such as *Leotiomycetes* (*Botrytis* spp.), *Dothideomycetes* (*Cladosporium* spp.), and *Sordariomycetes* (*Magnaporthe* spp.).

The two most significant groups of plant pathogens are the smuts (found in large populations within the subphylum *Ustilaginomycotina*) and the rusts (*Pucciniomycetes*), which belong to the basidiomycetes.

4.4.2.2 Oomycetes

Water molds, or oomycetes, are eukaryotic microorganisms that are saprophytes. It is a pathogen for plants, insects, crustaceans, aquatic organisms, vertebrates, and microorganisms. Oomycetes infect many native, agricultural, and ornamental plants, causing devastating diseases. They consist of more than 50 *Phytophthora* class, the most destructive dicotyledonous plant pathogens. *Phytophthora* infects almost all dicots and several monocots. The most famous among all is the Irish potato famine pathogen caused by *P. infestans*. This species causes late blight, which destroys potatoes and tomatoes. Other than *Phytophthora* genus, many oomycete pathogens cause grapevine downy mildew (obligate biotrophs *Plasmopara viticola)*, and white rust and downy mildew in a variety of crops (*Albugo*, *Bremia*, and *Peronospora* species).

4.4.2.3 Nematodes

Nematodes are elongated cylindrical worms that inhabit various habitats, such as soil, freshwater, and marine environments. Various species of nematodes exhibit a food source that includes fungi to plants. Parasitic nematodes (PPN), which consume plant components, are widely present in agricultural fields. Root-knot nematodes (RKN) and soybean cyst nematodes (SCN) are two highly economically unfavorable nematodes. After entering the root, the root-knot nematode (RKN) juvenile migrates down to the root tip intercellularly. Juveniles enter the vascular cylinder base and climb to the root.

The formation of a persistent feeding site in the differentiation zone of the root develops via the stimulation of nuclear division in host cells, without subsequent cytokinesis, producing giant cells. This leads to the development of galls known as "root knots." Stylets of cyst nematodes enter roots and move to the vascular cylinder, which disrupts the cells. They inject stylet secretions to feed at the vascular cylinder. A multinucleate syncytium is formed when the cell walls of adjacent feeding site cells and neighboring cells break down. Before becoming adults, cyst nematodes molt three times in the root. Sexual reproduction produces a female full of eggs. Female bodies become protective cysts for eggs after death.

The plant rhizosphere serves as a battleground for both beneficial and harmful organisms. Table 4.1 shows some PGPMs with their mode of action. The rhizosphere harbors microorganisms that form highly organized communities, characterized by specialized roles and distinct functions in their interactions with plants. Efficient plant health management is critical for generating sustainable food production and eliminating hunger. As a result, achieving safe and sustainable plant health requires outstanding integration of the rhizosphere microbiome throughout plant development, disease control, and a sustainable practice of biofertilizers and biopesticides.

4.5 Omics Insights Into Plants–Soil Microbiota

Naturally, microbes and plants engage in continuous interactions throughout various life cycle stages. Plant–microbe interaction, compared to other microbial interactions, holds paramount significance in maintaining ecosystem balance. The biotic and abiotic interactions influence the makeup of the plant microbiome. Plant-related microbial communities exhibit dependence on specific microenvironments, such as carposphere, spermosphere, phyllosphere, and endorhiza, with these microenvironments governed by environmental parameters. Microbial populations' survival mechanisms influence their impact on the plant system (Olanrewaju et al., 2019). The relationship between plants and microbes is complex and involves dynamic interactions that categorize the roles of microbes as mutualistic, commensal, and pathogenic (Agler et al., 2016). Plants have evolved variety of strategies in response to these interactions. These adaptations include specific defense responses, structural changes, and the release of secondary metabolites. These adaptive mechanisms serve to optimize plant growth and reproduction amid fluctuating environmental conditions (Pieterse et al., 2012). This molecular interplay, marked by reciprocal adaptations between plants and microbial communities, not only shapes the immediate responses but also propels the evolutionary trajectories of both parties. This dynamic coevolution plays a significant role in establishing beneficial allelic

Table 4.1 Examples of PGPR with mode of actions

S. No.	Type of bacteria	Name of the bacteria	Associated plant	Mechanism	References
1	Gram negative	*Acinetobacter genospecies A28*	*Triticum aestivum*	IAA production	Chopade et al. (2008)
2	Gram negative	*Acinetobacter haemolyticus 19*	*Triticum aestivum*		Chopade et al. (2008)
3	Gram negative	*Acinetobacter baumannii A16*	*Triticum aestivum*		Chopade et al. (2008)
4	Gram negative	*Acinetobacter junii A6*	*Triticum aestivum*		Chopade et al. (2008)
5	Gram negative	*Acinetobacter baumannii A13*	*Triticum aestivum*		Chopade et al. (2008)
6	Gram negative	*Acinetobacter genospecies A15,*	*Triticum aestivum*		Chopade et al. (2008)
7	Gram negative	*Acineto bacter baumannii A18,*	*Triticum aestivum*		Chopade et al. (2008)
8	Gram negative	*Acinetobacter baumannii A30*	*Triticum aestivum*		Chopade et al. (2008)
9	Gram negative	*Acinetobacter* strains PUCM1001	*Pennisetum glaucum*	Siderophore production, zinc oxide, and phosphate solubilization	Rokhbakhsh-Zamin et al. (2011)
10	Gram negative	*Acinetobacter* strains PUCM1002	*Pennisetum glaucum*	Siderophore production, zinc oxide solubilization	Rokhbakhsh-Zamin et al. (2011)
11	Gram negative	*Acinetobacter* strains PUCM1003	*Pennisetum glaucum*	Siderophore production, zinc oxide, and phosphate solubilization	Rokhbakhsh-Zamin et al. (2011)
12	Gram negative	*Acinetobacter* strains *PUCM1004*	*Pennisetum glaucum*		Rokhbakhsh-Zamin et al. (2011)
13	Gram negative	*Acinetobacter* strains *PUCM1005*	*Pennisetum glaucum*		Rokhbakhsh-Zamin et al. (2011)
14	Gram negative	*Acinetobacter* strains *PUCM1006*	*Pennisetum glaucum*		Rokhbakhsh-Zamin et al. (2011)
15	Gram negative	*Acinetobacter* strains *PUCM1007*	*Pennisetum glaucum*	Siderophores production, IAA production	Rokhbakhsh-Zamin et al. (2011)

(continued)

Table 4.1 (continued)

S. No.	Type of bacteria	Name of the bacteria	Associated plant	Mechanism	References
16	Gram negative	*Acinetobacter* strains *PUCM1008*	*Pennisetum glaucum*	Siderophore production, zinc oxide, and phosphate solubilization	Rokhbakhsh-Zamin et al. (2011)
17	Gram negative	*Acinetobacter* strains *PUCM1009*	*Pennisetum glaucum*	Siderophore production and zinc oxide solubilization	Rokhbakhsh-Zamin et al. (2011)
18	Gram negative	*Acinetobacter* strains *PUCM1011*	*Pennisetum glaucum*	Siderophore production, zinc oxide, and phosphate solubilization	Rokhbakhsh-Zamin et al. (2011)
19	Gram negative	*Acinetobacter* strains *PUCM1012*	*Pennisetum glaucum*	Siderophore production and phosphate solubilization	Rokhbakhsh-Zamin et al. (2011)
20	Gram negative	*Acinetobacter* strains *PUCM1013*	*Pennisetum glaucum*	Siderophore production and phosphate solubilization	Rokhbakhsh-Zamin et al. (2011)
21	Gram negative	*Acinetobacter* strains *PUCM1014*	*Pennisetum glaucum*	Siderophore production, zinc oxide, and phosphate solubilization	Rokhbakhsh-Zamin et al. (2011)
22	Gram negative	*Acinetobacter* strains *PUCM1015*	*Pennisetum glaucum*		Rokhbakhsh-Zamin et al. (2011)
23	Gram negative	*Acinetobacter* strains *PUCM1016*	*Pennisetum glaucum*		Rokhbakhsh-Zamin et al. (2011)
24	Gram negative	*Acinetobacter* strains *PUCM1017*	*Pennisetum glaucum*	Siderophore production	Rokhbakhsh-Zamin et al. (2011)
25	Gram negative	*Acinetobacter* strains *PUCM1018*	*Pennisetum glaucum*	Siderophore production, zinc oxide, and phosphate solubilization	Rokhbakhsh-Zamin et al. (2011)
26	Gram negative	*Acinetobacter* strains *PUCM1019*	*Pennisetum glaucum*		Rokhbakhsh-Zamin et al. (2011)
27	Gram negative	*Acinetobacter* strains *PUCM1020*	*Pennisetum glaucum*		Rokhbakhsh-Zamin et al. (2011)
28	Gram negative	*Acinetobacter* strains *PUCM1021*	*Pennisetum glaucum*		Rokhbakhsh-Zamin et al. (2011)

(continued)

Table 4.1 (continued)

S. No.	Type of bacteria	Name of the bacteria	Associated plant	Mechanism	References
29	Gram negative	*Acinetobacter* strains *PUCM1022*	*Pennisetum glaucum*	Siderophore production and zinc oxide solubilization	Rokhbakhsh-Zamin et al. (2011)
30	Gram negative	*Acinetobacter* strains *PUCM1023*	*Pennisetum glaucum*	Siderophore production, zinc oxide, and phosphate solubilization	Rokhbakhsh-Zamin et al. (2011)
31	Gram negative	*Acinetobacter* strains *PUCM1024*	*Pennisetum glaucum*		Rokhbakhsh-Zamin et al. (2011)
32	Gram negative	*Acinetobacter* strains *PUCM1025*	*Pennisetum glaucum*		Rokhbakhsh-Zamin et al. (2011)
33	Gram negative	*Acinetobacter* strains *PUCM1026*	*Pennisetum glaucum*		Rokhbakhsh-Zamin et al. (2011)
34	Gram negative	*Acinetobacter* strains *PUCM1027*	*Pennisetum glaucum*		Rokhbakhsh-Zamin et al. (2011)
35	Gram negative	*Acinetobacter* strains *PUCM1028*	*Pennisetum glaucum*		Rokhbakhsh-Zamin et al. (2011)
36	Gram negative	*Acinetobacter* strains *PUCM1029*	*Pennisetum glaucum*		Rokhbakhsh-Zamin et al. (2011)
37	Gram negative	*Acinetobacter* strains *PUCM1029*	*Pennisetum glaucum*		Rokhbakhsh-Zamin et al. (2011)
38	Gram negative	*Acinetobacter haemolyticus* A19	*Triticum aestivum*	Antibiotic and chitinase production	Mujumdar et al. (2014)
39	Gram negative	*Azospirillum brasilense Sp245*	*Phaseolus vulgaris*	Root growth increased	Remans et al. (2008)
40	Gram negative	*Achromobacter xylosoxidans*	*Vigna radiata*	Influence plant homeostasis	Ma et al. (2009)
41	Gram negative	*Azospirillum brasilence*	*Saccharum officinarum*	Increases root growth	Orlandini et al. (2014)
42	Gram negative	*Acinetobacter calcoaceticus*	*Cucumus sativus*	Produces different gibberellins and phosphate solubilizing properties	Kang et al. (2009)
43	Gram negative	*Achromobacter piechaudii ARV8*	*Solanum lycopersicum and Piper nigrum*	Water stress tolerance	Mayak et al. (1999)

(continued)

Table 4.1 (continued)

S. No.	Type of bacteria	Name of the bacteria	Associated plant	Mechanism	References
44	Gram negative	*Azospirillum brasilense Sp6*	*Atriplex lentiformis*	Growth in acidic, metalliferous desert	de-Bashan et al. (2010)
45	Gram negative	*Azoarcus* sp.	*Kallar grass*	Nitrogen fixation, enhances stress tolerance	Hurek et al. (2002)
46	Gram negative	*Aeromonas punctata PNS-1*	*Arabidopsis thaliana*	Increases root growth	Iqbal (2013)
47	Gram negative	*Azotobacter chroococcum HKN-5*	*Brassica juncea*	Soil bioremediation of Pb and zinc	Wu et al. (2006)
48	Gram negative	*Achromobacter xylosoxidans AUM54*	*Catharanthus roseus*	IAA production, phosphate solubilization, siderophore	Karthikeyan et al. (2012)
49	Gram negative	*Azospirillum lipoferum KYR F6*	*Haplopappus sp*	IAA production, nitrogen fixation, phosphate solubilization	Navarro-Noyaa et al. (2012)
50	Gram negative	*Acinetobacter* sp. *ALEB16*	*Atractylodes lancea*	ABA, SA production	Wang et al. (2015)
51	Gram negative	*Alcaligenes faecalis. Faecalis. S8*	*Withania somnifera*	Phosphate solubilization, IAA production	Abdallah et al. (2016)
52	Gram negative	*Azospirilium*	*Cynara scolymus*	Phosphate solubilization, nitrogen-fixing bacteria l	Jahanian et al. (2012)
53	Gram negative	*Azotobacter*	*Cynara scolymus*	Phosphate solubilization, nitrogen-fixing bacteria l	Jahanian et al. (2012)
54	Gram positive	*Arthrobacter* sp. *SMR3*	*Papaver somniferum*	IAA production	Pandey et al. (2016)
55	Gram negative	*Burkholderia ambifaria*	*Daucus carota*	ACC production	Pellegrini et al. (2021)
56	Gram negative	*Bradyrhizobium MRM6*	*Vigna radiata*	Herbicide tolerance improvement	Ahemad and Khan (2012a, b)
57	Gram negative	*Bradyrhizobium MRM6*	*Vigna radiata*	Increased the growth	Ahemad and Khan (2011, 2012a, b)
58	Gram negative	*Bradyrhizobium* sp. *(vigna) RM8*	*Vigna radiata*	Enhanced the nodule numbers	Wani et al. (2008)

(continued)

Table 4.1 (continued)

S. No.	Type of bacteria	Name of the bacteria	Associated plant	Mechanism	References
59	Gram negative	*Burkholderia caribensis XV*	*Mangifera indica*	Mango growth promotion, floral fate, and increases flowers	de los Santos-Villalobos et al. (2013)
60	Gram negative	*Bradyrhizobium elkanii*	*Glycine max*	Nodule formation	Fukuhara et al. (1994)
61	Gram negative	*Enterobacter agglomerans*	*Solanum lycopersicum*	Phosphate solubilization	Oteino et al. (2015)
62	Gram negative	*Enterobacter* sp. *TAP02*	*T. amplexicaulis*	IAA production, nitrogen fixation, phosphate solubilization	El-Sayed et al. (2014)
63	Gram negative	*Enterobacter aerogenes*	*Zea mays*	Promoting growth and yield at high salt stress	Nadeem et al. (2007)
64	Gram negative	*Enterobacter cloacae Rs-35*	*Gossypium*	Encourage the establishment of systemic tolerance, and balance cell ion concentration	Yue et al. (2023)
66	Gram negative	*Herbaspirillum seropedicae*	*Oryza sativa*	Enhance production of gibberellins	Babalola (2010)
67	Gram negative	*Klebsiella* sp.	*Sorghum bicolor*	ACCD activity	Singh et al. (2015)
68	Gram negative	*Klebsiella* strains	*Triticum aestivum*	IAA production	Sachdev et al. (2009)
69	Gram negative	*Lysobacter antibioticus*	*Oryza sativa*	Produces antibiotics, controls bacterial leaf blight	Jha et al. (2005)
70	Gram negative	*Mesorhizobium strain MRC4*	*Cicer arietinum*	Enhance overall growth of plants	Ahemad and Khan (2011)
71	Gram negative	*Methylobacterium radiotolerans*	*Oryza sativa and Solanum lycopersicum*	Enhanced root and shoot length	Chinnadurai et al. (2009)
72	Gram negative	*Methylobacterium mesophilicum*	*Oryza sativa*	Effect on N-Acyl-homoserine lactone	Sanders et al. (2000)
73	Gram negative	*Methylobacterium oryzae*	*Oryza sativa*	Methanol utilization, growth promotion	Madhaiyan et al. (2007)

(continued)

Table 4.1 (continued)

S. No.	Type of bacteria	Name of the bacteria	Associated plant	Mechanism	References
74	Gram negative	*Mesorhizobium huakuii B3*	*Astragalus sinicus*	Soil bioremediation of cadmium heavy metal	Sriprang et al. (2003)
75	Gram negative	*Ochrobactrum intermedium*	*Cucurbita*	Salt-tolerant PGPR strains from salt-contaminated soil	Abdelrahman et al. (2021)
76	Gram negative	*PD-P6 (Enterobacter cloacae)*	*Phoenix dactylifera*	Enhance root elongation and facilitate nutrient uptake in roots under salinity stress	Yaish et al. (2015)
77	Gram negative	*Pseudomonas fluorescens*	*Setaria italica*	Seed germination and seedling growth in drought conditions	Niu et al. (2018)
78	Gram negative	*Pseudomonas alcaliphila*	*Phaseolus vulgaris*	Enhanced the root growth parameters	AlAli et al. (2021)
79	Gram negative	*Pseudomonas mallei (RBG4, ET17)*	*Eggplant*	Reduced wilt incidence by 83% caused by *Ralstonia solanacearum*	Ramesh and Phadke (2012)
80	Gram negative	*Pseudomonas* sp. *5Vm1K*	*Allium cepa*	Bulb size and weight were increased	Novello et al. (2021)
81	Gram positive	*Paenibacillus polymyxa*	*Phaseolus vulgaris*	Increased drought stress	Ngumbi and Kloepper (2016)
82	Gram negative	*Pseudomonas cepacia*	*Cucumis sativus*	Inhibit the growth of *Pythium ultimum*	Pérez-Montaño et al. (2014)
83	Gram negative	*Pseudomonas cepacia*	*Phaseolus vulgaris*	Inhibit the growth *Sclerotium rolfsii*	Pérez-Montaño et al. (2014)
84	Gram negative	*Pseudomonas cepacia*	*Gossypium hirsutum*	Inhibit the growth *Rhizoctonia solani*	Pérez-Montaño et al. (2014)
85	Gram negative	*Pseudomonas* sp.	*Phaseolus vulgaris*	Prevent Fusarium wilt	Rathore (2015)
86	Gram negative	*Pseudomonas aeruginosa GPS 21*	*Arachis hypogaea*	Increased plant growth and pod production.	Kishore et al. (2005)

(continued)

Table 4.1 (continued)

S. No.	Type of bacteria	Name of the bacteria	Associated plant	Mechanism	References
87	Gram negative	*Pseudomonas aeruginosa RTE4*	*Camellia sinensis*	Bio-control activity against two foliar fungal pathogens	Chopra et al. (2020)
88	Gram negative	*Pseudomonas mallei (RBG4, ET17)*	*Solanum melongena*	Reduced wilt incidence by 83% caused by *Ralstonia solanacearum*	Ramesh and Phadke (2012)
89	Gram negative	*Pseudomonas* sp. *5VmIK*	*Allium cepa*	Bulb size and weight were increased	Novello et al. (2021)
90	Gram negative	*Pseudomonas putida*	*Cynara scolymus*	Phosphate solubilizing, nitrogen-fixing bacteria l	Jahanian et al. (2012)
91	Gram negative	*Pseudomonas* sp. *PS1*	*Vigna radiata*	Overall growth production	Ahemad and Khan (2011, 2012a, b)
92	Gram negative	*Pseudomonas* sp. *A3R3*	*Alyssum serpyllifolium, Brassica junce*	Increased biomass	Ma et al. (2011)
93	Gram negative	*Pseudomonas* sp.	*Triticum aestivum*	Increased nutrient uptake	Sharma et al. (2011)
94	Gram negative	*Psychrobacter* sp. *SRS8*	*Ricinus communis, Helianthus annuus*	Stimulated plant growth	Ma et al. (2011)
95	Gram negative	*Pseudomonas*	*Litchi chinensis*	Stimulated plant growth	Kumari and Khare (2020)
96	Gram negative	*Pseudomonas fluorescens APF1*	*Solanum lycopersicum*	Increased fresh and dry weight	Lemessa and Zeller (2007)
97	Gram negative	*Pseudomonas alcaliphila*	*Phaseolus vulgaris*	Enhanced the root growth parameters	AlAli et al. (2021)
98	Gram negative	*Pseudomonas*	*Litchi chinensis*	Effect on root length, shoot length, and leaves	Kumari and Khare (2020)
99	Gram negative	*Pseudomonas fluorescens*	*Arabidopsis thaliana*	Effect on iron inside plant tissues and plant growth	Vansuyt et al. (2007)
100	Gram negative	*Pseudomonas* sp. *PS1*	*Vigna radiata*	Improved plant dry weight, nodule formation	Ahemad and Khan (2012)
101	Gram negative	*Pseudomonas* sp. *NBRI 4014*	*Glycine max*	Heavy metal resistance with root elongation	Gupta et al. (2002)

(continued)

Table 4.1 (continued)

S. No.	Type of bacteria	Name of the bacteria	Associated plant	Mechanism	References
102	Gram negative	*Pseudomonas extremorientalis*	*Triticum aestivum*	Increased overall growth	Egamberdieva and Kucharova (2009)
103	Gram negative	*Pseudomonas stutzeri CSP03*	*Capparis spinosa*	IAA production, Nitrogen fixation, P solubilization, Siderophore production	El-Sayed et al. (2014)
104	Gram negative	*Pseudomonas* sp.	*Glycine max*	Increased soil enzyme activities	Sharma et al. (2011)
105	Gram negative	*Pseudomonas* sp. *A3R3*	*Alyssum serpyllifolium, Brassica juncea*	Increased biomass	Ma et al. (2011)
106	Gram negative	*Rhizobium phaseoli*	*Vigna radiata L*	Effect on salinity and overall growth	Zahir et al. (2010)
107	Gram negative	*Rhizobium strain MRP1*	*Pisum sativum*	Effect overall growth	Ahemad and Khan (2011)
108	Gram negative	*Rhizobium strain MRL3*	*Lens esculentus*	Effect overall growth	Ahemad and Khan (2011)
109	Gram negative	*Rhizobium* sp. *RP5*	*Pisum sativum*	Intrinsic capacities of nitrogen fixation, growth stimulation, and reduction of nickel and zinc toxicity.	Wani et al. (2008)
110	Gram negative	*Rhizobium leguminosarum*	*Phaseolus vulgaris*	P solubilization	Ahemad and Kibret (2014)
111	Gram negative	*Serratia marcescens EB 67*	*Pennisetum glaucum*	Increase in plant weight	Hameeda et al. (2006)
112	Gram negative	*Sinorhizobium meliloti* strain	*Medicago truncatula*	Enhanced salt tolerance	Bianco and Defez (2009)
113	Gram negative	*Sphingomonas sp*	*Arabidopsis thaliana*	Inhibit the growth of *Pseudomonas syringae* pv. tomato	Innerebner et al. (2011)
114	Gram negative	*Serratia ureilytica Bac5*	*Ocimum sanctum*	Siderophore, ACC deaminase, Phosphate solubilization	Barnawal et al. (2012)
115	Gram negative	*Xanthomonas* sp. *RJ3*	*Brassica napus*	Increased the growth of other microorganisum	Rajkumar et al. (2006)
116	Gram positive	*Bacillus subtilis LK14*	*Moringa peregrina*	Phosphate solubilization, IAA production	Khan et al. (2016)

(continued)

Table 4.1 (continued)

S. No.	Type of bacteria	Name of the bacteria	Associated plant	Mechanism	References
117	*Gram positive*	*Bacillus cereus BC1AW*	*Solanum lycopersicum*	Phosphate solubilization, production of siderophore, HCN and IAA, reduce bacterial growth	Kurabachew and Wydra (2013)
118	Gram positive	*Bacillus megaterium*	*Capsicum annuum*	Phosphate solubilization, promotes growth	Rodriguez and Fraga (1999)
119	Gram positive	*Bacillus RC23*	*Actinidia deliciosa*	Root induction in kiwifruit stem cuttings	Erturk et al. (2010)
120	Gram positive	*Bacillus velezensis 83*	*Zea mays (Maize)*	Increase in height and root biomass in maize seedlings	Karina et al. (2020)
121	Gram positive	*Bacillus circulans*	*Vigna radiata*	Phosphate solubilization	Oteino et al. (2015)
122	Gram positive	*Bacillus mucilaginosus*	*Piper nigrum, Cucumis sativus*	Improve potassium intake capacity	Liu et al. (2012)
123	Gram positive	*Bacillus subtilis*	*Hordeum vulgare*	Inhibit powdery mildew	Prathap and Kumari (2015); Oyedele and Ogunbanwo, (2014)
124	Gram positive	*Bacillus species PSB10*	*Cicer arietinum*	Significantly improved growth, nodulation	Wani and Khan (2010)
125	Gram positive	*Bacillus RC23*	*Actinidia deliciosa*	Root induction in kiwifruit stem cuttings	Erturk et al. (2010)
126	Gram positive	*Clavibacter michiganensis*	*Curcuma longa*	IAA production	Kumar et al. (2016)
127	Gram positive	*Curtobacterium plantarum*	*Malus pumila*	Induces resistance to pathogens, enhances growth	Sessitsch et al. (2004)
128	Gram positive	*Micromonospora aurantiaca*	Various crops	Induces plant defense mechanisms, promotes root and shoot growth	Martinez-Hidalgo and Hirsch (2017)
129	Gram positive	*Paenibacillus graminis BR 35*	*Aster gymnocephalus*	Nitrogen fixation, P solubilization	Navarro-Noyaa et al. (2012)
130	Gram positive	*Streptomyces coelicolor*	*Solanum lycopersicum*	Produces antifungal compounds, enhances growth	Watkinson et al. (2001)

variants within genomic pools (Frantzeskakis et al., 2018). The continual refinement of these genomic pools underlines the intricate and evolving nature of the symbiotic relationship between plants and microbes, with far-reaching implications for ecological dynamics and adaptation in response to environmental challenges. Recent advancements in data analysis facilitate the correlation of candidate microbes with specific phenotypes such as plant growth, yield, nutrient uptake efficiency, and disease tolerance (Edwards et al., 2018; Moroenyane et al., 2021; Wallace et al., 2018).

In order to choose plants that efficiently attract beneficial microbes, it is crucial to uncover the molecular mechanisms behind microbiome recruitment. The emergence of omics technologies facilitates a thorough characterization of plants. Advancements in molecular biology contribute to unraveling the structural, functional, and interactive aspects of genes, microbial communities, transcript sequences, protein-protein interactions, and metabolic networks. These technological strides enhance our understanding of both positive and negative effects within joint metabolic networks. Despite the complexity, "Omics" science offers an integrative analysis approach to unravel complex plant–microbe interactions.

4.5.1 Genomics Insights Into Plants–Soil Microbiota

Genomics is a multidisciplinary field dedicated to the comprehensive study of an organism's entire genome, drawing on concepts from genetics. This detailed process uses a collaborative approach that combines DNA sequencing, rDNA technique, and computational studies. These tools work together to facilitate the sequencing, assembly, and analysis of both the structural and functional aspects of genomes. Important advances in our understanding of several aspects of plant pathosystems are emphasized by groundbreaking contributions from genomics (Huang et al., 2023; Joshi et al., 2023; Krishna et al., 2024; Reyes-Ardila et al., 2024; Straub et al., 2021; Xu & Wang, 2019).

The genetic exploration of plant–microbe interactions has undergone a transformative shift, moving beyond the analysis of whole-genome sequences in diverse plant species and the annotation of enzymes, proteins, and metabolic networks (Fig. 4.1). This evolution extends to a more refined scrutiny of intra- and interspecific genetic variation at a comprehensive genomic scale. The advent of various genomic techniques has played a pivotal role in catalyzing this paradigm shift. These advanced methods have significantly enhanced our capacity to discern the intricacies of evolutionary and epidemiological mechanisms governing the inception and dissemination of diseases. The increasing accessibility of diverse genomic tools has notably facilitated the reliable identification of these underlying processes (Diwan et al., 2022). Moreover, it has made it possible to compare and correlate potential genes and pathways with certain phenotypic qualities, such as occupying a specific niche, having advantageous or detrimental traits, and adapting to surroundings with root versus shoot environments (Beilsmith et al., 2019; Kimotho & Maina, 2023; Levy et al., 2018; Sahu et al., n.d.; Wang et al., 2021).

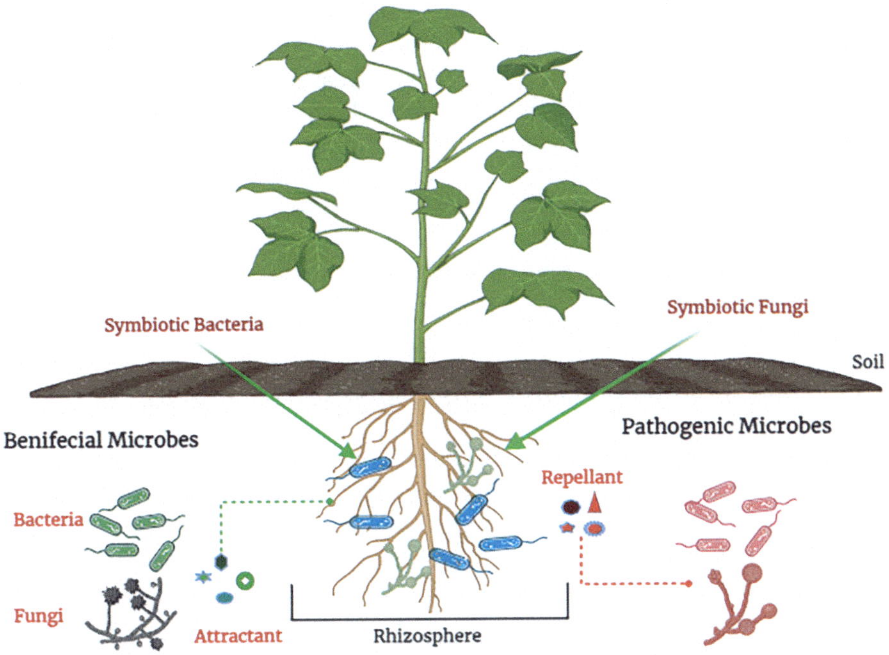

Fig. 4.1 Dialogue delivery between plant–soil–microbe responsible for plant–microbe interactions in the soil

The feedforward and feedback mechanisms are coordinated at multiple levels by the intricate interconnections among biochemical pathways. There will be significant implications for agriculture if these fundamental aspects of plant–microbe interactions are fully understood. The use of beneficial chemicals produced by microbes to boost crop yields and create plants that are especially suitable for use in sustainable agriculture is now possible; thanks to these discoveries (Diwan et al., 2022).

The scientific community values genomic data on plant diseases in multiple essential ways: Investigation of uncovered microorganisms, knowing the basic molecular processes that underlie host range and specificity, looking into the variations in expression patterns and putative virulence mechanisms among strains, able to determine the cause of the development of a plant disease, examining the processes involved in the coevolution of bacteria and their hosts, compare and contrast genetic traits within and between populations, find out how the host's susceptibility and resistance work (Ailloud et al., 2015; Garita-Cambronero et al., 2016; Goettelmann et al., 2022; Jacques et al., 2016; Kankanala et al., 2019; Lapidus & Korobeynikov, 2021; Lv et al., 2021; Parks et al., 2017; Pecman et al., 2017; Rocha et al., 2020; Saville & Ristaino, 2021; Shapiro et al., 2016; Susič et al., 2020; Xu et al., 2021; Yang et al., 2022). Our understanding of the positive and negative interactions between microorganisms and plants is increased by the management of biotic and abiotic stress. This will help to develop crops resistant to pathogens and make it easier to identify and treat diseases that are reemerging and emerging (Farrar et al., 2014; Knief, 2014; Bartoli & Roux, 2017; Kankanala et al., 2019).

In conclusion, genomics has become a key player that has revolutionized our understanding of plant–microbe interactions while also upending long-held biological beliefs (Dolatabadian, 2021; Frantzeskakis et al., 2020; Imam et al., 2016; Kroll et al., 2017). In spite of these revolutionary advances, we still have a long way to go before we can fully utilize the data gathered to take a close look at the complex mechanisms controlling the dynamic interaction between plant systems and the microbiological cosmos.

4.5.2 Metabolomics Insights into Plants–Soil Microbiota

Comprehensive, objective, and highly efficient investigations of the biosynthetic products found in the plant metabolome constitute the novel approach known as metabolomics, which is utilized to study the interactions between microbes and plants. This strategy plays a crucial role in shaping plant development, responses to stress, and interactions with weeds, as well as other microorganisms and larger organisms in the ecosystem (Turner et al., 2016). Metabolomics techniques provide more than just conventional large-scale metabolite profiling; they facilitate the creation of a metabolomic network atlas for plants. This process can involve the use of mutant or transgenic lines, the exploration of gene functions, or the assessment of impacts on metabolic pathways. There are two primary categories of metabolomics studies: (1) Targeted studies, in which a limited number of compounds are chosen for examination and (2) untargeted investigations, which permit the examination of an unknown quantity of distinctive characteristics (Ribbenstedt et al., 2018).

Given the diverse chemical and physical properties inherent in various classes of plant metabolites, the detection and quantification of these compounds mandate the application of at least three orthogonal methods. Consequently, a well-structured metabolomics analysis becomes essential for embracing the detection of a wide array of compounds. This methodology promotes a deeper comprehension of exudate composition and its relevance in plant–microbe interactions, in addition to facilitating comparative study (Mishra et al., 2022).

We utilize a range of instruments and methods to detect, screen, and measure metabolites, each with its own strengths and weaknesses. By using multiple technologies at the same time, we can achieve a detailed profile of an extract. Consequently, overlapping data should be combined to generate a comprehensive profile (Sakurai, 2022). The development of effective extraction methods is crucial, as the selection of extraction methods—liquid-liquid or solid phase—has a considerable impact on metabolite detection parameters such as recovery, repeatability, and specificity (Fritsche-Guenther et al., 2021). Careful consideration of extraction solvents further contributes to the efficacy of metabolite analysis (ElNaker et al., 2021; Mushtaq et al., 2014).

In the realm of metabolomics, isotopic labeling, and tracer experiments stand out as elegant approaches, providing crucial insights into precursor metabolites and their derivatives. These methodologies play an indispensable role in elucidating biosynthetic pathways of metabolites. Notably, these developments in the field of plant–microbe studies have opened up new avenues for the investigation of specific components of biological systems and how they interact with their surroundings. These developments have significantly improved our knowledge of the diversity of metabolites and bioactive substances. They also offer rare chances to identify the metabolic routes of certain metabolites (van Dam & Bouwmeester, 2016).

Numerous bacterial genes, such as nodulation genes, actively generate Nod factors during the nodulation process in roots, impacting both the host plant and microbial metabolism. Certain treatments can help identify the varying levels of metabolites. These processes play a significant role in shaping the rhizospheric community, as the presence of bacteria in the soil enhances the consumption of the released metabolites, especially aromatic organic acids (Zhalnina et al., 2018).

Metabolomics studies, particularly non-targeted approaches, have played a pivotal role in unraveling specific metabolic markers associated with plant–pathogen interactions. Lipid markers suggestive of *Plasmopara viticola* infection in grapevines were found by Negrel et al. (2018). Significant harm can be caused by *Plasmopara viticola*, which causes downy mildew in grapevines. Their use of a metabolomic non-targeted technique based on mass spectrometry made it easier to identify putative *Plasmopara*-specific metabolites.

Furthermore, Zhu et al. (2018) utilized metabolomics to investigate the infection of soybean hypocotyls by *Phytophthora sojae*. They examined the metabolic variations between susceptible (S) and resistant (R) soybean lines at 12 hours post-inoculation (hpi) and 36 hpi, revealing 90 differentially accumulated metabolites and highlighting variations between the two lines. This exemplifies the utility of metabolomics in comprehensively studying diverse plant–microbe interactions (de Falco et al., 2019; Hong et al., 2012; Negrel et al., 2018; Scandiani et al., 2015; Su et al., 2018; William Allwood et al., 2010; Yogendra et al., 2014; Zhu et al., 2018) (Table 1).

Despite the notable contributions of metabolomic analysis to our understanding of plant–microbe systems, several limitations hinder its widespread acceptance. Foremost among these challenges is the considerable cost associated with the equipment and technical expertise required for conducting metabolic studies, rendering metabolomics a less accessible and economically viable approach for many researchers. Moreover, the scarcity of publicly available metabolomic reference databases poses a significant hurdle, complicating the attribution of a given metabolite to a specific organism. Notwithstanding these challenges, it is crucial to recognize that metabolomics remains a powerful and influential tool in the realm of plant–microbe interactions (Sharma et al., 2020). This analytical approach excels in the detection and quantification of small molecules, providing valuable insights into molecular variations occurring at the intricate interface of plant–microbe interactions. As technology advances and collaborative efforts contribute to expanding metabolomic databases, it is anticipated that these limitations will be progressively addressed, further enhancing the utility and accessibility of metabolomics in unraveling the complexities of plant–microbe systems.

4.6 Challenges and Future Directions

Plant–microbe interaction in the soil holds a crucial position in shaping the ecosystem dynamics as well as plant health. The associations between plant roots and soil microbes involve a complex back-and-forth of genetic, biochemical, and environmental factors. The intricate and multifaceted nature of plant–microbe interactions poses a significant challenge. The microbial diversity, the variability in plant responses, and the myriad signaling pathways involved make it challenging to unravel the complete picture. The signal exchange between the roots of plants and soils plays a very significant role in the life of plants. The variability of environmental conditions adds an additional layer of complexity. The dynamics of plant–microbe interaction depend on various conditions such as temperature, pH of the soil, soil type, the composition of soil, and geographical variability along with the humidity. Future research should explore how these interactions acclimatize to diverse environmental factors and evolve strategies to harness these adaptations for sustainable agriculture in a changing climate. The omics approach has been found to be very significant in understanding plant–microbe interactions. The understanding as well as harnessing of these relationships holds noteworthy insinuations for the conservation of the environment, sustainable agriculture, and food security.

This field holds a good opportunity for the application of knowledge in various fields such as microbiome engineering. One of the primary challenges in understanding complex plant–bacteria interactions is the vast diversity of microbial communities present in the soil. Soil microorganisms are dynamic and heterogeneous communities of bacteria, fungi, archaea, and bacteria among others. Deciphering the functions and interactions of these diverse cells is a daunting task. Furthermore, the involvement of many chemicals during multifaceted signaling in-between microbes and plants makes it difficult to identify specific molecular interactions. The impact of environmental stressors on these interactions further increases the complexity of the system. Composition and thus the functioning of the microbial communities can be altered by soil physiological properties, soil degradation, and climate change as well as pollution. Understanding how these stressors influence the delicate balance between beneficial and pathogenic microbes is crucial for forecasting and extenuating their influences on plant health and productivity. Translating research findings into practical applications faces hurdles. Implementation in real-world agricultural settings involves deliberations such as scalability, cost-effectiveness, and compatibility with existing farming practices. Future directions should include research on developing feasible and scalable applications of plant–microbe interactions for sustainable agriculture.

Excessive use of chemical fertilizers in agriculture is responsible for the shift or change of the microbial diversity by disturbing the plant–root–soil microbiome interactions as well as the release of harmful chemicals to the soil also causing soil pollution. These chemical fertilizers can lead to the depletion of beneficial microbes from the interactions negatively impacting soil fertility and resilience. However, restoration of healthy microbial communities in the soil might be beneficial in sustainable agriculture. As the research in this area advances, ethical considerations associated with genetically modified organisms (GMOs) and the utilization of engineered microbes in the environment need careful attention. Future research should include thorough risk assessments and ethical frameworks to guide the accountable application of plant–soil–microbe interactions.

Looking ahead, the future directions of research in plant–soil–microbe interactions holds potential for innovative solutions. Developments in omics technologies—such as genomics, transcriptomics, and metabolomics—provide unparalleled insights into the molecular mechanisms underlying these interactions. Integrating big data and computational approaches will enable researchers to unravel complex microbial networks and forecast the outcomes of specific plant–microbe associations. Harnessing the potential of PGPRs and mycorrhizal fungi represents a promising approach to improving plant health and nutrient mobilization. Bioinoculants containing beneficial microbes have shown promise for improved crop production as well as reduction of chemical fertilizers. Future research should emphasize on optimizing these microbial formulations for different soil types, climates, and crops. The concept of "microbiome engineering" holds immense potential for tailoring microbial communities to enhance plant traits such as biotic and abiotic stress tolerance and nutrient uptake. This transformative approach requires a deeper understanding of the functional roles of specific microbial taxa and their interactions within the soil ecosystem. Advancements in synthetic biology can lead to the development of custom-designed microbes with specific functionalities. This opens doors to creating tailored solutions for plant–microbe interactions, providing precise control over desired outcomes.

4.7 Conclusion

Plant–microbe interaction in the soil is not only responsible for shaping the soil ecosystem but also it is involved in the growth of plants via various mechanisms and in return plants secrete molecules that are utilized by microbes for their colonization and growth. Omics approaches are involved in the study of these delicate, but targeted plant–microbe interactions have increased our understanding of the mechanism underlining this interaction. The microbial diversity present in the soil is related to the wealth of the soil and its productivity. The relation between the microbes in the soil might be beneficial or harmful depending on the types of interactions that are involved between them. The metabolites as well as the transcriptomics profile of the plant and microbes are also influenced by the interaction between them. Surprisingly, we found that 84% of PGPR are Gram-negative bacteria. Future research in the field of synthetic biology and AI-based modeling could boost the uses of the plant–microbe interaction in uplifting crop production, which will be of priority in the next decade to fulfill the world's food demand. Deep knowledge of plant–microbe interaction might be very beneficial for sustainable agriculture as well as it might be helpful in precise farming.

References

Abdallah, R. A. B., Trabelsi, B. M., Nefzi, A., Khiareddine, H. J., & Remadi, M. D. (2016). Isolation of Endophytic bacteria from Withania somnifera and assessment of their ability to suppress Fusarium Wilt disease in tomato and to promote plant growth. *Journal of Plant Pathology and Microbiology, 7*, 352.

Abdelrahman, H., Zaghloul, R. A., Hassan, E., El-Zehery, H. R. A., & Salem, A. A. (2021). New strains of plant growth-promoting rhizobacteria in combinations with humic acid to enhance squash growth under saline stress. *Egyptian Journal of Soil Science, 61*(1), 129–146.

Agler, M. T., Ruhe, J., Kroll, S., Morhenn, C., Kim, S.-T., Weigel, D., & Kemen, E. M. (2016). Microbial Hub Taxa link host and abiotic factors to plant microbiome variation. *PLoS Biology, 14*(1), e1002352. https://doi.org/10.1371/journal.pbio.1002352

Ahemad, M., & Khan, M. S. (2011). Functional aspects of plant growth-promoting rhizobacteria: Recent advancements. *Insight Microbiology, 1*, 39–54. https://doi.org/10.5567/IMICRO-IK.2011.39.54

Ahemad, M., & Khan, M. S. (2012a). Influence of selective herbicides on plant growth promoting traits of greengram-symbiont Bradyrhizobium. *International Journal of Agronomy, 86*(4), 384–388.

Ahemad, M., & Khan, M. S. (2012b). Plant growth promoting activities of phosphate-solubilizing Enterobacter asburiae as influenced by fungicides. *International Journal of Agronomy*.

Ahemad, M., & Kibret, M. (2014). Mechanisms and applications of plant growth promoting rhizobacteria: current perspective. *Journal of King Saud University-Science, 26*(1), 1–20.

Ailloud, F., Lowe, T., Cellier, G., Roche, D., Allen, C., & Prior, P. (2015). Comparative genomic analysis of *Ralstonia solanacearum* reveals candidate genes for host specificity. *BMC Genomics, 16*(1), 270. https://doi.org/10.1186/s12864-015-1474-8

AlAli, H. A., Khalifa, A., & Almalki, M. (2021). Plant growth-promoting rhizobacteria from Ocimum basilicum improve growth of Phaseolus vulgaris and Abelmoschus esculentus. *South African Journal of Botany, 139*, 200–209.

Antoun, H., & Prevost, D. (2005). Ecology of plant growth promoting rhizobacteria. In: Z. A. Siddiqui (Ed.), *PGPR: Biocontrol and biofertilization*. Springer, Berlin, 1–38. https://doi.org/10.1007/1-4020-4152-7_1

Araújo, A. S., Leite, L. F., Santos, V. B., & Carneiro, R. F. (2009). Soil microbial activity in conventional and organic agricultural systems. *Sustainability, 1*(2), 268–276.

Babalola, O. O. (2010). Beneficial bacteria of agricultural importance. *Biotechnology Letters, 32*, 1559–1570.

Backer, R., Rokem, J. S., Ilangumaran, G., Lamont, J., Praslickova, D., Ricci, E., Subramanian, S., & Smith, D. L. (2018). Plant growth-promoting rhizobacteria: Context, mechanisms of action, and roadmap to commercialization of biostimulants for sustainable agriculture. *Frontiers in Plant Science, 9*, 1473.

Bagyaraj, D. J. (2014). Mycorrhizal Fungi. *Proceedings of the Indian National Science Academy, 80*, 415–428.

Bailey, V. L., McCue, L. A., Fansler, S. J., Boyanov, M. I., DeCarlo, F., Kemner, K. M., & Konopka, A. (2013). Micrometer-scale physical structure and microbial composition of soil macroaggregates. *Soil Biology and Biochemistry, 65*, 60–68.

Banerjee, S., & van der Heijden, M. G. A. (2023). Soil microbiomes and one health. *Nature Reviews, 21*, 6.

Barnawal, D., Bharti, N., Maji, D., Chanotiya, C. S., & Kalra, A. (2012). 1-Aminocyclopropane-1 carboxylic acid (ACC) deaminase-containing rhizobacteria protect Ocimum sanctum plants during waterlogging stress via reduced ethylene. *Plant Physiology and Biochemistry, 58*, 227–235.

Barns, S. M., Delwiche, C. F., Palmer, J. D., & Pace, N. R. (1996). Perspectives on archaeal diversity, thermophily and monophily from environmental rRNA sequences. *Proceedings of the National Academy of Sciences of the United States of America, 93*, 9188–9193.

Bartoli, C., & Roux, F. (2017). Genome-wide association studies in plant pathosystems: Toward an ecological genomics approach. *Frontiers in Plant Science, 8*. https://www.frontiersin.org/articles/10.3389/fpls.2017.00763

Beare, M. H., Coleman, D. C., Crossley, D. A., Jr., Hendrix, P. F., & Odum, E. P. (1995). A hierarchical approach to evaluating the significance of soil biodiversity to biogeochemical cycling. *Plant and Soil, 170*, 5–22.

Beilsmith, K., Thoen, M. P. M., Brachi, B., Gloss, A. D., Khan, M. H., & Bergelson, J. (2019). Genome-wide association studies on the phyllosphere microbiome: Embracing complexity in host–microbe interactions. *The Plant Journal, 97*(1), 164–181. https://doi.org/10.1111/tpj.14170

Benizri, E., Dedourge, O., Di Battista-Leboeuf, C., Nguyen, C. S., & Piutti, G. A. (2002). Effect of maize rhizodeposits on soil microbial community structure. *Applied Soil Ecology, 21*, 261–265.

Berg, G., Grube, M., Schloter, M., & Smalla, K. (2014). Unraveling the plant microbiome: looking back and future perspectives. *Frontiers in Microbiology, 5*, 148.

Berg, G., Köberl, M., Rybakova, D., Müller, H., Grosch, R., & Smalla, K. (2017). Plant microbial diversity is suggested as the key to future biocontrol and health trends. *FEMS Microbiology Ecology, 93*(5), fix050.

Bianco, C., & Defez, R. (2009). Medicago truncatula improves salt tolerance when nodulated by an indole-3-acetic acidoverproducing Sinorhizobium melilotistrain. *Journal of Experimental Botany, 60*(11), 3097–3107.

Borriss, R. (2011). Use of plant-associated Bacillus strains as biofertilizers and biocontrol agents in agriculture. In D. K. Maheshwari (Ed.), *Bacteria in agrobiology: Plant growth responses* (pp. 41–76). Springer.

Bulgarelli, D., Schlaeppi, K., Spaepen, S., et al. (2013). Structure and functions of the bacterial microbiota of plants. *Annual Review of Plant Biology, 64*, 807–838.

Byrnes, J. E. K., Gamfeldt, L., Isbell, F., Lefcheck, J. S., Griffin, J. N., Hector, A., Cardinale, B. J., Hooper, D. U., Dee, L. E., & Emmett Duffy, J. (2014). Investigating the relationship between biodiversity and ecosystem multifunctionality: challenges and solutions. *Methods in Ecology and Evolution, 5*, 111–124.

Casarrubia, S., Daghino, S., Kohler, A., Morin, E., Khouja, H. R., Daguerre, Y., & Martino, E. (2018). The hydrophobin-like OmSSP1 may be an effector in the ericoid mycorrhizal symbiosis. *Frontiers in Plant Science, 9*, 546.

Chinnadurai, D., Balachandar, S., & Sundaram, P. (2009). Characterization of 1-aminocyclopr opanc-1-carboxylate deaminase producing methylobacteria from phyllosphere of rice and their role in ethylene regulation. *World Journal of Microbiology and Biotechnology, 2009*(25), 1403–1411.

Chopade, B. A., Huddedar, S. B., Shete, A. M., Tilekar, J. N., Dhavale, D. D., & Gore, S. D. (2008). *Plasm encooing aa and a method thereof.* (US patent No. 7,341,868 B2).

Chopra, A., Bobate, S., Rahi, P., Banpurkar, A., Mazumder, P. B., & Satpute, S. (2020). Pseudomonas aeruginosa RTE4: A tea rhizobacterium with potential for plant growth promotion and biosurfactant production. *Frontiers in Bioengineering and Biotechnology, 8*, 861.

Compant, S., Samad, A., Faist, H., & Sessitsch, A. (2019). A review on the plant microbiome: Ecology, functions, and emerging trends in microbial application. *Journal of Advanced Research, 19*, 29–37.

Cooper, J. E. (2007). Early interactions between legumes and rhizobia: Disclosing complexity in a molecular dialogue. *Journal of Applied Microbiology, 103*, 1355–1365. https://doi.org/10.1111/j.1365-2672.2007.03366.x

Dastogeer, K. M. G., Tumpa, F. H., Sultana, A., Akter, M. A., & Chakraborty, A. (2020). Plant microbiome–an account of the factors that shape community composition and diversity. *Current Plant Biology*, 23100161. https://doi.org/10.1016/j.cpb.2020.100161

de Falco, B., Manzo, D., Incerti, G., Garonna, A. P., Ercolano, M., & Lanzotti, V. (2019). Metabolomics approach based on NMR spectroscopy and multivariate data analysis to explore the interaction between the leafminer Tuta absoluta and tomato (Solanum lycopersicum). *Phytochemical Analysis, 30*(5), 556–563. https://doi.org/10.1002/pca.2850

de los Santos-Villalobos, S., de Folter, S., Délano-Frier, J. P., et al. (2013). Growth promotion and flowering induction in mango (Mangifera indica L. cv "Ataulfo") trees by Burkholderia and rhizobium inoculation: Morphometric, biochemical, and molecular events. *Journal of Plant Growth Regulation, 32*, 615–627.

de-Bashan, L. E., Hernandez, J. P., Nelson, K. N., Bashan, Y., & Maier, R. M. (2010). Growth of quailbush in acidic, metalliferous desert mine tailings: effect of Azospirillum brasilense Sp6 on biomass production and Rhizosphere community structure. *Microbial Ecology, 60*(4), 915–927.

Dhakephalkar, P. K., & Chopade, B. A. (1994). High levels of multiple metal resistance and its correlation to antibiotic resistance in environmental isolates of Acinetobacter. *Biometals*, 7, 67–74. https://doi.org/10.1007/BF00205197

Dick, W. (2009). *Lecture on biochemistry process in soil microbiology*, Personal collection of W. Dick, The Ohio State University School of Environment and Natural Resources.

Diwan, D., Rashid, M. M., & Vaishnav, A. (2022). Current understanding of plant-microbe interaction through the lenses of multi-omics approaches and their benefits in sustainable agriculture. *Microbiological Research, 265*, 127180. https://doi.org/10.1016/j.micres.2022.127180

Dolatabadian, A. (2021). Plant–microbe interaction. *Biology, 10*(1), Article 1. https://doi.org/10.3390/biology10010015

Dong, C. J., Wang, L. L., Li, Q., & Shang, Q. M. (2019). Bacterial communities in the rhizosphere, phyllosphere and endosphere of tomato plants. *PLoS One, 14*(11), e0223847.

Edwards, A. P., & Bremner, J. M. (1967). Microaggregates in soils. *Journal of Soil Science, 18*, 64–73.

Edwards, J. A., Santos-Medellín, C. M., Liechty, Z. S., Nguyen, B., Lurie, E., Eason, S., Phillips, G., & Sundaresan, V. (2018). Compositional shifts in root-associated bacterial and archaeal microbiota track the plant life cycle in field-grown rice. *PLoS Biology, 16*(2), e2003862. https://doi.org/10.1371/journal.pbio.2003862

Egamberdieva, D., & Kucharova, Z. (2009). Selection for root colonising bacteria stimulating wheat growth in saline soils. *Biology and Fertility of Soils, 45*, 563–571. https://doi.org/10.1007/s00374-009-0366-y

ElNaker, N. A., Daou, M., Ochsenkühn, M. A., Amin, S. A., Yousef, A. F., & Yousef, L. F. (2021). A metabolomics approach to evaluate the effect of lyophilization versus oven drying on the chemical composition of plant extracts. *Scientific Reports, 11*(1), Article 1. https://doi.org/10.1038/s41598-021-02158-6

El-Sayed, W. S., Akhkha, A., El-Naggar, M. Y., & Elbadry, M. (2014). In vitro antagonistic activity, plant growth promoting traits and phylogenetic affiliation of rhizobacteria associated with wild plants grown in arid soil. *Frontiers in Microbiology, 5*, 651.

Yasar Erturk, Sezai Ercisli, Ayhan Haznedar And Ramazan Cakmakci (2010). Effects of plant growth promoting rhizobacteria (PGPR) on rooting and root growth of kiwifruit (Actinidia deliciosa) stem cuttings. Biological Research 43: 91–98.

Farrar, K., Bryant, D., & Cope-Selby, N. (2014). Understanding and engineering beneficial plant–microbe interactions: Plant growth promotion in energy crops. *Plant Biotechnology Journal, 12*(9), 1193–1206. https://doi.org/10.1111/pbi.12279

Fierer, N., & Jackson, R. B. (2006). The diversity and biogeography of soil bacterial communities, *Proceedings of the National Academy of Sciences, 103*(3), 626–631. https://doi.org/10.1073/pnas.0507535103

Fierer, N. (2017). Embracing the unknown: disentangling the complexities of the soil microbiome. *Nature Reviews Microbiology, 15*(10), 579–590.

Firmin, J. L., Wilson, K. E., Rossen, L., & Johnston, A. W. B. (1986). Flavonoid activation of nodulation genes in Rhizobium reversed by other compounds present in plants. *Nature, 324*, 90–92. https://doi.org/10.1038/324090a0

Floudas, D., Binder, M., Riley, R., Barry, K., Blanchette, R. A., Henrissat, B., Martínez, A. T., Otillar, R., Spatafora, J. W., & Yadav, J. S. (2012). The Paleozoic origin of enzymatic lignin decomposition reconstructed from 31 fungal genomes. *Science, 336*, 1715–1719.

Foster, R. C. (1988). Microenvironment of soil microorganisms. *Biology and Fertility of Soils, 6*, 189–203.

Frantzeskakis, L., von Dahlen, J. K., Panstruga, R., & Rose, L. E. (2018). Rapid evolution in the tug-of-war between microbes and plants. *New Phytologist, 219*(1), 12–14. https://doi.org/10.1111/nph.15220

Frantzeskakis, L., Di Pietro, A., Rep, M., Schirawski, J., Wu, C.-H., & Panstruga, R. (2020). Rapid evolution in plant–microbe interactions – a molecular genomics perspective. *New Phytologist, 225*(3), 1134–1142. https://doi.org/10.1111/nph.15966

Fritsche-Guenther, R., Gloaguen, Y., Bauer, A., Opialla, T., Kempa, S., Fleming, C. A., Redmond, H. P., & Kirwan, J. A. (2021). Optimized workflow for on-line derivatization for targeted metabolomics approach by gas chromatography-mass spectrometry. *Metabolites, 11*(12), Article 12. https://doi.org/10.3390/metabo11120888

Fukuhara, H., Minakawa, Y., Akao, S., & Minamisawa, K. (1994). The involvement of indole-3-acetic acid produced by Bradyrhizobium elkanii in nodule formation. *Plant and Cell Physiology, 35*(8), 1261–1265.

Garita-Cambronero, J., Palacio-Bielsa, A., López, M. M., & Cubero, J. (2016). Draft genome sequence for virulent and avirulent strains of Xanthomonas arboricola isolated from Prunus spp. In Spain. *Standards in Genomic Sciences, 11*(1), 12. https://doi.org/10.1186/s40793-016-0132-3

Genre, A., & Russo, G. (2016). Does a common pathway transduce symbiotic signals in plant-microbe interactions? *Frontiers in Plant Science, 7*, 96. https://doi.org/10.3389/fpls.2016.00096

Geurts, R., Xiao, T. T., & Reinhold-Hurek, B. (2016). What does it take to evolve a nitrogen-fixing endosymbiosis? *Trends in Plant Science, 21*, 199–208. https://doi.org/10.1016/j.tplants.2016.01.012

Giri, B., Giang, P. H., Kumari, R., Prasad, R., & Varma, A. (2005). Microbial diversity in soils. In F. Buscot & A. Varma (Eds.), *Soil biology, Volume 3, Microorganisms in soils: Roles in genesis and function*. © Springer Berlin Heidelberg.

Glick, B. R. (1995). The enhancement of plant growth by free-living bacteria. *Canadian Journal of Microbiology, 41*, 109–117. https://doi.org/10.1139/m95-015

Goettelmann, F., Roman-Reyna, V., Cunnac, S., Jacobs, J. M., Bragard, C., Studer, B., Koebnik, R., & Kölliker, R. (2022). Complete genome assemblies of all Xanthomonas translucens pathotype strains reveal three genetically distinct clades. *Frontiers in Microbiology, 12*. https://www.frontiersin.org/articles/10.3389/fmicb.2021.817815

Griesmann, M., Chang, Y., Liu, X., Song, Y., Haberer, G., Crook, M. B., et al. (2018). Phylogenomics reveals multiple losses of nitrogen-fixing root nodule symbiosis. *Science, 361*, eaat1743. https://doi.org/10.1126/science.aat1743

Gupta, A., Meyer, J. M., & Goel, R. (2002). Development of heavy metal-resistant mutants of phosphate solubilizing Pseudomonas sp. NBRI 4014 and their characterization. *Current Microbiology, 45*, 323–327.

Hameeda, B., Rupela, O. P., Reddy, G., et al. (2006). Application of plant growth-promoting bacteria associated with composts and macrofauna for growth promotion of Pearl millet (Pennisetum glaucum L.). *Biology and Fertility of Soils, 43*, 221–227.

Hawksworth, D. L. (1991). The fungal dimension of diversity: magnitude, significance, and conservation. *Mycological Research, 95*, 641–655.

Hong, Y.-S., Martinez, A., Liger-Belair, G., Jeandet, P., Nuzillard, J.-M., & Cilindre, C. (2012). Metabolomics reveals simultaneous influences of plant defence system and fungal growth in Botrytis cinerea-infected Vitis vinifera cv. Chardonnay berries. *Journal of Experimental Botany, 63*(16), 5773–5785. https://doi.org/10.1093/jxb/ers228

Hoorman, J. J. (2016). *Role of soil bacteria, assistant professor and extension educator, agriculture and natural resources*. Ohioline, Ohio State University Extension.

Huang, J., Liu, S., & Cook, D. E. (2023). Dynamic genomes—Mechanisms and consequences of genomic diversity impacting plant-fungal interactions. *Physiological and Molecular Plant Pathology, 125*, 102006. https://doi.org/10.1016/j.pmpp.2023.102006

Huddedar, S. B., Shete, A. M., Tilekar, J. N., et al. (2002). Isolation, characterization, and plasmid pUPI126-mediated indole-3-acetic acid production in Acinetobacter strains from rhizosphere of wheat. *Applied Biochemistry and Biotechnology, 102*, 21–39. https://doi.org/10.1385/ABAB:102-103:1-6:021

Huey, C. J., Gopinath, S. C. B., Uda, M. N. A., Zulhaimi, H. I., Jaafar, M. N., Kasim, F. H., & Yaakub, A. R. W. (2020). Mycorrhiza: A natural resource assists plant growth under varied soil conditions. *3 Biotech, 10*(5), 204. https://doi.org/10.1007/s13205-020-02188-3. Epub 2020 Apr 18. PMID: 32337150; PMCID: PMC7165205.

Hurek, T., Reinhold-Hurek, B., Van Montagu, V., & Kellenberger, M. (2002). Root colonization and systemic spreading of Azoarcus sp. strain BH72 in grasses. *Journal of Bacteriology, 176*(7), 1913–1923.

Imam, J., Singh, P. K., & Shukla, P. (2016). Plant microbe interactions in post genomic era: Perspectives and applications. *Frontiers in Microbiology, 7*, 209205. https://doi.org/10.3389/fmicb.2016.01488

Innerebner, G., Knief, C., & Vorholt, J. A. (2011). Protection of Arabidopsis thaliana against leaf-pathogenic Pseudomonas syringae by Sphingomonas strains in a controlled model system. *Applied and Environmental Microbiology, 77*(10), 3202–3210.

Iqbal, S. (2013). Hasnain Aeromonas punctata PNS-1: a promising candidate to change the root morphogenesis of A. thaliana in MS and sand system. *Acta Physiologiae Plantarum, 35*, 657–665.

Jacques, M.-A., Arlat, M., Boulanger, A., Boureau, T., Carrère, S., Cesbron, S., Chen, N. W. G., Cociancich, S., Darrasse, A., Denancé, N., Fischer-Le Saux, M., Gagnevin, L., Koebnik, R., Lauber, E., Noël, L. D., Pieretti, I., Portier, P., Pruvost, O., Rieux, A., et al. (2016). Using ecology, physiology, and genomics to understand host specificity in Xanthomonas. *Annual Review of Phytopathology, 54*(1), 163–187. https://doi.org/10.1146/annurev-phyto-080615-100147

Jahanian, A., Chaichi, M. R., Rezaei, K., Rezayazdi, K., & Khavazi, K. (2012). The effect of plant growth promoting rhizobacteria (PGPR) on germination and primary growth of artichoke (Cynara scolymus). *International Journal of Agriculture and Crop Sciences, 4*, 923–929.

Jha, C. K., Aeron, M., Patel, A., Maheshwari, B., & Saraf, L. (2005). Enterobacter: role in plant growth promotion. *Journal of Microbiology and Biotechnology*.

Jia, T., Yao, Y., Wang, R., Wu, T., & Chai, B. (2020). Dynamics relationship of phyllosphere and rhizosphere bacterial communities during the development of Bothriochloa ischaemum in copper tailings. *Frontiers in Microbiology, 11*, 869.

Jiménez-Guerrero, I., Acosta-Jurado, S., Del Cerro, P., Navarro-Gómez, P., López-Baena, F. J., Ollero, F. J., Vinardell, J. M., & Pérez-Montaño, F. (2017). Transcriptomic studies of the effect of nod gene-inducing molecules in Rhizobia: different weapons, one purpose. *Genes (Basel), 9*(1), 1.

Joshi, N., Ruparelia, J. A., Saraf, M., & Jha, C. K. (2023). Techniques to study plant–microbe interactions that lead to efficient sustainable agriculture. In S. Chhabra, R. Prasad, N. R. Maddela, & N. Tuteja (Eds.), *Plant microbiome for plant productivity and sustainable agriculture* (pp. 401–421). Springer Nature. https://doi.org/10.1007/978-981-19-5029-2_17

Jung, J., Jun-Seob, K., Julian, T., Gabriele, B., & Choong-Min, R. (2020). Archaea, tiny helpers of land plants. *Computational and Structural Biotechnology Journal, 18*, 2494–2500.

Kang, S. M., Joo, G. J., Hamayun, M., et al. (2009). Gibberellin production and phosphate solubilization by newly isolated strain of Acinetobacter calcoaceticus and its effect on plant growth. *Biotechnology Letters, 31*(2), 277–281.

Kankanala, P., Nandety, R. S., & Mysore, K. S. (2019). Genomics of plant disease resistance in legumes. *Frontiers in Plant Science, 10*. https://www.frontiersin.org/articles/10.3389/fpls.2019.01345

Kaplan, H. B. & Greenberg, E. P. (1985). Diffusion of autoinducer is involved in regulation of the Vibrio fischeri luminescence system. *J Bacteriol, 163*(3), 1210–1214. https://doi.org/10.1128/jb.163.3.1210-1214.1985

Karina, K., Rosliana, I., Rosadi, I., et al. (2020). Safety of technique and procedure of stromal vascular fraction therapy: From liposuction to cell administration. *Scientifica (Cairo)*. 2863624. https://doi.org/10.1155/2020/2863624

Karthikeyan, B., Joe, M. M., Islam, R., & Sa, T. (2012). ACC deaminase containing diazotrophic endophytic bacteria ameliorate salt stress in Catharanthus roseus through reduced ethylene levels and induction of antioxidative defense systems. *Symbiosis, 56*, 77–86.

Khan, A. L., Halo, B. A., Elyassi, A., Ali, S., Al-Hosni, K., Hussain, J., et al. (2016). Indole acetic acid and ACC deaminase from endophytic bacteria improves the growth of Solanum lycopersicum. *Electronic Journal of Biotechnology, 21*, 58–64.

Kimotho, R. N., & Maina, S. (2023). Unraveling plant–microbe interactions: Can integrated omics approaches offer concrete answers? *Journal of Experimental Botany, erad448*, 1289. https://doi.org/10.1093/jxb/erad448

Kim, H. M., Chae, N., Jung, J. Y., & Lee, Y. K. (2013). Isolation of facultatively anaerobic soil bacteria from Ny-A° lesund, Svalbard. *Polar Biology, 36*, 78796.

Kishore, G. K., Pande, S., & Podile, A. R. (2005). Phylloplane bacteria increase seedling emergence, growth and yield of field-grown groundnut (Arachis hypogaea L.). *Applied Microbiology, 40*(4), 260–268.

Knief, C. (2014). Analysis of plant microbe interactions in the era of next generation sequencing technologies. *Frontiers in Plant Science, 5*. https://www.frontiersin.org/articles/10.3389/fpls.2014.00216

Kobayashi, H., Naciri-Graven, Y., Broughton, W. J., & Perret, X. (2004). Flavonoids induce temporal shifts in gene-expression of nod-box controlled loci in Rhizobium sp. NGR234. *Molecular Microbiology, 51*, 335–347. https://doi.org/10.1046/j.1365-2958.2003.03841.x

Kowalenko, C. G., & Ivarson, K. C. (1978). Effect of moisture content, temperature and nitrogen. *Soil Biology and Biochemistry, 10*(5), 417–423.

Krishna, S., Bhutia, D. D., Chaubey, R. K., & Sudhir, I. (2024). Chapter 4 – Exploring plant microbiome: A holistic approach to sustainable agriculture. In J. A. Parray, N. Shameem, & D. Egamberdieva (Eds.), *Microbiome drivers of ecosystem function* (pp. 61–77). Academic Press. https://doi.org/10.1016/B978-0-443-19121-3.00013-2

Kroll, S., Agler, M. T., & Kemen, E. (2017). Genomic dissection of host–microbe and microbe–microbe interactions for advanced plant breeding. *Current Opinion in Plant Biology, 36*, 71–78. https://doi.org/10.1016/j.pbi.2017.01.004

Kumar, A., Singh, V., Singh, M., Singh, P. P., Singh, S. K., Singh, P. K., et al. (2016). Isolation of plant growth promoting rhizobacteria and their impact on growth and curcumin content in Curcuma longa L. *Biocatalysis and Agricultural Biotechnology, 8*, 1–7.

Kumari, N., & Khare, P. (2020). Effect of indigenous beneficial rhizobacterial isolates on growth of Litchi seedlings. *FLORA AND FAUNA, 26*(2), 213–217.

Kurabachew, H., & Wydra, K. (2013). Characterization of plant growth promoting rhizobacteria and their potential as bioprotectant against tomato bacterial wilt caused by Ralstonia solanacearum. *Biological Control, 67*(1), 75–83.

Lamont, J. R., Wilkins, O., Bywater-Ekegard, M., & Smith, D. L. (2017). From yogurt to yield: potential applications of lactic acid bacteria in plant production. *Soil Biology and Biochemistry, 111*, 1–9.

Lapidus, A. L., & Korobeynikov, A. I. (2021). Metagenomic data assembly—The way of decoding unknown microorganisms. *Frontiers in Microbiology, 12*. https://www.frontiersin.org/articles/10.3389/fmicb.2021.613791

Lea, U. S., Slimestad, R., Smedvig, P., & Lillo, C. (2007). Nitrogen deficiency enhances expression of specific MYB and bHLH transcription factors and accumulation of end products in the flavonoid pathway. *Planta, 225*, 1245–1253. https://doi.org/10.1007/s00425-006-0414-x

Lemessa, F., & Zeller, W. (2007). Screening rhizobacteria for biological control of Ralstonia solanacearum in Ethiopia. *Biological Control, 42*(3), 336–344.

Levy, A., Salas Gonzalez, I., Mittelviefhaus, M., Clingenpeel, S., Herrera Paredes, S., Miao, J., Wang, K., Devescovi, G., Stillman, K., Monteiro, F., Rangel Alvarez, B., Lundberg, D. S., Lu, T.-Y., Lebeis, S., Jin, Z., McDonald, M., Klein, A. P., Feltcher, M. E., Rio, T. G., et al. (2018).

Genomic features of bacterial adaptation to plants. *Nature Genetics, 50(1)*, Article 1. https://doi.org/10.1038/s41588-017-0012-9
Liu, C. W., & Murray, J. D. (2016). The role of flavonoids in nodulation hostrange specificity: An update. *Plants (Basel), 5*(3), 33.
Liu, D., Lian, B., & Dong, H. (2012). Isolation of Paenibacillus sp. and assessment of its potential for enhancing mineral weathering. *Geomicrobiology Journal, 29*(5), 413–421.
Liu, N., Shen, G., Xu, Y., Liu, H., Zhang, J., et al. (2020). Extensive inter-plant protein transfer between Cuscuta parasites and their host plants. *Molecular Plant, 13*, 573–585.
Lowenfels, J., & Lewis, W. (2006). Teaming with microbes: A Gardener's guide to the soil food web, Chapter 3: Bacteria, Timber Press, Portland, Oregon. *Microbiology, 21*, 6–20.
Lugtenberg, B., & Kamilova, F. (2009). Plant-growth-promoting rhizobacteria. *Annual Review of Microbiology, 63*, 541–556.
Lv, Z., He, Z., Hao, L., Kang, X., Ma, B., Li, H., Luo, Y., Yuan, J., & He, N. (2021). Genome sequencing analysis of Scleromitrula shiraiana, a causal agent of mulberry sclerotial disease with narrow host range. *Frontiers in Microbiology, 11*. https://www.frontiersin.org/articles/10.3389/fmicb.2020.603927
Ma, Y., Rajkumar, M., & Freitas, H. (2009). Inoculation of plant growth promoting bacterium Achromobacter xylosoxidans strain Ax10 for the improvement of copper phytoextraction by Brassica juncea. *Journal of Environmental Management, 90*(2), 831–837.
Ma, Y., Rajkumar, R., & Freitas, H. (2011). Isolation and characterization of nickel uptake by nickel resistant bacterial isolate (Pseudomonas sp. A3R3). *Microbial Ecology*.
MacLean, A. M., Bravo, A., & Harrison, M. J. (2017). Plant signaling and metabolic pathways enabling arbuscular mycorrhizal symbiosis. *Plant Cell, 29*, 2319–2335. https://doi.org/10.1105/tpc.17.00555
Madhaiyan, M., Poonguzhali, S., Lee, J., Senthilkumar, M., Lee, K., & Sa, T. (2007). Methylobacterium oryzae sp. nov., an aerobic, pink-pigmented, facultatively methylotrophic, methanol-utilizing bacterium isolated from rice. *International Journal of Systematic and Evolutionary Microbiology, 57*, 326.
Maindad, D. V., Kasture, V. M., Chaudhari, H., Dhavale, D. D., Chopade, B. A., & Sachdev, D. P. (2014). Characterization and fungal inhibition activity of siderophore from wheat rhizosphere associated Acinetobacter calcoaceticus strain HIRFA32. *Indian Journal of Microbiology, 54*(3), 315–322. https://doi.org/10.1007/s12088-014-0446-z
Marschner, H., & Dell, B. (1994). Nutrient uptake in mycorrhizal symbiosis. *Plant and Soil, 159*, 89–102.
Martinez-Hidalgo, P., & Hirsch, A. M. (2017). The nodule microbiome: N2-fixing rhizobia do not live alone. *Phytobiomes Journal, 1*, 70.
Markmann, K., & Parniske, M. (2009). Evolution of root endosymbiosis with bacteria: How novel are nodules? *Trends Plant Sci, 14*(2):77–86. https://doi.org/10.1016/j.tplants.2008.11.009
Mayak, S., Tirosh, T., & Glick, B. R. (1999). Effect of wild-type and mutant plant growth-promoting rhizobacteria on the rooting of mung bean cuttings. *Journal of Plant Growth Regulation, 18*(2), 49–53.
Meding, S. M., & Zasoski, R. J. (2008). Hyphal-mediated transfer of nitrate, arsenic, cesium, rubidium, and strontium between arbuscular mycorrhizal forbs and grasses from a California oak woodland. *Soil Biology and Biochemistry, 40*, 126–134.
Mendes, R., Garbeva, P., & Raaijmakers, J. M. (2013). The rhizosphere microbiome: significance of plant beneficial, plant pathogenic, and human pathogenic microorganisms. *FEMS Microbiology Reviews, 37*(5), 634–663.
Mishra, A. K., Sudalaimuthuasari, N., Hazzouri, K. M., Saeed, E. E., Shah, I., & Amiri, K. M. A. (2022). Tapping into plant–microbiome interactions through the lens of multi-omics techniques. *Cells, 11*(20), 3254. https://doi.org/10.3390/cells11203254
Moroenyane, I., Tremblay, J., & Yergeau, É. (2021). Temporal and spatial interactions modulate the soybean microbiome. *FEMS Microbiology Ecology, 97*(1), fiaa206. https://doi.org/10.1093/femsec/fiaa206

Mujumdar, S. S., Bashetti, S. P., & Chopade, B. A. (2014). Plasmid pUPI126-encoded pyrrolnitrin production by Acinetobacter haemolyticus A19 isolated from the rhizosphere of wheat. *World Journal of Microbiology and Biotechnology, 30*, 495–505.

Mushtaq, M. Y., Choi, Y. H., Verpoorte, R., & Wilson, E. G. (2014). Extraction for metabolomics: Access to the metabolome. *Phytochemical Analysis, 25*(4), 291–306. https://doi.org/10.1002/pca.2505

Nadarajah, K., & Abdul Rahman, N. S. N. (2021). Plant–microbe interaction: Aboveground to belowground, from the good to the bad. *International Journal of Molecular Sciences, 22*(19), 10388.

Nadeem, S. M., Zahir, Z. A., Naveed, M., & Arshad, M. (2007). Preliminary investigations on inducing salt tolerance in maize through inoculation with rhizobacteria containing ACC deaminase activity. *Canadian Journal of Microbiology, 53*(10), 1141–1149.

Nannipieri, P., Kandeler, E., & Ruggiero, P. (2002). Enzyme activities and microbiological and biochemical processes in soil. In R. G. Burns & R. P. Dick (Eds.), *Enzymes in the environment: Activity, ecology, and applications* (pp. 1–33). Marcel Dekker. https://doi.org/10.1201/9780203904039-8

Navarro-Noyaa, Y. E., Hernández-Mendoza, E., Morales-Jiménez, J., Jan-Robleroa, J., Martínez-Romero, E., & Hernández-Rodríguez, C. (2012). Isolation and characterization of nitrogen fixing heterotrophic bacteria from the rhizosphere of pioneer plants growing on mine tailings. *Applied Soil Ecology, 62*, 52–60.

Negrel, L., Halter, D., Wiedemann-Merdinoglu, S., Rustenholz, C., Merdinoglu, D., Hugueney, P., & Baltenweck, R. (2018). Identification of Lipid markers of Plasmopara viticola infection in grapevine using a non-targeted metabolomic approach. *Frontiers in Plant Science, 9*. https://www.frontiersin.org/articles/10.3389/fpls.2018.00360

Ng, W. L., & Bassler, B. L. (2009). Bacterial quorum-sensing network architectures. *Annual Review of Genetics, 43*, 197–222.

Ngumbi, E., & Kloepper, J. (2016). Bacterial-mediated drought tolerance: current and future prospects. *Applied Soil Ecology, 105*, 109–125.

Niu, X., Song, L., Xiao, Y., & Ge, W. (2018). Drought-tolerant plant growth-promoting rhizobacteria associated with foxtail millet in a semi-arid agroecosystem and their potential in alleviating drought stress. *Frontiers in Microbiology, 8*, 2580.

Novello, G., Cesaro, P., Bona, E., Massa, N., Gosetti, F., Scarafoni, A., Todeschini, V., Berta, G., Lingua, G., & Gamalero, E. (2021). The effects of plant growth-promoting bacteria with biostimulant features on the growth of a local onion cultivar and a commercial zucchini variety. *Agronomy, 1*(5), 888.

Novick, R. P., & Geisinger, E. (2008). Quorum sensing in staphylococci. *Annual Review of Genetics, 42*, 541–564.

Novick, R. P., Projan, S. J., Kornblum, J., Ross, H. F., Ji, G., Kreiswirth, B., Vandenesch, F., & Moghazeh, S. (1995). The agr P2 operon: An autocatalytic sensory transduction system in Staphylococcus aureus. *Molecular & General Genetics, 248*, 446–458.

Oades, J. M., & Waters, A. G. (1991). Aggregate hierarchy in soils. *Soil Research, 29*, 815–828.

Olanrewaju, O. S., Ayangbenro, A. S., Glick, B. R., & Babalola, O. O. (2019). Plant health: Feedback effect of root exudates-rhizobiome interactions. *Applied Microbiology and Biotechnology, 103*(3), 1155–1166. https://doi.org/10.1007/s00253-018-9556-6

Oldroyd, G. E. (2013). Speak, friend, and enter: signalling systems that promote beneficial symbiotic associations in plants. *Nature Reviews. Microbiology, 11*, 252–263. https://doi.org/10.1038/nrmicro2990

Orlandini, V., Emiliani, G., Fondi, M., Maida, I., Perrin, E., & Fani, R. (2014). Network analysis of plasmidomes: the Azospirillum brasilense Sp245 case. *International Journal of Evolutionary Biology, 2014*, 951035.

Oteino, N., Lally, R. D., Kiwanuka, S., Lloyd, A., Ryan, D., Germaine, K. J., & Dowling, D. N. (2015). Plant growth promotion induced by phosphate solubilizing endophytic Pseudomonas isolates. *Frontiers in Microbiology, 6*, 745.

Oyedele, A. O., & Ogunbanwo, T. S. (2014). Antifungal activities of Bacillus subtilis isolated from some condiments and soil. *African Journal of Microbiology Research, 8*(18), 1841–1849.

Pandey, S. S., Singh, S., Babu, C. S. V., Shanker, K., Shrivastava, N. K., & Kalra, A. (2016). Endophytes of opium poppy differentially modulate host plant productivity and genes for the biosynthetic pathway of benzylisoquinoline alkaloids. *Planta, 243*, 1097–1114.

Parks, D. H., Rinke, C., Chuvochina, M., Chaumeil, P.-A., Woodcroft, B. J., Evans, P. N., Hugenholtz, P., & Tyson, G. W. (2017). Recovery of nearly 8,000 metagenome-assembled genomes substantially expands the tree of life. *Nature Microbiology, 2*(11), Article 11. https://doi.org/10.1038/s41564-017-0012-7

Pecman, A., Kutnjak, D., Gutiérrez-Aguirre, I., Adams, I., Fox, A., Boonham, N., & Ravnikar, M. (2017). Next generation sequencing for detection and discovery of plant viruses and viroids: Comparison of two approaches. *Frontiers in Microbiology, 8*. https://www.frontiersin.org/articles/10.3389/fmicb.2017.01998

Pellegrini, M., Pagnani, G., Rossi, M., D'Egidio, S., Del Gallo, M., & Forni, C. (2021). Daucus carota L. Seed Inoculation with a consortium of bacteria improves plant growth, soil fertility status and microbial community. *Applied Sciences, 11*(7), 3274.

Pérez-Montaño, F., Alías-Villegas, C., Bellogín, R. A., Del Cerro, P., Espuny, M. R., Jiménez-Guerrero, I., et al. (2014). Plant growth promotion in cereal and leguminous agricultural important plants: From microorganism capacities to crop production. *Microbiological Research, 169*(5–6), 325–336.

Pieterse, C. M. J., Van der Does, D., Zamioudis, C., Leon-Reyes, A., & Van Wees, S. C. M. (2012). Hormonal modulation of plant immunity. *Annual Review of Cell and Developmental Biology, 28*(1), 489–521. https://doi.org/10.1146/annurev-cellbio-092910-154055

Pieterse, C. M. J., Zamioudis, C., Berendsen, R. L., Weller, D. M., Van Wees, S. C. M., & Bakker, P. A. H. M. (2014). Induced systemic resistance by beneficial microbes. *Annual Review of Phytopathology, 52*, 347–375.

Pimentel, D. C., Wilson, C., McCullum, C., Hunag, R., Dwen, P., Flack, J., Tran, Q., Saltman, T., & Cliff, B. (1997). Economic and environmental benefits of biodiversity. *BioScience, 47*, 747–757.

Prathap, M., & Kumari, B. R. (2015). A critical review on plant growth promoting rhizobacteria. *Journal of Plant Pathology and Microbiology, 6*(4).

Prescott, L. M., Harley, J. P., & Klein, D. A. (1996). The diversity of the microbial world. In L. M. Prescott, J. P. Harley, & D. A. Klein (Eds.), *Microbiology*. WCB Publishers.

Rajkumar, M., Nagendran, R., Lee, K., Lee, W., & Kim, S. (2006). Influence of plant growth promoting bacteria and Cr^{6+} on the growth of Indian mustard. *Chemosphere, 62*(5), 741–748. https://doi.org/10.1016/j.chemosphere.2005.04.117

Ramesh, R., & Phadke, G. S. (2012). Rhizosphere and endophytic bacteria for the suppression of eggplant wilt caused by Ralstonia solanacearum. *Crop Protection, 37*, 35–41.

Rathore, P. (2015). A review on approaches to develop plant growth promoting rhizobacteria. *International Journal of Recent Science Research, 5*(2), 403–407.

Reyes-Ardila, W. L., Rugeles-Silva, P. A., Duque-Zapata, J. D., Vélez-Martínez, G. A., Tarazona Pulido, L., Cardona Tobar, K. M., Díaz Gallo, S. A., Muñoz Flórez, J. E., Díaz-Ariza, L. A., & López-Alvarez, D. (2024). Exploring genomics and microbial ecology: Analysis of Bidens pilosa L. genetic structure and soil microbiome diversity by RAD-Seq and metabarcoding. *Plants, 13*(2), Article 2. https://doi.org/10.3390/plants13020221

Ribbenstedt, A., Ziarrusta, H., & Benskin, J. P. (2018). Development, characterization and comparisons of targeted and non-targeted metabolomics methods. *PLoS One, 13*(11), e0207082. https://doi.org/10.1371/journal.pone.0207082

Rocha, J., Shapiro, L. R., & Kolter, R. (2020). A horizontally acquired expansin gene increases virulence of the emerging plant pathogen Erwinia tracheiphila. *Scientific Reports, 10*(1), Article 1. https://doi.org/10.1038/s41598-020-78157-w

Rodriguez, H., & Fraga, R. (1999). Phosphate solubilizing bacteria and their role in plant growth promotion. *Biotechnology Advances, 17*, 319.

Rokhbakhsh-Zamin, F., Rokhbakhsh-Zamin, D., Sachdev, N., Kazemi-Pour, A., Engineer, K. R., Pardesi, S., Zinjarde, P. K., & Dhakephalkar, B. A. (2011). Chopade Characterization of plant-growth-promoting traits of Acinetobacter species isolated from rhizosphere of Pennisetum glaucum. *Journal of Microbiology and Biotechnology, 21*, 556–566.

Sachdev, D. P., Chaudhari, H. G., Kasure, V. M., Dahavale, D. D., & Chopade, B. A. (2009). Isolation and characterization of indole acetic acid (IAA) producing Klebsiella pneumoniae strains from rhizosphere of wheat (Triticum aestivum) and their effect on plant growth Indian. *The Journal of Experimental Biology, 47*(2009), 993–1000.

Sahu, J., Vaishnav, A., & Singh, H. B. (n.d.). Insights in plant-microbe interaction through genomics approach (Part III). *Current Genomics, 21*(6), 399–400.

Sakurai, N. (2022). Recent applications of metabolomics in plant breeding. *Breeding Science, 72*(1), 56–65. https://doi.org/10.1270/jsbbs.21065

Sanders, J. W., Martin, J. W., Hooke, M., & Hooke, J. (2000). Methylobacterium mesophilicum infection: Case report and literature review of an unusual opportunistic pathogen. *Clinical Infectious Diseases, 30*(6), 936–938.

Santoyo, G., Orozco-Mosqueda, M. D., & Govindappa, M. (2012). Mechanisms of biocontrol and plant growth-promoting activity in soil bacterial species of Bacillus and Pseudomonas: a review. *Biocontrol Science and Technology, 22*, 855–872.

Saville, A. C., & Ristaino, J. B. (2021). Global historic pandemics caused by the FAM-1 genotype of Phytophthora infestans on six continents. *Scientific Reports, 11*(1), Article 1. https://doi.org/10.1038/s41598-021-90937-6

Scandiani, M. M., Luque, A. G., Razori, M. V., Ciancio Casalini, L., Aoki, T., O'Donnell, K., Cervigni, G. D. L., & Spampinato, C. P. (2015). Metabolic profiles of soybean roots during early stages of Fusarium tucumaniae infection. *Journal of Experimental Botany, 66*(1), 391–402. https://doi.org/10.1093/jxb/eru432

Seed, P. C., Passador, L., & Iglewski, B. H. (1995). Activation of the Pseudomonas aeruginosa lasI gene by LasR and the Pseudomonas autoinducer PAI: An autoinduction regulatory hierarchy. *Journal of Bacteriology, 177*, 654–659.

Sessitsch, A., Reiter, B., Pfeifer, U., & Wilhelm, E. (2004). Cultivation-independent population analysis of bacterial endophytes in three potato varieties based on eubacterial and Actinomycetes-specific PCR of 16S rRNA genes. *FEMS Microbiology Ecology, 39*(1), 23–32.

Shapiro, L. R., Scully, E. D., Straub, T. J., Park, J., Stephenson, A. G., Beattie, G. A., Gleason, M. L., Kolter, R., Coelho, M. C., De Moraes, C. M., Mescher, M. C., & Zhaxybayeva, O. (2016). Horizontal gene acquisitions, mobile element proliferation, and genome decay in the host-restricted plant pathogen Erwinia Tracheiphila. *Genome Biology and Evolution, 8*(3), 649–664. https://doi.org/10.1093/gbe/evw016

Sharma, S., Aneja, R., Mayer, M. K., Munch, A., & Schloter, M. (2011). Characterization of bacterial communities in rhizosphere soil of grain legumes. *Microbial Ecology, 49*(3), 407–415.

Sharma, M., Sudheer, S., Usmani, Z., Rani, R., & Gupta, P. (2020). Deciphering the omics of plant-microbe interaction: Perspectives and new insights. *Current Genomics, 21*(5), 343–362. https://doi.org/10.2174/1389202921999200515140420

Sharma, V., Salwan, R., Moliszewska, E., Ruano-Rosa, D., & Jędryczka, M. (2023). *The chemical dialogue between plants and beneficial microorganisms*. Academic Press.

Shivlata, L., & Satyanarayana, T. (2017). Actinobacteria in agricultural and environmental sustainability. In J. S. Singh & G. Seneviratne (Eds.), *Agro-environmental sustainability* (pp. 173–218). Springer.

Singh, R. P., Jha, P., & Jha, P. N. (2015). The plant-growth-promoting bacterium Klebsiella sp. SBP-8 confers induced systemic tolerance in wheat (Triticum aestivum) under salt stress. *Journal of Plant Physiology, 184*, 57–67.

Sivasakthi, S., Usharani, G., & Saranraj, P. (2014). Biocontrol potentiality of plant growth promoting bacteria (PGPR)-Pseudomonas fluorescens and Bacillus subtilis: A review. *African Journal of Agricultural Research, 9*, 1265–1277.

Soil Quality Institute, Grazing Lands Technology Institute, and National Soil Survey Center, Natural Resources Conservation Service, USDA. Rangeland Soil Quality-Soil Biota USDA, Natural Resources Conservation Service, 2001.

Song, G. C., Im, H., Jung, J., Lee, S., Jung, M. Y., & Rhee, S. K. (2019). Plant growth-promoting archaea trigger induced systemic resistance in Arabidopsis thaliana against Pectobacterium carotovorum and Pseudomonas syringae. 2019. *Environmental Microbiology, 21*(3), 940–948.

Sriprang, R., Hayashi, M., Ono, H., Takagi, M., Hirata, K., & Murooka, Y. (2003). Enhanced accumulation of Cd21 by a Mesorhizobium sp. transformed with a gene from Arabidopsis thaliana coding for phytochelatin synthase. *Applied and Environmental Microbiology, 69*(3), 1791–1796.

Stotzky, G. (1997). Soil as an environment for microbial life. In J. D. van Elsas, J. T. Trevors, & E. M. H. Wellington (Eds.), *Modern soil microbiology* (pp. 1–20). Marcel Dekker.

Straub, C., Colombi, E., & McCann, H. C. (2021). Population genomics of bacterial plant pathogens. *Phytopathology, 111*(1), 23–31. https://doi.org/10.1094/PHYTO-09-20-0412-RVW

Su, X., Lu, G., Guo, H., Zhang, K., Li, X., & Cheng, H. (2018). The dynamic transcriptome and metabolomics profiling in Verticillium dahliae inoculated Arabidopsis thaliana. *Scientific Reports, 8*(1), Article 1. https://doi.org/10.1038/s41598-018-33743-x

Suman, J., Rakshit, A., Ogireddy, S. D., Singh, S., Gupta, C., & Chandrakala, J. (2022). Microbiome as a key player in sustainable agriculture and human health. *Frontiers in Soil Science, 2*, 821589.

Susič, N., Janežič, S., Rupnik, M., & Stare, B. G. (2020). Whole genome sequencing and comparative genomics of two nematicidal Bacillus strains reveals a wide range of possible virulence factors. *G3 Genes|Genomes|Genetics, 10*(3), 881–890. https://doi.org/10.1534/g3.119.400716

Tate, R. L., III. (1995). *Soil microbiology*. Wiley.

Tikhonovich, I. A., & Provorov, N. A. (2007). Beneficial plant–microbe interactions. In Y. T. Dyakov, V. G. Dzhavakhiya, & T. Korpela (Eds.), *Comprehensive and molecular phytopathology* (pp. 365–420). Elsevier. https://doi.org/10.1016/B978-044452132-3/50018-3

Turner, M. F., Heuberger, A. L., Kirkwood, J. S., Collins, C. C., Wolfrum, E. J., Broeckling, C. D., Prenni, J. E., & Jahn, C. E. (2016). Non-targeted metabolomics in diverse sorghum breeding lines indicates primary and secondary metabolite profiles are associated with plant biomass accumulation and photosynthesis. *Frontiers in Plant Science, 7*. https://www.frontiersin.org/articles/10.3389/fpls.2016.00953

van Dam, N. M., & Bouwmeester, H. J. (2016). Metabolomics in the rhizosphere: Tapping into belowground chemical communication. *Trends in Plant Science, 21*(3), 256–265. https://doi.org/10.1016/j.tplants.2016.01.008

van Velzen, R., Holmer, R., Bu, F., Rutten, L., van Zeijl, A., Liu, W., et al. (2018). Comparative genomics of the nonlegume Parasponia reveals insights into evolution of nitrogen-fixing rhizobium symbioses. *Proceedings. National Academy of Sciences. United States of America, 115*, E4700–E4709. https://doi.org/10.1073/pnas.1721395115

van Velzen, R., Doyle, J. J., & Geurts, R. (2019). A resurrected scenario: Singlegain and massive loss of nitrogen-fixing nodulation. *Trends in Plant Science, 24*, 49–57. https://doi.org/10.1016/j.tplants.2018.10.005

Vansuyt, G., Robin, A., Briat, J. F., Curie, C., & Lemanceau, P. (2007). Iron acquisition from Fe-pyoverdine by Arabidopsis thaliana. *Molecular Plant-Microbe Interactions, 20*(4), 441–447.

Wallace, J. G., Kremling, K. A., Kovar, L. L., & Buckler, E. S. (2018). Quantitative Genetics of the Maize Leaf Microbiome. *Phytobiomes Journal, 2*(4), 208–224. https://doi.org/10.1094/PBIOMES-02-18-0008-R

Wang, B., & Qiu, Y. L. (2006). Phylogenetic distribution and evolution of mycorrhizas in land plants. *Mycorrhiza, 16*, 299–363. https://doi.org/10.1007/s00572-005-0033-6

Wang, Y., Xu, Z., Bach, S. J., & McAllister, T. A. (2008). Effects of phlorotannins from Ascophyllum nodosum (brown seaweed) on in vitro ruminal digestion of mixed forage or barley grain. *Animal Feed Science and Technology, 145*(1/4), 375–395

Wang, X. M., Yang, B., Ren, C. G., Wang, H. W., Wang, J. Y., & Dai, C. C. (2015). Involvement of abscisic acid and salicylic acid in signal cascade regulating bacteria. *Plant, 153*, 30–42.

Wang, Q., Liu, J., & Zhu, H. (2018). Genetic and molecular mechanisms underlying symbiotic specificity in legume-rhizobium interactions. *Frontiers in Plant Science, 9*, 313.

Wang, K., Wu, Y., Ye, M., Yang, Y., Asiegbu, F. O., Overmyer, K., Liu, S., & Cui, F. (2021). Comparative Genomics Reveals Potential Mechanisms of Plant Beneficial Effects of a Novel Bamboo-Endophytic Bacterial Isolate Paraburkholderia sacchari Suichang626. *Frontiers in Microbiology, 12*, 686998. https://doi.org/10.3389/fmicb.2021.686998

Wani, P. A., Khan, M. S., & Zaidi, A. (2008). Effect of metal-tolerant plant growth-promoting rhizobium on the performance of pea grown in metal-amended soil. *Archives of Environmental Contamination and Toxicology, 55*(1), 33–42.

Wani, P. A., & Khan, M. S. (2010). Bacillus species enhance growth parameters of chickpea (Cicerarietinum L.) in chromium stressed soils. *Food and Chemical Toxicology, 48*, 3262–3267. https://doi.org/10.1016/j.fct.2010.08.035

Watkinson, S. C., Boddy, L., & Money, N. P. (2001). *The fungi*. Academic Press.

Wei, Y., Perez, L. J., Ng, W. L., Semmelhack, M. F., & Bassler, B. L. (2011). Mechanism of Vibrio cholerae autoinducer-1 biosynthesis. *ACS Chemical Biology, 6*, 356–365.

William Allwood, J., Clarke, A., Goodacre, R., & Mur, L. A. J. (2010). Dual metabolomics: A novel approach to understanding plant–pathogen interactions. *Phytochemistry, 71*(5), 590–597. https://doi.org/10.1016/j.phytochem.2010.01.006

Williams, P., & Camara, M. (2009). Quorum sensing and environmental adaptation in Pseudomonas aeruginosa: A tale of regulatory networks and multifunctional signal molecules. *Current Opinion in Microbiology, 12*, 182–191.

Wilpiszeski, R. L., Aufrecht, J. A., Retterer, S. T., Sullivan, M. B., Graham, D. E., Pierce, E. M., Zablocki, O. D., Palumbo, A. V., & Elias, D. A. (2019). Soil aggregate microbial communities: towards understanding microbiome interactions at biologically relevant scales. *Applied and Environmental Microbiology, 85*, 324–319.

Wu, C. H., Wood, T. K., Mulchandani, A., & Chen, W. (2006). Engineering plant-microbe symbiosis for rhizoremediation of heavy metals. *Applied and Environmental Microbiology, 72*(2), 1129–1134.

Xu, J., & Wang, N. (2019). Where are we going with genomics in plant pathogenic bacteria? *Genomics, 111*(4), 729–736. https://doi.org/10.1016/j.ygeno.2018.04.011

Xu, P., Wang, H., Qin, C., Li, Z., Lin, C., Liu, W., & Miao, W. (2021). Analysis of the taxonomy and pathogenic factors of Pectobacterium aroidearum L6 using whole-genome sequencing and comparative genomics. *Frontiers in Microbiology, 12*. https://www.frontiersin.org/articles/10.3389/fmicb.2021.679102

Yaish, M. W., Antony, I., & Glick, B. R. (2015). Isolation and characterization of endophytic plant growth-promoting bacteria from date palm tree (Phoenix dactylifera L.) and their potential role in salinity tolerance. *Antonie van Leeuwenhoek, 107*, 1519–1532.

Yang, B., Liu, Q., & Dong, T. (2017). Variation of Soil nutrients and microbial community diversity of different land use types in the Jialing River, Sichuan Province. *Research Soil and Water Conservation, 24*, 26.

Yang, S., Johnson, M. A., Hansen, M. A., Bush, E., Li, S., & Vinatzer, B. A. (2022). Metagenomic sequencing for detection and identification of the boxwood blight pathogen Calonectria pseudonaviculata. *Scientific Reports, 12*(1), Article 1. https://doi.org/10.1038/s41598-022-05381-x

Yogendra, K. N., Kushalappa, A. C., Sarmiento, F., Rodriguez, E., & Mosquera, T. (2014). Metabolomics deciphers quantitative resistance mechanisms in diploid potato clones against late blight. *Functional Plant Biology, 42*(3), 284–298. https://doi.org/10.1071/FP14177

Yue, H., Sun, S., Wang, R., et al. (2023). Study on the mechanism of salt relief and growth promotion of Enterobacter cloacae on cotton. *BMC Plant Biology, 23*, 656.

Zahir, Z. A., Shah, M. K., Naveed, M., Akhtar, M. J. (2010). Substrate dependent auxin production by Rhizobium phaseoli improve the growth and yield of Vigna radiate L. under salt stress conditions. *Journal of Microbiology and Biotechnology, 20*, 1288–1294

Zhalnina, K., Louie, K. B., Hao, Z., Mansoori, N., da Rocha, U. N., Shi, S., Cho, H., Karaoz, U., Loqué, D., Bowen, B. P., Firestone, M. K., Northen, T. R., & Brodie, E. L. (2018). Dynamic root exudate chemistry and microbial substrate preferences drive patterns in rhizosphere microbial community assembly. Nature. *Microbiology, 3*(4), Article 4. https://doi.org/10.1038/s41564-018-0129-3

Zhou, J., Deng, Y. E., Shen, L., Wen, C., Yan, Q., Ning, D., et al. (2016). Temperature mediates continental-scale diversity of microbes in forest soils. *Nature Communications, 7*(1), 12083.

Zhu, L., Zhou, Y., Li, X., Zhao, J., Guo, N., & Xing, H. (2018). Metabolomics analysis of soybean hypocotyls in response to Phytophthora sojae infection. *Frontiers in Plant Science, 9*. https://www.frontiersin.org/articles/10.3389/fpls.2018.01530

PGPR Consortia: Micro Solution to Major Challenges

5

Nishra Joshi, Janki Ruparelia, M. Tofazzal Islam, and Chaitanya Kumar Jha

Abstract

The world is grappling with pressing environmental concerns that are adversely impacting the agricultural sector's ability to meet global food demands. A significant driver of this crisis is climate change, creating extreme weather events that disrupt established agricultural systems and the natural water cycle. Consequently, this upheaval in the environmental equilibrium has become a pivotal factor contributing to reduced food production and escalated food prices. In response to these challenges, it becomes imperative to adopt farming practices that are not only sustainable but also environmentally friendly, particularly in the face of increasingly unpredictable weather conditions. One noteworthy side of exploration involves harnessing the potential of rhizosphere microbiota in agriculture. These microorganisms have evolved intricate mechanisms enabling them to thrive amid adverse abiotic conditions, such as fluctuating temperatures and varying pH levels. Their positive attributes serve as valuable tools in mitigating the consequences of climate change on agricultural productivity. This discourse primarily underscores the critical importance and urgency of embracing sustainable and eco-friendly agricultural techniques, with a specific emphasis on the application of climate-smart biofertilizers tailored to navigate the challenges

N. Joshi · J. Ruparelia
Department of Microbiology and Biotechnology, University School of Science, Gujarat University, Navrangpura, Ahmedabad, India

Microbiology Department, Gujarat Arts and Science College, Ellisbridge, Ahmedabad, India

M. T. Islam
Institute of Biotechnology and Genetic Engineering (IBGE), Bangabandhu Sheikh Mujibur Rahman Agricultural University, Gazipur, Bangladesh

C. K. Jha (✉)
Microbiology Department, Gujarat Arts and Science College, Ellisbridge, Ahmedabad, India

M. Saraf et al. (eds.), *Climate Change and Soil Microorganisms for Environmental Sustainability*, Microorganisms for Sustainability 28,
https://doi.org/10.1007/978-981-96-3425-5_5

posed by evolving climate conditions. Additionally, the examination of microbial consortia, comprising diverse microorganisms with promising functionalities, is also discussed as a strategic approach to address the multifaceted impacts of climate change on agriculture.

Keywords

Plant growth-promoting rhizobacteria (PGPR) · PGPR consortia · Crop stress management · Climate-smart agriculture

5.1 Introduction

The projected rise in global population, coupled with the challenges posed by climate change, underscores the urgency of finding sustainable solutions to ensure food security for future generations. As atmospheric CO_2 concentrations continue to increase, the resulting changes in temperature and precipitation patterns are expected to have profound implications for agricultural systems worldwide. Climate models predict more frequent and severe weather events, such as droughts, floods, and heatwaves, which can devastate crops and lead to food shortages (Shah et al., 2021).

Furthermore, the degradation of agricultural soils due to salinization, erosion, and nutrient depletion exacerbates the impact of climate change on crop productivity. Soil fertility is essential for supporting healthy plant growth and mitigating the effects of environmental stresses. However, conventional farming practices, including the heavy use of synthetic fertilizers and pesticides, have contributed to soil degradation and biodiversity loss (Jha & Saraf, 2015). This degradation reduces the soil's ability to sequester carbon and regulate water and nutrient cycles, further exacerbating climate change (Agnihotri & Mitra, 2023).

In light of these challenges, there is a growing recognition of the need to transition toward more sustainable agricultural practices that promote soil health, biodiversity, and resilience to climate change. Plant–microbe interactions, particularly those involving plant growth-promoting rhizobacteria (PGPR), offer promising avenues for enhancing crop productivity while reducing the environmental impact of agriculture. PGPR have been shown to improve soil structure, enhance nutrient uptake, and boost plant tolerance to various abiotic stresses, including drought, salinity, and heavy metal toxicity (Ansari & Ahmad, 2019).

Moreover, PGPR-mediated biopriming represents an innovative approach to enhance crop resilience under changing climate conditions. Biopriming involves treating seeds with beneficial microbes before planting, thereby priming them to better withstand environmental stresses and promote vigorous growth (Kumar et al., 2019). By harnessing the natural abilities of PGPR to enhance plant–microbe interactions, biopriming offers a sustainable alternative to traditional agrochemicals and can help mitigate the adverse effects of climate change on crop production.

In conclusion, the integration of PGPR-based solutions into sustainable agriculture practices holds great promise for addressing the dual challenges of feeding a growing population while mitigating the impacts of climate change on food security. By leveraging the natural abilities of PGPR to enhance plant resilience and soil health, we can build more resilient and sustainable food systems that are better equipped to withstand the challenges of a changing climate.

5.2 Scenario of Climate Change and Its Impact on Agriculture Sector of India

Climate change is a critical global issue that poses considerable hazards to ecosystems, economies, and communities around the world. Agriculture is one of the most sensitive industries to the effects of climate change, since it is critical for food security, livelihoods, and economic development, especially in countries like India, where a big proportion of the population relies on agriculture for survival (Adams et al., 1998; Yohannes, 2016).

In recent decades, India's climate has changed noticeably, with rising temperatures, altered precipitation patterns, increased frequency of extreme weather events, and changes in monsoon rain distribution (Redondo-Gómez et al., 2021). These changes have far-reaching ramifications for agricultural output, food security, and rural lives throughout the country (Nelson et al., 2009).

Temperature and precipitation fluctuations are likely to have different effects on regional and temporal distribution, changing the time and duration of growing seasons. In this case, high temperatures combined with insufficient precipitation may accelerate plant growth, limit grain filling, lower nutrient efficiency, and increase crop water use. As a result, considerable shifts in agroecological zones are projected. Furthermore, fluctuations in soil organic matter, which regulates agricultural landform and fertility, are predicted. As a result, improvements to nutrient levels and agricultural techniques required for cultivating certain crops in the altered agroecological zones are impending. Temperature and precipitation patterns are expected to shift, increasing potential evapotranspiration and perhaps exacerbating drought (Anwar et al., 2013). The substantial depletion of nutrients, as well as the widespread use of synthetic agrochemicals such as chemical fertilizers and growth regulators, are directly related to the climate change scenario. For instance, methane (CH_4), a strong greenhouse gas (GHG) that contributes to climate change, has been linked to increased emissions when nitrogen fertilizer is used in intensive agricultural systems. This influences the stratospheric ozone (O_3) layer in addition to enhancing the effects of carbon dioxide (CO_2) Future agricultural intensification is expected to increase CO_2 and CH_4 emissions, which could worsen anthropogenic climate change. Furthermore, because of the stratospheric O_3 depletion caused by increased CH_4 emissions, surface ultraviolet (UV-B) radiation levels may rise. In turn, exposure to UV-B light may indirectly prevent the absorption of photosynthetic carbon. Crop productivity and growth may suffer as a result of these widespread consequences in a variety of geographical areas (Anwar et al., 2013). Hence, these practices lead to soil degradation, compounding climate change's impact on agricultural sustainability and resilience (Sarker et al., 2021).

The development of essential food crops and medicinal plants in India is facing serious challenges due to climate change. A vast array of plant species are being impacted by changing environmental circumstances, ranging from common crops like rice and wheat to therapeutic herbs like ashwagandha and tulsi. Extreme weather, water scarcity, and erratic monsoon patterns all have an impact on rice and wheat production, which lowers yields and jeopardizes food security (Lobell et al., 2012; Rani et al., 2022; Soora et al., 2013). The growth, availability, and therapeutic qualities of plants like ashwagandha and tulsi are also susceptible to variations in temperature and precipitation (Sharma et al., 2020). Secondary compounds produced by plants, which are essential for their therapeutic qualities, can be impacted by climate change (Sharma et al., 2020). The production of cotton is also impacted by climate change, which poses threats to crop yields and farmer incomes through variations in temperature, rainfall, and insect prevalence (Iqbal et al., 2022).

Developing adaptive solutions to promote sustainable agriculture, food security, and healthcare practices in India requires an understanding of the sensitivities of medicinal plants and crops to climate change. Mitigation strategies can lessen the negative effects of climate change on agricultural output and the availability of medicinal plants. Examples of these strategies include the development of climate-resilient crop types and sustainable cultivation practices.

Rhizobacteria that promote plant growth (PGPR) may become micro solution to the major even enormous problem that climate change is posing to agriculture. Within the plant's rhizosphere, these advantageous microorganisms are essential for improving agricultural yield and resilience in the face of changing environmental conditions. Climate change-related nutritional deficits are mitigated by PGPR, which solubilizes critical elements like iron and phosphorus to promote plant uptake. Furthermore, crops are able to flourish in unfavorable circumstances because PGPR-produced phytohormones and enzymes strengthen plants against abiotic challenges like salinity, drought, and temperature swings. Furthermore, some PGPR strains have the ability to act as biocontrol agents, inhibiting the development of plant pathogens and lowering the frequency of disease, protecting the health of crops (Kumar et al., 2019). In addition to these immediate advantages, PGPR improves soil health by encouraging the decomposition of organic matter, the cycling of nutrients, and the improvement of soil structure, all of which support the establishment of robust plants in healthy soils. One important benefit of PGPR is that it can reduce the need for chemical inputs like pesticides and fertilizers, which will lessen the impact on the environment and save production expenses (Joshi et al., 2023a).

Ultimately, the use of PGPR emphasizes the importance of sustainable agriculture and adaptation techniques by showing potential for raising crop yields, enhancing crop quality, and guaranteeing food security in the face of problems brought on by climate change (Fig. 5.1).

Several factors can influence the effectiveness of using PGPR to improve plant development, including the plant's growth stage, species, cultivation techniques, soil features, and experimental circumstances. Given recent advances in understanding plant–microbe interactions, this short review digs into the processes by

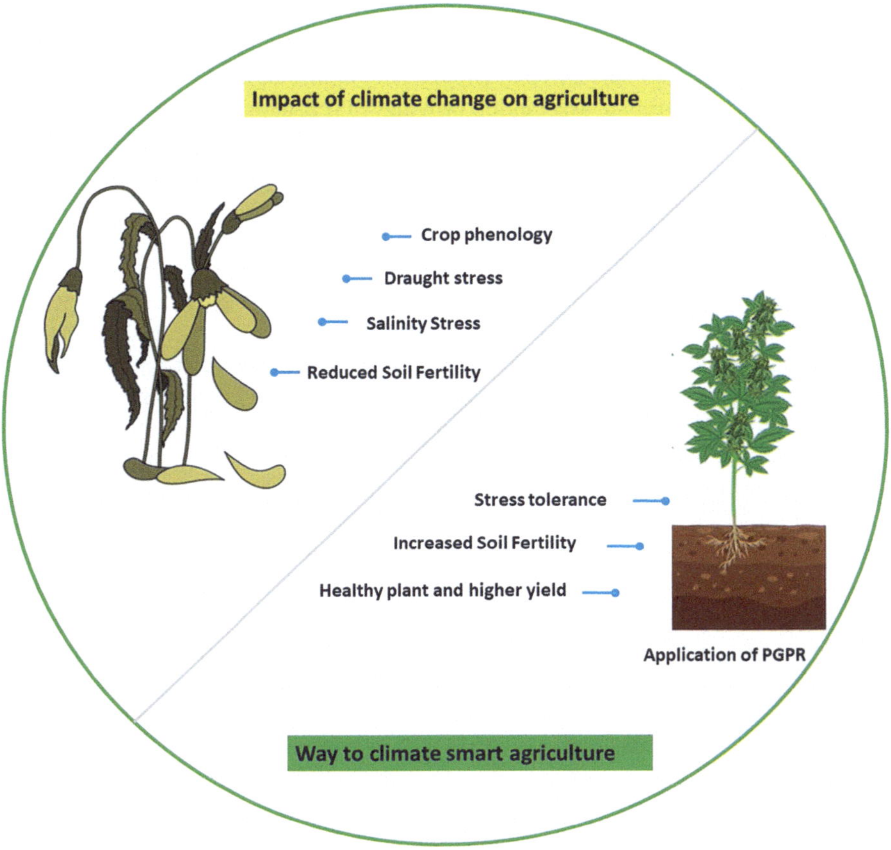

Fig. 5.1 Role of PGPR in climate-smart agriculture

which PGPR promote tolerance to abiotic stress. Furthermore, it discusses the challenges of using PGPR as biopriming agents to improve crop output in the face of changing climate circumstances.

5.3 Beneath the Surface: Unveiling PGPR's Stress Alleviation Strategies

In the rhizosphere, the dynamic contact between soil and plant roots, live an exceptional community of helpful microorganisms known as *plant growth-promoting rhizobacteria (PGPR)* (Goswami et al., 2016). It is therefore imperative for scientists to learn how PGPR interact with plants and reduce stress in light of the difficulties presented by climate change and rising environmental pressures. This section explores the intricate mechanisms employed by PGPR to mitigate abiotic stress, delving into the scientific underpinnings of their strategies.

The rhizosphere serves as a lively hotspot of microbial activity, where PGPR establishes symbiotic relationships with plants. Beneath the surface of the soil, PGPR colonize the root surfaces and engage in complex signaling interactions with the plant host. This intimate association provides PGPR with access to plant-derived carbon compounds and facilitates the exchange of signals that trigger stress response pathways in both the microbe and the plant. There are several elements that influence the effectiveness of plant growth-promoting rhizobacteria (PGPR), including the soil ecology, the type and age of the plant or crop that the bacteria interact with, and their growth phase. The chemistry of the plant-soil is constantly being adjusted by these helpful bacteria, which promotes plant health and growth. Plant growth promoting rhizobacteria (PGPR) uses a variety of strategies to reduce plant stress, such as increasing soil nutrient availability and producing phytohormones, osmoprotectants, and antioxidant enzymes (Jha & Saraf, 2015). Gaining insight into the mechanisms by which PGPR affect plant health and guarantee sustainable agricultural practices is essential to realizing the full potential of PGPR and successfully applying them in agriculture. This requires comprehensive knowledge of PGPR-mediated direct and indirect mechanisms that optimize plant benefits while minimizing environmental impacts. Solubilization of inorganic phosphate, auxins (specifically, 3-indole acetic acid [3 IAA]) production of ammonia, the enzyme 1-aminocyclopropane-1-carboxylate (ACC) deaminase, and the synthesis of siderophore molecules are the most significant and researched direct mechanisms. On the other hand, indirect mechanisms are defined as inhibitory activities in which growth-promoting microorganisms counteract the detrimental effects of pathogenic organisms (biotic stress) (Jha & Saraf, 2015). Some of the most prevalent indirect strategies are the formation of hydrogen cyanide, antibiotics, and enzymes capable of attacking and destroying pathogen cell walls (Vocciante et al., 2022).

Phosphorus (P) is essential for plant growth, comprising about 0.2% of a plant's dry weight and ranking second only to nitrogen in limiting mineral nutrients for crops (Jha & Saraf, 2015). However, traditional approaches to addressing soil phosphorus deficiency through fertilizers often lead to environmental concerns such as groundwater contamination and eutrophication. To mitigate these issues, there is a growing interest in enhancing phosphorus fertilization efficiency while reducing environmental pollution (Goswami et al., 2016). Soil microorganisms, particularly phosphorus solubilizing microorganisms (PSM), play a crucial role in enhancing plant-available phosphorus through solubilization and mineralization of insoluble phosphorus compounds in soil. A diverse array of microbial organisms, including bacteria, fungi, actinomycetes, and algae, contribute to phosphorus solubilization and mineralization in soils. Notable species like *Azotobacter*, *Bacillus*, *Burkholderia*, *Enterobacter*, *Erwinia*, *Kushneria*, *Paenibacillus*, *Ralstonia*, *Rhizobium*, *Rhodococcus*, *Serratia*, *Bradyrhizobium*, and *Thiobacillus* are reported phosphorus solubilizers and mineralizers (Goswami et al., 2016; Patel & Goswami, 2020). Bacterial-mediated phosphorus (P) solubilization generally entails soil acidification caused by bacteria's production of organic acids. These rhizobacteria can create organic acids such as gluconic acid and 2-ketogluconic acid (Sharma et al., 2013). These organic acids have chelating capabilities, which allow them to attach to cations linked to phosphate, making the phosphate more available for plant roots to absorb.

Plant growth-promoting rhizobacteria (PGPR) hold significance also for improving iron acquisition in plants because they produce siderophores, which are lower molecular weight organic compounds that chelate iron when there is an iron shortage (Hayat et al., 2012). By preventing pathogens from obtaining iron, these siderophores not only assist plants absorb iron but also prevent the development of disease. Numerous bacterial species possess the ability to produce siderophores, such as *Azospirillum*, *Klebsiella*, *Pantoea*, *Pseudomonas*, *Azotobacter*, *Bacillus*, and *Serratia* (Ansari & Ahmad, 2019; Jha & Saraf, 2015). By decreasing metal ion availability and minimizing the buildup of hazardous metals in the soil, PGPR-producing siderophores help plants survive times of metal stress. Additionally, the presence of plant growth-promoting strains further enhances the uptake of other metals such as zinc. Overall, PGPR-mediated siderophore production represents an essential mechanism for improving plant iron uptake, enhancing plant health, and mitigating metal stress in the soil (Kumar et al., 2019).

Indole acetic acid (IAA), a common natural auxin found in plants, is synthesized by up to 80% of rhizobacteria and plays a crucial role in stimulating root growth and enhancing mineral and nutrient uptake from the soil (Mehmood et al., 2018). It influences various aspects of plant growth and development, including cell proliferation, root and xylem development, seed germination, and stress resistance. Tryptophan, an amino acid present in root exudates, serves as the main precursor molecule for IAA biosynthesis in bacteria (Mehmood et al., 2018). The biosynthesis of IAA by plant growth-promoting rhizobacteria (PGPR) typically involves the conversion of tryptophan to indole-3-pyruvic acid and then to indole-3-acetic aldehyde. This mechanism is common in bacteria like *Pseudomonas*, *Bacillus*, *Burkholderia*, and *Rhizobium*, contributing to root growth promotion (Jha & Saraf, 2015; Joshi et al., 2023b). However, microbially produced phytohormones are more effective than chemically produced ones due to their continuous slow release and greater efficacy in promoting plant growth.

PGPR employ various mechanisms to indirectly promote plant growth while counteracting the detrimental effects of pathogens. These bacteria can produce antibiotics or lytic enzymes that degrade the cell wall of phytopathogens (Bhattacharyya & Jha, 2012; Goswami et al., 2016). Moreover, PGPR compete with pathogens for nutrients and root colonization sites, thereby reducing the proliferation of pathogens. Furthermore, PGPR may produce small amounts of hydrogen cyanide (HCN), which collaborates with other biocontrol mechanisms to further inhibit pathogen growth and enhance plant health.

5.4 Harnessing PGPR for Climate-Smart Agriculture

Plant growth-promoting rhizobacteria (PGPR) play a fundamental role in climate-smart agriculture by enhancing plant resilience to environmental stresses and promoting sustainable crop production (Agnihotri & Mitra, 2023). Through various mechanisms, PGPR mitigate the adverse effects of climate change on agricultural systems. These beneficial microbes facilitate nutrient uptake by plants, particularly under conditions of nutrient deficiency exacerbated by climate change-induced soil degradation. Moreover, PGPR enhances plant tolerance to abiotic stresses such as drought, salinity, and temperature extremes, thereby improving crop yield stability in the face of changing climatic conditions. Additionally, PGPR-mediated biofertilization reduces the dependency on chemical fertilizers, mitigating greenhouse gas emissions associated with their production and application (Joshi et al., 2023a). Furthermore, PGPR contribute to soil health and carbon sequestration through the promotion of root growth, organic matter decomposition, and nutrient cycling. Overall, harnessing the potential of PGPR in climate-smart agriculture offers a sustainable approach to enhance agricultural productivity, resilience, and environmental sustainability in the face of climate change challenges.

5.4.1 Role of PGPR in Alleviating Abiotic Stress Factors

In recent years, the agricultural sector has faced increasingly frequent and severe challenges due to climate change-induced extreme environmental conditions such as drought, intense rainfall, salinity, temperature extremes, heavy metal contamination, and nutrient deficiencies. These stressors have led to substantial reductions in crop yields and overall quality worldwide. Traditional breeding methods and genetic engineering approaches for stress tolerance are time-consuming, resource-intensive, and face ethical and social acceptance challenges (Fiodor et al., 2021). Consequently, there is growing recognition of the vital role played by beneficial microorganisms in stress management and the development of climate-resilient agriculture. Microorganisms offer a sustainable and environmentally friendly approach to improving crop resilience and ensuring food security in a changing world.

Salinity is recognized as one of the primary abiotic stressors affecting crop productivity. Salinity stress hampers essential plant processes like photosynthesis, respiration, and protein synthesis, leading to oxidative stress through the generation of harmful reactive oxygen species (ROS) (Agnihotri & Mitra, 2023). To combat oxidative stress, plants activate antioxidant defense systems comprising enzymes such as peroxidase, superoxide dismutase, catalase, ascorbate peroxidase, and glutathione reductase. Halophilic/halotolerant microorganisms offer an influential and eco-friendly method to sustain productivity under salt stress conditions by enhancing plant immunity, increasing antioxidant enzyme activity, and improving nodule formation in legumes (Sood et al., 2019). PGPR have been shown to enhance the growth of various crops, including eggplant, wheat, and quinoa, under salt stress conditions, while biostimulants protect photosynthetic pigments and improve chlorophyll content in crops like corn and rice (Lobell et al., 2012; Soora et al., 2013).

Extreme temperatures, whether cold or warm, can also significantly impact crop yields. Beneficial microbes have the potential to mitigate temperature-induced stress in crops. For instance, *Pseudomonas* sp. and *Acinetobacter oleivorans* IRS14 have been found to promote the growth of wheat plants under low-temperature stress conditions by regulating key physiological and biochemical factors. These microbes modulate biochemical and metabolic pathways in plants, reducing chilling stress and increasing plant growth rate and biomass. Overall, harnessing the potential of beneficial microorganisms offers promising solutions for enhancing crop resilience and ensuring food security in the face of climate change-induced stresses (Singh et al., 2024).

5.4.2 Role of PGPR in Alleviating Biotic Stress Factors

Pathogenic microorganisms such as bacteria, fungi, viruses, nematodes, and protists pose significant challenges to agricultural productivity. Plant growth-promoting rhizobacteria (PGPR) offer a promising solution through their multifaceted mechanisms of action. PGPR effectively suppresses the proliferation of phytopathogens and other deleterious microorganisms through a range of sophisticated strategies. These include the production of antagonistic substances like bacteriocin, antibiotics, hydrogen cyanide, and antimicrobial metabolites, which inhibit the growth of pathogens (Hamid et al., 2021). Furthermore, PGPR can engage in parasitism, compete with pathogens for essential nutrients and space, and secrete a variety of lytic enzymes such as chitinases, glucanases, and proteases that degrade the cell walls of pathogenic organisms (Hamid et al., 2021). Additionally, PGPR have the ability to induce systemic resistance in plants, augmenting their innate defense mechanisms against diseases. This intricate interplay of mechanisms underscores the pivotal role of PGPR in bolstering plant health and mitigating the impact of biotic stresses on agricultural systems.

Bacteriocins, peptide secretions with narrow-spectrum antimicrobial action, are potent weapons produced by both Gram-negative (e.g., colicin) and Gram-positive (e.g., nisin) bacteria (Zimina et al., 2020). These toxins exhibit a high degree of specificity in targeting and eliminating competing bacterial species (Rooney et al., 2020). Notably, bacteriocins have demonstrated promising efficacy in vitro against bacterial spot disease in tomatoes (Príncipe et al., 2018). Notably, some specific bacteriocins synthesized by *Bacillus* species may exhibit a broader spectrum of inhibitory activity. Antibiotics manufactured by PGPR exhibit enhanced efficiency owing to their multifaceted properties, encompassing antibacterial, insecticidal, antiviral, phytotoxic, cytotoxic, and anthelminthic attributes (Fernando et al., 2007). Notably, *Pseudomonas* species are renowned for their production of a diverse array of antifungal antibiotics, including but not limited to 2,4-diacetylphloroglucinol (2,4-DAPG), butyrolactones, rhamnolipids, and N-butylbenzene sulfonamide (Ramadan et al., 2016). Similarly, *Bacillus* species are prolific producers of antibiotics, such as bacilysin, bacillaene, and mycobacillin, complemented by the secretion of various lipopeptide biosurfactants like bacillomycin, renowned for their potent antibacterial activity (Mahmood et al., 2019). Furthermore, plants inoculated with the PGPR strain boost their resistance potential against phytopathogens. Shen et al. reported that PGPR strains *P. stutzeri*, *B. subtilis*, *B. amyloliquefaciens*, and *Stenotrophomonas maltophilia* isolated from the root zone of cucumber successfully suppressed *Phytophthora capsica*, causing Phytophthora crown rot in cucumber, by inoculating cucumber seeds with tested PGPR (Shen et al., 2013).

5.5 Synergistic Strategies: Utilizing PGPR Consortia for Climate-Smart Agriculture

Climate-smart agriculture has become immensely necessary due to the multiple challenges imparted by climate change that threaten the global production of food crops and medicinally important crops, yield production, and environmental sustainability. Here, PGPR emerge as a one-stop solution for climate-smart agriculture as it offers various benefits like enhanced agricultural productivity, resilience, and adaptation to climate change while minimizing greenhouse gas emissions and environmental pollution. PGPR being a potential candidate for climate-smart agriculture provide multiple benefits to plants but it also come with limitations that hinder its efficacy in promoting sustainable agriculture. These limitations include their narrow functional diversity, inconsistent performance under field conditions, reduced resilience to environmental fluctuations, and narrow-spectrum activity against specific stressors (Kaushal et al., 2023). Furthermore, single strains may also exhibit limitations in nutrient cycling and fertility. To overcome these limitations, PGPR consortia represent a more versatile and sustainable approach to climate-smart agriculture. PGPR consortia consist of a different two or more microbial strains showing compatibility with each other (Joshi et al., 2023b; Panwar et al., 2014). They exhibit greater resilience, adaptability, and efficacy in promoting plant growth and yield, suppressing pathogens, and enriching soil health through enhanced nutrient cycling (Panwar et al., 2014).

In agriculture, it is crucial for members of consortia to engage in positive interactions, fostering mutualistic growth that ensures stable performance over extended periods of cultivation, thereby achieving the anticipated beneficial effects when applied to crops. Synergistic behavior among members of a consortium in agriculture is intricately governed by various molecular signaling mechanisms, with quorum sensing being particularly pivotal (Mukherjee & Bassler, 2019). Quorum sensing is a fundamental process that allows bacteria to transition between two distinct gene expression programs: one at low cell density, and another at high cell density, facilitating social and group behaviors, which are advantageous for consortia dynamics (Ng & Bassler, 2009). This response to changes in cell density occurs through the synthesis of low-molecular-weight molecules known as autoinducers within bacterial cells. Subsequently, these autoinducers are released or actively secreted into the extracellular environment as the cell population grows. As the concentration of autoinducers increases extracellularly, they reach a threshold level necessary for detection. Upon reaching this threshold, the autoinducers bind to associated receptors, initiating signal transduction cascades that lead to population-wide changes in gene expression. Consequently, quorum sensing enables synchronized cellular behavior within a population, facilitating collective actions that are beneficial for consortia function, without compromising the desired effects of compatible plant growth-promoting bacteria (PGPB) in consortia (Schikora et al., 2016). Acyl homoserine lactone (AHL) emerges as a predominant signal molecule facilitating cell-to-cell communication (Liu et al., 2012). AHLs, produced notably by bacterial species such as *Serratia liquefaciens* and *S. plymuthica*, exert profound effects on root development and plant biomass (Pérez-Montaño et al., 2013). Similarly, species like *Sinorhizobium fredii* and *Pantoea ananatis* induce biofilm formation within the roots of *Oryza sativa* and *Phaseolus vulgaris*, fostering plant growth (Pérez-Montaño et al., 2013). AHLs also modulate intracellular Ca2+ levels and primary root growth (Liu et al., 2012; Pérez-Montaño et al., 2013). Beyond AHLs, volatile organic compounds (VOCs) represent another vital signaling molecule within bacterial consortia, facilitating interplay between plants and bacteria, as well as among bacterial species (Schulz-Bohm et al., 2017). Comprising terpenoids, alkenes, alkanes, sulfur-containing compounds, and alcohols, VOCs exert beneficial effects on nearby or distant bacterial species within the consortium, bolstering crop enhancement. Notably, VOCs possess antimicrobial properties, effectively managing plant pathogens, and regulate genes associated with hormonal signaling, thereby orchestrating plant growth (Schulz-Bohm et al., 2017). Certain bacterial VOCs also engage in direct antagonism against other bacteria, endowing them with potent biocontrol capabilities against plant pathogens, particularly exemplified by *Pseudomonas* and *Bacillus* species (Santoyo et al., 2021; Schulz-Bohm et al., 2017). These multifaceted compounds play indispensable roles in modulating gene expression linked to hormonal signaling, defensive pathways, microbial stress tolerance, virulence, and biofilm formation. In sunflowers, the bacterial consortium comprising *Azotobacter* sp. and *Azospirillum* sp. optimized nitrogen fertilization by 50%, while achieving the highest grain production, oil, and protein levels. Jha and Saraf (2012) demonstrated the co-inoculation efficacy of *Micrococcus* sp., *Acinetobacter calcoaceticus, Brevibacillus brevis*, and *Bacillus licheniformis* in enhancing the yield of *Jatropha curcas L.* These bioinoculant consortia exhibit the higher efficacy over single inoculum in IAA production, ACC deaminase production, production of siderophores, and solubilize inorganic phosphorus (Jha & Saraf, 2012). Bacterial consortia of four taxa *Enterobacter cloacae*, *Pseudomonas putida*, *Citrobacter freundii, and Comamonas testosterone* exhibited an effective role in solubilizing the phosphate, therefore improving crop productivity by twofold, particularly in tomatoes (Baas et al., 2016). Several studies involving the inoculation of plants with bacterial consortia have shown significant increases in yield and biomass as well as stress alleviation (Table 5.1) (Raklami et al., 2021; Schröder et al., 2018).

5.5.1 Role of PGPR Consortia Under Abiotic and Biotic Stress Conditions

Amid the flourishing challenges posed by climate change, characterized by a plethora of biotic and abiotic stressors, researchers have underscored the pivotal role of certain microorganisms in augmenting plant resilience. These microorganisms exhibit the capacity to impart tolerance and resistance to plants, thereby enhancing

Table 5.1 Beneficial effect of PGPR consortia on crops

PGPR	Crop	Beneficial effect	Reference
Pantoea cypripedii + *Enterobacter aerogenes*	Chickpea	Increased uptake of phosphorus	Singh et al. (2014)
Bacillus pumilus + *Pseudomonas pseudoalcaligenes*	Rice	Improved plant growth and yield	Jha and Subramanian (2014)
Bacillus cereus Y5 + *Bacillus* sp. *Y14* + *Bacillus subtilis Y16*	Wheat	Salinity stress tolerance and higher grain yield	Shahzad et al. (2017)
Bacillus cereus BC1AW + *Pseudomonas putida PP3WT*	Tomato	Reduced bacterial wilt disease	Kurabachew and Wydra (2013)
Lactobacillus plantarum PM411 + *Lactobacillus plantarum TC92*	Strawberry	Disease prevention in strawberry	Daranas et al. (2018)
Bacillus cereus AR156 + *Bacillus subtilis SM21* + *Serratia* sp. *XY21*	Cucumber	Drought and significant increment in leaf proline content	Wang et al. (2012)
Bacillus spp. + *Pseudomonas aeruginosa*	Sunflower	Sunflower necrosis disease suppression	Srinivasan and Mathivanan (2011)

their ability to withstand diverse stressors. Tolerance refers to the inherent ability of a microbe to endure adverse environmental conditions, while resistance denotes its capability to withstand stress through specific physiological mechanisms. Empirical evidence supports the notion that microorganisms endowed with these attributes can function as beneficial inoculants, facilitating plant adaptation to stressful environments. Leveraging the synergistic interactions of multiple plant growth-promoting rhizobacteria (PGPR) consortia-based strategies offer distinct advantages over single-strain inoculants in ameliorating stress-induced crop losses.

Abiotic stressors such as drought, temperature extremes, salinity, heavy metals, and pollutants disrupt plant growth and metabolism, leading to reduced yields, impaired photosynthesis, and cellular damage (Fiodor et al., 2021). Study of *Ocimum basilicum* (basil) inoculated with a bacterial consortium *Pseudomonas* sp., *Bacillus lentus*, and *Azospirillum brasilense* exhibited enhanced levels of chlorophyll content and antioxidant activity in plants under water stress circumstances (Heidari & Golpayegani, 2012). Nadeem et al. (2013) noted prominent improvements in the germination rate, growth, yield, and nutritional content of wheat when treated with a consortium of *Enterobacter cloacae*, *Pseudomonas putida*, *P. fluorescens*, and *Serratia ficaria* particularly under saline stress conditions (Nadeem et al., 2013). Some plant growth-promoting bacteria also exhibit metal-tolerance capacity. For example, *Flavobacterium* sp., *Rhodococcus* sp., and *Variovorax paradoxus* enhanced the root development of Indian mustard, irrespective of the presence or absence of cadmium (Cd), thus affirming their role as growth promoters even under conditions of metal stress (Belimov et al., 2005). In conditions of abiotic stress, elevated levels of ethylene in plants inhibit both root and shoot growth. However, bacteria possessing 1-aminocyclopropane-1-carboxylate (ACC) deaminase can degrade excessive 1-aminocyclopropane-1-carboxylate (ACC) produced by the plant, thereby regulating the total ACC content and promoting plant growth (Gamalero & Glick, 2015). For instance, Saikia et al. (2018) demonstrated that the introduction of a consortium comprising three ACC-deaminase-producing rhizobacteria (*Ochrobactrum pseudogrignonense RJ12*, *Pseudomonas* sp. *RJ15*, and *Bacillus subtilis RJ46*) significantly improved seed germination percentage, root length, shoot length, and dry weight of legumes *Vigna mungo* and *Pisum sativum* under drought stress conditions.

The triggering of plant defense mechanisms is typically more potent and efficient when stimulated by microbial consortia rather than a single microbial inoculant. For instance, Berendsen et al. (Berendsen et al., 2018) investigated the microbiota of Arabidopsis during infection with the phytopathogenic fungus *Hyaloperonospora arabidopsidis*. They identified three predominant bacterial genera—*Xanthomonas* spp., *Stenotrophomonas* spp., and *Microbacterium* spp. It was observed that these bacteria synergistically formed biofilms in vitro, and while individually they had minimal impact on the plant, together they induced systemic resistance against downy mildew and enhanced plant growth. Hence, the strains within this bacterial consortium demonstrate clear compatibility for controlling phytopathogens. The synergistic interaction of *Pseudomonas putida* CRN-09 and *Bacillus subtilis* CRN-16 was tested in *Vigna radiate* (mung bean), where the consortium developed systemic resistance to the phytopathogenic fungus *Macrophomina phaseolina*. This led to higher plant levels of peroxidase (PO), polyphenol oxidase (PPO), phenylalanine ammonia-lyase (PAL), β-1, 3 glucanase, and chitinase (Sharma et al., 2018). Thus, the integration of various biocontrol agents within the consortium is supposed to offer heightened protection and holds promise for effectively managing numerous plant diseases.

5.6 Conclusion

In the agriculture sector, the application of PGPR, sole or as consortia presents a promising avenue for sustainable agriculture, particularly in the face of escalating challenges posed by climate change. By harnessing the collective capabilities of multiple microbial strains, PGPR consortia offer a holistic approach to enhancing plant resilience and productivity under challenging environmental conditions. Through the secretion of beneficial metabolites, such as phytohormones, enzymes, and antimicrobial compounds, these consortia contribute to improved nutrient uptake, stress tolerance, and disease suppression in plants. Furthermore, their ability to modulate plant defense mechanisms and promote systemic resistance highlights their potential as eco-friendly alternatives to conventional agrochemicals. Moving forward, continued research efforts are essential to unraveling the complex interactions within PGPR consortia and optimizing their application in diverse agroecosystems. By integrating PGPR consortia into climate-smart agricultural practices, we can pave the way toward more sustainable and resilient food production systems capable of meeting the challenges of a changing climate while ensuring global food security.

References

Adams, R. M., Hurd, B. H., Lenhart, S., & Leary, N. (1998). Effects of global climate change on agriculture: An interpretative review. *Climate Research, 11*(1), 19–30.

Agnihotri, P., & Mitra, A. K. (2023). Understanding the impact of global climate change on abiotic stress in plants and the supportive role of PGPR. In *Abiotic stress in plants-adaptations to climate change*. IntechOpen.

Ansari, F. A., & Ahmad, I. (2019). Alleviating drought stress of crops through PGPR: Mechanism and application. In *Microbial interventions in agriculture and environment: Volume 2: Rhizosphere, microbiome and agro-ecology* (pp. 341–358). Springer.

Anwar, M. R., Liu, D. L., Macadam, I., & Kelly, G. (2013). Adapting agriculture to climate change: A review. *Theoretical and Applied Climatology, 113*, 225–245.

Baas, P., Bell, C., Mancini, L. M., Lee, M. N., Conant, R. T., & Wallenstein, M. D. (2016). Phosphorus mobilizing consortium Mammoth P™ enhances plant growth. *PeerJ, 4*, e2121.

Belimov, A. A., Hontzeas, N., Safronova, V. I., Demchinskaya, S. V., Piluzza, G., Bullitta, S., & Glick, B. R. (2005). Cadmium-tolerant plant growth-promoting bacteria associated with the roots of Indian mustard (Brassica juncea L. Czern.). *Soil Biology and Biochemistry, 37*(2), 241–250.

Berendsen, R. L., Vismans, G., Yu, K., Song, Y., de Jonge, R., Burgman, W. P., Burmølle, M., Herschend, J., Bakker, P. A., & Pieterse, C. M. (2018). Disease-induced assemblage of a plant-beneficial bacterial consortium. *The ISME Journal, 12*(6), 1496–1507.

Bhattacharyya, P. N., & Jha, D. K. (2012). Plant growth-promoting rhizobacteria (PGPR): Emergence in agriculture. *World Journal of Microbiology and Biotechnology, 28*, 1327–1350.

Daranas, N., Badosa, E., Francés, J., Montesinos, E., & Bonaterra, A. (2018). Enhancing water stress tolerance improves fitness in biological control strains of Lactobacillus plantarum in plant environments. *PLoS One, 13*(1), e0190931.

Fernando, W., Nakkeeran, S., Zhang, Y., & Savchuk, S. (2007). Biological control of Sclerotinia sclerotiorum (Lib.) de Bary by Pseudomonas and Bacillus species on canola petals. *Crop Protection, 26*(2), 100–107.

Fiodor, A., Singh, S., & Pranaw, K. (2021). The contrivance of plant growth promoting microbes to mitigate climate change impact in agriculture. *Microorganisms, 9*(9), 1841.

Gamalero, E., & Glick, B. R. (2015). Bacterial modulation of plant ethylene levels. *Plant Physiology, 169*(1), 13–22.

Goswami, D., Thakker, J. N., & Dhandhukia, P. C. (2016). Portraying mechanics of plant growth promoting rhizobacteria (PGPR): A review. *Cogent Food & Agriculture, 2*(1), 1127500.

Hamid, B., Zaman, M., Farooq, S., Fatima, S., Sayyed, R., Baba, Z., Sheikh, T., Reddy, M., El Enshasy, H., & Gafur, A. (2021). Bacterial plant biostimulants: A sustainable way towards improving growth, productivity, and health of crops. *Sustainability, 13*(2856), 1–24. https://doi.org/10.3390/su13052856

Hayat, R., Ahmed, I., & Sheirdil, R. A. (2012). An overview of plant growth promoting rhizobacteria (PGPR) for sustainable agriculture. In *Crop production for agricultural improvement* (pp. 557–579). Springer.

Heidari, M., & Golpayegani, A. (2012). Effects of water stress and inoculation with plant growth promoting rhizobacteria (PGPR) on antioxidant status and photosynthetic pigments in basil (Ocimum basilicum L.). *Journal of the Saudi Society of Agricultural Sciences, 11*(1), 57–61.

Iqbal, A., Iqbal, M., Alamzeb, M., Qiang, D., Xiangru, W., Huiping, G., Hengheng, Z., Nianchang, P., Fahad, S., & Anwar, S. (2022). Climate change and cotton production. In *Improvement of plant production in the era of climate change* (pp. 95–112). CRC Press.

Jha, C. K., & Saraf, M. (2012). Evaluation of multispecies plant-growth-promoting consortia for the growth promotion of Jatropha curcas L. *Journal of Plant Growth Regulation, 31*, 588–598.

Jha, C. K., & Saraf, M. (2015). Plant growth promoting rhizobacteria (PGPR): A review. *Journal of Agricultural Research and Development, 5*(2), 108–119.

Jha, Y., & Subramanian, R. (2014). PGPR regulate caspase-like activity, programmed cell death, and antioxidant enzyme activity in paddy under salinity. *Physiology and Molecular Biology of Plants, 20*, 201–207.

Joshi, N., Ruparelia, J. A., Saraf, M., & Jha, C. K. (2023a). Techniques to study plant–microbe interactions that lead to efficient sustainable agriculture. In *Plant microbiome for plant productivity and sustainable agriculture* (pp. 401–421). Springer.

Joshi, N., Saraf, M., Jha, C. K., Sudha, A., Alharbi, S. A., Alfarraj, S., & Datta, R. (2023b). Harnessing the efficacy of multifunctional rhizobacterial consortia for promoting the growth of Anethum graveolens L. *Frontiers in Sustainable Food Systems, 7*, 1126621.

Kaushal, M., Devi, S., Kumawat, K. C., & Kumar, A. (2023). Microbial consortium: A boon for a sustainable agriculture. In *Climate change and microbiome dynamics: Carbon cycle feedbacks* (pp. 15–31). Springer.

Kumar, A., Patel, J. S., Meena, V. S., & Srivastava, R. (2019). Recent advances of PGPR based approaches for stress tolerance in plants for sustainable agriculture. *Biocatalysis and Agricultural Biotechnology, 20*, 101271.

Kurabachew, H., & Wydra, K. (2013). Characterization of plant growth promoting rhizobacteria and their potential as bioprotectant against tomato bacterial wilt caused by Ralstonia solanacearum. *Biological Control, 67*(1), 75–83.

Liu, F., Bian, Z., Jia, Z., Zhao, Q., & Song, S. (2012). The GCR1 and GPA1 participate in promotion of Arabidopsis primary root elongation induced by N-acyl-homoserine lactones, the bacterial quorum-sensing signals. *Molecular Plant-Microbe Interactions, 25*(5), 677–683.

Lobell, D. B., Sibley, A., & Ivan Ortiz-Monasterio, J. (2012). Extreme heat effects on wheat senescence in India. *Nature Climate Change, 2*(3), 186–189.

Mahmood, I., Rizvi, R., Sumbul, A., & Ansari, R. A. (2019). Potential role of plant growth promoting Rhizobacteria in alleviation of biotic stress. In *Plant health under biotic stress: Volume 2: Microbial interactions* (pp. 177–188). Springer.

Mehmood, U., Inam-ul-Haq, M., Saeed, M., Altaf, A., Azam, F., & Hayat, S. (2018). A brief review on plant growth promoting rhizobacteria (PGPR): A key role in plant growth promotion. *Plant Protection, 2*(2), 77–82.

Mukherjee, S., & Bassler, B. L. (2019). Bacterial quorum sensing in complex and dynamically changing environments. *Nature Reviews Microbiology, 17*(6), 371–382.

Nadeem, S. M., Zahir, Z. A., Naveed, M., & Nawaz, S. (2013). Mitigation of salinity-induced negative impact on the growth and yield of wheat by plant growth-promoting rhizobacteria in naturally saline conditions. *Annals of Microbiology, 63*, 225–232.

Nelson, G. C., Rosegrant, M. W., Koo, J., Robertson, R., Sulser, T., Zhu, T., Ringler, C., Msangi, S., Palazzo, A., & Batka, M. (2009). *Climate change: Impact on agriculture and costs of adaptation* (Vol. 21). International Food Policy Research Institute.

Ng, W.-L., & Bassler, B. L. (2009). Bacterial quorum-sensing network architectures. *Annual Review of Genetics, 43*, 197–222.

Panwar, M., Tewari, R., & Nayyar, H. (2014). Microbial consortium of plant growth-promoting rhizobacteria improves the performance of plants growing in stressed soils: An overview. In *Phosphate solubilizing microorganisms: principles and application of microphos technology* (pp. 257–285). Springer.

Patel, D., & Goswami, D. (2020). Phosphorus solubilization and mobilization: Mechanisms, current developments, and future challenge. In *Advances in plant microbiome and sustainable agriculture: Functional annotation and future challenges* (pp. 1–20). Springer.

Pérez-Montaño, F., Jiménez-Guerrero, I., Sánchez-Matamoros, R. C., López-Baena, F. J., Ollero, F. J., Rodríguez-Carvajal, M. A., Bellogín, R. A., & Espuny, M. R. (2013). Rice and bean AHL-mimic quorum-sensing signals specifically interfere with the capacity to form biofilms by plant-associated bacteria. *Research in Microbiology, 164*(7), 749–760.

Príncipe, A., Fernandez, M., Torasso, M., Godino, A., & Fischer, S. (2018). Effectiveness of tailocins produced by Pseudomonas fluorescens SF4c in controlling the bacterial-spot disease in tomatoes caused by Xanthomonas vesicatoria. *Microbiological Research, 212*, 94–102.

Raklami, A., Bechtaoui, N., Tahiri, A.-I., Slimani, A., Bargaz, A., Meddich, A., & Oufdou, K. (2021). Co-inoculation with rhizobacteria and mycorrhizae can improve wheat/faba bean intercrop performance under field conditions. *Frontiers in Agronomy, 3*, 734923.

Ramadan, E. M., AbdelHafez, A. A., Hassan, E. A., & Saber, F. M. (2016). Plant growth promoting rhizobacteria and their potential for biocontrol of phytopathogens. *African Journal of Microbiology Research, 10*(15), 486–504.

Rani, N., Joshi, M., Sagar, A., & Sharma, H. R. (2022). Potential impacts of environmental pollution on the growth and metabolism of medicinal plants: An overview. In *Environmental pollution and medicinal plants* (pp. 1–16). CRC Press.

Redondo-Gómez, S., Mesa-Marín, J., Pérez-Romero, J. A., López-Jurado, J., García-López, J. V., Mariscal, V., Molina-Heredia, F. P., Pajuelo, E., Rodríguez-Llorente, I. D., & Flowers, T. J. (2021). Consortia of plant-growth-promoting rhizobacteria isolated from halophytes improve response of eight crops to soil salinization and climate change conditions. *Agronomy, 11*(8), 1609.

Rooney, W. M., Grinter, R. W., Correia, A., Parkhill, J., Walker, D. C., & Milner, J. J. (2020). Engineering bacteriocin-mediated resistance against the plant pathogen Pseudomonas syringae. *Plant Biotechnology Journal, 18*(5), 1296–1306.

Saikia, J., Sarma, R. K., Dhandia, R., Yadav, A., Bharali, R., Gupta, V. K., & Saikia, R. (2018). Alleviation of drought stress in pulse crops with ACC deaminase producing rhizobacteria isolated from acidic soil of Northeast India. *Scientific Reports, 8*(1), 3560.

Santoyo, G., Guzmán-Guzmán, P., Parra-Cota, F. I., de los Santos-Villalobos, S., del Carmen Orozco-Mosqueda, M., & Glick, B. R. (2021). Plant growth stimulation by microbial consortia. *Agronomy, 11*(2), 219.

Sarker, A., Ansary, M. W. R., Hossain, M. N., & Islam, T. (2021). Prospect and challenges for sustainable management of climate change-associated stresses to soil and plant health by beneficial rhizobacteria. *Stress, 1*(4), 200–222.

Schikora, A., Schenk, S. T., & Hartmann, A. (2016). Beneficial effects of bacteria-plant communication based on quorum sensing molecules of the N-acyl homoserine lactone group. *Plant Molecular Biology, 90*, 605–612.

Schröder, P., Beckers, B., Daniels, S., Gnädinger, F., Maestri, E., Marmiroli, N., Mench, M., Millán, R., Obermeier, M. M., & Oustriere, N. (2018). Intensify production, transform biomass to energy and novel goods and protect soils in Europe—A vision how to mobilize marginal lands. *Science of the Total Environment, 616*, 1101–1123.

Schulz-Bohm, K., Martín-Sánchez, L., & Garbeva, P. (2017). Microbial volatiles: Small molecules with an important role in intra-and inter-kingdom interactions. *Frontiers in Microbiology, 8*, 289291.

Shah, A., Nazari, M., Antar, M., Msimbira, L. A., Naamala, J., Lyu, D., Rabileh, M., Zajonc, J., & Smith, D. L. (2021). PGPR in agriculture: A sustainable approach to increasing climate change resilience. *Frontiers in Sustainable Food Systems, 5*, 667546.

Shahzad, S., Khan, M. Y., Zahir, Z. A., Asghar, H. N., & Chaudhry, U. K. (2017). Comparative effectiveness of different carriers to improve the efficacy of bacterial consortium for enhancing wheat production under salt affected field conditions. *Pakistan Journal of Botany, 49*(4), 1523–1530.

Sharma, S. B., Sayyed, R. Z., Trivedi, M. H., & Gobi, T. A. (2013). Phosphate solubilizing microbes: Sustainable approach for managing phosphorus deficiency in agricultural soils. *Springerplus, 2*, 1–14.

Sharma, C. K., Vishnoi, V. K., Dubey, R., & Maheshwari, D. (2018). A twin rhizospheric bacterial consortium induces systemic resistance to a phytopathogen Macrophomina phaseolina in mung bean. *Rhizosphere, 5*, 71–75.

Sharma, M., Thakur, R., Sharma, M., Sharma, A. K., & Sharma, A. K. (2020). Changing scenario of medicinal plants diversity in relation to climate change: A review. *Plant Archives, 20*(2), 4389–4400.

Shen, X., Hu, H., Peng, H., Wang, W., & Zhang, X. (2013). Comparative genomic analysis of four representative plant growth-promoting rhizobacteria in Pseudomonas. *BMC Genomics, 14*, 1–20.

Singh, O., Gupta, M., Mittal, V., Kiran, S., Nayyar, H., Gulati, A., & Tewari, R. (2014). Novel phosphate solubilizing bacteria 'Pantoea cypripedii PS1'along with Enterobacter aerogenes PS16 and Rhizobium ciceri enhance the growth of chickpea (Cicer arietinum L.). *Plant Growth Regulation, 73*, 79–89.

Singh, R., Kaur, S., Bhullar, S. S., Singh, H., & Sharma, L. K. (2024). Bacterial biostimulants for climate smart agriculture practices: Mode of action, effect on plant growth and roadmap for commercial products. *Journal of Sustainable Agriculture and Environment, 3*(1), e12085.

Sood, G., Kaushal, R., & Gupta, S. (2019). Role of PGPR in sustainable agriculture under changing scenario of climate change. In *Agricultural impacts of climate change* (Vol. 1, pp. 109–118). CRC Press.

Soora, N. K., Aggarwal, P. K., Saxena, R., Rani, S., Jain, S., & Chauhan, N. (2013). An assessment of regional vulnerability of rice to climate change in India. *Climatic Change, 118*, 683–699.

Srinivasan, K., & Mathivanan, N. (2011). Plant growth promoting microbial consortia mediated classical biocontrol of sunflower necrosis virus disease. *Journal of Biopesticides, 4*(1), 65.

Vocciante, M., Grifoni, M., Fusini, D., Petruzzelli, G., & Franchi, E. (2022). The role of plant growth-promoting rhizobacteria (PGPR) in mitigating plant's environmental stresses. *Applied Sciences, 12*(3), 1231.

Wang, C.-J., Yang, W., Wang, C., Gu, C., Niu, D.-D., Liu, H.-X., Wang, Y.-P., & Guo, J.-H. (2012). Induction of drought tolerance in cucumber plants by a consortium of three plant growth-promoting rhizobacterium strains. *PLoS One, 7*(12), e52565.

Yohannes, H. (2016). A review on relationship between climate change and agriculture. *Journal of Earth Science & Climatic Change, 7*(2), 335.

Zimina, M., Babich, O., Prosekov, A., Sukhikh, S., Ivanova, S., Shevchenko, M., & Noskova, S. (2020). Overview of global trends in classification, methods of preparation and application of bacteriocins. *Antibiotics, 9*(9), 553.

The Role of Plant–Microbe Partnerships in Ecosystem Functionality

6

Shri Kant Tripathi, Somnath Chakrobortty, Ayushi Tyagi, P. K. Toushif, Salam Suresh Singh, Keshav Kumar Upadhyay, Ngangbam Somen Singh, and Balu Ananda Chopade

Abstract

Plants and their associated microbiota form a plant holobiont, and interactions among these organisms determine the functioning of the holobiont and the performance of the plant. Furthermore, the degree of interaction between these organisms and various species strongly influences ecosystem functioning. Therefore, any switch in these interactions may profoundly alter the functionality of the ecosystem and its ability to provide goods and services to society. These interactions are highly complex in nature and are less known because of the methodological difficulties associated with understanding the temporal dynamics of these interactions occurring at plant–soil interface "rhizosphere." As per the available information, plant microbes and their interactions prevent common diseases, fix atmospheric nitrogen, increase the resilience of plants to stress tolerance, enhance physical vigor and robustness, boost the efficiency of nutrient consumption, and provide tolerance to abiotic challenges. Research is progressing to unravel the complexity of the microorganisms involved and their interactions to develop sustainable crop production practices for different regions. However, further studies are needed to understand microbiome interactions and holobiont functioning to manipulate them more effectively to develop sustainable crop production models and overcome environmental consequences. This chapter primarily focuses on plant–microbe interactions and their role in the functioning and sustainability of ecosystems.

S. K. Tripathi (✉) · S. Chakrobortty · A. Tyagi · P. K. Toushif · S. S. Singh · K. K. Upadhyay
N. S. Singh
Department of Forestry, School of Earth Science and Natural Resources Management, Mizoram University, Aizawl, Mizoram, India

B. A. Chopade
Department of Biotechnology, AKS University, Satna, Madhya Pradesh, India

M. Saraf et al. (eds.), *Climate Change and Soil Microorganisms for Environmental Sustainability*, Microorganisms for Sustainability 28,
https://doi.org/10.1007/978-981-96-3425-5_6

Keywords

Plant–microbe partnership · Nitrogen fixation · Ecosystem functionality · Sustainability

6.1 Introduction

The plant holobiont encompasses a plant and all its associated microorganisms, along with their interactions that determine the fitness of plants through the entire set of genes and the hologenome (genome of the host plant and all epi-and endophytes). These interactions among plants and their associated microbes are highly complex and important for the proper functioning of any ecosystem (Mitter et al., 2016). Plants create specific environments "rhizosphere" for their associated microbes by releasing an array of organic compounds from the roots, amounting to about 10–40% of the total photosynthate fixed to fuel the associated microbes in the rhizosphere, and in turn, these microbes release nutrients by degrading organic matter and growth-promoting hormones to facilitate the plant growth (Newman, 1985; Rokhbakhsh-Zamin et al., 2011; Sachdev et al., 2009; Huddedar et al., 2002; Jones et al., 2009; McNear Jr., 2013). However, the relationship between plants and microbes is highly complex and has become a focus of current research because of the inability to understand these interactions under different climatic and anthropogenic gradients, and to develop balanced hologenetic combinations to sustain future ecosystem production under changing climate scenarios (Jefferson, 1994). Microbes associated with plants promote their growth and development through various processes. For example, adding nitrogen to the soil through the fixation of atmospheric nitrogen gas increases the availability of phosphorus and zinc, produces plant growth regulators such as auxins, cytokinins, and gibberellins, and suppresses ethylene synthesis through activities such as 1-aminocyclopropane-1-carboxylate (ACC) deaminase (Hakim et al., 2021; Olanrewaju & Babalola, 2021). Furthermore, these microbes suppress the growth and activity of plant pathogens (Maindad et al., 2014). These microbial activities can avoid competition for space and nutrients, antibacterial action, secretion of hydrolytic enzymes, and suppression of pathogen-producing enzymes or toxins that induce plant defense mechanisms.

Therefore, plant–microbe partnerships have the potential to enhance biomass production for food, feed, and industrial processes under different ecological regimes. Furthermore, these partnerships can be used for the remediation of contaminated soil and groundwater. Certain plant-associated microbes with metal-sequestration abilities can reduce metal toxicity and help move metals to aboveground plant parts through the phytoremediation of metal-contaminated soils and groundwater. In the process of phytoremediation of certain organic contaminants, plants depend on the presence of specific microorganisms in their vicinity that possess the necessary degradation pathways. These microorganisms play vital roles in the efficient breakdown of organic contaminants. This breakdown reduces the phytotoxicity of these contaminants and minimizes the release of volatile

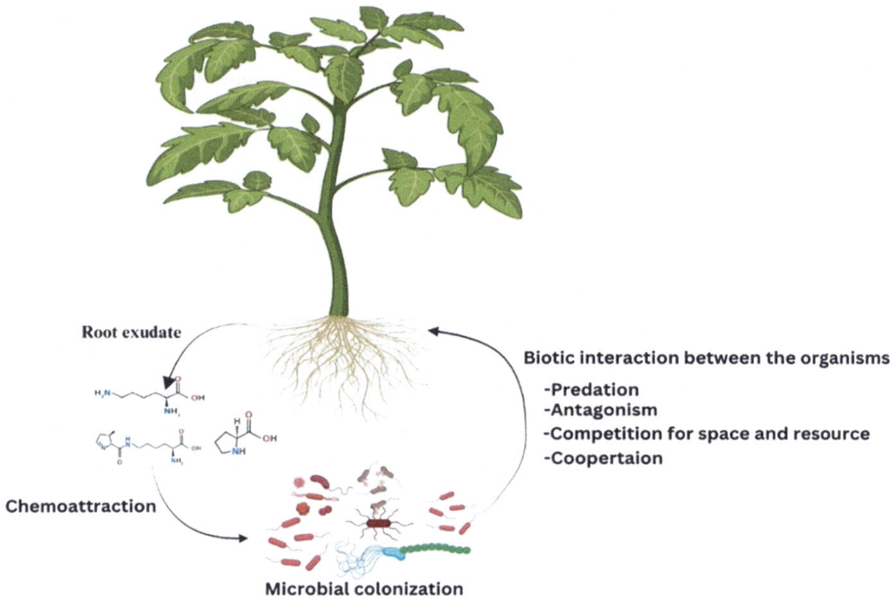

Fig. 6.1 Diagrammatic representation of plant–microbe interaction

contaminants or their degradation intermediates into the atmosphere. Consequently, there is growing interest in utilizing systems biology approaches to comprehensively understand the synergistic interactions between plants and their beneficial microbiota (Fig. 6.1).

The rhizosphere, which is one of the most complex microhabitats, consists of an integrated network of plant roots, soil, and a diverse consortium of bacteria, fungi, eukaryotes, and archaea (Hauchhum & Tripathi, 2019a; Hakim et al., 2021). The rhizosphere environment directly influences the growth and yield of crops, and vice versa, and thus has wide applicability in sustainable crop production practices in the future (Hauchhum & Tripathi, 2019b). This is because most agricultural soils are exhausted due to the heavy use of chemical fertilizers and pesticides to gain maximum yield from the agroecosystem (Kumar & Dubey, 2020), and they need to be nurtured through a balanced holobiont for their future stability. Additionally, most of the crop varieties used have low nutrient uptake efficiencies (ca. 20–30% of applied fertilizers), leading to the loss of the remaining (ca. 70–80%) nutrients from agroecosystems (Fageria, 2014), causing severe consequences in other ecosystems (Tripathi, 2009). Furthermore, the plant varieties developed in the 1960s during the "green revolution" are becoming nonresponsive to fertilizers with stagnant yield (Kumar & Dubey, 2020). Therefore, developing balanced rhizospheres at scales of ecosystems where a variety of organisms interact with each other and their environment to form a sphere of life, including plants and other life forms comprising both biotic and abiotic components, is a prerequisite to sustain ecosystem functioning. Ecosystem functioning involves a number of ecological processes, such as the flow of energy and cycling of nutrients within the ecosystem through photosynthesis and organic matter decomposition (Tripathi & Singh, 1994). These properties are sensitive to changes in biotic and abiotic variables within the ecosystem (Tripathi & Singh, 1992; Wapongnungsang et al., 2017). Therefore, developing a healthy ecosystem through balanced holobionts would be important in developing a healthy human society and coping with the increasing human population and biotic and abiotic stresses due to climate change, land scarcity, and food insecurity in the future.

In recent years, sustainable agricultural production (SAP) has become one of the world's most debated topics in food security. This approach maximizes productivity by increasing the efficiency of resource use and minimizing adverse effects on the environment by restoring the quality of soil and water (Walters et al., 2008). This can be achieved through 4R strategies, that is, the use of the right source of nutrients at the right rate, right time, and right place. It can be further expanded to make it more efficient by including two components, for example, the plant using the nutrient and their rhizosphere to maximize the interaction. This can be achieved through the selection of low-fertilizer-responsive crop germplasm and inoculation of appropriate plant growth-promoting bacteria through rhizosphere engineering (Hakim et al., 2021).

This approach represents a significant and evolving field of research and is briefly outlined in this chapter. The topics covered in this chapter include plant–microbe interactions in the ecosystem, their effects on productivity and the control of diseases, the factors that influence plant–microbe interactions and ecosystem functionality, as well as recent developments and upcoming difficulties in achieving the goals of ecosystem sustainability.

6.2 Plant–Microbe Interactions, Disease Management, and Ecosystem Productivity

The interaction between plants and microbes is a complex and ongoing process that dates back millions of years, coinciding with the colonization of plants on Earth. Over this extensive period, plants have established associations with various microorganisms, leading to the creation of a distinct ecological unit known as the "holobiont" (Jefferson, 1994). Within ecosystems, plants frequently encounter both beneficial and pathogenic microorganisms, predominantly bacteria and fungi. Throughout their life cycle, microorganisms typically engage with host plants in one of three ways: pathogens, parasites, and mutualists (Mitter et al., 2016). Understanding these three interactions is crucial for achieving ecosystem sustainability through microbial management. These relationships involve various processes that positively or negatively affect plant life. For example, rhizobia and mycorrhizal cling in plant roots add atmospheric nitrogen and mineral nutrients to plants (Hakim et al., 2021). They also directly stimulate plant growth through phytohormones, exhibit antagonistic effects against pathogenic microorganisms, and help plants cope with stressors (Walters et al., 2008; Maindad et al., 2014; Sachdev, 2010, Chopade et al., 2008). In contrast, when microorganisms invade plants, they can act as saprophytes and cause necrotrophic effects in colonized plants. This implies that negative interactions can be hazardous to plants. Understanding the interactions between microorganisms and plants is therefore essential to understand the benefits and drawbacks of bacteria on plants, as well as their function in ecosystem dynamics.

6.2.1 Plant–Microbe Interaction on Ecosystem Productivity

Productive agriculture is vital for ensuring the survival of the global population and securing food for the growing population, which is expected to reach approximately 9 billion by 2050 (Kumar & Dubey, 2020). Conventional agriculture often relies heavily on chemical interventions to boost production, without compromising environmental and health considerations (Tal, 2018). The excessive use of synthetic chemical fertilizers and pesticides and the low nutrient use efficiencies (20–30%) of agricultural crops have led to many environmental consequences (Fageria, 2014; Hakim et al., 2021). For example, erosion of biodiversity, emission of greenhouse gases, water pollution, soil contamination, deterioration of soil health and biology, and health risks for humans. Therefore, future agricultural sustainability requires the management of food security to meet the increasing demand, mitigate climate change effects, and improve soil fertility. A sustainable approach to address these challenges involves harnessing the power of plant growth-promoting microorganisms (PGPMs), particularly bacteria and fungi. PGPM is a subset of efficient microbes used to boost crop yield through various mechanisms (e.g., nitrogen fixation, siderophore production, ACC deaminase activity, phytohormone secretion, phosphorus and mineral solubilization, and improvement of soil characteristics) (Malgioglio et al., 2021). In addition, they facilitate the release of trace elements and secrete antioxidants, exopolysaccharides, and bioactive compounds (i.e., vitamins, hormones, and enzymes), enhance the water and nutrient absorption efficiency of crops, and stimulate crop growth and productivity. However, the intricate interactions that occur among crops, the environment, and soil microorganisms, particularly in the rhizosphere (the narrow zone around the roots), pose challenges to the effective use of PGPMs to enhance the functionality of ecosystems and their sustainability (Naik et al., 2019).

Over the past few years, extensive human activities, such as unsustainable intensive agricultural practices, overgrazing, and deforestation, have negatively affected soil fertility. This has resulted in soil degradation and deterioration of agricultural production. To address this issue, researchers have turned to microbial communities inhabiting the rhizosphere and phyllosphere (leaf surface) of plants as a potential solution for achieving sustainable soil restoration and agricultural production. These microbial communities offer a safe and eco-friendly approach through their inherent abilities to enhance soil quality by amplifying the availability of nutrients to host plants, fixing atmospheric nitrogen, mobilizing essential nutrients, such as phosphorus, potassium, and iron, and improving soil structure and equilibrium. In particular, Indian soils often have low organic phosphorus content, leading to a low bioavailability of this crucial nutrient. Phosphate-solubilizing microbes (PSM) play a critical role in solubilizing insoluble phosphate (PO_4^{3-}) and making them accessible to plants. While certain bacteria and fungi, such as *Bacillus*, *Pseudomonas*, *Aspergillus*, and *Penicillium*, are known as good phosphate solubilizes, their performance can sometimes be inconsistent under real-world conditions. Genetic modifications and co-inoculation techniques offer potential avenues for improving their effectiveness under field conditions. Plant growth-promoting bacteria from the rhizosphere and mycorrhizal fungi are subgroups of PGPMs that enhance soil productivity and plant growth. They can be applied alongside manure, chemical fertilizers, organic fertilizers, and molasses to provide a sustainable crop production alternative in the form of microbial consortia comprising the replacement of chemical fertilizers and pesticides and reducing the environmental impact without compromising crop yields.

The interplay between various species and kingdoms of organisms in the rhizospheric/phyllospheric interface is essential for the proper functioning, health, stability, and sustainability of ecosystems, including agroecosystems (De-la-Peña, & Loyola-Vargas, 2014). Plant–microbe interactions under high plant richness and diversity play a significant role in shaping the soil microbial communities in the rhizosphere, thus influencing ecosystem dynamics and stability (Schlatter et al., 2015). Abiotic factors, such as soil type, pH, temperature, geography, and climate conditions, exert a considerable influence on these microbial communities, which, in turn, affect plant growth and development. In brief, mutualistic relationships among plants, microbiological populations, and environmental factors in the soil are important for maintaining ecosystem functionality by enhancing soil nutrients and assisting plants to overcome environmental stress (Santoyo et al., 2017). This has led to the development of a sustainable ecosystem in the future under a changing climate.

6.2.2 Plant–Microbe Interactions and Disease Management

Various plant diseases occur because of harmful interactions with microbes, including viroids, viruses, bacteria, and fungi, which lead to an approximately 13% loss in crop productivity on a global scale. Approximately 11,000 microbes have been identified, including eight genera of bacteria, two mollicutes, 120 fungi, and 30 different types of viruses responsible for causing different plant diseases. Plant pathogenic bacteria, such as *Pseudomonas*, *Erwinia*, and *Xanthomonas*, have been extensively studied to understand the specificity of plant–microbe interactions. The effectiveness of biocontrol agents in managing plant diseases has been demonstrated and is commercially available (Whipps, 2004; Elmer & Reglinski, 2006; Whipps & Gerhardson, 2007). To combat plant pathogens, these biocontrol agents offer various mechanisms including competition for nutrients and space at the infection site, antibiosis, parasitism, production of enzymes that degrade cell walls, induction of resistance in the plant, and manipulation of bacterial signaling molecules. Compant et al. (2005) provided a comprehensive review of the mechanisms involved in plant disease biocontrol by plant growth-promoting bacteria. This study focused on two specific mechanisms: Induced resistance and manipulation of signaling molecules.

6.2.2.1 Induced Resistance

Bacterial interactions can develop two distinct plant defense mechanisms (i.e., systemic acquired resistance [SAR] and induced systemic resistance [ISR]), which play a crucial role in protecting plants against further infections. The first initiates a localized increase in phytohormone accumulation and the production of phloem mobile signals, whereas the second involves non-*Venturia. inaequalis* pathogenic free-living rhizosphere bacteria and endophytes. An important difference between SAR and ISR is the involvement of salicylic acid. ISR is activated via a salicylic acid-independent pathway. Plant disease control involves the use of *Pseudomonas* strains with biocontrol activity to induce resistance against the pathogenic fungus *Venturia inaequalis* in apple trees (Malus domestica). This was developed by a study of *Pseudomonas* movement on apple leaves and its antagonistic interactions (Gau et al., 2004). Confocal laser scanning microscopy revealed that *Pseudomonas fluorescens* Bk3 localized near stomatal openings and penetrated the cuticle by secreting cutinases and obtaining nutrients from the fluid within the apoplastic space. Isolated leaf cuticles stimulated the production of extracellular bacterial proteins (Singh et al., 2004). The interactions between *V. inaequalis* and apple trees were analyzed by constructing cDNA libraries using suppression subtractive hybridization for resistant and susceptible cultivars (Degenhardt et al., 2005). Many transcripts related to plant defense, such as those encoding β-1,3-glucanase, cysteine protease inhibitors, and metallothionein, were more abundant in the resistant cultivar. Conversely, susceptible cultivars had higher levels of RuBisCo transcripts. Apple trees treated with the biocontrol strains *V. inaequalis* and *P. fluorescens* Bk3 expressed similar proteins (Kürkcüoglu et al., 2007). The presence of *P. fluorescens* Bk3 has been shown to enhance defense mechanisms in apple trees, making it an effective biocontrol agent. The authors suggested that low levels of RuBisCo and high levels of metallothionein resembled the characteristics of "old leaves," making them less attractive to *V. inaequalis* (Degenhardt et al., 2005). The study did not investigate whether *P. fluorescens* Bk3 induced SAR- or ISR-type plant resistance.

6.2.2.2 Manipulation of Bacterial Signaling Molecules

Certain bacteria rely on signaling molecules for the development of pathogenicity in the same environment, and other microbes may degrade these signaling molecules. In a study by Newman et al. (2008), recombinant biocontrol strains producing carAB genes, which are essential for the degradation of a fatty acid signaling molecule, were found to reduce the virulence of *Xanthomonas* sp. and *Xylella fastidiosa*. Plants also play a role in this intricate interplay. The secondary metabolites produced by plants have been shown to mimic or inhibit quorum-sensing (QS) molecules. For example, *Medicago truncatula* produces numerous compounds that either stimulate or inhibit QS, as demonstrated by Gao et al. (2003). Plants have been engineered to produce molecules that mimic or block QS signals as well as enzymes that degrade QS molecules or interfere with QS strategies. Scott et al. (2006) genetically modified tobacco plants to synthesize acyl-homoserine lactones (AHL) in chloroplasts. These AHL molecules are transported and secreted in the phyllosphere (leaf surface) and the rhizosphere (root environment). Pretreatment of potato slices with *Bacillus thuringiensis* results in reduced maceration caused by the virulence of *Erwinia carotovora* (Dong et al., 2004; Cloutier & Bauduin, 1995). The reduced pathogenesis was attributed to the disruption of *the QS system by E. carotovora owing* to the degradation of QS molecules by AHL lactonase. In contrast, pretreatment with *B. thuringiensis* lacking the ability to produce AHL lactonase did not reduce maceration. Manipulating plants to express AHL lactonase has shown promise in protecting plants against pathogens. More recently, a gene called quorum-sensing signal degradation (qsdA) from *Rhodococcus erythropolis*, which encodes a new type of AHL lactonase, has been characterized. This gene was found to confer quorum-quenching capabilities to *Pseudomonas fluorescens*, leading to enhanced protection of potato tubers against the soft rot pathogen, *Pectobacterium carotovorum* (Uroz et al., 2008).

6.2.3 Plant–Microbe Partnership and Pest Management

Plant–microbe partnerships hold great potential for pest management in both natural and managed ecosystems (Usta, 2013). These partnerships involve beneficial microorganisms that can either directly suppress pests or enhance the inherent defense mechanisms of plants to cope with pest attacks. One promising avenue for pest management involves plant endophytes, which inhabit the same ecological niche within plants as the pathogens (Ruiu, 2018). These endophytes exhibit antimicrobial activities that suppress the growth of pathogens within plant tissues and increase the resistance of plants to pest attacks. Previous studies have demonstrated the effectiveness of endophytes in controlling plant diseases and pests through inoculation (Ryan et al. 2008). Additionally, endophytes can induce plant disease and pest resistance through a mechanism known as induced systemic resistance (ISR), which upregulates the salicylic acid, jasmonate, and ethylene pathways in plants (Yi et al., 2013; White et al., 2019). For example, *Colletotrichum tropicale* was found to regulate gene expression in *Theobroma cacao*, resulting in enhanced plant immunity (Mejía et al., 2014). Other endophytes, including *Burkholderia*, *Rahnella*, *Pseudomonas*, and *Curtobacterium*, found in Salicaceae plants, exhibit biocontrol activity against various plant pathogens such as *Rhizoctonia solani* AG-8, *Fusarium culmorum*, *Gaeumannomyces graminis*, and *Pythium ultimum* (Kandel et al., 2017). Endophytic *Phomopsis cassia*, isolated from *Cassia spectabilis*, produces substances similar to 3,11,12-trihydroxycadalene and cadinene sesquiterpenes, exhibiting antifungal activity against *Cladosporium cladosporioides* and *Cladosporium sphaerospermum* (Silva et al., 2006). Certain root-inhabiting microorganisms, including *Trichoderma*, mycorrhizal species, and *Pseudomonas*, can trigger the immune response of plants against various plant pathogens (Pieterse et al., 2014).

Fungal strains, such as *Acremonium implicatum*, have been shown to negatively impact root galls caused by *Meloidogyne incognita* (Lin et al., 2013).

Certain pest management strategies using plant–microbe partnerships are as follows:

(i) Endophytic microorganisms that inhabit plant tissues without harming the host plant produce several compounds that deter pests. For example, certain fungal endophytes can synthesize alkaloids that discourage herbivory.
(ii) Symbiotic relationships between mycorrhizal fungi and plant roots enhance plant health by increasing nutrient uptake and developing resistance to pests.
(iii) Nitrogen-fixing bacteria such as rhizobia enrich plant growth by converting atmospheric nitrogen into a reactive form that can be used by plants. Plants with a high nitrogen content are less appealing to certain herbivorous pests.
(iv) Beneficial microbes trigger the systemic defense response of plants through ISR, making them more resistant to pests and pathogens.
(v) Microbe-based biofertilizers, such as compost tea and microbial inoculants, enhance soil health and plant vigor and exhibit greater resilience against pest attacks.
(vi) Microbial pathogens, such as *Bacillus thuringiensis* (Bt), serve as insecticides to plant leaves by producing toxins that target specific insect pests without affecting beneficial organisms.
(vii) Pest-repelling microbial associations in crops or cover crops can deter pests from the primary crops. For example, marigolds release root exudates that inhibit nematodes.
(viii) Beneficial microbes protect plants by suppressing pathogenic microorganisms that attract pests. For example, mycorrhizal fungi compete with pathogenic fungi for resources.
(ix) Targeted inoculation of crops with specific beneficial microorganisms tailored to pest pressure and crop type can also be an effective strategy.
(x) Combining different beneficial microorganisms, such as a blend of mycorrhizal fungi, rhizobia, and plant growth-promoting bacteria, can provide comprehensive benefits to plants and boost their overall health and resilience against pests.
(xi) Microbial-based biostimulants enhance plant growth and development, resulting in more robust and pest-resistant plants.
(xii) The entomopathogenic potential of microbes has been exploited as a biocontrol agent against pests. For instance, the inoculation of *Azospirillum* in *Oryza sativa* induces the production of hydroxycinnamic acids (HCAs), such as N-feruloyl putrescine and N-p-coumaroyl putrescine, which play crucial roles in plant defense against *Burkholderia glumae* (Valette et al., 2020). Numerous studies have reported the efficacy of commercially available microbial biopesticides.
(xiii) Microbial herbicides are preparations based on phytopathogenic microbes that can be used for biological control of weeds. Prolonged use of chemical herbicides has resulted in the emergence of herbicide-resistant weeds.

Targeting herbicide-resistant weeds can help reduce environmental herbicide usage (Abbas et al., 2018). Microbial herbicides consist of fungal and bacterial pathogens that grow on or within a target weed, causing their death or suppressing their growth (Kremer, 2019). For example, the bacterial pathogen, *Xanthomonas campestris* pv.*poae*, known to cause lethal wilting in *Poa annua*, colonizes the vascular cylinder and produces a significant amount of xanthan polysaccharide, leading to xylem occlusion and death of the host weed (Imaizumi et al., 1999). However, ensuring the survival of these herbicidal microbes and considering their growth phase can affect the effectiveness of the formulations. Phytotoxic metabolites produced by microbes can also be used for weed control because they are highly effective against target weeds and do not raise concerns about microbial survival. For example, *Pseudomonas syringae* pv. *tagetis* (Pst), a bacterial phytopathogen, causes apical chlorosis in ragweed (*Ambrosia* spp.), *Canada thistle* (*Cirsium arvense*), and common cocklebur (*Xanthium strumarium*) by producing the phytotoxic metabolite tagetitoxin, ultimately leading to death of the host plant (Johnson et al., 1996). Therefore, the extraction and application of phytotoxic metabolites hold promise for effective weed control.

A deep understanding of the specific pests and plant–microbe interactions within a given ecosystem is important for effective pest management. Furthermore, selecting the most suitable microbes for the target crop and pest species is critical. Integrated pest management (IPM) strategies that combine various approaches, including plant–microbe partnerships, can contribute to sustainable and environmentally friendly pest management practices (Table 6.1).

6.3 Plant–Microbe Interactions for Restoration Ecology

Ecosystem degradation and habitat loss are recognized as critical issues around the globe, which keep pressing threats to global biodiversity, ecosystem services, and human well-being. In response, restoration ecology plays a pivotal role in coping with changing environments and in improving biodiversity and ecosystem services (Hobbs & Cramer, 2008; Perring et al., 2015). Ecosystem restoration can be achieved through both active and passive interventions. However, active intervention is applied in severely degraded ecosystems, and this technique requires intensive labor and is expensive. However, Passive intervention is eco-friendly, cost-effective, and takes a longer time to bring back the degraded ecosystem in service. Among passive interventions, the intricate relationship of plant microbes plays a critical role in the restoration of ecosystems by enhancing plant growth and soil quality, and promoting biodiversity (Alma et al. 2018).

Microbes associated with plants, such as rhizobacteria and fungi, play important roles in solubilizing inorganic nutrients and converting them into plants to enhance plant growth. Some bacteria, such as *Rhizobium*, have symbiotic relationships with plants (leguminous), absorb atmospheric nitrogen, convert it into the available form

Table 6.1 Table showing plant–microbe interaction for sustainable agriculture

S.No.	Plant	Microbes associated with plant	Effect	References
1.	Maize	*Bacillus amyloliquefaciens* SQR9	Root colonization	Bhatt et al. (2022)
2.	Wheat	*Azospirillum brasilense*	Heat Stress Tolerance	Bhatt et al. (2022)
3.	Wheat	*Acinetobacter calcoaceticus HIRFA32*	Inhibit growth of *F. oxysporum (Biocontrol)*	Maindad et al. (2014)
4.	Grapevine	*Trichoderma harzianum*	Biosuppression of *Botrytis cinerea*	Elmer and Reglinski (2006)
5.	Mung bean	*Rhizobium* and *Pseudomonas*	Increased biomass and yield	Hakim et al., 2021
6.	Soyabean	*Bradyrhizobium* and *Streptomyces griseoflavus*	Enhanced nodulation and N_2 fixation	Hakim et al. (2021)
7.	Wheat	*Methylobacterium phyllosphaerae* strain IARI-HHS2-6	Cold stress tolerance	Ho et al. (2017)
8.	Rice	*Pseudomonas pseudoalcaligenes*	Salinity stress tolerance	Ho et al. (2017)
9.	Asparagus bean	*Photobacterium halotolerans* MELD1	Photoprotection against Hg	Ho et al. (2017)
10.	Wheat	*Acinetobacter* spp.	IAA production	Huddedar et al. (2002)
11.	Mustard plant	*Tetrathiobacter* and *Bacillus* strains	Improved phosphorus content and plant growth	Hakim et al. (2021)
12.	Wheat	*Pseudomonas* spp., *Paraburkholderia* spp., and *Pantoea* spp.	P solubilization, N_2 fixation, IAA production	Compant et al. (2019)
13.	Messina creeper	*Fusarium oxysporum* and *Fusarium fujikuroi*	IAA production	Bhatt et al. (2022)
14.	Wheat	*Klebsiella pneumoniae*	IAA Production and Growth Enhancement	Sachdev et al. (2009)
15.	*Sugarcane*	*Novosphingobium sediminicola* and *Ochrobactrum intermedium*	Nitrogen fixation	Bhatt et al. (2022)

for the plant, and support plant growth. For instance, *Rhizobium* inoculation in the native leguminous plants of the Prairies helps to increase the N content in the soil and helps other plants (Beyhaut et al., 2014). Similarly, inoculation of phosphorus-solubilizing bacteria (PSB) in degraded forests helps to convert the insoluble form of phosphorus (inorganic phosphorus) into soluble phosphate. The PSB-inoculated plant produces organic acids, such as citric acid, gluconic acid, and lactic acid, which help to chelate the cations bound to phosphate, resulting in the formation of phosphate ions in the rhizospheric soil, thus helping the availability of phosphorus to the surrounding native plants (Kumar et al., 2023).

Thus, in restoration ecology, the idea is to establish ecosystems that work in a restoration-oriented manner, are healthy and diverse, and can deliver ecosystem services. Plant–microbe associations play vital roles in the restoration process because they affect nutrient cycling, soil formation, and plant health. Knowledge of such interactions can be utilized by practitioners involved in ecosystem restoration to improve the efficiency of microbial-driven restoration approaches.

Case Studies of Microbial Contributions to Ecosystem Restoration

(i) *Application of Mycorrhizal Fungi in Reforestation*

Mycorrhizae are a type of soil fungi that usually develop on or in the plant roots, enhancing their ability to absorb nutrients and minerals. Using mycorrhizal fungi-inoculated tree seedlings in reforestation programs has been documented to increase the chances of speedy recovery of degraded lands. For instance, research in degraded tropical forests revealed that mycorrhizal inoculation enhanced tree standing volume and biomass density, as well as the number of tree species per hectare, compared to the non-inoculated control (Holste & Kobe, 2017).

(ii) *Nitrogen-Fixing Bacteria in Grassland Ecosystem Restoration*

Some efficient bacteria, such as *Rhizobium* and *Azospirillum*, have great importance in remediating nitrogen-deficient areas, such as grasslands. Of these rhizosphere bacteria, those that live in mutualistic relationships with grasses and other herbs contribute to the supply of nitrogen, which is required for the growth and overall productivity of the ecosystem (Bauer et al., 2012). The use of nitrogen-fixing bacteria in a degraded grassland restoration project was demonstrated to quickly increase vegetation cover and enhance soil nitrogen status, making restoration easier.

(iii) *Endophytic Bacteria in Wetland Restoration*

Wetland habitats support many plants and animals, yet they can be severely affected by activities such as drainage or pollution. Although some endophytic bacteria dwell intracellularly within plant tissues, these microorganisms can also improve wetland plant stress tolerance and facilitate their successful colonization of disturbed site conditions (Ashraf et al., 2018). During wetland restoration, the use of endophytic bacteria as a biofertilizer to enhance the growth of native wetland plants was effective in dense saline soils, thus helping to restore wetland vegetation and ecosystem functions.

(iv) *Fungal Consortia in Degraded Soils*

Soils degraded by toxic metals or industrial pollutants represent difficult cases for restoration of the ecosystem (Singh et al., 2024a). Assembling metal-tolerant and pollutant-degrading fungal communities will assist in restoring contaminated sites and the subsequent planting of native plants. In a restoration project within an ex-mining area, disturbed soils were effectively bioaugmented with a mycorrhizal fungal community and native plant propagules, which subsequently supported the growth of vegetation and degradation of soil pollutants (Hassan et al., 2020; Singh et al., 2024b; Huidrom et al., 2024).

The intricate plant–microbe interaction plays a crucial role to bring back degraded ecosystems in service, facilitating the establishment of native plants, and enhancing ecosystem resilience. This chapter illustrates how plant–microbe partnership helps restore the degraded ecosystem by harnessing the role of microbes. So, a deep understanding of the intricate relationship between microbes and plant is very important to restore the degraded ecosystem.

6.4 Plant-Microbial Interactions and Ecosystem Functionality in a Changing Environment

Abiotic factors (i.e. temperature, precipitation, and edaphic variables) potentially affect the growth and productivity of the species within the ecosystem, and subsequently, the functioning of the ecosystem through altered plant–microbe interactions and plant physiological processes. Other soil physicochemical characteristics (i.e. temperature, nutrient availability, organic matter, pH, and soil texture) driven by temperature and precipitation patterns directly influence the composition and activities of microbes in the soil, and in turn, affect the growth and activity of plant roots. Furthermore, the changing pattern of roots strongly affects the amount and composition of root exudates, and thus, the rhizosphere biotic communities. Root-microbe interactions in the rhizosphere are highly sensitive to temperature and precipitation patterns, along with their associated variables in the soil; hence, future climate change may have profound effects on rhizosphere interactions and ecosystem functioning. The roles of abiotic, biotic, and geographical factors are discussed in the following sections.

6.4.1 Abiotic Factors

6.4.1.1 Temperature

The increasing frequency of extreme heat and cold events due to climate change has led to noticeable changes in crop yields worldwide. Changes in crop yield have been advocated as a result of altered plant–microbe interactions in the rhizosphere because of their high sensitivity to biotic variables. Hydroponically grown Japanese cucumbers in a controlled growth chamber at various temperatures have been reported to show high organic acid content at higher temperatures; hence, with an increase in temperature variation, plant–microbe interactions are likely to be altered, which will further affect ecosystem functions (MAFF, 2020). By growing strawberry plants at two temperature ranges (5–10 °C and 20–30 °C), it has been found that plants exposed to lower temperatures have higher amino acid concentrations in the rhizosphere, which significantly affects *Rhizoctonia fragariae* pathogenicity in the plant (Husain & McKeen, 1963). In addition, increasing temperatures are believed to predominantly influence plant physiological processes (photosynthesis, transpiration, respiration, and yield). High temperatures cause declining yields in many crops because of altered physiological processes and have been a serious concern for decreasing crop productivity at the global level (Hatfield & Prueger, 2015; Schauberger et al., 2017). Changes in temperature and sunshine hours have been reported to alter the supply of host-plant carbon, which may be the driving force regulating the temporal dynamics of arbuscular mycorrhizal (AM) fungal communities. Therefore, changing temperatures due to climate change may substantially alter future AM fungal community dynamics and ecosystem functioning (Dumbrell et al., 2011). However, our understanding of many spatiotemporal processes that regulate the diversity of functionally important taxa, especially microbes, is poor. Therefore, an understanding of the mechanisms that regulate the diversity of functionally important taxa would be important for sound predictions of the way ecosystems function under changing environments (McGuire & Treseder, 2010).

6.4.1.2 Soil Moisture

Bacterial communities have been reported to be significantly altered due to seasonal variability in rainfall and soil moisture content, and even rapid changes in soil bacterial abundance of ecologically important phyla were noticed within hours to days after seasonal drought (Cruz-Martínez et al., 2012; Placella et al., 2012; Šťovíček et al., 2017). Floods and droughts triggered by shifting climate patterns have considerably affected global cereal harvest (Lipiec et al., 2013). The synergistic effects of both heat and drought have been widely reported to significantly affect the yield of many crops more strongly than the effects of stress alone (Dreesen et al., 2012; Santos-Medellín et al., 2018; Rollins et al., 2013). Temporary wilting was found to cause plants to release more amino acids from their roots, possibly increasing the number of pathogens in the rhizosphere. Water deficits arising from both insufficient rainfall and soil water during the growing season are crucial for sustaining a high crop yield (Armada et al., 2014; Vadez et al., 2011, 2012; Wahid, 2007). An increase in intense rain events, along with a reduction in the number of rainy days, has been reported to increase the risk of drought (Trenberth, 2011; Vadez et al., 2011), which will considerably affect productivity under rainfed conditions due to water scarcity as a result of changing climate because dry weather is injurious to crops. Therefore, a balanced soil moisture content is important for sustaining crop yield.

6.4.1.3 Soil pH and Nutrient Status

The functioning of any terrestrial ecosystem is driven by the diversity of microbes (Wang et al., 2019), the availability of carbon and essential nutrients (i.e. nitrogen, phosphorus, and potassium), and chemical niches developed in soil (Sharma et al., 2013, 2021; Delgado-Baquerizo et al., 2016). Soil microbial diversity is considered to be important for driving elemental cycling within ecosystems through the exchange of material and energy (Delgado-Baquerizo et al., 2017a, b). Soil pH, a measure of acidity or alkalinity, controls many chemical and biochemical properties of the soil. Soil pH influences many important soil properties, such as the bioavailability of nutrients or pollutants to plant roots and the activity of soil microbes (Brady & Weil, 2014). The soil pH is strongly affected by the soil parent material, and the leaching of cations through high rainfall promotes acidification in humid regions. In contrast, most soils in arid and semiarid regions are alkaline (Brady & Weil, 2014). Furthermore, plant roots and microorganisms contribute to soil acidification through ion uptake, respiration, and nitrification. In the rhizosphere, root activity can change local soil pH by up to two units (Philippot et al., 2013). Plants growing in nutrient-poor soils have been reported to have an increased rate of exudation of amides and amino acids from their roots. In particular, these two groups of compounds (amides and amino acids) were twice as prevalent in the exudates of plants grown in phosphate-deficient soils. pH gradients have been shown to influence microbial populations in small field experiments, where the effects were stronger on bacterial diversity than on fungal diversity (Rousk et al., 2009). Dumbrell et al. (2011) in their study of 425 unique plants from 28 different species have provided convincing evidence in support of the claim that soil pH affected niche differentiation within the AM fungal community.

In a long-term fertilization trial on maize, Toljander et al. (2008) found that the community composition of AM fungi and bacteria was significantly influenced by the application of different fertilizers. Furthermore, the authors reported that changes in pH, phosphate concentration, and soil carbon content significantly altered arbuscular mycorrhizal (AM) fungal communities. Additionally, it was discovered that soils with low iron levels stimulated phenolic root secretion, which changed the microbial ecology of the rhizosphere. Changes in the microbial community composition were mainly correlated with changes in pH induced by the fertilization regime. However, factors such as phosphate and soil carbon content also significantly contributed to these changes. The results of this study demonstrated that long-term fertilizer application in the soil significantly affected specific groups of fungi and bacteria and the quality of soil in arable lands. In general, long-term fertilizer application significantly altered the bacterial community composition with a reduction in bacterial taxon richness in the soil (Moreira et al., 2016).

6.4.1.4 Ultraviolet (UV) Radiation

Ultraviolet radiation (UVR) is an important abiotic factor that affects both plants and microorganisms and their interactions. The distinct bands (UV-A, UV-B, and UV-C) of UV radiation affect plants and their associated microorganisms differently; for example, UV-A and UV-B affect morphogenesis and phototropism, whereas UV-B and UV-C strongly influence secondary metabolite production. The direct effects of UV-B radiation on soil microorganisms include changes in pigment content, growth, and the induction of carbon assimilation during amino acid synthesis (Vanhaelewyn et al., 2020). Furthermore, prolonged or intense exposure to UV radiation can directly damage DNA. Consequently, to cope with UV stress, plants can alter their root metabolism which strongly affects the release of carbon and nutrients in the rhizosphere. Furthermore, it affects bacterial communities in the rhizosphere. Plants have a significant effect on the microbial composition of the rhizosphere, with root exudates acting as the main source of energy and carbon for fungi and bacteria.

In contrast, phyllosphere (plant surface) communities are more exposed to environmental factors such as UVR than soil or rhizospheric communities (Chen et al.,

2022). There are various mechanisms underlying UVR tolerance in bacterial species. For example, a plasmid encoding the DNA repair rulAB operon confers UVR tolerance to the plant pathogenic bacterium *Pseudomonas syringae*. The ability of *Xanthomonas campestris* to tolerate UV radiation has also been linked to the production and absorption of extracellular polysaccharides. Pigmentation is another defense mechanism used by phyllospheric bacteria against UV-A radiation (320–400 nm). Microorganisms belonging to the archaea domain are resistant to UV-B radiation, and environments with higher UV-B exposure show higher species diversity. Gram-negative bacterial communities in Arctic soils experience nutrient shortages and environmental stresses. The selection of various bacterial communities in the rhizosphere can be influenced by changes in root metabolism caused by UVR exposure. Overall, UVR exposure can decrease root biomass, leading to less microbial colonization, lower soil nutrient levels, potential nutrient cycles, and soil community disruptions.

6.4.1.5 Carbon Dioxide (CO_2)

Terrestrial ecosystems are greatly affected by the enrichment of atmospheric CO_2, which also interacts with the soil carbon cycle. Changes in the dynamics of organic carbon were primarily responsible for these effects. According to previous research (Van Ginkel et al., 2000; Hu et al., 2001), exposure to high CO_2 concentrations can result in slower microbial decomposition rates in pastures. According to Hu et al. (2001), increased plant growth may result in lower nitrogen availability for microorganisms, thereby decreasing the capacity of the microbial community for degradation. The dynamics of soil organic matter may be indirectly affected by higher atmospheric CO_2 concentrations.

6.4.2 Biotic Factors

6.4.2.1 Microorganisms

In addition to several abiotic factors, plant and microbial species also influence root exudation. Therefore, biotic factors, mainly plant and microorganism species, strongly influence the exudation process (i.e. release of organic and inorganic carbon and nutrients) by altering the permeability of root cells and root metabolism (Neumann & Ludewig, 2023). The interaction between plants and soil microbes in the rhizosphere is crucial, as it affects the growth of both plants and soil microbes. Carbon in the rhizosphere serves as a source of energy for microbial metabolism, and nutrients are used by plants for their growth and activity. Thus, they play crucial roles in plant growth and exudate production. Since they provide plant hosts with compounds derived from microbes that can affect plant signaling and metabolism, soil microbes are frequently referred to as the "plant secondary genome" (Hakim et al., 2021). Some microbes, penicillin, and polymyxin-like antibiotics can alter cell permeability, leading to increased leakage and subsequent increase in organic material exudation. Soil microbes can induce exudation of phenolic compounds to enhance plant iron absorption in cases of low iron availability in the soil.

6.4.3 Geographical Factors (Altitude, Latitude, and Longitude)

Geographical factors, including latitude (angular distance from the equator), longitude (angular distance from the prime meridian), and altitude (height above sea level), play an important role in species distribution and diversity within ecosystems. The variations in species with respect to geographical factors mainly occur due to altered abiotic factors, which affect the biota and plant–microbe interactions within the rhizosphere (Dessaux et al., 2016). However, studies on these aspects are limited. Van Horn et al. examined abiotic factors such as pH, sulfates, and organic matter controlling soil biodiversity in Antarctica. They found that sites rich in organic carbon at lower elevations were frequently populated by members of the phyla *Actinobacteria* and *Acidobacteria*, while *Firmicute*s and *Proteobacteria* dominated sites at higher elevations with lower humidity. Although soil microbial parameters were correlated with soil water content, pH, organic matter, and sulfates, the strength and direction of these relationships varied among regions, indicating significant contextual effects at the local and regional levels. This study emphasized the significance of sampling at the regional scale to comprehend the particular geographic factors influencing soil microbial communities. Although some studies have found that bacterial diversity and richness altered with elevation and different land use patterns due to changing soil physicochemical properties indicating that lower diversity at different locations are not single modulators (Manpoong et al., 2020). Further large-scale studies are needed to clarify these relationships and their impact on microorganisms.

6.5 Application of Plant–Microbe Interactions in Improving Ecosystem Functioning

The rhizosphere is a thin envelope around the roots that harbors diverse microbial groups, which positively affect plant growth by performing a number of functions. For instance, they regulate the availability of nutrients through nutrient cycling, protect plants under biotic and abiotic stresses, and subsequently influence plant growth and productivity, as discussed previously. Therefore, plant–microbe relationships within the rhizosphere can possibly be exploited to optimize plant nutrient uptake, growth, and production through rhizosphere engineering in an eco-friendly manner for sustainable production practices. Furthermore, it can be used in other cases, such as the recovery of degraded ecosystems. The possible applications of rhizosphere engineering are as follows.

(i) Detecting potential microbes that can be used as sources of biofertilizer for specific plants or for the combination of crops using recent advances in DNA sequencing technology and bioinformatics by performing detailed studies on the composition and functions of plant-associated microbial communities. This knowledge holds the potential to be harnessed for the selection of microbial communities and the engineering of rhizosphere processes to enhance plant growth and health.

(ii) Identification and use of PGPMs as biofertilizers and biopesticides to sustain ecosystem productivity, particularly in agriculture, by characterizing rhizosphere microbes with the potential for growth promotion and facilitating nutrient uptake by producing plant growth-promoting compounds and providing protection against plant pathogens.

(iii) Identification and application of nitrogen-fixing bacteria as biofertilizers to develop a certain level of independence in the use of synthetic N fertilizers in cropland and to minimize negative environmental impacts.

(iv) Identification and application of region/crop-specific potential mycorrhizal fungi as biofertilizers for solubilization and the continuous availability of phosphorus in the soil is particularly important in tropical regions where the majority of the soil is phosphorus deficient.

(v) Targeting the application of beneficial microbes with the potential for plant defense against pathogens is an eco-friendly strategy for disease management in agriculture.

(vi) Microbes play a pivotal role in managing soil health and hence can be exploited for their potential use in the remediation of polluted soils using microbe-based solutions for environmental cleanup and ecosystem restoration.

(vii) Application of engineered microbes with specific functions to enhance plant–microbe interactions by producing specific metabolites that benefit plants.

(viii) Targeting and application of plant/region-specific microbes from different ecosystems can be later used for the restoration of degraded ecosystems by re-establishing plant–microbe interactions by accelerating the process of recovery.

6.6 Challenges and Opportunities for Future

The exponential increase in the human population, coupled with biotic and abiotic stresses due to climate change and land scarcity, has severely impacted global agricultural production, which is likely to worsen due to the expected increase in the human population by approximately 9 billion by 2050. Many varieties of crop plants developed during "green revolution" have become nonresponsive to fertilizers with stagnation in their yield under agricultural intensification to maximize agriculture production using heavy doses of chemical fertilizers and pesticides. Furthermore, most crop varieties have low nutrient use efficiencies (20–30%) of the applied chemical fertilizers, and thus most of the nutrients applied are lost from the system, causing several environmental consequences. These include erosion of biodiversity, emission of greenhouse gases, water pollution, soil contamination, and deterioration of soil biology and health. Therefore, future agricultural sustainability requires the management of food security to meet the increasing demand for food, mitigate climate change effects, and improve soil fertility.

The rhizosphere engineering is one of the most important tools that be used to potentially meet all the requirements. Therefore, exploiting region-specific interactions between plant and microbial communities would be an appropriate approach to meet the demands of the growing world population. Plant growth-promoting rhizobacteria (PGPR) hold promise for sustainable agriculture by providing substitutes for chemical fertilizers, fungicides, and chemical-based stimulators. In recent years, PGPR has replaced chemical fertilizers in silviculture, horticulture, agriculture, and environmental remediation processes. The major obstacle in PGPR research is its restricted applicability to small-scale experiments (i.e. at the laboratory and greenhouse levels); therefore, there is an urgent need to transform the success of these research outcomes into larger field experiments in the natural environment. Further, challenges in implementing such research include the screening, production, marketing, and commercialization of region-specific microbial inoculants. Lastly, plant ecology-based solutions for sustainable agriculture will require the development of practical approaches in consultation with farmers, policymakers, breeders, and agricultural scientists, keeping in mind the social and economic constraints to facilitate wider adoption of the technology developed.

References

Abbas, T., Zahir, Z. A., Naveed, M., & Kremer, R. J. (2018). Limitations of existing weed control practices necessitate development of alternative techniques based on biological approaches. *Advances in Agronomy, 147*, 239–280.

Armada, E., Portela, G., Roldán, A., & Azcón, R. (2014). Combined use of beneficial soil microorganism and agro-waste residue to cope with plant water limitation under semiarid conditions. *Geoderma, 232*, 640–648.

Ashraf, S., Afzal, M., Naveed, M., Shahid, M., & Ahmad Zahir, Z. (2018). Endophytic bacteria enhance remediation of tannery effluent in constructed wetlands vegetated with Leptochloa fusca. *International Journal of Phytoremediation, 20*(2), 121–128.

Bauer, J. T., Kleczewski, N. M., Bever, J. D., Clay, K., & Reynolds, H. L. (2012). Nitrogen-fixing bacteria, arbuscular mycorrhizal fungi, and the productivity and structure of prairie grassland communities. *Oecologia, 170*, 1089–1098.

Beyhaut, E., Larson, D. L., Allan, D. L., & Graham, P. H. (2014). Legumes in prairie restoration: Evidence for wide cross-nodulation and improved inoculant delivery. *Plant and Soil, 377*, 245–258.

Bhatt, K., Suyal, D., Kumar, S., Singh, K., & Goswami, P. (2022). New insights into engineered plant-microbe interactions for pesticide removal. *Chemosphere, 309*, 136635.

Brady, N. C., & Weil, R. R. (2014). *The nature and properties of soils*. Pearson Education Limited.

Chen, Z., Dong, Y., & Huang, X. (2022). Plant responses to UV-B radiation: Signaling, acclimation and stress tolerance. *Stress Biology, 2*(1), 51. https://doi.org/10.1007/s44154-022-00076-9

Chopade, B. A., Huddedar, S. B., Shete, A. M., Tilekar, J. N., Dhavale, D. D., & Gore, S. D. (2008). *Plasmid encoding IAA and a method thereof*. US Patent 7,341,868.

Cloutier, C., & Bauduin, F. (1995). Biological control of the colorado potato beetle *leptinotarsa decemlineata* (coleoptera: chrysomelidae) in quebec by augmentative releases of the two-spotted stinkbug *perillus bioculatus* (hemiptera: pentatomidae). *Canadian Entomologist, 127*, 195–212. https://doi.org/10.4039/Ent127195-2

Compant, S., Samad, A., Faist, H., & Sessitsch, A. (2019). A review on the plant microbiome: Ecology, functions, and emerging trends in microbial application. *Journal of Advanced Research, 19*, 29–37.

Compant, S., Duffy, B., Nowak, J., Clément, C., & Barka, E. A. (2005). Use of plant growth-promoting bacteria for biocontrol of plant diseases: Principles, mechanisms of action, and future prospects. *Applied and Environmental Microbiology, 71*(9), 4951–4959.

Cruz-Martínez, K., Rosling, A., Zhang, Y., Song, M., Andersen, G. L., & Banfield, J. F. (2012). Effect of rainfall-induced soil geochemistry dynamics on grassland soil microbial communities. *Applied and Environmental Microbiology, 78*(21), 7587–7595.

Degenhardt, J., Al-Masri, A. N., Kürkcüoglu, S., Szankowski, I., & Gau, A. E. (2005). Characterization by suppression subtractive hybridization of transcripts that are differentially expressed in leaves of apple scab-resistant and susceptible cultivars of Malus domestica. *Molecular Genetics and Genomics, 273*, 326–335.

Delgado-Baquerizo, M., Maestre, F. T., Reich, P. B., Jeffries, T. C., Gaitan, J. J., Encinar, D., Berdugo, M., Campbell, C. D., & Singh, B. K. (2016). Microbial diversity drives multifunctionality in terrestrial ecosystems. *Nature Communications, 7*, 10541.

Delgado-Baquerizo, M., Reich, P. B., Khachane, A. N., Campbell, C. D., Thomas, N., Freitag, T. E., Abu Al-Soud, W., Sorensen, S., Bardgett, R. D., & Singh, B. K. (2017a). It is elemental: Soil nutrient stoichiometry drives bacterial diversity. *Environmental Microbiology, 19*(3), 1176–1188.

Delgado-Baquerizo, M., Eldridge, D. J., Ochoa, V., Gozalo, B., Singh, B. K., & Maestre, F. T. (2017b). Soil microbial communities drive the resistance of ecosystem multifunctionality to global change in drylands across the globe. *Ecology Letters, 20*(10), 1295–1305.

De-la-Peña, C., & Loyola-Vargas, V. M. (2014). Biotic interactions in the rhizosphere: A diverse cooperative enterprise for plant productivity. *Plant Physiology, 166*(2), 701–719.

Dessaux, Y., Grandclément, C., & Faure, D. (2016). Engineering the rhizosphere. Special issue: Unravelling the secrets of the rhizosphere. *Trends in Plant Science, 21*(3), 266–278. https://doi.org/10.1016/j.tplants.2016.01.002

Dolatabadian, A. (2020). Plant-microbe interaction. *Biology (Basel), 10*(1), 15. https://doi.org/10.3390/biology10010015. PMID: 33396600; PMCID: PMC7823258.

Dong, Y. H., Zhang, X. F., Xu, J. L., & Zhang, L. H. (2004). Insecticidal Bacillus thuringiensis silences Erwinia carotovora virulence by a new form of microbial antagonism, signal interference. *Applied and Environmental Microbiology, 70*(2), 954–960.

Dreesen, P. E., De Boeck, H. J., Janssens, I. A., & Nijs, I. (2012). Summer heat and drought extremes trigger unexpected changes in productivity of a temperate annual/biannual plant community. *Environmental and Experimental Botany, 79*, 21–30.

Dumbrell, A. J., Ashton, P. D., Aziz, N., Feng, G., Nelson, M., Dytham, G., Fitter, A. H., & Helgason, T. (2011). Distinct seasonal assemblages of arbuscular mycorrhizal fungi revealed by massively parallel pyrosequencing. *The New Phytologist, 190*(3), 794–804.

Elmer, P. A. G., & Reglinski, T. (2006). Biosuppression of *Botrytis cinerea* in grapes. *Plant Pathology, 55*, 155–177.

Fageria, N. K. (2014). *Nitrogen Management in Crop Production* (p. 436). CRC Press. https://doi.org/10.1201/b17101

Gao, M., Teplitski, M., Robinson, J. B., & Bauer, W. D. (2003). Production of substances by Medicago truncatula that affect bacterial quorum sensing. *Molecular Plant-Microbe Interactions, 16*(9), 827–834.

Gau, A. E., Koutb, M., Piotrowski, M., & Kloppstech, K. (2004). Accumulation of pathogenesis-related proteins in the apoplast of a susceptible cultivar of apple (Malus domestica cv. Elstar) after infection by Venturia inaequalis and constitutive expression of PR genes in the resistant cultivar Remo. *European Journal of Plant Pathology, 110*, 703–711.

Hakim, S., Naqqash, T., Nawaz, M. S., Laraib, I., Siddique, M. J., Zia, R., Mirza, M. S., & Imran, A. (2021). Rhizosphere engineering with plant growth-promoting microorganisms for agriculture and ecological sustainability. *Frontiers in Sustainable Food Systems, 5*, 617157. https://doi.org/10.3389/fsufs.2021.617157

Hassan, A., Periathamby, A., Ahmed, A., Innocent, O., & Hamid, F. S. (2020). Effective bioremediation of heavy metal–contaminated landfill soil through bioaugmentation using consortia of fungi. *Journal of Soils and Sediments, 20*, 66–80.

Hatfield, J. L., & Prueger, J. H. (2015). Temperature extremes: Effect on plant growth and development. *Weather and Climate Extremes, 10*, 4–10.
Hauchhum, R., & Tripathi, S. K. (2019a). Impact of rhizosphere microbes of three early colonizing annual plants on improving soil fertility during vegetation establishment under different fallow periods following shifting cultivation. *Agricultural Research, 9*(2), 213–221. https://doi.org/10.1007/s40003-017-0286-2535
Hauchhum, R., & Tripathi, S. K. (2019b). Carbon and nitrogen differences in rhizosphere soil of annual plants in abandoned lands following shifting agriculture in northeast India. *Nutrient Cycling in Agroecosystems*. https://doi.org/10.1007/s10705-019-099725
Ho, Y. N., Mathew, D. C., & Huang, C. C. (2017). Plant-microbe ecology: Interactions of plants and symbiotic microbial communities. In *Plant ecology-traditional approaches to recent trends* (p. 93–119). https://doi.org/10.5772/intechopen.69088
Hobbs, R. J., & Cramer, V. A. (2008). Restoration ecology: Interventionist approaches for restoring and maintaining ecosystem function in the face of rapid environmental change. *Annual Review of Environment and Resources, 33*, 39–61.
Holste, E. K., & Kobe, R. K. (2017). Tree species and soil nutrients drive tropical reforestation more than associations with mycorrhizal fungi. *Plant and Soil, 410*, 283–297.
Hu, S., Chapin Iii, F. S., Firestone, M. K., Field, C. B., & Chiariello, N. R. (2001). Nitrogen limitation of microbial decomposition in a grassland under elevated CO2. *Nature, 409*(6817), 188–191.
Huddedar, S. B., Shete, A. M., Tilekar, J. N., Gore, S. D., Dhavale, D. D., & Chopade, B. A. (2002). Isolation, characterization, and plasmid pUPI126-mediated indole-3-acetic acid production in Acinetobacter strains from rhizosphere of wheat. *Applied Biochemistry and Biotechnology, 102*, 21–39.
Huidrom, M., Singh, S. S., & Meitei, M. D. (2024). Role of Naga Nullah tributary as a major polluter of Loktak: A Ramsar site from Manipur, India with 3 decades in the Montreux records of the wetland's convention. *Sustainable Water Resources Management, 10*(2), 56.
Husain, S. S., & McKeen, W. E. (1963). Rhizoctonia fragariae sp. nov. in relation to Strawberry degeneration in Southwestern Ontario. *Phytopathology, 53*(5), 532–540.
Imaizumi, S., Honda, M., & Fujimori, T. (1999). Effect of temperature on the control of annual bluegrass (Poa annua L.) with Xanthomonas campestris pv. poae (JT-P482). *Biological Control, 16*(1), 13–17.
Jefferson, R. (1994). *The Hologenome. Agriculture, environment and the developing world: A future of PCR*. Cold Spring Harbor or Cold Spring Harbor Laboratory Press.
Johnson, D. R., Wyse, D. L., & Jones, K. J. (1996). Controlling weeds with phytopathogenic bacteria. *Weed Technology, 10*(3), 621–624.
Jones, D. L., Nguyen, C., & Finlay, R. D. (2009). Carbon flow in the rhizosphere: Carbon trading at the soil-root interface. *Plant and Soil, 321*, 5–33.
Kandel, S. L., Firrincieli, A., Joubert, P. M., Okubara, P. A., Leston, N. D., McGeorge, K. M., ... & Doty, S. L. (2017). An in vitro study of bio-control and plant growth promotion potential of Salicaceae endophytes. *Frontiers in Microbiology, 8*, 386.
Kremer, R. J. (2019). Bioherbicides and nanotechnology: Current status and future trends. *Nano-biopesticides Today and Future Perspectives*, 353–366.
Kumar, A., & Dubey, A. (2020). Rhizosphere microbiome: Engineering bacterial competitiveness for enhancing crop production. *Journal of Advanced Research, 24*, 337–352.
Kumar, M., Shankar, A., Chaudhary, S., & Prasad, V. (2023). Phosphate solubilizing microorganisms: Multifarious applications. In *Plant microbiome for plant productivity and sustainable agriculture* (pp. 245–262). Springer Nature Singapore.
Kürkcüoglu, S., Degenhardt, J., Lensing, J., Al-Masri, A. N., & Gau, A. E. (2007). Identification of differentially expressed genes in Malus domestica after application of the non-pathogenic bacterium Pseudomonas fluorescens Bk3 to the phyllosphere. *Journal of Experimental Botany, 58*(3), 733–741.

Lin, S., Wu, X., Cao, J. Z., & Wang, F. L. (2013). Biocontrol potential of chitinase-producing nematophagous fungus Acremonioum implicatum against Meloidogyne incognita. *Acta Phytopathol Sin, 43*(5), 509–517.

Lipiec, J., Doussan, C., Nosalewicz, A., & Kondracka, K. (2013). Effect of drought and heat stresses on plant growth and yield: A review. *International Agrophysics, 2013*(27), 463–477. https://doi.org/10.2478/intag-2013-0017

MAFF. (2020). https://www.maff.go.jp/j/tokei/kouhyou/sakumotu/sakkyou_yasai/

Maindad, D. V., Kasture, V. M., Chaudhari, H., Dhavale, D. D., Chopade, B. A., & Sachdev, D. P. (2014). Characterization and fungal inhibition activity of siderophore from wheat rhizosphere associated Acinetobacter calcoaceticus strain HIRFA32. *Indian Journal of Microbiology, 54*, 315–322.

Malgioglio, G., Rizzo, G. F., Nigro, S., Lefebvre du Prey, V., Herforth-Rahmé, J., Catara, V., & Branca, F. (2021). Plant-microbe interaction in sustainable agriculture: The factors that may influence the efficacy of PGPM application. *Sustainability, 14*, 2253. https://doi.org/10.3390/su14042253

Manpoong, C., De Mandal, S., Bangaruswamy, D. K., Perumal, R. C., Benny, J., Beena, P. S., et al. (2020). Linking rhizosphere soil biochemical and microbial community characteristics across different land use systems in mountainous region in Northeast India. *Meta Gene, 23*, 100625.

McGuire, K. L., & Treseder, K. K. (2010). Microbial communities and their relevance for ecosystem models: Decomposition as a case study. *Soil Biology and Biochemistry, 42*, 529–535.

McNear, D. H., Jr. (2013). The rhizosphere - roots, soil and everything in between. *Nature Education Knowledge, 4*(3), 1.

Mejía, L. C., Herre, E. A., Sparks, J. P., Winter, K., García, M. N., Van Bael, S. A., ... & Maximova, S. N. (2014). Pervasive effects of a dominant foliar endophytic fungus on host genetic and phenotypic expression in a tropical tree. *Frontiers in Microbiology, 5*, 479.

Mitter, B., Pfaffenbichler, N., & Sessitsch, A. (2016). Plant–microbe partnerships in 2020. *Microbial Biotechnology, 9*(5), 635–640. https://doi.org/10.1111/1751-7915.12382

Moreira, H., Pereira, S. I., Marques, A. P., Rangel, A. O., & Castro, P. M. (2016). Mine land valorization through energy maize production enhanced by the application of plant growth-promoting rhizobacteria and arbuscular mycorrhizal fungi. *Environmental Science and Pollution Research, 23*, 6940–6950.

Naik, K., Mishra, S., Srichandan, H., Singh, P. K., & Sarangi, P. K. (2019). Plant growth promoting microbes: Potential link to sustainable agriculture and environment. *Biocatalysis and Agricultural Biotechnology, 21*, 101326.

Neumann, G., & Ludewig, U. (2023). Rhizosphere chemistry influencing plant nutrition. In Z. Rengel, I. Cakmak, & P. J. White (Eds.), *Marschner's mineral nutrition of plants* (4th ed., pp. 545–585). Academic Press. https://doi.org/10.1016/B978-0-12-384905-2.00014-5

Newman, E. I. (1985). The rhizosphere: Carbon sources and microbial populations. In A. H. Fitter (Ed.), *Ecological interactions in soil* (p. 107). Blackwell Scientific Publications.

Newman, K. L., Chatterjee, S., Ho, K. A., & Lindow, S. E. (2008). Virulence of plant pathogenic bacteria attenuated by degradation of fatty acid cell-to-cell signaling factors. *Molecular Plant-Microbe Interactions, 21*(3), 326–334.

Olanrewaju, O. S., & Babalola, O. O. (2021). The rhizosphere microbial complex in plant health: A review of interaction dynamics. *Journal of Integrative Agriculture, 21*(8), 2168–2182.

Perring, M. P., Standish, R. J., Price, J. N., Craig, M. D., Erickson, T. E., Ruthrof, K. X., et al. (2015). Advances in restoration ecology: Rising to the challenges of the coming decades. *Ecosphere, 6*(8), 1–25.

Philippot, L., Raaijmakers, J. M., Lemanceau, P., & Van Der Putten, W. H. (2013). Going back to the roots: the microbial ecology of the rhizosphere. *Nature Reviews Microbiology, 11*(11), 789–799.

Pieterse, C. M., Zamioudis, C., Berendsen, R. L., Weller, D. M., Van Wees, S. C., & Bakker, P. A. (2014). Induced systemic resistance by beneficial microbes. *Annual Review of Phytopathology, 52*(1), 347–375.

Placella, S. A., Brodie, E. L., & Firestone, M. K. (2012). Rainfall-induced carbon dioxide pulses result from sequential resuscitation of phylogenetically clustered microbial groups. *Proceedings of the National Academy of Sciences of the United States of America, 109*, 10931–10936.
Rokhbakhsh-Zamin, F., Sachdev, D., Kazemi-Pour, N., Engineer, A., Pardesi, K. R., Zinjarde, S., et al. (2011). Characterization of plant-growth-promoting traits of Acinetobacter species isolated from rhizosphere of Pennisetum glaucum. *Journal of Microbiology and Biotechnology, 21*(6), 556–566.
Rollins, J. A., Habte, E., Templer, S. E., Colby, T., Schmidt, J., & von Korff, M. (2013). Leaf proteome alterations in the context of physiological and morphological responses to drought and heat stress in barley (Hordeum vulgare L.). *Journal of Experimental Botany, 64*(11), 3201–3212.
Rousk, J., Brookes, P. C., & Bååth, E. (2009). Contrasting soil pH effects on fungal and bacterial growth suggest functional redundancy in carbon mineralization. *Applied and Environmental Microbiology, 75*(6), 1589–1596.
Ruiu, L. (2018). Microbial biopesticides in agroecosystems. *Agronomy, 8*, 235. https://doi.org/10.3390/agronomy8110235
Ryan, R. P., Germaine, K., Franks, A., Ryan, D. J., & Dowling, D. N. (2008). Bacterial endophytes: Recent developments and applications. *FEMS Microbiology Letters, 278*(1), 1–9.
Sachdev, S. (2010). Holographic metals and the fractionalized Fermi liquid. *Physical Review Letters, 105*(15), 151602.
Sachdev, D. P., Chaudhari, H. G., Kasture, V. M., Dhavale, D. D., & Chopade, B. A. (2009). Isolation and characterization of indole acetic acid (IAA) producing Klebsiella pneumoniae strains from rhizosphere of wheat (Triticum aestivum) and their effect on plant growth. *Indian Journal of Experimental Biology, 47*(12), 993.
Santos-Medellín, C., Edwards, J., Liechty, Z., Nguyen, B., & Sundaresan, V. (2018). Drought stress results in a compartment-specific restructuring of the rice root-associated microbiomes. *MBio, 8*(4), 10–1128.
Santoyo, G., Hernández-Pacheco, C., Hernández-Salmerón, J., & Hernández-León, R. (2018). The role of abiotic factors modulating the plant-microbe-soil interactions: Toward sustainable agriculture. A review. *Spanish Journal of Agricultural Research, 15*(1), e03R01. https://doi.org/10.5424/sjar/2017151-9990
Schauberger, B., Archontoulis, S., Arneth, A., Balkovic, J., Ciais, P., Deryng, D., et al. (2017). Consistent negative response of US crops to high temperatures in observations and crop models. *Nature Communications, 8*(1), 13931.
Schlatter, D. C., Bakker, M. G., Bradeen, J. M., & Kinkel, L. L. (2015). Plant community richness and microbial interactions structure bacterial communities in soil. *Ecology, 96*(1), 134–142.
Scott, R. A., Weil, J., Le, P. T., Williams, P., Fray, R. G., von Bodman, S. B., & Savka, M. A. (2006). Long-and short-chain plant-produced bacterial N-acyl-homoserine lactones become components of phyllosphere, rhizosphere, and soil. *Molecular Plant-Microbe Interactions, 19*(3), 227–239.
Sharma, S. B., Sayyed, R. Z., Trivedi, M. H., & Gobi, T. A. (2013). Phosphate solubilizing microbes: Sustainable approach for managing phosphorus deficiency in agricultural soils. *Springerplus, 2*, 1–14.
Sharma, P., Kumar, T., Yadav, M., Gill, S. S., & Chauhan, N. S. (2021). Plant-microbe interactions for sustainable agriculture and food security. *Plant Gene, 28*, 100325.
Silva, G. H., Teles, H. L., Zanardi, L. M., Young, M. C. M., Eberlin, M. N., Hadad, R., ... & Araújo, Â. R. (2006). Cadinane sesquiterpenoids of Phomopsis cassiae, an endophytic fungus associated with Cassia spectabilis (Leguminosae). *Phytochemistry, 67*(17), 1964–1969.
Singh, P., Piotrowski, M., Kloppstech, K., & Gau, A. E. (2004). Investigations on epiphytic living Pseudomonas species from Malus domestica with an antagonistic effect to Venturia inaequalis on isolated plant cuticle membranes. *Environmental Microbiology, 6*(11), 1149–1158.
Singh, S. S., Chanda, R., Singh, N. S., Ramtharmawi, Devi, N. R., Devi, K. V., ... & Tripathi, S. K. (2024a). Microplastic pollution: Exploring trophic transfer pathways and ecological impacts. *Discover Environment, 2*(1), 103.

Singh, S. S., Meitei, M. D., Upadhyay, K. K., Chanda, R., Mawi, R., Singh, N. S., ... & Tripathi, S. K. (2024b). Bioremediation of toxic metals using aquatic macrophytes: Challenges and opportunities. *CLEAN–Soil, Air, Water, 52*(11), e202400273.
Štovíček, A., Kim, M., Or, D., et al. (2017). Microbial community response to hydration-desiccation cycles in desert soil. *Scientific Reports*. https://doi.org/10.1038/srep45735
Tal, A. (2018). Making conventional agriculture environmentally friendly: Moving beyond the glorification of organic agriculture and the demonization of conventional agriculture. *Sustainability, 10*(4), 1078.
Toljander, J. F., Santos-González, J. C., Tehler, A., & Finlay, R. D. (2008). Community analysis of arbuscular mycorrhizal fungi and bacteria in the maize mycorrhizosphere in a long-term fertilization trial. *FEMS Microbiology Ecology, 65*(2), 323–338.
Trenberth, K. E. (2011). Changes in precipitation with climate change. *Climate Research, 47*(1-2), 123–138.
Tripathi, S. K. (2009). Human influences on mobility of nitrogen in the environment: Needs for research and management. *Acta Ecologica Sinica, 29*(2), 130–135.
Tripathi, S. K., & Singh, K. P. (1992). Abiotic and litter quality control during the decomposition of different plant parts in a dry tropical bamboo savanna in India. *Pedobiologia, 36*, 241–256.
Tripathi, S. K., & Singh, K. P. (1994). Productivity and nutrient cycling in recently harvested and mature bamboo savannas in the dry tropics. *Journal of Applied Ecology, 31*, 109–124.
Usta, C. (2013). Microorganisms in biological pest control — A review (bacterial toxin application and effect of environmental factors). In M. Silva-Opps (Ed.), *Current progress in biological research*. InTech. https://doi.org/10.5772/55786
Uroz, S., Oger, P. M., Chapelle, E., Adeline, M. T., Faure, D., & Dessaux, Y. (2008). A Rhodococcus qsdA-encoded enzyme defines a novel class of large-spectrum quorum-quenching lactonases. *Applied and Environmental Microbiology, 74*(5), 1357–1366.
Vadez, V., Kholova, J., Choudhary, S., Zindy, P., Terrier, M., Krishnamurth, L., Kumar, P. R., & Turner, N. C. (2011). Whole plant response to drought under climate change. In S. S. Yadav, R. Redden, J. L. Hatfield, H. Lotze-Campen, & A. E. Hall (Eds.), *Crop adaptation to climate change*. Chichester-Wiley-Blackwell.
Vadez, V., Berger, J. D., Warkentin, T., Asseng, S., Ratnakumar, P., Rao, K. P. C., Gaur, P. M., Munier-Jolain, N., Larmure, A., Voisin, A. S., Sharma, H. C., Pande, S., Sharma, M., Krishnamurthy, L., & Zaman, M. A. (2012). Adaptation of grain legumes to climate change: A review. *Agronomy for Sustainable Development, 32*, 31–44.
Valette, M., Rey, M., Gerin, F., Comte, G., & Wisniewski-Dyé, F. (2020). A common metabolomic signature is observed upon inoculation of rice roots with various rhizobacteria. *Journal of Integrative Plant Biology, 62*(2), 228–246.
Van Ginkel, J. H., Gorissen, A., & Polci, D. (2000). Elevated atmospheric carbon dioxide concentration: Effects of increased carbon input in a Lolium perenne soil on microorganisms and decomposition. *Soil Biology and Biochemistry, 32*(4), 449–456.
Vanhaelewyn, L., Straeten, D. V. D., De Coninck, B., & Vandenbussche, F. (2020). Ultraviolet radiation from a plant perspective: The plant-microorganism context. *Frontiers in Plant Science*. PMID: 33384704 PMCID: PMC7769811. https://doi.org/10.3389/fpls.2020.597642
Wahid, A. (2007). Physiological implications of metabolites biosynthesis in net assimilation and heat stress tolerance of sugarcane sprouts. *Journal of Plant Research, 120*, 219–222.
Walters, D. R., Paterson, L., Walsh, D. J., & Havis, N. D. (2008). Priming for plant defense in barley provides benefits only under high disease pressure. *Physiological and Molecular Plant Pathology, 73*(4-5), 95–100.
Wang, C., Zhou, X., Guo, D., Zhao, J., Li, Y., Feng, G., Gao, Q., Yu, H., & Zhao, L. (2019). Soil pH is the primary factor driving the distribution and function of microorganisms in farmland soils in Northeastern China. *Annales de Microbiologie, 69*, 1461–1473. https://doi.org/10.1007/s13213-019-01529-9
Wapongnungsang, Hauchhum, R., & Tripathi, S. K. (2017). Litter decomposition Vis-a-Vis carbon and nitrogen dynamics of Tephrosia candida components in different fallow periods following shifting cultivation in Mizoram. *Indian Journal of Ecology, 44*(4), 791–796.

Whipps, J. M. (2004). Prospects and limitations for mycorrhizas in biocontrol of root pathogens. *Canadian Journal of Botany, 82*, 1198–1227.

Whipps, J. M., & Gerhardson, B. (2007). Biological pesticides for control of seed- and soil-borne plant pathogens. In J. D. Van Elsas, J. D. Jansson, & J. T. Trevors (Eds.), *Modern soil microbiology edition* (2nd ed., pp. 479–501). CRC Press.

White, J. F., Kingsley, K. L., Zhang, Q., Verma, R., Obi, N., Dvinskikh, S., ... & Kowalski, K. P. (2019). Endophytic microbes and their potential applications in crop management. *Pest Management Science, 75*(10), 2558–2565.

Yi, H. S., Yang, J. W., & Ryu, C. M. (2013). ISR meets SAR outside: Additive action of the endophyte Bacillus pumilus INR7 and the chemical inducer, benzothiadiazole, on induced resistance against bacterial spot in field-grown pepper. *Frontiers in Plant Science, 4*, 122.

Microbes to the Rescue: The Future of Plant Protection

7

Kinjal Bhatt, Avni Divatia, Sudeshna Menon, and Sebastian Vadakan

Abstract

With every passing decade, anthropogenic pursuits have intensified unwarranted cycles of meeting the demands of the overpopulous world, affecting the environment, which climate change has made living difficult. Within this cycle, there is a focus on increasing the yield of plants especially agricultural crops to meet the growing food demand across the globe. To attain more yield in less time, incidences of overharvesting arable land and unsuitable farming methods have led to a decline in environmental health. This decline comprises diminishing biodiversity, deteriorating soil, water, and air quality, burgeoning biotic and abiotic stress, and a non-sustainable ecosystem. The zealous efforts on plant protection by overuse of chemical fertilizers, pesticides, etc., have been detrimental as it has resulted in accumulation of toxic compounds in soil and emission of greenhouse gases. The realization that declining environmental health needs to be reversed has affected strong policies in countries, including India. One of the policies is to reduce the use of recalcitrant fertilizers and pesticides. Plant protection approaches now include establishing phytomicrobiomes to enhance plant growth, mitigate biotic and abiotic stress, and enhance biocontrol and plant immunity. This intervention may be slow in comparison to the use of chemicals; however, it is predicted to be long-lasting, sustainable, and eco-friendly.

Keywords

Environment and climate change · Agriculture and horticulture · Plant growth-promoting microbes · Phytomicrobiomes · Biotic and abiotic stress · Biocontrol and plant immunity

K. Bhatt · A. Divatia · S. Menon (✉) · S. Vadakan
Department of Biochemistry – Biotechnology, St. Xavier's College, Ahmedabad, India
e-mail: sudeshna.menon@sxca.edu.in

M. Saraf et al. (eds.), *Climate Change and Soil Microorganisms for Environmental Sustainability*, Microorganisms for Sustainability 28,
https://doi.org/10.1007/978-981-96-3425-5_7

7.1 Introduction

"A rapidly warming planet poses an existential threat to all life on earth" is a statement based on data analysis by the Natural Resources Defence Council (NRDC) (www.nrdc.org). With the growing demand for energy and food, there has been a substantial increase in the use of fossil fuels (Fig. 7.1). The fossil fuel consumption has increased by eightfold since 1950, with coal being the main energy source earlier, and now it has shifted more to use of oil and gas. The fossil energy consumption by India is about 9000 TWh, of which about 28% comes from the use of oil and only about 6% comes from gas. The consumption of fossil energy per person is reported to be highest for the United States of America. The use of fossil fuels has contributed majorly to greenhouse gas emissions. (Richie, 2022).

Accumulation of carbon dioxide in the atmosphere has been the major reason for global warming. The rise in temperature has led to unprecedented incidences of floods, droughts, hurricanes, wildfires, etc. These catastrophes have been disturbing the ecosystem. According to an NRDC report there has been a gradual increase in the sea levels, melting of glaciers; warmer and acidified oceans affecting aquatic life; substantial effect on land ecosystem with increase in outbreaks of pests, plant diseases and invasive species; difficulties in farming and raising livestock; deterioration in soil quality due to extreme heat and increased precipitation, especially in regions of industrialization and monoculture farming; losses in agriculture leading to food shortage and inflations; extinction of several animal and plant species as the climate changes the habitat conditions; migration of species from existing habitats

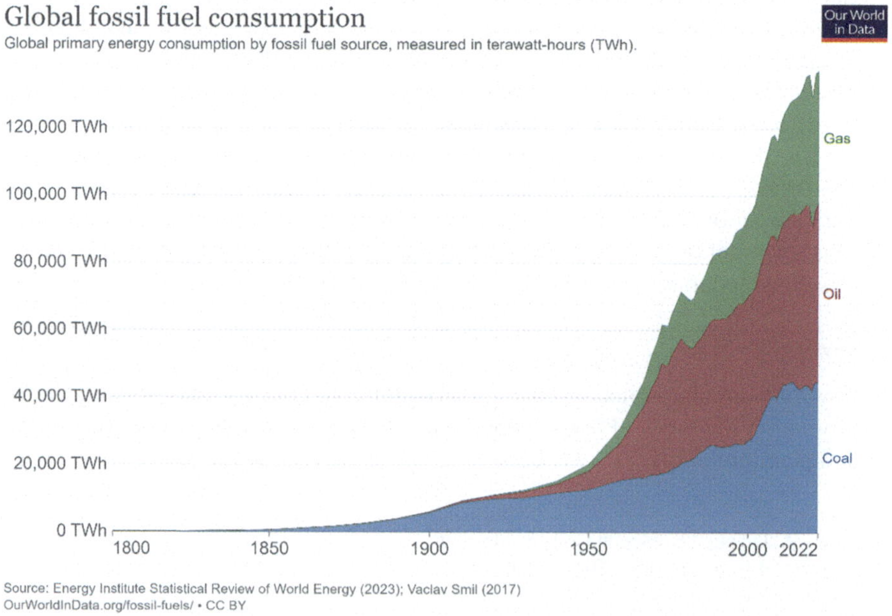

Fig. 7.1 Global utilization of fossil fuel with the increase in demand for food and energy. (Richie, 2022)

to new habitats; effecting on health of human due to the decline in air quality, water quality and soil quality, which in turn has led to an increase in incidences of respiratory disorders, viral pandemics, insect borne diseases such as Malaria and Zika, mental health; establishing inequities between developed, developing and poor countries with the latter finding it very difficult to adapt to the climatic changes; faltering of economic sustainability. All these entail curbing greenhouse gas emissions as a policy, and this has found a place in the list of sustainable goals of the United Nations reports. There has been a considerable effort in exploring how natural communities of microbes can interact with plants to decrease the use of fertilizers and pesticides that are synthesized chemicals to make plants and the environment healthy. This chapter discusses the putative role of microbes in maintaining the health and growth of the plant, as well as mitigating stress and in developing plant's immunity for its own protection.

7.2 Climate Change, Agriculture, and Horticulture Crops

One of the most pressing challenges of our time is climate change, with the rise in global temperatures, ecosystems are threatened. Changes in climatic conditions in different regions of the world have negatively affected agricultural productivity, food security, various ecosystem services, and general composition, as well as the quality of flora and fauna (Field, 2012; Leal Filho et al., 2021). Some of the most important consequences of climate change have been temperature extremes and uncertainty or irregularity in precipitation patterns, which ultimately affect agricultural crops (Agnihotri & Mitra, 2023). It has been reported that the total agricultural productivity has been reduced by 21% due to the increase in extreme weather disasters caused by climate change. Since 1961, tropical regions such as Africa is having a greater impact on reduced agricultural productivity than temperate regions such as Europe and Central Asia (Ortiz-Bobea et al., 2021). It is perilous to meet the global food needs with the rapidly increasing world population. The global agricultural production must enhance by 50% to sustain approximately 9 billion people by 2050. But as food production increases, the exploitation of arable land and overuse of chemical fertilizers further contribute to greenhouse gas (GHG) emissions and climate change (Shah et al., 2021; Kelbessa et al., 2023). Changing climate not only lowers crop productivity but also increases the total cost of agricultural products. Major weather-dependent crops such as wheat, rice, barley, maize, sorghum, and millet have seen large production losses due to climate change (Toulotte et al., 2022).

Governments and farmers are also working to reduce chemical inputs into horticultural systems as awareness of their importance has grown over the past few years. In recent years it has been observed that there is enhanced chemical contamination of rivers and underground water as well as the salinization of soils brought on by horticultural activities. One of the most awaited challenges in the field of horticulture is to guarantee high production levels for the plants by limiting the use of environmentally harmful synthetic chemicals. Therefore, solutions are required to maintain crop productivity and to simultaneously reduce chemical inputs in terms of chemical fertilizers and pesticides (Ruzzi & Aroca, 2015).

7.2.1 Abiotic Stress

Agricultural and horticultural crops are subject to different abiotic stresses (salinity, nutrient deficiency, drought, heat, cold, flooding, ultraviolet radiation, etc.) due to climate change, which finally causes oxidative stress in plant cells. Various abiotic factors negatively affect the crop yield and quality of agricultural crops which undermines global food security (Kumar & Verma, 2018; Qiao et al., 2022). Abiotic stresses account for around 50–70% of the world's food losses according to a recent report (Francini & Sebastiani, 2019). Various abiotic stresses transform cultivated areas into uncultivated regions. One such devastating abiotic stress is drought. It affects 64% of all land area and threatens agricultural productivity worldwide (Ma et al., 2020). It disrupts some normal functions of plants such as respiration, nutrient uptake, antioxidant defense mechanism, and photosynthesis. It can also lead to wilting and eventual death of plants due to insufficient water supply (Vishnupradeep et al., 2022). Plants close their stomatal pores under drought conditions, which adversely affects their metabolism mainly photosynthesis as it lowers CO_2 uptake and fixation. Drought also represents an adverse effect on the soil microorganisms, as it reduces the availability of nutrients. In addition, drought stress can lead to the development of secondary stress like oxidative stress, which may result in a dramatic rise in reactive oxygen species (ROS) formation (Abdelaal et al., 2021).

Salinity stress, on the other hand, can cause nutrient deficiency, ion imbalance, and water uptake issues in plants, which may result in restricted growth and reduced productivity. The primary cause of soil salinity is the accumulation of various salts like sodium chloride, sodium carbonate, sodium nitrate, and magnesium chloride in plants. Similar to drought stress, long-term exposure to salt causes oxidative stress, resulting in an excessive accumulation of reactive oxygen species (ROS) in plant tissues (Gupta et al., 2022). It was estimated that salinity was responsible for more than 50% of the average output loss worldwide. Numerous scientists have created improved methods for growing key crops such as potato, eggplant, cucumber, onion, carrot, cabbage, spinach, pepper, cauliflower, and lettuce (Shivakumar & Bhaktavatchalu, 2017).

Extreme temperatures, both high and low, can also cause significant damage to plant tissues and impede their growth and development. Heat stress negatively affects photosynthesis, membrane permeability, and water content of plants. Moreover, heat-induced stress has been observed to reduce the accumulation of proline, which is essential for the growth and flowering of plants (Khan et al., 2020; Mukhtar et al., 2023).

Heavy metal accumulation in the environment is a cause for concern and is another destructive stress factor for plants. Certain metals are necessary for a variety of plant metabolic processes at very little concentrations; however, at higher concentrations, these metals can have a detrimental effect on plants. Metals at their high concentrations disrupt soil structure, reduce microbial diversity, obstruct root growth, and cause water and nutrient imbalances, all of which reduce plant growth (Mohanty et al., 2021; Barman et al., 2022).

The effects of the aforementioned abiotic stresses are expected to increase as a result of climatic changes (Saijo & Loo, 2020).

7.2.2 Biotic Stress

Biotic stress, in combination with indirect losses due to changes in climate, has become a major obstacle to achieve sustainable food production. Current estimates of the losses to major food crops due to biotic stress have reached 20%–40% (Velásquez et al., 2018). Biotic stress can be induced in plants by various plant pathogens such as viruses, bacteria, fungi, insects, nematodes, and weeds, which directly affect plant health and also reduce crop productivity (Madani et al., 2019). Climate fluctuations can result in a shift in the geographic distribution of some plant pathogens like pests and disease-causing microbes, therefore increasing the vulnerability of plants to these threats. It should be noted that the impact of climate change may differ on the basis of the pathogen, as they may have positive, neutral, or negative consequences for crop disease development. Biotic stress can have a direct impact on host nutrients, resulting in the death of plants, both before and after harvest. The detrimental effects of plant diseases on food crops are bound to increase as a result of climate change due to the increased emergence and virulence of pathogens. Abiotic and biotic stresses have a significant impact on crop production, and climate change aggravates these stress factors (Yang et al., 2021).

At this critical juncture, there is an urgent need to shift to more sustainable production of crop processes that focus more on promoting sustainable mechanisms that allow crops to thrive in a wider range of environmental conditions and also produce optimal sustainable yields. Several strategies have been developed to promote sustainable agriculture, which includes the use of smart irrigation systems, agrochemicals, and the development of stress-tolerant strains, etc. but most of these techniques are time consuming, expensive, increase GHG emissions, and also lead to residual crop contamination that threatens human and animal health. In this context, the use of plant growth-promoting rhizobacteria (PGPR) and their compounds are found as an economic and environment-friendly approach to improve crop productivity under a wide range of adverse environmental conditions (Fiodor et al., 2021). A shift to organic and sustainable farming in agriculture, as well as horticulture, is to reduce dependence on chemical products and to raise plants' natural activities.

7.3 Role of PGPR in Enhancing the Growth and Mitigation of Various Biotic–Abiotic Stresses

PGPR inhabit the rhizosphere and are potentially beneficial for sustainable agriculture and horticulture. They improve the health of plant health and also promote the growth of the plant by using various direct and indirect mechanisms (Fig. 7.2).

Commonly fruit species are heterozygous and self-sterile, which has made PGPR inoculation crucial for fruit multiplication. This is due to the fact that fruit cultivars grown for commercial purposes do not grow true to type when reproduced from seeds. Thus, techniques for seedling propagation are useless. Many bacteria in the genera of *Agrobacterium*, *Alcali*, *Bacillus*, *Pseudomonas*, *Streptomyces* induces the formation of roots in stem cuttings (Bassil et al., 1991; Rinallo et al., 1999). These bacteria produce indole acetic acid (IAA) (Goto, 1990). Exogenous indole-3-butyric acid (IBA) can be used to speed up the roots of bacteria-infused cuttings. (Falasca et al., 2000).

As a sustainable approach, PGPR could be employed in agricultural systems to reduce the detrimental effects of various biotic and abiotic stresses and climate change conditions (Khan et al., 2020). Figures 7.3a and b shows different strategies used by PGPR against various biotic and abiotic stresses. Plants are typically less tolerant to nutritional shortages, osmotic shifts, or salinity than soil microbes

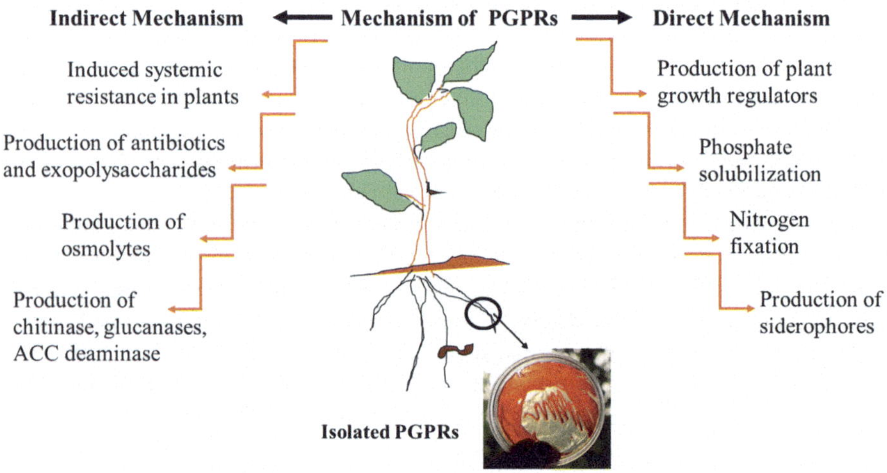

Fig. 7.2 Various mechanisms employed by PGPRs in plant growth and stress management

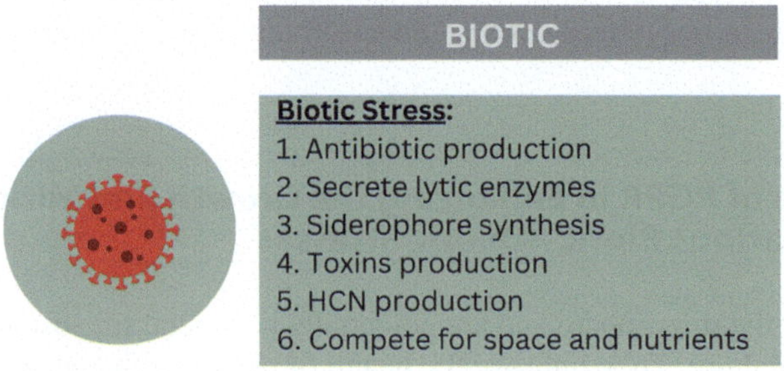

Fig. 7.3 (**a**) Role of PGPR against biotic stresses, (**b**) Role of PGPR against various abiotic stresses

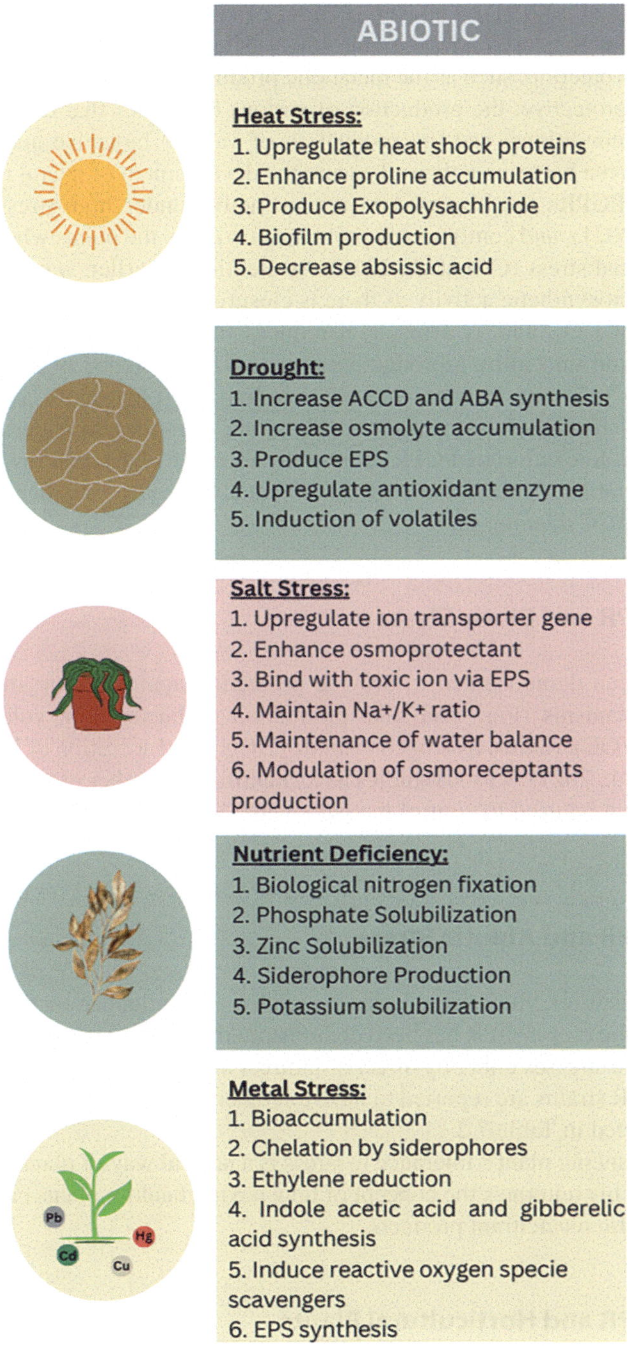

Fig. 7.3 (continued)

(Chanratana et al., 2019). As a result, PGPRs can play a protective role in supporting plants when they are under extreme stress. Microorganisms can provide different types of protection, such as the metabolic production of various compounds that can be osmoprotective, the production of various hormones that can act on plants under stress conditions, and the production of various bioactive and antioxidant molecules. Some bacteria in the soil ensure the development of lateral roots in cereals. Several PGPRs are observed to remain active under high-stress conditions (Shah et al., 2021), and continuously provide nutrition to the plants when they suffer from nutritional stress (Chu et al., 2019). As mentioned earlier, water stress results in loss of photosynthetic activity as there is closure of stomata resulting in reduced availability of CO_2, and as a result, it reduces the vegetative development, leaf senescence, and unhealthy reproductive organs. The existence of microbes in the soil increases tolerance to water stress due to enhanced hydraulic conductivity of the roots, which enables the extraction of the water from the soil under low gradient conditions. Colonization of PGPR around the roots of plants can lower the water stress in vegetables, cereals, and ornamental plants by controlling the level of ethylene through ACC deaminase.

7.3.1 PGPR and Biotic Stress

PGPR has been demonstrated to provide protection against biotic stress through various mechanisms (Fig. 7.3a) like production of bacteriocin, volatile organic compound (VOC), toxins, hydrogen cyanide (HCN), and secretion of lytic enzymes (Mohanty et al., 2021). Various studies have identified a number of promising PGPR strains that can be used to control the pathogens. Some of them are represented in Table 7.1.

7.3.2 PGPR and Abiotic Stress

As discussed earlier, all of the abiotic stresses have an influence on the overall performance of the crop. PGPR has been demonstrated to improve crop productivity by reducing these stresses either by direct or indirect mechanisms (Batool et al., 2020). Various PGPR strains are reported to have tolerance against different abiotic factors and are depicted in Table 7.2.

Thus, enhancing plant's tolerance to stress is a natural way of plant protection by the PGPRs. This reiterates the concept of how having healthy plants can reduce the use of synthetic recalcitrant products.

7.3.3 PGPR and Horticultural Plants

The interactions of growth promoting bacteria, mycorrhiza and plants play a very crucial role in enhancing the fertility of soil and its quality, biocontrol, decreasing

Table 7.1 Biotic stress tolerance in crops mediated by PGPR

No.	Biotic stress	PGPR strains	Crops	Reference
1	*Aphis gossypii glover*	*Rhizobium etli strain G12*	*Cucurbita pepo*	Martinuz et al. (2012)
2	*Fusarium oxysporum*	*Tricoderma koningiopsis Th003 WP*	*Physalis peruviana*	Díaz et al. (2013)
3	*Fusarium graminearum*	*Pseudomonas fluorescens*	*Triticum aestivum (wheat)*	Moussa et al. (2013)
4	*Bean yellow mosaic virus (BYMV)*	*Rhizobium leguminosarum*	*Vicia faba*	Al-Ani and Adhab (2013)
5	*Rhizoctonia solani*	*Bacillus amyloliquefaciens (SN13)*	*Oryza sativa*	Srivastava et al. (2016)
6	*Cucumber mosaic virus*	*Peanibacillus lentimorbus B-30488*	*Nicotiana tabacum*	Kumar et al. (2016)
7	*Spodoptera litura*	*Pseudomonas putida and Rothia sp*	*Solanum lycopersicum*	Bano and Muqarab (2017)
8	*Peronophythora litchii*	*Bacillus amyloliquefaciens LY-1*	*Litchi (Litchi chinensis Sonn.)*	Wu et al. (2017)
9	*Clavibacter michiganensis*	*Pseudomonas* sp. *23S*	*Solanum lycopersicum L*	Takishita et al. (2018)
10	*Fusarium wilt*	*Streptomyces thermocarboxydus*	*Solanum lycopersicum L*	Passari et al. (2019)

heavy metal toxicity, protection and propagation of plants as well as in horticulture (Table 7.3).

7.3.4 PGPR and Soil Bioremediation

Microbes can reduce the uptake of heavy metals by plants. In vegetable crops with mycorrhizal association, it is observed that there is a reduction in copper uptake and an increase in antioxidant activity (Prisa, 2023). Plants grown in soils polluted with heavy metals such as mercury adapted better when colonized in the root system by symbiotic microorganisms (El-Esawi et al., 2019). By enhancing plant growth and biomass in polluted areas, endophytic strains can increase the movement of heavy metals in the soil and promote phytoextraction. Many microorganisms can carry out the bioaccumulation of heavy metals in the cell biomass by reducing their mobility in the plant-soil system. Several strains of plant growth-promoting microorganisms promote physiological adaptation in heavy metal-polluted soils without leading to the accumulation of toxic products in leaves and roots (Bhise & Dandge, 2019).

Table 7.2 PGPR-mediated abiotic stress tolerance

No.	Crop	Stress type	PGPR strain	Mechanism	Reference
1	Maize (*Zea mays*)	Drought	*Klebsiella variicola* F2 (KJ465989); *Raoultella planticola* YL2 (KJ465991); *Pseudomonas fluorescens* YX2	Induced accumulation of betaine, choline, and glycine resulting in water loss	Gou et al. (2015)
2	Wheat (*Triticum aestivum L.*)	Drought and salinity	*Arthrobacter protophormiae* (SA3) and *Dietzia natronolimnaea* (STR1)	Increased IAA, reduced ABA and ACC levels, adjusted expression of ethylene signaling regulatory compartment (CTR1) pathway and DREB2 transcription factor	Barnawal et al. (2017)
3	Soybean (*Glycine max*)	Salinity	*Paenibacillus mucilaginosus*	Volatile organic compounds produced by bacteria decreased concentration of sodium ions in shoot and root as well as enhanced concentration of proline in root	Ma et al. (2018)
4	Pea (*Pisum sativum L.*)	Drought and heavy metals	*Rhizobium leguminosarum bv. Viciae*	ACC deaminase increased Nodulation, enhanced shoot biomass, higher water, and nutrient uptake	Belimov et al. (2019)
5	Maize (*Zea mays*)	Drought	*Bacillus licheniformis* (FMCH001)	Enhanced activity of the antioxidant enzyme catalase in roots; enhanced water use efficiency	Akhtar et al. (2020)
6	*Arabidopsis*	Drought	*Bacillus amyloliquefaciens* SN13	Increased concentrations of various osmolytes; upregulated expression of the AP2/ERF, GST, ERD4, and ARF2 genes, which are related to phytohormone production	Tiwari et al. (2021)

(continued)

Table 7.2 (continued)

No.	Crop	Stress type	PGPR strain	Mechanism	Reference
7	Wheat (*Triticum aestivum L.*)	Heat	*Bacillus safensis*	Increased expression of heat shock proteins (HSPs) and small HSPs	Sarkar et al. (2021)
8	Tomato (*Lycopersicon esculentum*)	Drought	*Pseudomonas fluorescens* G20–18	Improved chlorophyll content of leaves; enhanced accumulation of ABA in the roots; activity of various enzymes involved in carbohydrate metabolism increases; improved antioxidant capacity	Mekureyaw et al. (2022)
9	Pepper (*Piper nigrum*)	Drought	*Bacillus butanolivorans* KJ40	Activated peroxidase and glutathione peroxidase; increased expression of the sHSP, Cadhn, bZIP1, and NCED1 genes	Kim et al. (2022)
10	Tomato (*Lycopersicon esculentum*)	Heat	*Bacillus safensis* SCAL1	Increased activity of ACC deaminase, IAA, GA3, and kinetin; increased exopolysaccharide production; stimulated the synthesis of antioxidant enzymes; improved chlorophyll content	Mukhtar et al. (2023)

7.4 Plant Protection Products

The United Nations Environment Programme, has one of its major foci on plant protection. Plant diseases, weeds, and pests have led to plant damage, thereby affecting livelihoods, crisis in the supply of food, green cover, and carbon footprint. Countries across the world through their Ministries of Agriculture have initiated programs to ensure plant protection. These programs majorly focus on Integrated Pest Management by using products, especially pesticides and chemical fertilizers, for improving crop yield as well as eliminating new pests and pathogens. Plant protection products comprise fungicides, insecticides, plant growth regulators, preservatives, herbicides, etc., which have to be authorized by specific regulatory bodies. These plant protection products especially synthetic products have been instrumental in enhancing the production of plants and crops, enabling meeting the increasing demands of growing populations of countries, including India. However, long-term use of these products has led to detrimental impacts on the environment, biodiversity, and as well as health of humans. The environmental risk is dependent on the

Table 7.3 Beneficial effects of plant growth-promoting rhizobacteria on horticultural plants

Name of PGPR	Plant	Beneficial effect	Reference
G. mosseae *Rhizobium* *Sphingomonas* *P.agglomerans strain* FFd *Bacillus* strain *Bacillus megateriuma* *B. subtilis* strain TN Vel-35a *Serratia marcescens* *Rhizoglomus irregularis*	Tomato	Growth Protein production Plant height, gibberellin synthesis Increased mineral content, length of plant length as well as dry matter, weight, number and width of fruits Increased fresh root weight, stem diameter, and dry shoot weight Increased shoot and total plant dry weight Increased fresh and dry shoot and root weights; enhanced chlorophyll content and nutrient uptake, increased maximum fruit weight Increased shoot dry weight, plant height, number of fruits, and weight of yield per plant Delayed plant senescence and increased root mycorrhization, flowering, increased phosphorus content in the fruit, seedling height, stem diameter, and epigeal and hypogean biomass	Gamalero et al. (2004) Garcia-Fraile et al. (2012) Khan et al. (2014) Dursun et al. (2010) Kokalis-Burelle et al. (2002) Porcel et al. (2014) Babu et al. (2015) Almaghrabi et al. (2013) Conversa et al. (2013)
P. fluorescens *Rhizobium etli strain G12b* *Bacillus, Actinomycetes, Azotobacter strains*	Potato	Enhanced stolon length Increased resistance against disease stress Increased tuber weights and tuber numbers	Kloepper and Schroth (1981) Reitz et al. (2000) Oswald et al. (2010)
Rhizobium	Carrot	Siderophore production, protein production Increased dry matter of shoots and roots, increased Root length and root hair number	Garcia-Fraile et al. (2012) Flores-Félix et al. (2013)
Rhizobium *Rhizophagus irregularis* *Gigaspora margarita* *Bacillus licheniformis*	Pepper	Siderophore production, protein production Enhanced dry biomass Increased in the root and shoot length, and dry weight under drought stress	Garcia-Fraile et al. (2012) Tullio et al. (2007) Lim and Kim (2013)
A. brasilense strain Sp7-S *R. leguminosarum*	Lettuce	Increase in the number of leaves, seedling height, and root length Nutrient uptake	Mangmang et al. (2015) Cardarelli et al. (2010)

Agrobacterium *Alcaligenes sp*	Strawberry	Root inoculation increased Fruit yield Increased fruit yield, number, and weight under high Calcareous soil conditions	Ipek et al. (2011) Ipek et al. (2014)
B. subtilis	Apple	Increased trunk diameter Enhanced growth of apple leaves and improved fruit sweetness and moisture content	Utkhede and Li (1989) Ryu et al. (2011)
Pseudomonas putida	Grape	Increased graft callusing, scion shoot growth, cane hardening	Sabir (2013)
Pseudomonas chlororaphisd *Arthrobacter*	Walnut	Increased plant height, shoot and root dry weight, as well as increased phosphorus and nitrogen uptake	Yu et al. (2012)
P. fluorescens	Pear	Decreased bacteria population and freeze injury + INA	Lindow et al. (1996)
R. tropici	Field bean	Growth	Camacho et al. (2001)
Pisolithus albus	*Acacia spirorbis and Eucalyptus globulus*	Growth through uptake of N, P, K	Sousa et al. (2012)
Pseudomonas fluorescens	Broccoli	Enhanced plant growth, nutrient uptake	Tanwar et al. (2014)
R. leguminosarum	Peas	Root length, shoot height and dry weight	Dileep Kumar et al. (2001)
R. leguminosarum	Courgette	Nutrient uptake	Cardarelli et al. (2010)
Azospirillium	Rice	Product enhancement	Ji et al. (2020)
B. amyloliq-uefaciens	Wheat	Increased root growth	Talboys et al. (2009)

active substance of the product, its toxicity, and its degradability potential. These products can contaminate the environment through spray drifts, leaching, vaporization, sorption, surface run-offs, etc. (Tudi et al., 2021).

In India, the pesticide market has been increasing, as the demand for agricultural products has been increasing, and yet at the same time, plant diseases have been increasing. The fruit borer has led to a lot of loss of tomatoes in the past few years, thereby increasing its cost. According to IMARC's Market Research Report, in India, the pesticide market size that was 229.4 billion rupees will reach 342.3 billion rupees by 2028. In another report by Mordor Intelligence, there has been a steady rise in the India Crop Protection Chemicals Market, as the crop yield in India has been steadily decreasing. To combat this, there is an increased demand for plant protection products. Integrated Pest Management Practises in India have been adopted to suit the needs of the farmers as well as extensive detriments of the use of chemicals. There has been a steady shift to using biopesticides, including microbial pesticides, plant-based products like neem formulations, essential oils, and biofertilizers.

The concept of healthy plants and environment as a counter-action against pests and pathogens is a burgeoning sustainable practice. Green materials for pest management have been growing research across the globe. There have been several studies of new pesticides being developed from natural monoterpenes, rhamnolipids, thiolactomycin, and RNAi pesticides (Song et al., 2020). The RNAi pesticides are probably a robust method of protecting crops. The dsRNA is capable of binding to the target DNA of the pest and operates like an RNAi when applied (Hashiro & Yasueda, 2022).

This gradual shift to biobased plant protection products has been because the chemical-based products have led to a decrease in the soil quality, germination of seeds, root as well as shoot length of plants, content of chlorophyll, and photosynthesis. Moreover, pesticides have also been directly associated with several chronic diseases in humans, such as asthma, cancer, neurodegenerative disorders, reproductive disorders, and coronary artery diseases. (Pathak et al., 2022).

7.4.1 Plant Protection Products and Climate Change

A report released by Pesticide Action Network North America (PANNA) in January 2023 observed that the production and synthesis of pesticides emit several greenhouse gases contributing to global warming. Moreover, sulfuryl fluoride a type of pesticide is a greenhouse gas with a far too high potency than carbon dioxide. Some pesticides give rise to volatile organic compounds that can produce ozone upon reaction with nitrous oxide. The increased temperature will make the plants more susceptible to diseases, further increasing the demand for pesticides. Therefore, the recommendation is for a reduction in the use of synthetic pesticides and adopting agroecology, which supports pest-resistant crops.

Change in climate has resulted in a shift in the distribution of pests as well as pesticide resistance. A meta-analysis study linking pesticide resistance to the overwintering type of insect, Diamondback moth showed that climate warming can lead to pesticide resistance (Ma et al., 2021). Studies have shown that persistent use of pesticides is a threat to earthworms, and climate change such as increased temperature can result in changes in the moisture content of soil and pH, which can enhance the toxicity of chemicals and pesticides. Moreover, hypoxia too can directly affect the soil organisms. These changes can further affect the dynamic network between species that are growing over the ground and those that are growing underground. It has been predicted that these altered ecosystems due to environmental stress and toxic compounds can lead to a gradual loss of diversity and species (Kaka et al., 2021). In a study carried out in a protected Mediterranean coastal wetland, pesticides such as difenocon and azoxystrobin can pose unacceptable ecological risks for aquatic life, especially in conditions of increased temperature followed by heavy precipitation events (Martínez-Megías et al., 2023). Fusarium head blight is a major disease that has affected the wheat crop yield in many countries. Several studies have shown that air pollution and climate changes, such as increased temperature, increase the severity of the Fusarium head blight. Mycotoxin levels are found to be increased in the grains, and the grain yield is several folds lower due to the combination effect of the disease and climate change such as increased rain days and increased carbon dioxide content in the air (Chakraborty & Newton, 2011).

India being a tropical country has to bear the brunt of changes in climate, especially a rise in temperature and CO_2 levels. Having a large C3 plant cultivation, an increase in temperature is bound to affect productivity and biomass. Pests and diseases associated with a change in climate have been a major threat to chickpeas, cotton, soybean, chili, areca nut, paddy, etc. A warm, rainless humid climate seems to be favorable for pest multiplication by overconsumption and by a reduction in the overall nitrogen content in the plants and disadvantage for host plant resistance.

A paradigm shift to the use of biochemicals that enhance the immune system of plants has been envisaged to enable a reduction in the use of synthetic plant protection products. Natural elicitors targeting the plant defense mechanism are a great promise as they will enhance the plant's innate ability to resist diseases, as well as elicit the production of beneficial phytochemicals such as abscisic acid, jasmonic acid, and jasmonoyl isoleucine in the plants making them naturally healthy (Iriti & Vitalini, 2020; Xiong et al., 2023).

7.5 Plant Immunity and Its Elicitation

Plants have an innate immune system, which is able to identify invasion by pathogens. The ability to identify invasion is due to the presence of several receptors, termed pattern recognition receptors (PRR). The PRRs are able to recognize pathogen/molecular patterns associated with microbes (Fig. 7.4). PRRs are vital for plant development, resistance to stress associated with abiotic factors, and symbiotic interactions. Flagellin in bacteria may be recognized by the plant immune system, and yet it may be feasible that different receptors may be present in different plants to recognize different flagellin epitopes. An immunogenic epitope in bacteria may

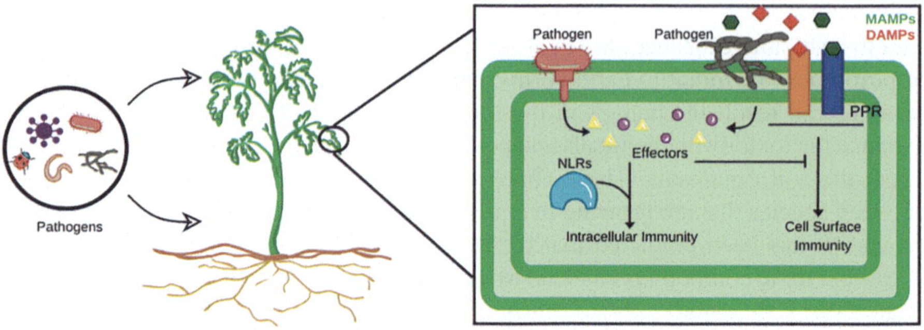

Fig. 7.4 Elicitation of plant immunity by cell surface receptors of pathogens and intracellular receptors in plants mediating PAMP-triggered immunity (PTI) and effector-triggered immunity (ETI), respectively

elicit an immune response in rice but not in Brassica or Arabidopsis. The plant immune system was also able to respond to effectors, released or present in the microbes. The resistance to disease has thus been attributed to the presence of Resistance (R) genes in the plant genome, which are triggered in response to effectors (Han, 2019). PRRs possess highly variable extracellular domains, such as leucine-rich repeat (LRR), lysine motif (LysM), lectin, and epidermal growth factor (EGF)-like domains, enabling them to recognize a wide range of ligands (Yu et al., 2017). The recognition of these ligands or effectors, produced by pathogens to colonize or proliferate, by the PRRs triggers an activated immune response that leads to programmed cell death that localizes and restricts the spread of the infection. *Capsicum chinense* plants that contain the PR gene, L^3, when infected with pepper mild mottle virus (PMMoV), were found to give rise to a hypersensitive reaction with the production of several PR proteins, such as glucanases, chitinases, peroxidases, osmotin–like proteins (Elvira et al., 2008).

At least two forms of defense pattern have been observed in several species of plants; pathogen-associated molecular pattern (PAMP) triggered immunity (PTI) and effector-triggered immunity (ETI) (Fig. 7.4). As mentioned earlier, the former can be triggered by a certain conserved molecular pattern such as flagellin, chitin from fungus, etc. It can also be resulting from plant materials such as cellobiose or oligogalacturonides arising from any damage to plant cells (Sowden et al., 2017). These patterns or biomolecules are recognized and stimulated by the pattern recognition receptors that are present on the cell surface of the plant. This mechanism is effective against non-adapted pathogens. Several pathogens survive and multiply by suppressing PAMP-triggered immunity by producing effector molecules (Dodds & Rathjen, 2010).

The effector-riggered immunity is based on the identification of pathogen-related molecules called effectors by internal receptors. These "effectors" are recognized through intracellularly located nucleotide-binding receptors that are rich in leucine. The subsequent series of responses focus on the production of proteins that resist the invasion of pathogens as well as secondary metabolites that may induce the formation of ROS, nitric oxide, and other plant defense molecules. It may also include the physical closing of stomata and strengthening of the cell wall and plasmodesmata. Thus, the effector-triggered immunity is targeted toward adapted pathogens. Studies have been reported on the synergistic behavior of these two mechanisms for a better and effective plant defense response (Liu et al., 2023).

There are three main signal transduction pathways that have been observed to be involved in the immune response to different pathogens. These are Jasmonic acid-dependent response, ethylene-dependent response, and salicylic acid-dependent response. Induction of these signaling pathways results in several responses such as the production of reactive oxygen species, phytoalexins; callose deposition and change in cell wall composition, synthesis of defense enzymes, and build-up of PR proteins (Thakur & Sohal, 2013; Sowden et al., 2017). In a study on *Arabidopsis thaliana*, it was observed that MAMP perception triggered a salicylic acid-dependent response leading to the accumulation of amino acids such as glutamine and serine. These accumulated amino acids suppress the growth and virulence of bacteria. Exogenous application of these amino acids too resulted in resistance against the bacteria, *Pseudomonas syringae* (Zhang et al., 2023). In another study, pyroglutamic acid derivatives from sugar beet were tested against *Zymoseptoria tritici*, a common wheat pathogen. Foliar application of these molecules reduced disease severity by 63%. These molecules activated plant defense dependent on lipooxygenases, peroxidase, and other related based pathways (Mejri et al., 2023). External application of salicylic acid and its derivative methyl salicylic acid in low concentrations has been shown to be effective in controlling pathogens and in improving photosynthesis (Sowden et al., 2017).

Based on their chemical nature and properties, plant immunity inducers have been classified into different classes (Fig. 7.5). (Yang et al., 2022).

Fungal elicitors have been classified into saccharine elicitors and protein elicitors. Chitins and chitosans found in the cell walls of fungi are able to elicit plant immune response by increasing chitinase and phytoalexin production, modifying ion fluxes and ROS, and altering gene expression. Glucans present in the cell wall of fungi can be recognized by receptors in plants eliciting an immune response. A glycoprotein isolated from *Colletotrichum falcatum* induced an increase in total phenolics, phenylalanine ammonia-lyase and peroxidase in sugarcane leaves as well as in vitro cell suspensions of sugarcane (Sundar et al., 2002). The fungal protein elicitors include cell wall-related enzymes such as carbohydrate esterases, polysaccharide lyases, and glycoside hydrolases (GH). Several GH class elicitors have been identified in the non-pathogenic *Trichoderma* spp., making them effective biocontrol agents (Guo & Cheng, 2022). The causative fungi, *Eutypa lata*, which is responsible for grapevine trunk disease, secretes secondary metabolites, Eutypine, and its derivatives, which are able to induce defense response genes in the grapevine (Guan et al., 2022).

Thus, it is to be understood that PAMP-triggered immunity and effector-triggered immunity have the capacity to induce systemic acquired resistance using salicylic acids. NPR1 (nonexpressor of pathogenesis-related genes) and PR genes (pathogenesis related) are actively involved in resisting disease using jasmonic acid and

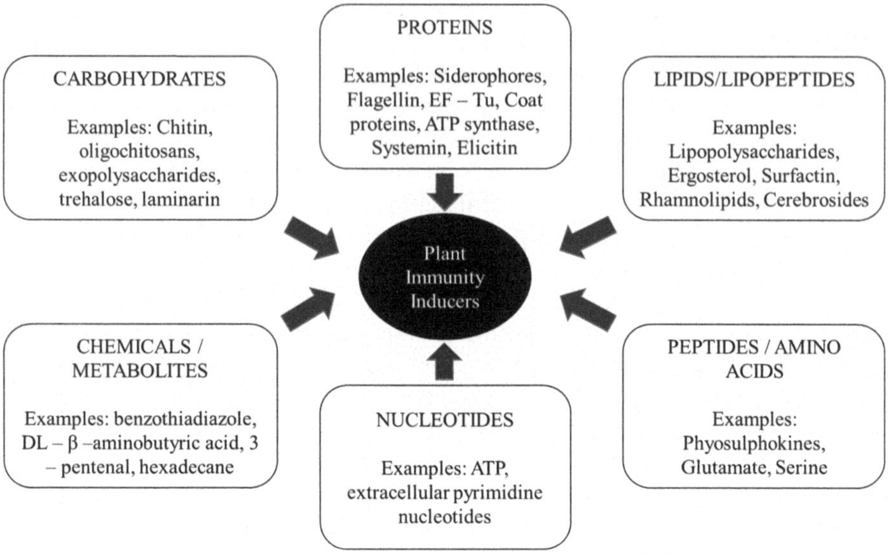

Fig. 7.5 Classes of plant immunity inducers

salicylic acid signaling pathways. It has been reported that NPR1 the master key player is translocated to the nucleus to activate PR gene expression. Under normal circumstances, NPR1 is found in the cytoplasm in an oligomeric form. During stressful conditions, salicylic acid is reported to accumulate in plants. Shah et al. reported that salicylic acid may be reversibly binding to catalase and other hydrogen peroxide scavenging enzymes like ascorbate peroxidase to inhibit them (Shah et al., 1997). Thus a reducing condition is created in plants, which leads to the monomerization of the NPR1. Liu et al. reported that these monomeric forms are phosphorylated to facilitate its translocation to the nucleus and facilitate the expression of PR gene (Liu et al., 2023). It has been reported that PR gene expression is required for developing a systemic acquired response.

Chen et al. reported the role of N-hydroxypipecolic acid (NHP) acting synergistically with salicylic acid in systemic plant immunity (Chen et al., 2018). NHP which is synthesized from lysine is hypothesized to be facilitating salicylic acid synthesis and salicylic acid regulating the NHP biosynthesis in return—a double whammy (Sun et al., 2020)! How is salicylic acid participating in the amplification of receptors that are able to recognize the pattern and nucleotide binding and leucine - rich receptors (NLR)-mediated signalingproduced in plants? Salicylic acid is produced primarily by the isochorismate pathway (Fig. 7.6). Chorismate is converted in plastids to isochorismate-by-isochorismate synthase 1 (ICS1). With the signal for initiating defense, isochorismate is transported to the cytosol from plastids. It then reacts with glutamate to form isochorismate-9-glutamate. This spontaneously gets converted to salicylic acid. It is also important to take into account that the synthesis of salicylic acid is upregulated during stressful conditions in a plant. ISC1 is under

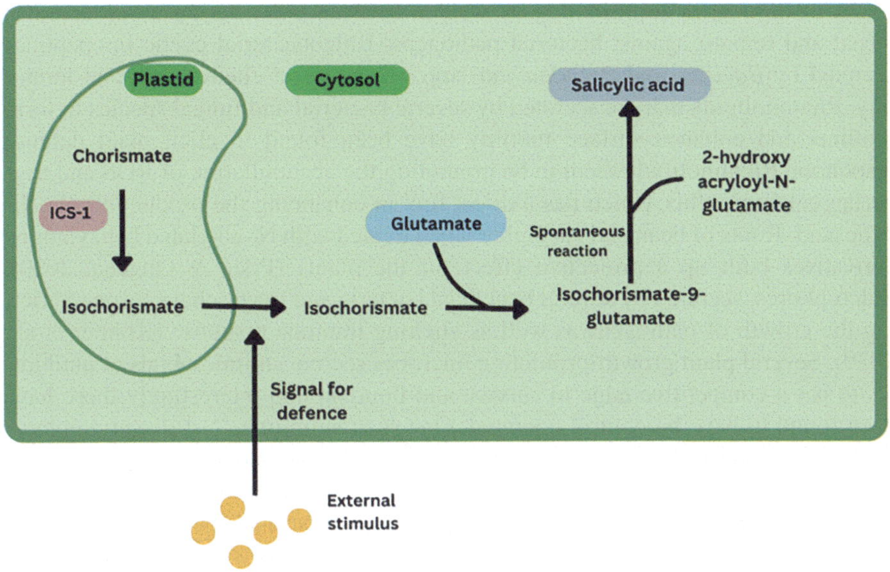

Fig. 7.6 Synthesis of salicylic acid through isochorismate pathway in response to a signal for defense

positive and negative regulation. Calmodulin binding protein-60 and calcium signaling are reported to have a key role in the activation of salicylic acid biosynthesis (Huang et al., 2020).

7.5.1 Plant Immunity and Plant–Microbe Interaction

Apart from the elicitation of plant immunity due to plant pathogens, pests, etc., there are various studies that have shown that interactions of plants with beneficial microbes lead to enhanced plant immunity. This gives the scope of using good microbes and their metabolites as plant immunizers or plant vaccines, as the plants are able to remember the infection. A mutualistic symbiotic association of plants with friendly microbes can create an environment within the plant that resists the growth of harmful pathogens. Colonization by Arbuscular mycorrhizal fungi and symbiosis with rhizobia bacteria have enabled the suppression of plant diseases and pests by elicitation of induced systemic resistance in the plants (Nishad et al., 2020).

Several fungi and bacteria have been identified that not only promote plant growth, they also act as biocontrol agents. These microbes have within them molecules that enable them to promote plant growth, improve root colonization, improve communication to enhance colonization, as well as help, induce systemic resistance in the plants. Plant rhizospheric microbiome is therefore a focal point of research. These molecules help the plants by directly interacting with the pathogen and also by triggering the plant's immune response. Acyl-homoserine lactones have been found to have growth-promoting as well as resistance elicitation as observed in wheat and tomato against bacterial pathogens. Rhizobacterial cyclic lipopeptides secreted by *Bacillus* and *Pseudomonas* spp. are effective elicitors of plant immunity. Rhamnolipids that are secreted by several bacterial and fungal species to form biofilms and enhance surface motility have been found to elicit plant defense responses. Rhamnolipids seem to be promoting the accumulation of ROS and triggering calcium influx which has a direct link in enhancing the production of salicylic acid. Roots of bean and cucumber when treated with N–alkylated benzylamine derivatives built up a protective effect for the plants (Pršić & Ongena, 2020). Siderophores, secreted by several beneficial bacteria and fungi, chelate iron restricting the growth of pathogens as well as eliciting immune response (Kramer et al., 2020). Several plant growth-promoting microbes secrete antimicrobials or antibiotics to get a competitive edge to survive and flourish, and interestingly these have been found to have biocontrol potential. One such antibiotic, 2, 4-diacetyl phloroglucinol produced by *P. fluorescens*, has been found to confer resistance in Arabidopsis against several pathogens (Chae et al., 2020). Several volatile organic compounds secreted by plant growth-promoting microbes have been identified to be immune elicitors. It has been reported that different stereoisomers of the volatile organic compound, 2, 3-butanediol, are able to induce systemic resistance in pepper against different viruses (Kong et al., 2018). This phenomenon provides long-lasting as well as broad-spectrum crop protection, although a clear understanding of the mode of action of these elicitors, will enable their use more effectively. Research across the globe is focusing on reducing pesticide use by incorporating biocontrol agents in agricultural practices. The dose and time of application of these biocontrol agents are very important in eliciting a good immune response.

7.6 Biocontrol Applications, Microbe Interactions, and Climate

Agroeconomy and green crop protection policies in several parts of the world have paved way for the use of biocontrol agents to improve crop health and suppress diseases. This has entailed a detailed study on the preparation of different formulations and their application success. An exogenous application may also affect the existing rhizosphere microbiome and hence studies are being carried out to understand the changes in communities under and below the ground and its correlation with climate change.

A biocontrol formulation prepared using *Pseudomonas chlororaphis*, *PcPCL1606* (isolated from rhizosphere of Avocado) was tested for its effect on rhizospheric microbial communities. A metagenomic analysis proved that the application of this formulation did not have any detectable effect on the microbial communities in soil without the pathogenic fungi, *R. necatrix.* In soil with the pathogen, there was an observed increase in species that had antifungal potential and there was a marked reduction in the proliferation of the pathogen. The antifungal potential of *PcPCL1606* has been attributed to the production of 2-hexyl, 5-propylresorcinol (Tienda et al., 2020).

Different formulations of *Trichoderma spp and Talaromyces* spp. have been explored to assess their biocontrol potential. These fungi have been observed to through different mechanisms suppress the growth of pathogenic fungi. In a study with peat-based and talc-based formulations of *T. flavus*, sugar beet seeds were coated with this formulation and sown in soil infected with *R.solani*. After 60 days, the increased number of healthy seedlings indicated that *T. flavus* was able to control the sugar beet damping off disease (Kakvan et al., 2013). Wilt caused by *Fusariam oxysporium f.* sp. *lycopersici* has been a very highly prevalent disease. Several biocontrol formulations have been successfully prepared to combat this plant pathogen. Water-dispersible granules were developed using *Bacillus subtilis* and *Trichoderma harzianum*, and these showed a 48% reduction in disease incidence and an increase in the plant growth promotion. The culture filtrates on analysis were found to have chitinases, glucanases, proteases, volatile organic compounds, and metabolites that significantly reduced the mycelial growth of the pathogen (Jangir et al., 2021). A study with a granular formulation comprising *T. asperellum* spores showed that corn seedlings were protected from *R. solani* with a 73% reduction in infected plants (Herrera et al., 2020). *Psuedomomas aureofacians* strains have been successfully used to control cotton damping off by *R. solani*. Organic formulations of this strain enhanced the phenazine production that was responsible for the antifungal potential (Samavat et al., 2014). *Serratia marcenses* strain ETR17, isolated from the rhizosphere has biocontrol potential as it showed significant in vitro antagonism against different foliar and root pathogens of tea plant (Dhar Purkayastha et al., 2018). *Bacillus subtilis AF1* bioformulation when supplemented with chitin showed improved biocontrol as well as plant-promoting potential. The formulation was observed to promote seedling growth of both groundnut and pea in soil infested with *A. niger* and *Fusariam udum* (Manjula & Podile, 2011).

These fungi through their repository of metabolites enable symbiotic associations with plants as well as reduce disease burden, and are contributing to alleviating the effects of climate change. Their carbon dioxide storing capacity is instrumental in reducing greenhouse gases. They have been reported to be harboring gigatons of carbon (Hawkins et al., 2023). Apart from their carbon sequestration, fungi can also be used to produce biodegradable and energy-economical products, thereby meeting the United Nations Sustainable Development Goals. Interestingly, several meta-analyses studies have shown that elevated carbon dioxide levels enhance the plant–microbe interactions, although more data is needed to support this phenomenon as a global phenomenon. As a contraindication, there have been studies that have also indicated that elevated levels of greenhouse gases have led to a reduction in certain species of both plants as well as microbes. The metabolite exudation from the roots of the plants has a direct effect on the composition of the rhizospheric microbial communities. Therefore, elevated carbon dioxide, increased temperature, and drought stress can lead to the establishment of different PGPBs and PGPFs in the soil (Compant et al., 2010). Fungal biotechnology should enable petroleum-based economy to move to a biobased economy mitigating all types of pollution and building a strong, healthy ecosystem (Antonelli et al., 2019). Due to the overuse of plastics, soil has microplastics that affect the microbiome, as well as enter the food web. Moreover, plastics life cycle also leads to the emission of greenhouse gases. Several fungi and bacteria have been studied that have plastic degrading potential, and these can mitigate the effect of plastics on the environment (Srikanth et al., 2022). Due to increased use of biocontrols, there should be a significant decline in the use of industrially manufactured chemicals, which in turn leads to less burning of fossil fuels and greenhouse gas emissions.

Change in climate has led to an increase in abiotic stress such as floods, droughts, and salinity. and these have direct implications on soil and plant health. Tolerant plant growth-promoting microbes to a great extent have been reported to increase the growth of plants in response to these abiotic stresses. Researchers have identified and characterized many such tolerant plant growth-promoting microbes (Naamala & Smith, 2020). Syntrophic microbial growth promotion is an intriguing phenomenon that can mitigate plant diseases and adverse effects of climate change. Syntrophy enables the utilization of a substrate by combined metabolic pathways in multiple microbial species. An example of syntrophy is an association of methanotrophs in the rhizosphere with *Hyphomicrobium* spp. Methnotrophs can be inhibited by methanol, but its association with *Huphomicrobium* spp., which removes methanol, enables it to sustain itself. Thus, this is a phenomenon by which harmful substances in the environment can be removed or depleted (Liechty et al., 2020).

Facilitated migration is another phenomenon in which microbes interact with other microbes in the community to migrate from one place to another place to enable root colonization. Fungal hyphae and mycelia create fungal highways that can be used by other species to migrate. Mycelia of *Phomopsis liquidambaris* is used by rhizobia to migrate to the rhizosphere of legumes for nodulation. This concept enables the development of multiple strain biocontrol agents to improve the survival of disease-suppressing microbes and adapt to changing environmental conditions (Niu et al., 2020). The soil microbiome is also instrumental in regulating biogeochemical cycles of nutrients and their sequestration and improving water retention. The addition of soil conditioners such as biochar, straw, etc. can augment the soil fertility, and the microbiome, especially biocontrol agents (Zonayet et al., 2023).

Numerous studies demonstrate that plants can engage in simultaneous communication with the rhizospheric and mycorrhizospheric microbial population through a sophisticated system of species- and strain-specific communication involving a highly specialized conversation at the germplasm strain level (Rodríguez et al., 2020). Recent research has shown that bacteria can communicate, which is essential for their survival and ability to compete (Sagar et al., 2021). This has a variety of effects, such as suppressing rivals through cooperative behavior that benefits both the individual and the group (Rocha et al., 2012). Without a doubt, the dynamics of the root microbiome depend heavily on bacterial communication (Atkinson & Williams, 2009; Rocha et al., 2012). Signal molecules must be created, dispersed, and then recognized by the other bacteria in the community in order for bacteria to communicate. After perception, signal molecules change gene transcription, which affects the physiology and activity of recipients (Atkinson & Williams, 2009). Communication is essential for community coordination and the continuation of effective bacterial functions. The formation of biofilm, adhesion, and motility has also been demonstrated to be associated with signal molecule-mediated communication (Rocha et al., 2012), control of virulence-associated factors (Yajima, 2014), and propagation (Atkinson & Williams, 2009). Regulation of various characteristics of bacterial behavior is frequently associated with population density. Quorum sensing (QS) is the term used to describe this phenomenon, which is caused by density-dependent stimuli and reactions (Sperandio et al., 2002; Weisskopf, 2022). Additionally, it has been discovered that these endophytic cell populations can be significantly more effective at enhancing the growth of plant growth than their non-endophytic and rhizospheric counterparts, improving plant physiology and adaptability to both biotic and abiotic environmental stressors.

The potential of plant–microbe interactions holds promise in mitigating the detrimental effects of climate change. However, the race between the speedy changes in global climate change and the efficacy of plant–microbe interactions to combat these changes will decide the tilt in the future. Studies have shown that elevated temperatures have increased microbial respiration rate, affected utilization of organic compounds, and increased virulence of soil-borne pathogens. Low soil moisture has a negative impact on the rhizosphere microbiome. UV radiation too can have an impact on the survival of microbiomes (Sharma et al., 2022). The incidences of plant diseases will increase with an increase in the severity of global climate change, which in turn will have an impact on food security, economy, and sociocultural aspects (Singh et al., 2023). With advancements in biotechnology and bioinformatics, the engineering of microbes, plants, and rhizospheres to withstand and strive under various environmental changes has been a progressive area of research (Dries et al., 2021). Understanding the fine details of plant–microbe interactions and microbe–microbe interactions will be key to building a healthy ecosystem and surviving the onslaught of the implications of climate change.

7.7 Concluding Remarks

Our current understanding of the various mechanisms that plants use for self-defense against pathogens is remarkable. The complex mechanisms of NLR, pathogen recognition receptors, and the promising role of salicylic and jasmonic acids are areas for further studies to develop resist varieties for enhancement of life on earth as well as unraveling the mechanism of plant immunity.

Using PGPR as a bio-fertilizer, bio-pesticide, or bio-reactive agent provides hope for improving crop yield in a changing climate, which is essential for global food security. Although several potential PGPR have been identified by researchers, their commercialization has not yet been successful. Overall, it can be concluded that researchers have a better understanding of the fundamental mechanisms of PGPR. However, more research is necessary to further understand the interactions between microbes and plants. On a commercial scale, a considerable amount of work remains to be done to ensure that PGPR is an effective tool in sustainable agriculture. In addition, future research should also focus on identifying new stress-resistant PGPR strains as well as identifying which genes and metabolites of PGPR play a role in abiotic and biotic stress tolerance in the context of climate change. Several advanced approaches like metagenomics, CRISPR, and metabolomics could be used in this context. In recent years, plant microbiome engineering has also attracted significant scientific interest due to its eco-friendly nature and its potential to increase crop productivity in changing climate conditions. However, further research into the microbiome of plants and its interactions with other microorganisms is necessary in order to gain a better scientific foundation from which to implement novel strategies for increasing crop productivity.

References

Abdelaal, K., AlKahtani, M., Attia, K., Hafez, Y., Király, L., & Künstler, A. (2021). The role of plant growth-promoting bacteria in alleviating the adverse effects of drought on plants. *Biology, 10*(6), 520.

Agnihotri, P., & Mitra, A. K. (2023). Understanding the impact of global climate change on abiotic stress in plants and the supportive role of PGPR. In M. Oliveira & A. Fernandes-Silva (Eds.), *Abiotic stress in plants – Adaptations to climate change*. IntechOpen; EBOOK (PDF) ISBN978-1-83768-498-4. https://doi.org/10.5772/intechopen.104274

Akhtar, S. S., Amby, D. B., Hegelund, J. N., Fimognari, L., Großkinsky, D. K., Westergaard, J. C., Müller, R., Moelbak, L., Liu, F., & Roitsch, T. (2020). *Bacillus licheniformis* FMCH001 increases water use efficiency via growth stimulation in both normal and drought conditions. *Frontiers in Plant Science, 11*, 297.

Al-Ani, R. A., & Adhab, M. A. (2013). Bean yellow mosaic virus (BYMV) on broadbean: Characterization and resistance induced by *Rhizobium leguminosarum. Journal of Pure and Applied Microbiology, 7*(1), 135–142.

Almaghrabi, O. A., Massoud, S. I., & Abdelmoneim, T. S. (2013). Influence of inoculation with plant growth promoting rhizobacteria (PGPR) on tomato plant growth and nematode reproduction under greenhouse conditions. *Saudi Journal of Biological Sciences, 20*, 57–61.

Antonelli, A., Smith, R. J., & Simmonds, M. S. J. (2019). Unlocking the properties of plants and fungi for sustainable development. *Nature Plants, 5*, 1100–1102.

Atkinson, S., & Williams, P. (2009). Quorum sensing and social networking in the microbial world. *The Journal of the Royal Society Interface, 6*(40), 959–978.

Babu, N. A., Jogaiah, S., Ito, S. I., Kestur Nagaraj, A., & Tran, L. S. P. (2015). Improvement of growth, fruit weight and early blight disease protection of tomato plants by rhizosphere bacteria is correlated with their beneficial traits and induced biosynthesis of antioxidant peroxidase and polyphenol oxidase. *Plant Science, 231*, 62–73.

Bano, A., & Muqarab, R. J. (2017). Plant defence induced by PGPR against *Spodoptera litura* in tomato (*Solanum lycopersicum L.*). *Plant Biology, 19*(3), 406–412.

Barman, D., Dutta, I., & Jha, D. K. (2022). Heavy metal resistant bacteria from coal dumping site with plant growth promoting potentials. *Biologia, 77*, 1–3.

Barnawal, D., Bharti, N., Pandey, S. S., Pandey, A., Chanotiya, C. S., & Kalra, A. (2017). Plant growth-promoting rhizobacteria enhance wheat salt and drought stress tolerance by altering endogenous phytohormone levels and TaCTR1/TaDREB2 expression. *Physiologia Plantarum, 161*(4), 502–514.

Bassil, N. V., Proebsting, W. M., Moore, L. W., & Lightfoot, D. A. (1991). Propagation of hazelnut stem cuttings using Agrobacterium *rhizogenes*. *Horticultural Science, 26*, 1058–1060.

Batool, T., Ali, S., Seleiman, M. F., Naveed, N. H., Ali, A., Ahmed, K., Abid, M., Rizwan, M., Shahid, M. R., Alotaibi, M., & Al-Ashkar, I. (2020). Plant growth promoting rhizobacteria alleviates drought stress in potato in response to suppressive oxidative stress and antioxidant enzymes activities. *Scientific Reports, 10*(1), 16975.

Belimov, A. A., Zinovkina, N. Y., Safronova, V. I., Litvinsky, V. A., Nosikov, V. V., Zavalin, A. A., & Tikhonovich, I. A. (2019). Rhizobial ACC deaminase contributes to efficient symbiosis with pea (*Pisum sativum L.)* under single and combined cadmium and water deficit stress. *Environmental and Experimental Botany, 167*, 103859.

Bhise, K. K., & Dandge, P. B. (2019). Alleviation of salinity stress in rice plant by encapsulated salt tolerant plant growth promoting bacteria *Pantoea agglomerans* strain KL and its root colonization ability. *Archives of Agronomy and Soil Science, 65*(14), 1955–1968.

Camacho, M., Santamaria, C., Temprano, F., Rodriguez-Navarro, D. N., & Daza, A. (2001). Coinoculation with Bacillus sp. CECT 450 improves nodulation in *Phaseolus vulgaris* L. *Canadian Journal of Microbiology, 47*, 1058–1062.

Cardarelli, M., Rouphael, Y., Rea, E., & Colla, G. (2010). Mitigation of alkaline stress by arbuscular mycorrhiza in zucchini plants grown under mineral and organic fertilization. *Journal of Plant Nutrition and Soil Science, 173*(5), 778–787.

Chae, D. H., Kim, D. R., Cheong, M. S., Lee, Y. B., & Kwak, Y. S. (2020). Investigating the induced systemic resistance mechanism of 2,4-diacetylphloroglucinol (Dapg) using dapg hydrolase-transgenic arabidopsis. *Plant Pathology Journal, 36*, 255–266. https://doi.org/10.5423/PPJ.OA.02.2020.0031

Chakraborty, S., & Newton, A. C. (2011). Climate change, plant diseases and food security: An overview. *Plant Pathology, 60*, 2–14. https://doi.org/10.1111/j.1365-3059.2010.02411.x

Chanratana, M., Joe, M. M., Roy Choudhury, A., Anandham, R., Krishnamoorthy, R., Kim, K., & Sa, T. (2019). Physiological response of tomato plant to chitosan-immobilized aggregated *Methylobacterium oryzae* CBMB20 inoculation under salinity stress. *3 Biotech, 9*, 1–13.

Chen, Y. C., Holmes, E. C., Rajniak, J., Kim, J. G., Tang, S., Fischer, C. R., Mudgett, M. B., & Sattely, E. S. (2018). N-hydroxy-pipecolic acid is a mobile metabolite that induces systemic disease resistance in Arabidopsis. *Proceedings of the National Academy of Sciences of the United States of America, 115*(21), E4920–E4929.

Chu, T. N., Tran, B. T. H., Van Bui, L., & Hoang, M. T. T. (2019). Plant growth-promoting rhizobacterium *Pseudomonas* PS01 induces salt tolerance in Arabidopsis thaliana. *BMC Research Notes, 12*(1), 1–7.

Compant, S., Van Der Heijden, M. G., & Sessitsch, A. (2010). Climate change effects on beneficial plant–microorganism interactions. *FEMS Microbiology Ecology, 73*(2), 197–214. https://doi.org/10.1111/j.1574-6941.2010.00900.x

Conversa, G., Lazzizera, C., Bonasia, A., & Elia, A. (2013). Yield and phosphorus uptake of a processing tomato crop grown at different phosphorus levels in a calcareous soil as affected by mycorrhizal inoculation under field conditions. *Biology and Fertility of Soils, 49*, 691–703.

Dhar Purkayastha, G., Mangar, P., Saha, A., & Saha, D. (2018). Evaluation of the biocontrol efficacy of a *Serratia marcescens* strain indigenous to tea rhizosphere for the management of root rot disease in tea. *PLoS One, 13*(2), e0191761. https://doi.org/10.1371/journal.pone.0191761

Díaz, A., Smith, A., Mesa, P., Zapata, J., Caviedes, D., & Cotes, A. M. (2013). Control of *Fusarium* wilt in cape gooseberry by *Trichoderma koningiopsis* and PGPR. *IOBC-WPRS Bulletins, 86*, 89–94.

Dileep Kumar, B. S., Berggren, I., & Martensson, A. M. (2001). Potential for improving pea production by co-inoculation with fluorescent *Pseudomonas* and *Rhizobium*. *Plant and Soil, 229*, 25–34.

Dodds, P. N., & Rathjen, J. P. (2010). Plant immunity: Towards an integrated view of plant–pathogen interactions. *Nature Reviews. Genetics, 11*(8), 539–548.

Dries, L., Hendgen, M., Schnell, S., Löhnertz, O., & Vortkamp, A. (2021). Rhizosphere engineering: Leading towards a sustainable viticulture. *OENO one, 55*(2), 353–363. https://doi.org/10.20870/oeno-one.2021.55.2.4534

Dursun, A., Ekinci, M., & Donmez, M. F. (2010). Effects of foliar application of plantgrowth promoting bacterium on chemical contents, yield and growth of tomato (Lycopersicon Esculentum L.) and cucumber (Cucumis sativus L.). *Pakistan Journal of Botany, 42*, 3349–3356.

El-Esawi, M. A., Al-Ghamdi, A. A., Ali, H. M., & Alayafi, A. A. (2019). *Azospirillum lipoferum* FK1 confers improved salt tolerance in chickpea (Cicer arietinum L.) by modulating osmolytes, antioxidant machinery and stress-related genes expression. *Environmental and Experimental Botany, 159*, 55–65.

Elvira, M. I., Galdeano, M. M., Gilardi, P., García-Luque, I., & Serra, M. T. (2008). Proteomic analysis of pathogenesis-related proteins (PRs) induced by compatible and incompatible interactions of pepper mild mottle virus (PMMoV) in Capsicum chinense L 3 plants. *Journal of Experimental Botany, 59*(6), 1253–1265. https://doi.org/10.1093/jxb/ern032

Falasca, G., Reverberi, M., Lauri, P., Caboni, E., De Stradis, A., & Altamura, M. M. (2000). How *Agrobacterium rhizogenes* triggers de novo root formation in a recalcitrant woody plant: An integrated histological; ultrastructural and molecular analysis. *The New Phytologist, 145*, 77–93.

Field, C. B. (Ed.). (2012). *Managing the risks of extreme events and disasters to advance climate change adaptation: Special report of the intergovernmental panel on climate change*. Cambridge University Press.

Fiodor, A., Singh, S., & Pranaw, K. (2021). The contrivance of plant growth promoting microbes to mitigate climate change impact in agriculture. *Microorganisms., 9*(9), 1841.

Flores-Félix, J. D., Menéndez, E., Rivera, L. P., Marcos-García, M., Martínez-Hidalgo, P., Mateos, P. F., Martínez-Molina, E., Velázquez, M. E., García-Fraile, P., & Rivas, R. (2013). Use of *Rhizobium leguminosarum* as a potential biofertilizer for Lactucasativa and Daucus carota crops. *Journal of Plant Nutrition and Soil Science, 176*, 876–882.

Francini, A., & Sebastiani, L. (2019). Abiotic stress effects on performance of horticultural crops. *Horticulturae, 5*(4), 67.

Gamalero, E., Trotta, A., Massa, N., Copetta, A., Martinotti, M. G., & Berta, G. (2004). Impact of two fluorescent *Pseudomonad*s and an arbuscular mycorrhizal fungus on tomato plant growth, root architecture and P acquisition. *Mycorrhiza, 14*, 185–192.

Garcia-Fraile, P., Carro, L., & Robledo, M. (2012). Rhizobium promotes non-legumes growth and quality in several production steps: Towards a biofertilization of edible raw vegetables healthy for humans. *PLoS One, 7*, 38122.

Goto, M. (1990). *Fundamentals of bacterial plant pathology*. Academic.

Gou, W. E., Tian, L. I., Ruan, Z. H., Zheng, P. E., Chen, F. U., Zhang, L., Cui, Z., Zheng, P., Li, Z., Gao, M. E., & Shi, W. E. (2015). Accumulation of choline and glycinebetaine and drought stress tolerance induced in maize (*Zea mays*) by three plant growth promoting rhizobacteria (PGPR) strains. *Pakistan Journal of Botany, 47*(2), 581–586.

Guan, P., Schmidt, F., Fischer, J., Riemann, M., Thines, E., & Nick, P. (2022). The fungal elicitor eutypine from Eutypa lata activates basal immunity through its phenolic side chains. *Horticulture Research, 9*, uhac120. https://doi.org/10.1093/hr/uhac120

Guo, J., & Cheng, Y. (2022). Advances in fungal elicitor-triggered plant immunity. *International Journal of Molecular Sciences, 23*(19), 12003. https://doi.org/10.3390/ijms231912003

Gupta, A., Mishra, R., Rai, S., Bano, A., Pathak, N., Fujita, M., Kumar, M., & Hasanuzzaman, M. (2022). Mechanistic insights of plant growth promoting bacteria mediated drought and salt stress tolerance in plants for sustainable agriculture. *International Journal of Molecular Sciences, 23*(7), 3741.

Han, G. Z. (2019). Origin and evolution of the plant immune system. *The New Phytologist, 222*, 70–83. https://doi.org/10.1111/nph.15596

Hashiro, S., & Yasueda, H. (2022). RNA interference-based pesticides and antiviral agents: Microbial overproduction systems for double-stranded RNA for applications in agriculture and aquaculture. *Applied Sciences, 12*(6), 2954.

Hawkins, H. J., Cargill, R. I., Van Nuland, M. E., Hagen, S. C., Field, K. J., Sheldrake, M., Soudzilovskaia, N. A., & Kiers, E. T. (2023). Mycorrhizal mycelium as a global carbon pool. *Current Biology, 33*(11), 560–573.

Herrera, W., Valbuena, O., & Pavone-Maniscalco, D. (2020). Formulation of *Trichoderma asperellum* TV190 for biological control of *Rhizoctonia solani* on corn seedlings. *Egyptian Journal of Biological Pest Control, 30*, 44. https://doi.org/10.1186/s41938-020-00246-9

Huang, W., Wang, Y., Li, X., & Zhang, Y. (2020). Biosynthesis and regulation of salicylic acid and N-hydroxypipecolic acid in plant immunity. *Molecular Plant, 13*(1), 31–41.

Ipek, M., Pirlak, L., Esitken, A., Donmez, M. F., Turan, M., & Sahin, F. (2011). Plant growth-promoting rhizobacteria (PGPR) increase yield, growth and nutrition of strawberry under high calcareous soil conditions. *Journal of Plant Nutrition and Soil Science, 37*(7), 990–1001.

Ipek, M., Pirlak, L., Esitken, A., Dönmez, M. F., Turan, M., & Sahin, F. (2014). Plant growth-promoting rhizobacteria (PGPR) increase yield, growth and nutrition of strawberry under high-calcareous soil conditions. *Journal of Plant Nutrition, 37*, 990–1001.

Iriti, M., & Vitalini, S. (2020). Sustainable crop protection, global climate change, food security and safety—Plant immunity at the crossroads. *Vaccine, 8*(1), 42.

Jangir, M., Sharma, S., & Sharma, S. (2021). Development of next-generation formulation against *Fusarium oxysporum* and unraveling bioactive antifungal metabolites of biocontrol agents. *Scientific Reports, 11*, 22895. https://doi.org/10.1038/s41598-021-02284-1

Ji, J., Yuan, D., Jin, C., Wang, G., Li, X., & Guan, C. (2020). Enhancement of growth and salt tolerance of rice seedlings (Oryza sativa L.) by regulating ethylene production with a novel halotolerant PGPR strain *Glutamicibacter* sp. YD01 containing ACC deaminase activity. *Acta Physiologiae Plantarum, 42*, 1–17.

Kaka, H., Opute, P. A., & Maboeta, M. S. (2021). Potential impacts of climate change on the toxicity of pesticides towards earthworms. *Journal of Toxicology, 2021*, 8527991.

Kakvan, N., Heydari, A., Zamanizadeh, H. R., Rezaee, S., & Naraghi, L. (2013). Development of new bioformulations using Trichoderma and Talaromyces fungal antagonists for biological control of sugar beet damping-off disease. *Crop Protection, 1*(53), 80–84. https://doi.org/10.1016/j.cropro.2013.06.009

Kelbessa, B. G., Dubey, M., Catara, V., Ghadamgahi, F., Ortiz, R., & Vetukuri, R. R. (2023). Potential of plant growth-promoting rhizobacteria to improve crop productivity and adaptation to a changing climate. *CABI Reviews, 13*, 2023.

Khan, A. L., Waqas, M., & Kang, S. M. (2014). Bacterial endophyte *Sphingomonas* sp. LK11 produces gibberellins and IAA and promotes tomato plant growth. *Journal of Microbiology, 52*, 689–695.

Khan, M. A., Asaf, S., Khan, A. L., Jan, R., Kang, S. M., Kim, K. M., & Lee, I. J. (2020). Thermotolerance effect of plant growth-promoting *Bacillus cereus* SA1 on soybean during heat stress. *BMC Microbiology, 20*(1), 1–4.

Kim, S. T., Yoo, S. J., Weon, H. Y., Song, J., & Sang, M. K. (2022). *Bacillus butanolivorans* KJ40 contributes alleviation of drought stress in pepper plants by modulating antioxidant and polyphenolic compounds. *Scientia Horticulturae, 301*, 111111.

Kloepper, J. W., & Schroth, M. N. (1981). Development of a powder formulation of rhizobacteria for inoculation of potato seed pieces. *Phytopathology, 71*, 590.

Kokalis-Burelle, N., Vavrina, C. S., Rosskopf, E. N., & Shelby, R. A. (2002). Field evaluation of plant growth-promoting rhizobacteria amended transplant mixes and soil solarization for tomato and pepper production in Florida. *Plant and Soil, 238*, 257–266.

Kong, H. G., Shin, T. S., Kim, T. H., & Ryu, C. M. (2018). Stereoisomers of the bacterial volatile compound 2, 3-butanediol differently elicit systemic defense responses of pepper against multiple viruses in the field. *Frontiers in Plant Science, 9*, 90. https://doi.org/10.3389/fpls.2018.00090

Kramer, J., Özkaya, Ö., & Kümmerli, R. (2020). Bacterial siderophores in community and host interactions. *Nature Reviews. Microbiology, 18*, 152–163. https://doi.org/10.1038/s41579-019-0284-4

Kumar, A., & Verma, J. P. (2018). Does plant—Microbe interaction confer stress tolerance in plants: A review. *Microbiological Research, 207*, 41–52.

Kumar, S., Chauhan, P. S., Agrawal, L., Raj, R., Srivastava, A., Gupta, S., Mishra, S. K., Yadav, S., Singh, P. C., Raj, S. K., & Nautiyal, C. S. (2016). *Paenibacillus lentimorbus* inoculation enhances tobacco growth and extenuates the virulence of Cucumber mosaic virus. *PLoS One, 11*(3), e0149980.

Leal Filho, W., Azeiteiro, U. M., Balogun, A. L., Setti, A. F., Mucova, S. A., Ayal, D., Totin, E., Lydia, A. M., Kalaba, F. K., & Oguge, N. O. (2021). The influence of ecosystems services depletion to climate change adaptation efforts in Africa. *Science of the Total Environment, 779*, 146414.

Liechty, Z., Santos-Medellín, C., Edwards, J., Nguyen, B., Mikhail, D., Eason, S., Phillips, G., & Sundaresan, V. (2020). Comparative analysis of root microbiomes of rice cultivars with high and low methane emissions reveals differences in abundance of methanogenic archaea and putative upstream fermenters. *Msystems, 5*(1), 10–128.

Lim, J. H., & Kim, S. D. (2013). Induction of drought stress resistance by multi-functional PGPR *Bacillus licheniformis* K11 in pepper plant. *The Journal of Pathology, 29*, 201–208.

Lindow, S. E., McGourty, G., & Elkins, R. (1996). Interactions of antibiotics with *Pseudomonas fluorescens* strain A506 in the control of fire blight and frost injury to pear. *Phytopathology, 86*, 841–848.

Liu, J., Wu, X., Fang, Y., Liu, Y., Bello, E. O., Li, Y., Xiong, R., Li, Y., Fu, Z. Q., Wang, A., & Cheng, X. (2023). A plant RNA virus inhibits NPR1 sumoylation and subverts NPR1-mediated plant immunity. *Nature Communications, 14*(1), 3580.

Ma, J., Sun, C., Bai, L., Dong, R., Yan, Y., Yu, X., He, C., Zou, Z., & Li, Y. (2018). Transcriptome analysis of cucumber roots reveals key cold-resistance genes induced by AM fungi. *Plant Molecular Biology Reporter, 36*, 135–148.

Ma, Y., Dias, M. C., & Freitas, H. (2020). Drought and salinity stress responses and microbe-induced tolerance in plants. *Frontiers in Plant Science, 11*, 591911.

Ma, C. S., Zhang, W., Peng, Y., Zhao, F., Chang, X. Q., Xing, K., & Rudolf, V. H. W. (2021). Climate warming promotes pesticide resistance through expanding overwintering range of a global pest. *Nature Communications, 12*, 5351. https://doi.org/10.1038/s41467-021-25505-7

Madani, B., Mirshekari, A., & Imahori, Y. (2019). Physiological responses to stress. In *Postharvest physiology and biochemistry of fruits and vegetables* (pp. 405–423). Woodhead Publishing. https://doi.org/10.1016/B978-0-12-813278-4.00020-8

Mangmang, J. S., Deaker, R., & Rogers, G. (2015). Optimal plant growth-promoting concentration of *Azospirillum brasilense* inoculated to cucumber, lettuce and tomato seeds varies between bacterial strains. *Israel Journal of Plant Sciences, 62*, 145–152.

Manjula, K., & Podile, A. R. (2011). Chitin-supplemented formulations improve biocontrol and plant growth promoting efficiency of *Bacillus subtilis* AF 1. *Canadian Journal of Microbiology, 47*(7), 618–625. https://doi.org/10.1139/w01-057

Martínez-Megías, C., Mentzel, S., Fuentes-Edfuf, Y., Moe, S. J., & Rico, A. (2023). Influence of climate change and pesticide use practices on the ecological risks of pesticides in a protected Mediterranean wetland: A Bayesian network approach. *Science of the Total Environment, 878*, 163018. https://doi.org/10.1016/j.scitotenv.2023.163018

Martinuz, A., Schouten, A., Menjivar, R. D., & Sikora, R. A. (2012). Effectiveness of systemic resistance toward Aphis gossypii (Hom., Aphididae) as induced by combined applications of the endophytes *Fusarium oxysporum* Fo162 and *Rhizobium etli* G12. *Biological Control, 62*(3), 206–212.

Mejri, S., Ghinet, A., Magnin-Robert, M., et al. (2023). New plant immunity elicitors from a sugar beet byproduct protect wheat against *Zymoseptoria tritici*. *Scientific Reports, 13*, 90. https://doi.org/10.1038/s41598-022-26800-z

Mekureyaw, M. F., Pandey, C., Hennessy, R. C., Nicolaisen, M. H., Liu, F., Nybroe, O., & Roitsch, T. (2022). The cytokinin-producing plant beneficial bacterium *Pseudomonas fluorescens* G20-18 primes tomato (*Solanum lycopersicum*) for enhanced drought stress responses. *Journal of Plant Physiology, 270*, 153629.

Mohanty, P., Singh, P. K., Chakraborty, D., Mishra, S., & Pattnaik, R. (2021). Insight into the role of PGPR in sustainable agriculture and environment. *Frontiers in Sustainable Food Systems, 5*, 667150.

Moussa, T. A., Almaghrabi, O. A., & Abdel-Moneim, T. S. (2013). Biological control of the wheat root rot caused by *Fusarium graminearum* using some PGPR strains in Saudi Arabia. *The Annals of Applied Biology, 163*(1), 72–81.

Mukhtar, T., Ali, F., Rafique, M., Ali, J., Afridi, M. S., Smith, D., Mehmood, S., Amna, S. A., Jellani, G., & Sultan, T. (2023). Biochemical characterization and potential of *Bacillus safensis* strain SCAL1 to mitigate heat stress in *Solanum lycopersicum* L. *Journal of Plant Growth Regulation, 42*(1), 523–538.

Naamala, J., & Smith, D. L. (2020). Relevance of plant growth promoting microorganisms and their derived compounds, in the face of climate change. *Agronomy, 10*, 1179. https://doi.org/10.3390/agronomy10081179

Nishad, R., Ahmed, T., Rahman, V. J., & Kareem, A. (2020). Modulation of plant defense system in response to microbial interactions. *Frontiers in Microbiology, 11*, 1298. https://doi.org/10.3389/fmicb.2020.01298

Niu, B., Wang, W., Yuan, Z., Sederoff, R. R., Sederoff, H., Chiang, V. L., & Borriss, R. (2020). Microbial interactions within multiple-strain biological control agents impact soil-borne plant disease. *Frontiers in Microbiology, 11*, 585404.

Ortiz-Bobea, A., Ault, T. R., Carrillo, C. M., Chambers, R. G., & Lobell, D. B. (2021). Anthropogenic climate change has slowed global agricultural productivity growth. *Nature Climate Change, 4*, 306–312.

Oswald, A., Calvo Velez, P., Zúñiga Dávila, D., & Arcos Pineda, J. (2010). Evaluating soil rhizobacteria for their ability to enhance plant growth and tuber yield in potato. *The Annals of Applied Biology, 157*, 259–271.

Passari, A. K., Upadhyaya, K., Singh, G., Abdel-Azeem, A. M., Thankappan, S., Uthandi, S., Hashem, A., Abd Allah, E. F., Malik, J. A., As, A., & Gupta, V. K. (2019). Enhancement of disease resistance, growth potential, and photosynthesis in tomato *(Solanum lycopersicum)* by inoculation with an endophytic actinobacterium, *Streptomyces thermocarboxydus* strain BPSAC147. *PLoS One, 14*(7), e0219014.

Pathak, V. M., Verma, V. K., Rawat, B. S., et al. (2022). Current status of pesticide effects on environment, human health and it's eco-friendly management as bioremediation: A comprehensive review. *Frontiers in Microbiology, 13*, 962619. https://doi.org/10.3389/fmicb.2022.962619

Porcel, R., Zamarreño, Á. M., García-Mina, J. M., & Aroca, R. (2014). Involvement of plant endogenous ABA in *Bacillus megaterium* PGPR activity in tomato plants. *BMC Plant Biology, 14*, 36.

Prisa, D. (2023). Role of microorganisms in horticulture to improve plant quality. *Karbala International Journal of Modern Science, 9*(1), 4.

Pršić, J., & Ongena, M. (2020). Elicitors of plant immunity triggered by beneficial bacteria. *Frontiers in Plant Science, 11*, 594530.

Qiao, L., Wang, X., Smith, P., Fan, J., Lu, Y., Emmett, B., Li, R., Dorling, S., Chen, H., Liu, S., & Benton, T. G. (2022). Soil quality both increases crop production and improves resilience to climate change. *Nature Climate Change, 6*, 574–580.

Reitz, M., Rudolph, K., Schroder, I., Hoffmann-Hergarten, S., Hallmann, J., & Sikora, R. A. (2000). Lipopolysaccharides of *Rhizobium etli* strain G12 act in potato roots as an inducing agent of systemic resistance to infection by the cyst nematode *Globodera pallida*. *Applied and Environmental Microbiology, 66*, 3515–3518.

Richie, C. (2022). Environmental sustainability and the carbon emissions of pharmaceuticals. *Journal of Medical Ethics, 48*(5), 334–337.

Rinallo, C., Mittempergher, L., Frugis, G., & Mariotti, D. (1999). Clonal propagation in the genus Ulmus: Improvement of rooting ability by *Agrobacterium rhizogenes* T-DNA genes. *The Journal of Horticultural Science and Biotechnology, 74*, 502–506.

Rocha, J., Flores, V., Cabrera, R., Soto-Guzmán, A., Granados, G., Juaristi, E., & De La Torre, M. (2012). Evolution and some functions of the NprR–NprRB quorum-sensing system in the *Bacillus cereus* group. *Applied Microbiology and Biotechnology, 94*, 1069–1078.

Rodríguez, M., Torres, M., Blanco, L., Béjar, V., Sampedro, I., & Llamas, I. (2020). Plant growth-promoting activity and quorum quenching-mediated biocontrol of bacterial phytopathogens by *Pseudomonas segetis* strain P6. *Scientific Reports, 10*(1), 4121.

Ruzzi, M., & Aroca, R. (2015). Plant growth-promoting rhizobacteria act as bio stimulants in horticulture. *Scientia Horticulturae, 196*, 124–134. https://doi.org/10.1016/j.scienta.2015.08.042
Ryu, C. M., Shin, J. N., Qi, W., Ruhong, M., Kim, E. J., & Pan, J. G. (2011). Potential for augmentation of fruit quality by foliar application of bacilli spores on apple tree. *Plant Pathology Journal, 27*, 164–169.
Sabir, A. (2013). Improvement of grafting efficiency in hard grafting grape Berlandieri hybrid rootstocks by plant growth-promoting rhizobacteria (PGPR). *Scientia Horticulturae, 164*, 24–29.
Sagar, A., Rathore, P., Ramteke, P. W., Ramakrishna, W., Reddy, M. S., & Pecoraro, L. (2021). Plant growth promoting rhizobacteria, arbuscular mycorrhizal fungi and their synergistic interactions to counteract the negative effects of saline soil on agriculture: Key macromolecules and mechanisms. *Microorganisms, 9*(7), 1491.
Saijo, Y., & Loo, E. P. (2020). Plant immunity in signal integration between biotic and abiotic stress responses. *New Phytologist., 225*(1), 87–104.
Samavat, S., Heydari, A., Zamanizadeh, H. R., Rezaee, S., & Aliabadi, A. A. (2014). Application of new bioformulations of *Pseudomonas aureofaciens* for biocontrol of cotton seedling damping-off. *Journal of Plant Protection Research, 54*(4), 334–339. https://doi.org/10.2478/jppr-2014-0050
Sarkar, J., Chakraborty, U., & Chakraborty, B. (2021). High-temperature resilience in Bacillus safensis primed wheat plants: A study of dynamic response associated with modulation of antioxidant machinery, differential expression of HSPs and osmolyte biosynthesis. *Environmental and Experimental Botany, 182*, 104315.
Shah, J., Tsui, F., & Klessig, D. F. (1997). Characterization of a s alicylic a cid-i nsensitive mutant (sai1) of Arabidopsis thaliana, identified in a selective screen utilizing the SA-inducible expression of the tms2 gene. *Molecular Plant-Microbe Interactions, 10*(1), 69–78.
Shah, A., Nazari, M., Antar, M., Msimbira, L. A., Naamala, J., Lyu, D., Rabileh, M., Zajonc, J., & Smith, D. L. (2021). PGPR in agriculture: A sustainable approach to increasing climate change resilience. *Frontiers in Sustainable Food Systems, 5*, 667546.
Sharma, B., Singh, B. N., Dwivedi, P., & Rajawat, M. V. S. (2022). Interference of climate change on plant-microbe interaction: Present and future prospects. *Frontiers in Agronomy, 3*, 725804. https://doi.org/10.3389/fagro.2021.725804
Shivakumar, S., & Bhaktavatchalu, S. (2017). Role of plant growth-promoting rhizobacteria (PGPR) in the improvement of vegetable crop production under stress conditions. In A. Zaidi & M. Khan (Eds.), *Microbial strategies for vegetable production*. Springer. https://doi.org/10.1007/978-3-319-54401-4_4
Singh, B. K., Delgado-Baquerizo, M., Egidi, E., et al. (2023). Climate change impacts on plant pathogens, food security and paths forward. *Nature Reviews. Microbiology, 21*, 640–656. https://doi.org/10.1038/s41579-023-00900-7
Song, B., Seiber, J. N., Duke, S. O., & Li, Q. X. (2020). Green plant protection innovation: Challenges and perspectives. *Engineering (Beijing), 6*(5), 483–484. https://doi.org/10.1016/j.eng.2020.04.001
Sousa, N. R., Franco, A. R., Oliveira, R. S., & Castro, P. M. (2012). Ectomycorrhizal fungi as an alternative to the use of chemical fertilisers in nursery production of Pinus pinaster. *Journal of Environmental Management, 95*, 269–274.
Sowden, R. G., Watson, S. J., & Jarvis, P. (2017). The role of chloroplasts in plant pathology. *Essays in Biochemistry, 62*(1), 21–39.
Sperandio, V., Torres, A. G., & Kaper, J. B. (2002). Quorum sensing *Escherichia coli* regulators B and C (QseBC): A novel two-component regulatory system involved in the regulation of flagella and motility by quorum sensing in *E. coli*. *Molecular Microbiology, 43*(3), 809–821.
Srikanth, M., Sandeep, T. S. R. S., Sucharitha, K., & Godi, S. (2022). Biodegradation of plastic polymers by fungi: A brief review. *Bioresources and Bioprocessing, 9*, 42.
Srivastava, S., Bist, V., Srivastava, S., Singh, P. C., Trivedi, P. K., Asif, M. H., Chauhan, P. S., & Nautiyal, C. S. (2016). Unraveling aspects of *Bacillus amyloliquefaciens* mediated enhanced production of rice under biotic stress of *Rhizoctonia solani*. *Frontiers in Plant Science, 7*, 587.

Sun, T., Huang, J., Xu, Y., Verma, V., Jing, B., Sun, Y., Orduna, A. R., Tian, H., Huang, X., Xia, S., & Schafer, L. (2020). Redundant CAMTA transcription factors negatively regulate the biosynthesis of salicylic acid and N-hydroxypipecolic acid by modulating the expression of SARD1 and CBP60g. *Molecular Plant, 13*(1), 144–156.

Sundar, A. R., Velazhahan, R., & Vidhyasekaran, P. (2002). A glycoprotein elicitor isolated from Colletotrichum falcatum induces defense mechanisms in sugarcane leaves and suspension-cultured cells/Ein Glycoprotein aus Colletotrichum falcatum induziert als Elicitor Abwehrmechanismen in Zuckerrohrblättern und Zellkulturen. *Zeitschrift für Pflanzenkrankheiten und Pflanzenschutz/Journal of Plant Diseases and Protection, 109*, 601–611.

Takishita, Y., Charron, J. B., & Smith, D. L. (2018). Biocontrol rhizobacterium *Pseudomonas* sp. 23S induces systemic resistance in tomato (Solanum lycopersicum L.) against bacterial canker Clavibacter michiganensis subsp. michiganensis. *Frontiers in Microbiology, 9*, 2119.

Talboys, P. J., Owen, D. W., Healey, J. R., Withers, P. J. A., & Jones, D. L. (2009). Auxin secretion by *Bacillus amyloliquefaciens* FZB42 both stimulates root exudation and limits phosphorous uptake in Triticum aestivum. *BMC Plant Biology, 14*, 51.

Tanwar, A., Aggarwal, A., & Parkash, V. (2014). Effect of bioinoculants and super phosphate fertilizer on the growth and yield of broccoli (Brassica oleraceaL. var. italica Plenck). *New Zealand Journal of Crop and Horticultural Science, 42*(4), 288–302.

Thakur, M., & Sohal, B. S. (2013). Role of elicitors in inducing resistance in plants against pathogen infection: A review. *ISRN Biochemistry, 2013*, 762412. https://doi.org/10.1155/2013/762412

Tienda, S., Vida, C., Lagendijk, E., de Weert, S., Linares, I., González-Fernández, J., Guirado, E., de Vicente, A., & Cazorla, F. M. (2020). Soil application of a formulated biocontrol *Rhizobacterium, Pseudomonas chlororaphis* PCL1606, induces soil Suppressiveness by impacting specific microbial communities. *Frontiers in Microbiology, 11*, 1874. https://doi.org/10.3389/fmicb.2020.01874

Tiwari, S., Gupta, S. C., Chauhan, P. S., & Lata, C. (2021). An OsNAM gene plays important role in root rhizobacteria interaction in transgenic Arabidopsis through abiotic stress and phytohormone crosstalk. *Plant Cell Reports, 40*, 143–155.

Toulotte, J. M., Pantazopoulou, C. K., Sanclemente, M. A., Voesenek, L. A., & Sasidharan, R. (2022). Water stress resilient cereal crops: Lessons from wild relatives. *Journal of Integrative Plant Biology, 64*(2), 412–430.

Tudi, M., Daniel Ruan, H., Wang, L., Lyu Sadler, R., Connell, D., Chu, C., & Phung, D. T. (2021). Agriculture development, pesticide application and its impact on the environment. *International Journal of Environmental, 18*, 1112. https://doi.org/10.3390/ijerph18031112

Tullio, M., Rea, E., Cardarelli, M., & Colla, C. G. (2007). L'inoculo micorrizico costa poco e aumenta la resa produttiva. *Informatore Agrario, 28*, 54–57.

Utkhede, R. S., & Li, T. S. C. (1989). Chemical and biological treatments for control of apple replant disease in British Columbia. *Canadian Journal of Plant Pathology, 11*, 143–147.

Velásquez, A. C., Castroverde, C. D., & He, S. Y. (2018). Plant–pathogen warfare under changing climate conditions. *Current Biology, 28*(10), R619–R634.

Vishnupradeep, R., Bruno, L. B., Taj, Z., Karthik, C., Challabathula, D., Kumar, A., Freitas, H., & Rajkumar, M. (2022). Plant growth promoting bacteria improve growth and phytostabilization potential of Zea mays under chromium and drought stress by altering photosynthetic and antioxidant responses. *Environmental Technology and Innovation, 25*, 102154.

Weisskopf, L. (2022). Microbes to the rescue–exploring the chemistry of microbial communication and using it to protect plant health. *Chimia, 76*(11), 939–939.

Wu, Y., Lin, H., Lin, Y., Shi, J., Xue, S., Hung, Y. C., Chen, Y., & Wang, H. (2017). Effects of biocontrol bacteria *Bacillus amyloliquefaciens* LY-1 culture broth on quality attributes and storability of harvested litchi fruit. *Postharvest Biology and Technology, 132*, 81–87.

Xiong, Y., Zhao, D., Chen, S., Yuan, L., Zhang, D., & Wang, H. (2023). Deciphering the underlying immune network of the potato defense response inhibition by Phytophthora infestans nuclear effector Pi07586 through transcriptome analysis. *Frontiers in Plant Science, 14*, 1269959.

Yajima, A. (2014). Recent progress in the chemistry and chemical biology of microbial signaling molecules: Quorum-sensing pheromones and microbial hormones. *Tetrahedron Letters, 55*(17), 2773–2780.

Yang, N., Nesme, J., Røder, H. L., Li, X., Zuo, Z., Petersen, M., Burmølle, M., & Sørensen, S. J. (2021). Emergent bacterial community properties induce enhanced drought tolerance in Arabidopsis. *npj Biofilms and Microbiomes, 7*(1), 82.

Yang, B., Yang, S., Zheng, W., & Wang, Y. (2022). Plant immunity inducers: From discovery to agricultural application. *Stress Biology, 2*(1), 5. https://doi.org/10.1007/s44154-021-00028-9

Yu, X., Liu, X., Zhu, T. H., Liu, G. H., & Mao, C. (2012). Co-inoculation with phosphate-solubilizing and nitrogen-fixing bacteria on solubilization of rock phosphate and their effect on growth promotion and nutrient uptake by walnut. *European Journal of Soil Biology, 50*, 112–117.

Yu, X., Feng, B., He, P., & Shan, L. (2017). From chaos to harmony: Responses and signalling upon microbial pattern recognition. *Annual Review of Phytopathology, 55*, 109–137.

Zhang, X., Tubergen, P. J., Agorsor, I. D., et al. (2023). Elicitor-induced plant immunity relies on amino acids accumulation to delay the onset of bacterial virulence. *Plant Physiology, 192*(1), 601–615. https://doi.org/10.1093/plphys/kiad048

Zonayet, M., Paul, A. K., Faisal-E-Alam, M., Syfullah, K., Castanho, R. A., & Meyer, D. (2023). Impact of Biochar as a soil conditioner to improve the soil properties of saline soil and productivity of tomato. *Sustainability, 15*, 4832. https://doi.org/10.3390/su15064832

Plants, Microbes, and Carbon: The Interplay of Carbon Cycling and Sequestration

8

Dinesh Kumar, Meenu Saraf, and Bernard R. Glick

Abstract

Plants and microbes are key players in carbon cycling and sequestration, which has critical implications for the Earth's climate and the global carbon balance. The carbon cycle provides a balance between carbon fixation processes such as photosynthesis, and carbon release processes such as respiration. Carbon cycling involves the exchange of carbon dioxide (CO_2) between the atmosphere, plants, soil, and the oceans. Similarly, carbon sequestration involves long-term storage of carbon in various natural reservoirs, preventing its release into the atmosphere. The organic matter from the plant vegetation such as leaves, branches, and roots is incorporated into the soil. Some of it decomposes rapidly; however, a substantial portion is transformed into a more stable form of organic matter, such as humus. Cumulative evaluation and understanding their interplay are crucial in addressing climate change and developing sustainable environmental solutions for the mitigation of global warming and its impact. Over the past decades, there has been an unprecedented rise in the global average temperature of the Earth, critically threatening the livelihood of marginal communities and displacement of human population across the globe. A minimal increase in the average global

D. Kumar
Department of Biological Sciences, Birla Institute of Technology and Science (BITS) Pilani, Hyderabad Campus, Secunderabad, Telangana, India

Department of Microbiology and Biotechnology, University School of Sciences, Gujarat University, Navrangpura, Ahmedabad, Gujarat, India

M. Saraf (✉)
Department of Microbiology and Biotechnology, University School of Sciences, Gujarat University, Navrangpura, Ahmedabad, Gujarat, India
e-mail: msaraf@gujaratuniversity.ac.in

B. R. Glick
Department of Biology, University of Waterloo, Waterloo, ON, Canada

M. Saraf et al. (eds.), *Climate Change and Soil Microorganisms for Environmental Sustainability*, Microorganisms for Sustainability 28,
https://doi.org/10.1007/978-981-96-3425-5_8

temperature is likely to lead to disastrous climate conditions along with a loss of biological diversity. Therefore, finding and implementing effective long-term sustainable mitigation measures are the keys to address these challenges which are mainly the outcome of anthropogenic activities. To achieve the long-term objectives of natural carbon sequestration, the role of soil microorganisms is paramount and can be central to the mitigation of climate change and global warming.

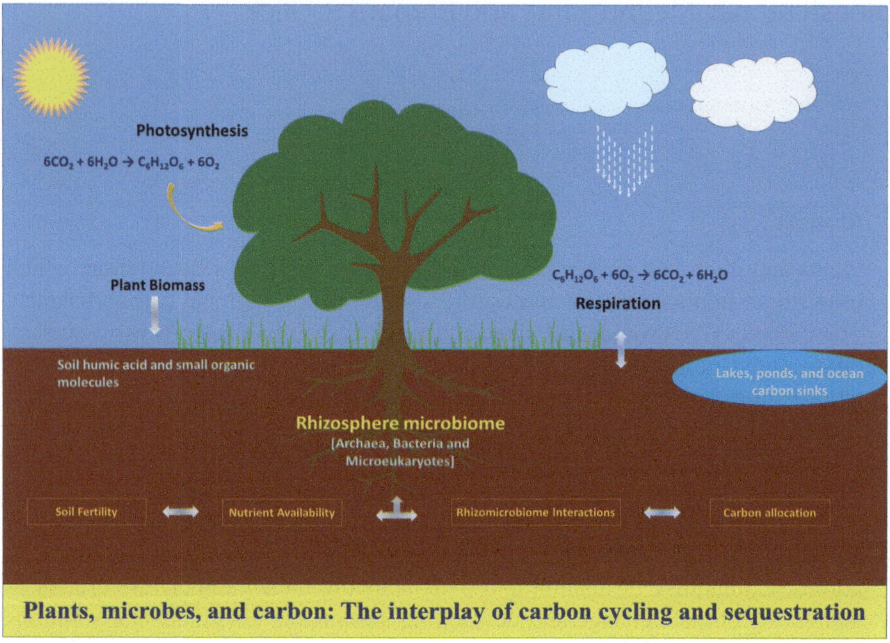

Keywords

Biogeochemical cycling · Greenhouse gases (GHGs) · Carbon sequestration · Biomass · Plant microbiome

8.1 Introduction

Plants and microbes are major players in carbon cycling and sequestration processes, which have significant implications for the Earth's climate and maintaining the global carbon balance (Raza et al., 2023). Understanding the interaction between plants, microbes, and soils in the uptake of atmospheric carbon is crucial in addressing climate change and developing sustainable environmental measures for the mitigation of global warming and its impact. Over the past decades, there has been an unprecedented rise in the global average temperature of the Earth due to an increase in greenhouse gases (GHGs), critically threatening the livelihood of many marginal communities and the potential displacement of numerous global populations (Letcher, 2021). Therefore, finding sustainable and long-term measures and solutions to this problem is the key to effectively addressing these critical global issues, which are mainly the consequence of anthropogenic activities. Natural ecosystems, e.g., forests, store a considerable amount of carbon and act as an important global carbon sink. Plants take up CO_2 and release oxygen (O_2) during photosynthesis, which transfers atmospheric carbon to them. Similarly, when plants decompose, the stored carbon in those plant parts is released and transferred back to the atmosphere and to the soil. This cyclic process supports various life forms by transferring carbon between living organisms and the environmental sinks (Raza et al., 2023; Sinsabaugh et al., 1991).

The largest carbon stock is stored within the continental crusts and upper mantle of the Earth formed millions of years ago, which is also the prime source of petrochemicals and crude oil. Similarly, carbon is stored in a range of different forms. For example, oceanic carbon is mainly present in the form of inorganic dissolved form (Garcia et al., 2020; Louca et al., 2016; Sunagawa et al., 2015; Zhang et al., 2018). The carbon stored in the Earth's atmosphere is tiny compared to the amount stored in global oceans, soils, and various geologic formations (Fig. 8.1). Furthermore, permafrost stores a large pool of carbon that is climatically protected from decomposition, although more of this carbon pool is becoming available as the global average temperature rises (Zhang et al., 2018). The different forest types store variable amounts of carbon and have different storage capacities, efficacy, and efficiency, although much of this variation is attributed to the fluctuating climatic conditions found in diverse ecological regions of the world. Warm and tropical regions tend to store higher levels of carbon in the aboveground components of plants compared to the amounts that they store belowground, while the boreal forest regions have enormous belowground carbon stores (Naylor et al., 2020).

The carbon exchange between different sinks and sources, whether global or local, can be a quite dynamically active form. This carbon that is exchanged between the source and sink is denoted as a flux, and the average rate at which carbon flows is termed as carbon turnover. An interesting parameter of global warming is the turnover of carbon within ecosystems across the globe, which provides an estimate of where the carbon might be most vulnerable to release as CO_2 into the atmosphere (Bernardini et al., 2022; McDaniel et al., 2019). Information on global carbon turnover within an ecosystem can influence informed management decisions that affect the rate of carbon turnover, ultimately influencing the carbon flux of ecosystems. The incremental additions of carbon to the atmosphere have an enormous effect on the global carbon cycle and global warming. Beginning of the Industrial Revolution, emissions from fossil fuel usage and other anthropogenic activities released huge amounts of carbon from deep sinks into the atmosphere. The anthropogenic fluxes caused a major shift of carbon from the Earth's crust in the form of crude oil and coal reservoirs, becoming a primary concern of air pollution. The inaccessible carbon that had been essentially locked away from the carbon cycle for millions of years and is now expected to remain in the atmosphere for decades and centuries

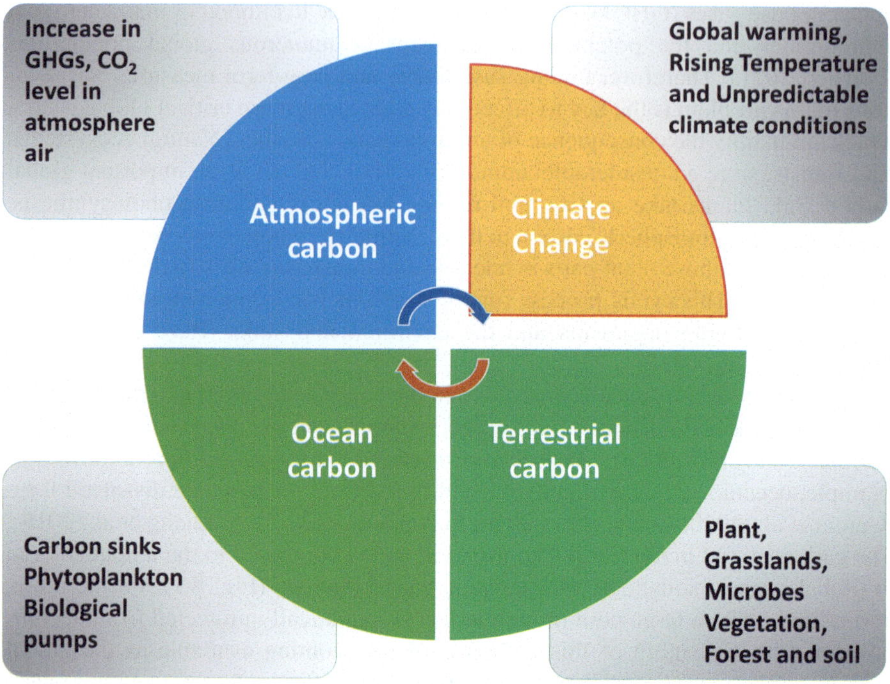

Fig. 8.1 Climate change and atmospheric, terrestrial, and ocean carbon sinks

(Pérez Castro et al., 2019). The addition of fossil fuel-based carbon to the atmosphere can be considered a net addition to the contemporary carbon cycle and an active driver of current climate change.

Carbon sequestration is the long-term storage of carbon in various reservoirs, preventing its release into the atmosphere (Zhang et al., 2018). Carbon sequestration process in oceans, plants, and soil helps to mitigate CO_2 levels in the atmosphere and the effects of climate change (Garcia et al., 2020). A significant fraction of the Earth's carbon is stored in soils. It is enriched with organic matter from plant leaves, branches, decomposing vegetation, and roots. Even though some of it decomposes quickly, a substantial amount of plant litter can be transformed into more stable forms of organic matter, such as humus (Sinsabaugh et al., 1991). It is a natural process of carbon sequestration in soil. The structure and composition of microbes in the soil are highly heterogeneous and diverse; functionally active communities of microorganisms and soil microbiomes play a crucial role in the cycling of soil organic carbon, nitrogen cycling, and other biogeochemical cycling processes supporting plant growth and development (Blagodatsky et al., 2010). The resilient, resistant, and stable microbial populations are of special interest to researchers since these microbial communities are sufficiently robust to withstand extreme environmental conditions and harbor genes encoding enzymes, biomolecules, and pathways that help them to survive extreme ecological conditions (Naylor et al., 2020). Additionally, plant roots and micro-eukaryotes in the soil play an important role in storing carbon in the form of root biomass and soil organic matter. The oxalate-carbonate pathway involves the conversion of atmospheric CO_2 fixed by plants into biomass and includes the biological recycling of calcium oxalate by soil microorganisms into calcium carbonate ($CaCO_3$) in terrestrial ecosystems. Recent research studies highlight findings of the phylogenetic characterization, metabolic plasticity, and physiological diversity of oxalotrophic bacteria in tropical ecosystem soils participating in the oxalate-carbonate pathway (Ballhausen & de Boer, 2016; Yu et al., 2021). Oxalotrophic bacteria could oxidize calcium oxalate. However, the diversity and alternative carbon sources of oxalotrophs participating in the oxalate-carbonate pathway are not fully characterized. The culture-independent approaches such as metagenomics, meta-transcriptomics, and multi-omics techniques could help to further elucidate these specialized group of microbes (Jansson & Hofmockel, 2018; McDaniel et al., 2019; Semenov, 2021).

A vast number of microorganisms are unexploited by cultivation-based approaches and thus remain accessible only through cultivation-independent methods. Genome-resolved metagenomics is one approach that enables the reconstruction of composite genomes from a microbial community (Kumar et al., 2022; Xu et al., 2021). Emerging computational methods and advanced sequencing technologies have now been applied at much larger scales and in numerous environments, including the global oceans, rumen, human microbiome, deep subsurface, and aquifers. These advancements have led to substantial understandings of evolutionary traits and functional metabolic properties of uncultivated microorganisms. As a result, researchers have expanded the microbial tree of life, a comprehensive genomic catalog of both cultivated and uncultivated bacteria and archaea for large-scale comparative genomic studies, thereby enabling mining for genes and functions of special interest and constructing genome-scale metabolic models (Beattie, 2018; Le Boulch et al., 2019; Mavrodi et al., 2021; Tatsis & O'Connor, 2016). Furthermore, recent genome reconstructions of uncultivated bacteria and archaea have yielded unique insights into the evolutionary trajectories of micro-eukaryotes and ancestral microbial traits and ecological drivers of carbon flux into soil (Lidbury et al., 2022).

In this chapter, we discuss the existing knowledge gaps in understanding plants, microbes, and carbon interactions, and how climate change impacts soil carbon cycling by the soil microbiome. More recent technical approaches should facilitate our understanding of the complex nature of carbon sequestration and soil microbiomes to better predict its future responses to climate change. In addition, we endeavor to comprehensively review existing literature on the role of microorganisms in carbon cycling and sequestration. Furthermore, identify key patterns, mechanisms, and gaps in the current understanding of the microbial processes involved in organic matter decomposition and CO_2 release, and the implications of microbial-mediated decomposition on the carbon cycle and its contribution to climate change are evaluated.

8.2 Feedback Loop System in Biological Carbon Cycles

There are two major types of carbon sequestration processes. First, the biological carbon sequestration process involves the capture and storage of carbon by living organisms. Plants, trees, forests, grasslands, oceans, and other ecosystems act as carbon sinks, absorbing more CO_2 through the photosynthesis process than they release through respiration or decomposition of organic matter (Zhang et al., 2017). Afforestation and reforestation efforts are important strategies to enhance biological carbon sequestration and natural carbon sinks. The additional process of carbon sequestration is the geological carbon sequestration process that involves storing CO_2 deep underground in geological formations, such as depleted crude oil and natural gas reservoirs or saline aquifers (Fagorite et al., 2023; Shaw & Mukherjee, 2022). Furthermore, carbon capture technologies aim to capture CO_2 emissions from industrial processes and inject them into suitable geological formations, preventing their immediate release into the atmosphere (Bahman et al., 2023; Bauer et al., 2023).

The interplay between plants, microbes, and carbon can lead to feedback loops that influence the overall carbon cycle process. For example, as global temperatures rise due to increased concentration of greenhouse gases (GHGs) in the atmosphere, it can influence and enhance microbial activity and decomposition rates, leading to more carbon being released into the atmosphere. This, in turn, contributes to supplementary global warming in a positive feedback loop. Similarly, carbon cycling plays a fundamental role in biotic and abiotic processes in the ocean, which intertwine to mediate the chemistry and redox status of carbon in the ocean and the atmosphere (Zhang et al., 2018). The increasing level of CO_2 gave rise to the phenomenon of ocean acidification, threatening the ocean's biodiversity, bleaching coral reefs, and disturbing the balanced ecosystem. The interactions between abiotic and biogenic organic carbon in the ocean are composite in nature, and there is a century-old enigma about the existence of a huge reservoir of recalcitrant dissolved organic carbon that is comparable in size to the magnitude of the pool of atmospheric CO_2 (Baumas & Bizic, 2023). The concepts of the biological carbon pump and the microbial loops shape our understanding of the natural carbon cycles. The concept of microbial carbon pump is closely connected to the biological carbon and the microbial loops, explicitly considers the consequence of the ocean's recalcitrant dissolved organic carbon reservoirs, and provides a mechanistic framework for the exploration of its formation and persistence in the context of global warming and climate change (Letcher, 2021). An understanding of the microbial loops has significantly benefited from advanced genomics, "multi-omics" and "meta-omics" research in microbial biogeochemistry. Therefore, predicting the ocean's response to climate change improves an integrative understanding of the recalcitrant global carbon sinks and microbial loops, is urgently needed (Zhang et al., 2018). The perception of a biological pump is a complex ecosystem process that transports particulate organic carbon and atmospheric CO_2 through vertical transportation of living biomass to marine sediments, focusing on sediment storage from the epipelagic zone to the deep interior of the oceans provides an important understanding of global ocean carbon sinks (Burd, 2023; Hilmi et al., 2023; Stirnimann, 2023). Furthermore, the microbial loop is a feedback mechanism of the primary environmental production of carbon in the form of dissolved organic matter and its utilization by microbial flora. Similarly, a microbial carbon pump is a theoretical framework for understanding the role of microbial processes involved in the making of recalcitrant dissolved organic matter in the marine ecosystem.

8.3 Plants, Microbes, and Carbon

Plants play a central role in the carbon cycle through the process of photosynthesis. During photosynthesis, plants absorb CO_2 from the atmosphere and, with the help of sunlight, convert it into organic compounds, such as simple sugars, and then subsequently complex biomolecules. Forests and plants remove CO_2 from the atmosphere, thereby acting as a natural carbon sink (Ali et al., 2023; Pessarrodona et al., 2023; Zhang et al., 2017). Microorganisms, including archaea, bacteria, and fungi, are vital components in the decomposition of organic matter. When plants or other organic materials die, microbes break them down releasing carbon back into the atmosphere in the form of CO_2 through the process of respiration and decomposition. Furthermore, specialized microbes, called oxalotrophs, use plant-derived oxalate as an exclusive carbon source. These microbes, which proliferate on oxalate, use the carbon atoms for their own metabolism, and in this process, one in every four carbon derived from oxalate is secreted into the soil as carbonate by the oxalotrophic microbes (Tolstanova et al., 2023). However, carbonates are not stable molecules and are rapidly converted to H_2O and $CO_{2,}$ especially in acidic and humid environmental conditions, explaining the sparseness of carbonates in such types of soils. However, under alkaline surroundings and in the presence of Ca^{2+}, the carbonate molecules precipitate as $CaCO_3$. Fortunately, massive $CaCO_3$ deposits are stable for hundreds to thousands of years. Furthermore, in this form of carbon sequestration photosynthetically fixed carbon atoms might also be sequestered into carbonates.

The key enzymes involved in oxalotrophy: *formyl-coenzyme A (CoA) transferase* (EC 2.8.3.16) and *oxalyl-CoA decarboxylase* (EC 4.1.1.8), are encoded by the *frc* and *oxc* genes, respectively. A research study of this phenomenon revealed that oxalate-degrading bacteria are dominated by *Actinobacteria*, *Firmicutes,* and *Proteobacteria* as identified from terrestrial and aquatic environments (Hervé et al., 2016). The active repertoire of microbial genes and receptors is associated with the signal transduction processes for carbohydrate utilization such as the TonB-dependent receptor found in gram-negative bacteria (Schauer et al., 2008). Interestingly, the bacterium *Ruminococcus flavefaciens* 007S has two type IV pili proteins that have been identified as cellulose degraders (Vodovnik et al., 2013). The bacterial groups *Burkholderia*, *Bradyrhizobium*, and *Azospirillum* are potential lignin-degrading microorganisms (Nacke et al., 2014). Similarly, methane (CH_4), a gaseous form of carbon under aerobic conditions, is consumed by certain forest soil methanotrophic bacteria, and such forest soils thus act as a methane sink. These bacteria consume both atmospheric methane and the methane produced by methanogenic archaea in waterlogged conditions (Vishwakarma & Kumar Dubey, 2007). *Alphaproteobacteria*, *Gammaproteobacteria*, and the acidophilic group *Verrucomicrobia* are the bacterial groups identified as methanotrophs, although there are others yet to be discovered and uncultured novel taxa (den Camp et al., 2009).

Microorganisms produce polysaccharide-degrading enzymes that can metabolize complex sugar molecules and break them into simple sugars and water molecules. Similarly, the fungal class *Basidiomycetes* is extremely effective at degrading dead woods, as the members of this class have the potential to depolymerize plant cell wall components like cellulose, hemicellulose, pectin, and lignin (albeit slowly). Furthermore, *Basidiomycota* members such as *Agaricus bisporus* and *Marasmius quercophilus* produce enzymes, *manganese peroxidase* and *laccase* both of which assist in plant litter decomposition (Meena et al., 2023). Several microorganisms, both bacteria and fungi, harbor genes encoding for *cellulase* enzymes as well as other enzymes necessary for plant cell wall degradation. Microbial enzymes such as *cellulase* and *glycoside hydrolases* (GHs) are present in all cellulose-degrading microbes (Khan et al., 2020). Some strains of *Actinobacteria* are also potential degraders of lignin which is mediated by oxidative enzymes, viz., *manganese peroxidases*, *oxidase*, *laccases*, and *hydrogen peroxidases* (Falade et al., 2017).

8.4 Photosynthesis, Carbon Uptake, and Storage by Plants

Plants are the primary mediators for carbon uptake from the atmosphere and conversion into organic carbon, which is used to build their tissues, growth, and development. As they grow, plants accumulate carbon in their stems, roots, and leaves (Askari-Khorasgani et al., 2019; Song et al., 2021). Therefore, plants and vegetation are essential for long-term carbon storage, as they can capture and store substantial amounts of carbon over their lifetimes (Fig. 8.2). Forests act as natural carbon sinks that absorb CO_2 from the atmosphere and store it in their biomass. Since the initial progress of agriculture by humans, there have been significant improvements in crop yields through increased use of fertilizers, irrigation, and strategic breeding of crops to develop new high-yielding varieties (Meadows-smith et al., 2017). This has both increased the amount of carbon fixed by growing plants and decreased the amount of fixed carbon by decaying plants.

During photosynthesis, most terrestrial plants convert only about 2% of the incoming light energy into chemical energy. Therefore, there is a need and opportunity to improve photosynthesis efficiency using synthetic biology, which might result in further increases in crop productivity by increasing the amount of carbon that is being assimilated by the plants. Improving the efficiency of photosynthesis would increase the amount of assimilated carbon, which could then be transported to the yield or to a plant organ of interest. Beyond increasing agricultural crop productivity, improving photosynthetic efficiency would capture higher CO_2 from the atmosphere, which can be stored in plant biomass and eventually sequestered in the soil or used to create secondary bioproducts (Terry, 1977). Removing carbon from

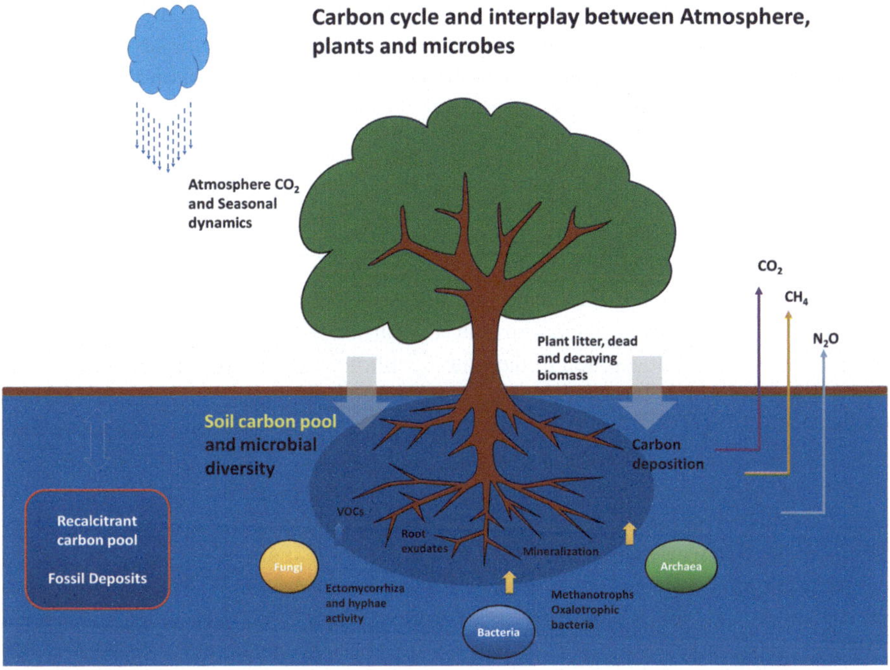

Fig. 8.2 Carbon cycle and microbial interaction between environment, soil, and plants

the atmosphere is an important climate change mitigation strategy, and photosynthesis is a natural way of capturing this atmospheric carbon. Furthermore, to improve photosynthetic efficiency for greater carbon capture, we need to modify large modular components of the photosynthesis machinery or even add novel mechanisms to the photosynthetic machinery. These complex changes are difficult to achieve through natural evolution or breeding techniques; however, they can be accomplished through beneficial microbe genetic manipulation techniques.

Energy is consumed at multiple steps throughout the photosynthetic process, preventing plants from efficiently capturing atmospheric carbon and converting the solar energy to chemical energy. When it comes to absorbing sunlight, the light-harvesting antenna in the chloroplasts only absorbs radiation that has wavelengths approximately between 400 and 700 nm, which is roughly equivalent to the visible light spectrum (Whatley & Whatley, 1981). The wavelengths in this spectrum account for less than 50% of the total energy in solar radiation, and some of it is reflected instead of absorbed, further decreasing the net amount of solar radiation that plants can convert to chemical energy. Additionally, chloroplasts frequently experience fluctuating light conditions due to variable cloud cover and winds. During these transitions, it can take several minutes for photosynthesis to return to maximal operational capacity under the new light conditions; these transitions have been estimated to cost about 10–40% of potential CO_2 assimilation (Long et al., 2022). The photosynthesis efficacy and efficiency, combined with the number of avenues for its improvement, underscores the potential need for engineering photosynthesis mechanisms to increase atmospheric carbon capture for crop yield and biomass storage. Researchers have demonstrated under greenhouse and field experiments how we can increase photosynthesis through engineering different components of the biological processes, including increasing how fast the photosynthetic machinery responds to fluctuating light, adding shorter and energy-efficient photorespiration recovery pathways, and deploying exogenously supplied photosynthetic enzymes to increase enzymatic reaction rates (Matthews, 2023).

There are various possibilities for photosynthesis engineering that are being explored, from expanding the range of solar wavelengths that can be absorbed by the light-harvesting antenna to introducing carboxysomes, carbon-concentrating compartments from blue-green algae, into C3 photosynthetic pathways (Matthews, 2023; Norena-Caro & Benton, 2018). Furthermore, given the complexity of photosynthetic machinery and the different avenues toward engineering it for increased efficiency and carbon capture, mathematical models have been powerful tools for identifying which strategies are promising for implementation. Predictive models shall continue to be important resources to identify and evaluate different engineering strategies for increasing carbon capture at the field scale, across developmental stages, and under future climate scenarios (Krevor et al., 2023). These models can also be used to investigate the predicted combined impact from stacking plants with multiple engineered traits, such as increased light absorption and increased enzymatic reaction rates or efficient photorespiration.

Despite successful greenhouse and field experiments demonstrating that we can engineer plant photosynthesis to increase carbon assimilation, there remain gaps in translating future engineering strategies from research labs and greenhouses into crops that can be planted in the fields by farmers around the world. For photosynthesis engineering to have a significant role in mitigating climate change and sustainably feeding our growing population, we need to accelerate the pipeline from discovery science to societal impact on a global scale. To address these challenges and meet the needs of future necessities, we must innovate and implement beyond our traditional approaches for increasing crop productivity. Engineering photosynthesis for improved efficiency and increased carbon capture, when used in conjunction with other scientific and societal advances, certainly helps us to solve significant global challenges of this century.

8.5 Microbial Activity and Biomass Decomposition

Microorganisms are essential components of soil ecosystems, performing crucial functions in biogeochemical cycling, carbon sequestration, and overall plant health and development. However, it remains indeterminate how microbial community structure, its functioning, and biogeochemical cycling, including GHG fluxes, would respond to climate change at different scales (Meena et al., 2023; Mukhtar et al., 2023; Naylor et al., 2020). Therefore, investigating the role of plant-microbe interactions in the context of global and regional climate change effects on the soil microbial community structure and functioning, as well as the climate-microbe-carbon feedback and plant-microbe interactions, is a critical aspect of the natural carbon sequestration process. The decomposition of organic matter is one of the most prominent biogeochemical processes directly affecting the resilience of ecosystems to climate change (Delgado-Baquerizo et al., 2017; Jousset et al., 2017; Schimel et al., 2007). Microorganisms including archaea, bacteria, and fungi, play a critical role in the process of organic matter decomposition. In this process of decomposition, CO_2 is released back into the atmosphere as a by-product of microbial respiration in the carbon cycle.

The microbial degradation of plant litter such as leaves, stems, roots, and dead matter is mediated by extracellular enzymatic activity. The specificity of this degradation makes microbial enzymes useful for comparative analysis of microbial communities, monitoring community succession, and evaluating the effects of ecosystem variables on microbial processes at the molecular level (Hirsch et al., 2003; Nobori et al., 2018). Thus, the application of microbial enzymatic assays to ecological studies of plant litter decomposition provides useful tools to ascertain and identify novel bacterial and microeukaryote communities. Furthermore, when applied in combination with general indices of microbial biomass and their activity and taxonomic analyses, specific enzyme activity assays offer a mechanistic approach to study the process of microbial breakdown of plant biomass (Khan et al., 2020). Similarly, soil nutrient cycling is linked with microbial biomass decomposition and influences the soil carbon dynamics. The decomposition of organic waste depends on several vital factors including soil microbial activity and the availability of nutrients. The stable components of soil organic matter play an obvious role in determining the dynamics of both micro- and macro-nutrients. However, there is a need to explore the functional role of composite soil microbiomes and environmental factors that regulate the decomposition of organic material such as climate, temperature, rainfall, and seasonal dynamics.

During the decomposition process, carbon is lost as CO_2 or sequestered into more recalcitrant forms, which are resistant to further biochemical degradation. Therefore, microbial enzymes such as *cellulase*, *reductase*, *laccase,* etc., have great potential in optimal plant litter degradation, however, need further investigation from an enzymatic perspective (López-Mondéjar et al., 2016). Organic matter breakdown is a biochemical process with consequences affecting climate change and ecosystem productivity. Microbial activities are among the largest CO_2 emission sources in the environment, next to anthropogenic activities; research investigations suggest that it has affected climate change over the past few decades. As the microbial respiration processes release CO_2 into the atmosphere, microbes act as custodians in the entire process of the carbon cycle. Therefore, it is crucial to note that microorganisms are one of the major contributors to the decomposition, conversion, and balance of the carbon cycle in the ecosystem (Zhang et al., 2017). Inconsistencies in the carbon cycle have implications for causing fluctuations in the entire carbon content of the natural ecosystems. Soil microbial communities in the terrestrial carbon cycle process require further attention and investigation. The key factors affecting the microbial breakdown processes are the types of substrates, temperature, pH, and moisture content. Therefore, it is essential to focus on the factors that affect the microbial environment and its behavior during the breakdown of organic constituents. Sustained efforts and new research initiatives are required to address global climate change and its effects on agricultural ecosystems, forests, oceans, and vice versa; there is a need to further evaluate the potential of functional microbial communities and their contribution to the global carbon sequestration process.

Among soil microbial communities, certain bacterial species are identified as highly resilient to extreme environments such as drought, pH, and hypersaline conditions. Previous studies reported that bacteria are effectively involved in breaking down organic matter and catabolizing undecomposed products (Blagodatsky et al., 2010; Raza et al., 2023; Sinsabaugh et al., 1991). Fungi are efficient at decomposition, but they often require a steady supply of nutrients and food to thrive under different environments. Therefore, overall, fungi might perform less extensive decomposition compared to bacteria but they might be more efficient litter decomposers (Ballhausen & de Boer, 2016; Poole et al., 2018). Additional research investigations are required for assessment and better understanding of the complex interactions between microbes, organic matter decomposition, CO_2 release, and climate change. These research studies are necessary for informing efforts for effective climate change mitigation and adaptation (Mukhtar et al., 2023; Pathak, 2023).

Due to their omnipresence in ecosystems, the diversity of soil microbes is essential in this key process. Soil organic carbon dynamics is an interaction between atmospheric CO_2 and the ecosystem. The carbon sequestration from atmospheric carbon to soil carbon occurs through "carbon-fixing" species, particularly autotrophic organisms, such as chemoautotrophic bacteria, and photosynthetic plants (De Deyn et al., 2008). Similarly, agricultural activities like soil tillage impact carbon sequestration potential and are influenced by microbial activity. Soil organic carbon fractions are strongly affected by abiotic and biotic factors, quality of soil surface plant debris such as leaf trash and crops, grassland residues, or the rate of decomposition of organic matter which might take days or years to completely decay (Barima et al., 2016). When organic matter is burned, it produces biochar that is resistant to further decomposition for a longer period. Predicting the consequences of climate change on ecosystem carbon fluxes, priming effects, and their feedback depends on understanding the limiting factors of the decomposition of soil organic matter and the biological activity of the microbial decomposers. According to Kuzyakov et al. (2000), priming effects gradually create an impact on the turnover of native soil organic matter. This change may include the addition of organic or biofertilizer to the soil, root exudation, or inputs of plant residues. Due to priming activity, large amounts of carbon (C), nitrogen (N), phosphorous (P), potassium (K), minerals, and other nutrients can be released very quickly into the soil (Blagodatsky et al., 2010). Any imbalance in the environmental conditions has the potential to change biogeochemical cycles, therefore increasing or decreasing the levels of atmospheric CO_2. Therefore, predicting the developments causing soil carbon imbalance is an important marker to understand global warming. Several processes and mechanisms, including respiration by autotrophic and heterotrophic living organisms, cause this carbon to return to the atmosphere.

8.6 Symbiotic Associations and Inter-Kingdom Interactions

Understanding microbial interactions is a vital strategy for colonizing and establishing microbes in a variety of environments and ecosystems. The establishment in the environmental condition depends on the species diversity, richness, and functional redundancy in the microbial community structure. These interactions involve numerous aspects of microbial ecology, including physiochemical changes, metabolite exchange, molecular conversions, signaling, chemotaxis, and genetic exchange resulting in genotype selection (Walitang et al., 2018). Therefore, these associations lead to adaptation and specialization, allowing the occupation of different niches by reducing biotic and abiotic stress or exchanging growth factors and signaling cascades. Microbial interactions occur through the exchange of molecular and genetic information, secondary metabolites, siderophores, plasmids, mobile genetic elements (MGEs), quorum sensing (QS), biofilm formation, cellular signaling, and transduction, among many others. The decisive unit of microbial communication and inter-kingdom alliances is the relative gene expression of each organism in response to the biotic and/or abiotic stimulus. Therefore, molecular mechanisms involved in the microbial interaction and strategy used by different microorganisms in the environment can modulate the establishment and structural variation of the microbial community structure (Bohan et al., 2017; Kalam et al., 2020; Richardson et al., 2018). Absolute quantification methods and precise measurements are theoretically feasible but hard to assess under transient conditions. Therefore, understanding the complexity of highly dynamic interactions of microbiomes requires minimal errors and absolute expertise. The microbiome systems are extremely dynamic and complex in nature due to the fluctuating environments that require decisive variable parameters in consideration, interchangeable components, data integration, and precision in planning and management to decipher keystone microbial communities (Glick & Gamalero, 2021).

The microbial nutrient preferences, metabolic substrate utilization, and root exudates shape the soil microbiome assembly for shared benefits. Microbiome engineering approaches using integrated quantitative systems and synthetic biology should guide strategies to engineer microbial consortiums fostering carbon sequestration, plant health, and development methods. Mycorrhiza forms symbiotic relationships with plant roots. Furthermore, fungi extend the reach of plant root systems, helping plants access nutrients and water in exchange for sugars and metabolites (Akiyama et al., 2005; van der Heijden et al., 2015). Mycorrhizal associations supplement plant growth and carbon uptake, promoting carbon sequestration in both plants and soils. Mycorrhizal hyphae growing through soil release exudates, including low-molecular-weight biomolecules, sugars, and organic acids. These carbon- and nitrogen-containing exudates are used and immobilized by soil microbial communities, and subsequently, this carbon forms one of the most stable soil carbon pools (Bais et al., 2006; Mavrodi et al., 2021; Preece & Peñuelas, 2020). There is increasing evidence that fungal residues play an important role in forming stable soil organic matter and might contribute effectively to the mineral-associated organic matter carbon sequestration. Large amounts of root exudates are released into the rhizosphere, where they can stimulate bacterial growth and activities to increase mineralization and nutrient availability (Chhabra et al., 2013). Plant exudates of energy-rich organic compounds are also redirected through arbuscular mycorrhizal hyphae in the soil. Therefore, mycorrhizal hyphae are a channel for a substantial number of plant-derived exudates in carbon cycle dynamics.

Beyond the terrestrial ecosystems, marine diatoms are the primary producers in global oceans, forming a massive annual bloom and covering the basis of the ocean food chain (Burd, 2023; Hay, 1996). Although these organisms are microscopic, their collective numbers are sufficient to influence the biogeochemistry of the planet, supplying a significant fraction of the Earth's oxygen. Marine microeukaryotes and bacteria are also highly abundant in the ocean and assimilate and decompose a significant portion of the organic carbon fixed by diatoms. Together, phototrophic microeukaryotes and heterotrophic bacteria establish components of the microbial loop in oceans where phytoplankton and bacteria contribute significantly to the cycling of carbon and other important nutrients. Bacteria are known to directly compete with phytoplankton for resource availability, such as light and dissolved organic carbon, which directly regulates ecological phytoplankton-bacterial interactions in the global oceans (Zhang et al., 2018). Emerging research studies revealed important mutually beneficial symbiotic relationships that allow both phytoplankton and marine bacteria to flourish in otherwise harsh environmental conditions. Although considerable past research work on phytoplankton-bacterial interactions focused on the temperate regions, these symbiotic associations are also critical in polar habitats, such as the Southern Ocean. The Southern Ocean accounts for approximately 40% of anthropogenic CO_2 uptake, but phytoplankton growth here is often limited by low iron availability (Stirnimann, 2023).

The inter-kingdom alliance between plants and associated microbiota should facilitate the development of novel precision agriculture systems and improvement in natural carbon capture efficiency needed to cope with the increasing global CO_2 level and unpredictable climatic conditions. The heritability of beneficial microbiota traits needs to be further investigated to explore genetic features that determine extended host plant phenotype in response to environmental variations such as light, temperature, cold, and seasonal influences (Theis et al., 2016). Finally, among other future benefits, employing the soil-plant microbiota should permit improved sustainable precision farming practices by deploying processes such as nutrient acquisition, drought resistance, salt tolerance, and disease resistance including the productive symbiosis in the enriched soil ecosystems in response to the abiotic and biotic factors (Glick, 2020; Reed & Glick, 2023).

8.7 Carbon Sequestration, Climate Change, and Its Impact

Carbon is mostly found in the lithosphere in the amount of approximately 66–100 million gigatonnes (Gt) (Hirt et al., 2023). The natural terrestrial carbon sequestration process requires capturing atmospheric CO_2 by crop vegetation and forests. Furthermore, a significant amount of soil carbon acts as the largest terrestrial carbon sink, comprising both soil organic carbon as 1500–1600 Gt carbon and soil inorganic carbon as 700–1700 Gt carbon, and principally contributes to the total terrestrial carbon stocks (3100 Gt) (Hirt et al., 2023). Soil carbon sinks undergo dynamic changes, influenced by various abiotic and biotic factors, temperature, atmospheric CO_2 concentrations, land usage patterns, and soil management practices. In contrast to soil carbon (3100 Gt), vegetation carbon (450 Gt) holds much smaller amounts of carbon. Deforestation, land-use changes, and burning of fossil fuels significantly altered the natural carbon cycle. The carbon that was previously trapped inside inaccessible sinks continues to accumulate in the atmosphere (Hirt et al., 2023). The increased release of CO_2 into the atmosphere from these activities contributes to the greenhouse effect, leading to global warming and climate change. Therefore, understanding the interplay of plants, microbes, and carbon in natural ecosystems is essential for developing effective strategies to mitigate climate change and improve the carbon sequestration process.

Among the several approaches for reducing the level of atmospheric GHGs, reforestation is the most common strategy for carbon sequestration. Trees are considered to be an ideal method to achieve this objective due to their natural ability to capture CO_2 through photosynthesis, converting it into biomass that can store carbon for an extended period (Raza et al., 2023). Approximately, one-third of the terrestrial surface of our planet is arid land that is not utilized for agricultural production and is in a state of significant degradation. However, many hectares of natural forests have been lost in recent years due to the conversion of forests to agricultural activities. Replantation and reclamation of arid lands by appropriate ecosystem engineering and vegetation would not only bring back infertile lands to vegetation but also provide ideal sites for carbon sequestration. The advantage of reclaiming arid and semi-arid regions for green forests and carbon sequestration is that they do not compete with lands used in agriculture and food production; in addition, they act as a reinforcement to reduce soil erosion and maintain soil fertility (Mutio et al., 2023).

Climate change and rising global average temperature have variable impacts on the microbial community structure and nutrient turnover. Such dramatic climate change responses, however, are difficult to generalize, even within an ecosystem, since they are subject to strong regional influences of current ambient environmental conditions. Finally, emerging tools, such as genetically engineered plants and microbes, provide mitigation strategies against global climate change impacts, particularly for agroecosystems. In a rapidly evolving world, addressing the knowledge gaps, assessments, and predictions of microbial responses requires the development of effective mitigation strategies. Real-time optimized high-throughput sequencing data analysis, interpretation, and visualization in the existing technologies using improvised scientific methods are a great opportunity to establish the core principles of microbiome and climate change research. The development and availability of spatial genomics, single-cell sequencing, long-read sequencing technologies with easier portability options, and computational modeling bring exciting times for researchers and the scientific fraternity. An augmented understanding of genetic interdependency and associated microbial community structure influences the function of microbiomes and will facilitate future efforts to obtain greater benefits for climate resilience, carbon sequestration, and global warming.

8.8 Conclusions

Overall, the intricate relationship between plants, microbes, and carbon is a crucial factor in regulating the carbon cycle. Sustainable land management practices, conservation of natural ecosystems, and innovative carbon capture technologies are essential to enhance carbon sequestration and mitigate the impacts of global climate change. It is essential to continue research and promote environmentally conscious policies to maintain a balanced carbon cycle and preserve the health of our planet.

To address the challenges of climate change, there are ongoing efforts to promote afforestation and reforestation, implement sustainable land management practices, and enhance soil carbon sequestration through agricultural practices like conservation tillage and cover cropping. These strategies aim to maximize carbon uptake by plants and promote the long-term storage of carbon in both plant biomass and soils. Therefore, in this regard, microbial genomics provides insights and recommendations for future research directions, potential mitigation strategies, and adaptation measures related to microbe's role in organic matter decomposition and CO_2 release in the context of climate change and mitigation actions.

Acknowledgment The authors would like to acknowledge Birla Institute of Technology and Science (BITS), Pilani, Hyderabad Campus, and Gujarat University, Ahmedabad for providing research support and resources.

Declaration The authors declare there are no financial interests or personal relationships which may be considered as potential competing interests.

References

Akiyama, K., Matsuzaki, K. I., & Hayashi, H. (2005). Plant sesquiterpenes induce hyphal branching in arbuscular mycorrhizal fungi. *Nature, 435*(7043), 824–827. https://doi.org/10.1038/nature03608

Ali, M., Javeed, H. M. R., Tariq, M., Khan, A. A., Qamar, R., Nawaz, F., Masood, N., Ditta, A., Abbas, T., Zamir, M. S. I., et al. (2023). Use of biochar for biological carbon sequestration. In *Climate change impacts on agriculture: Concepts, issues and policies for developing countries* (pp. 393–411). Springer.

Askari-Khorasgani, O., Hatterman-Valenti, H., Flores, F. B., & Pessarakli, M. (2019). Managing plant-environment-symbiont interactions to promote plant performance under low temperature stress. In *Journal of Plant Nutrition* (Vol. 42, Issue 16, pp. 2010–2027). Taylor and Francis Inc. https://doi.org/10.1080/01904167.2019.1648682

Bahman, N., Al-Khalifa, M., Al Baharna, S., Abdulmohsen, Z., & Khan, E. (2023). Review of carbon capture and storage technologies in selected industries: Potentials and challenges. *Reviews in Environmental Science and Bio/Technology, 22*, 1–20.

Bais, H. P., Weir, T. L., Perry, L. G., Gilroy, S., & Vivanco, J. M. (2006). The role of root exudates in rhizosphere interactions with plants and other organisms. *Annual Review of Plant Biology, 57*, 233–266. https://doi.org/10.1146/annurev.arplant.57.032905.105159

Ballhausen, M.-B., & de Boer, W. (2016). The sapro-rhizosphere: Carbon flow from saprotrophic fungi into fungus-feeding bacteria. *Soil Biology and Biochemistry, 102*, 14–17.

Barima, Y. S. S., Angaman, D. M., N'gouran, K. P., Bi, F. Z. T., & Samson, R. (2016). Involvement of leaf characteristics and wettability in retaining air particulate matter from tropical plant species. *Environmental Engineering Research*. https://www.koreascience.or.kr/article/JAKO201621650491549.page

Bauer, F., Tilsted, J. P., Pfister, S., Oberschelp, C., & Kulionis, V. (2023). Mapping GHG emissions and prospects for renewable energy in the chemical industry. *Current Opinion in Chemical Engineering, 39*, 100881.

Baumas, C. M. J., & Bizic, M. (2023). *Did you say marine snow? Zooming into different types of organic matter particles and their importance in the open ocean carbon cycle*.

Beattie, G. A. (2018). Metabolic coupling on roots. *Nature Microbiology, 3*(4), 396.

Bernardini, D., De Bernardini, N., Basile, A., Zampieri, G., Kovalovszki, A., Diaz, B. D. D., Offer, E., Wongfaed, N., Angelidaki, I., Kougias, P. G., Campanaro, S., & Treu, L. (2022). Integrating metagenomic binning with flux balance analysis to unravel syntrophies in anaerobic CO2 methanation. *Microbiome*, 1–18. https://doi.org/10.1186/s40168-022-01311-1

Blagodatsky, S., Blagodatskaya, E., Yuyukina, T., & Kuzyakov, Y. (2010). Model of apparent and real priming effects: Linking microbial activity with soil organic matter decomposition. *Soil Biology and Biochemistry, 42*(8), 1275–1283.

Bohan, D. A., Vacher, C., Tamaddoni-Nezhad, A., Raybould, A., Dumbrell, A. J., & Woodward, G. (2017). Next-generation global biomonitoring: Large-scale, automated reconstruction of ecological networks. *Trends in Ecology & Evolution, 32*(7), 477–487.

Burd, A. B. (2023). Modeling the vertical flux of organic carbon in the global ocean. *Annual Review of Marine Science, 16*, 135–161.

Chhabra, S., Brazil, D., Morrissey, J., Burke, J. I., O'Gara, F., Dowling, N., & D. (2013). Characterization of mineral phosphate solubilization traits from a barley rhizosphere soil functional metagenome. *MicrobiologyOpen, 2*(5), 717–724. https://doi.org/10.1002/mbo3.110

De Deyn, G. B., Cornelissen, J. H. C., & Bardgett, R. D. (2008). Plant functional traits and soil carbon sequestration in contrasting biomes. *Ecology Letters, 11*(5), 516–531.

Delgado-Baquerizo, M., Eldridge, D. J., Ochoa, V., Gozalo, B., Singh, B. K., & Maestre, F. T. (2017). Soil microbial communities drive the resistance of ecosystem multifunctionality to global change in drylands across the globe. *Ecology Letters, 20*(10), 1295–1305.

den Camp, H. J. M., Islam, T., Stott, M. B., Harhangi, H. R., Hynes, A., Schouten, S., Jetten, M. S. M., Birkeland, N.-K., Pol, A., & Dunfield, P. F. (2009). Environmental, genomic and taxonomic perspectives on methanotrophic Verrucomicrobia. *Environmental Microbiology Reports, 1*(5), 293–306.

Fagorite, V. I., Onyekuru, S. O., Opara, A. I., & Oguzie, E. E. (2023). The major techniques, advantages, and pitfalls of various methods used in geological carbon sequestration. *International Journal of Environmental Science and Technology, 20*(4), 4585–4614.

Falade, A. O., Nwodo, U. U., Iweriebor, B. C., Green, E., Mabinya, L. V., & Okoh, A. I. (2017). Lignin peroxidase functionalities and prospective applications. *MicrobiologyOpen, 6*(1), e00394.

Garcia, C. A., Hagstrom, G. I., Larkin, A. A., Ustick, L. J., Levin, S. A., Lomas, M. W., & Martiny, A. C. (2020). Linking regional shifts in microbial genome adaptation with surface ocean biogeochemistry. *Philosophical Transactions of the Royal Society B: Biological Sciences, 375*(1798), 20190254. https://doi.org/10.1098/rstb.2019.0254

Glick, B. R. (2020). *Beneficial plant-bacterial interactions* (2nd ed., p. 383). Springer.

Glick, B. R., & Gamalero, E. (2021). Recent developments in the study of plant microbiomes. *Microorganisms, 9*(7), 1533.

Hay, M. E. (1996). Marine chemical ecology: What's known and what's next? *Journal of Experimental Marine Biology and Ecology, 200*(1–2), 103–134.

Hervé, V., Junier, T., Bindschedler, S., Verrecchia, E., & Junier, P. (2016). Diversity and ecology of oxalotrophic bacteria. *World Journal of Microbiology and Biotechnology, 32*, 1–7.

Hilmi, N., Sutherland, M., Farahmand, S., Haraldsson, G., van Doorn, E., Ernst, E., Wisz, M. S., Claudel Rusin, A., Elsler, L. G., & Levin, L. A. (2023). Deep sea nature-based solutions to climate change. *Frontiers in Climate, 5*, 1169665.

Hirsch, A. M., Bauer, W. D., Bird, D. M., Cullimore, J., Tyler, B., & Yoder, J. I. (2003). Molecular signals and receptors: Controlling rhizosphere interactions between plants and other organisms. *Ecology, 84*(4), 858–868.

Hirt, H., Boukcim, H., Ducousso, M., & Saad, M. M. (2023). Engineering carbon sequestration on arid lands. *Trends in Plant Science, 28*, 1218. https://doi.org/10.1016/j.tplants.2023.08.009

Jansson, J. K., & Hofmockel, K. S. (2018). The soil microbiome—from metagenomics to metaphenomics. *Current Opinion in Microbiology, 43*, 162–168.

Jousset, A., Bienhold, C., Chatzinotas, A., Gallien, L., Gobet, A., Kurm, V., Küsel, K., Rillig, M. C., Rivett, D. W., Salles, J. F., Van Der Heijden, M. G. A., Youssef, N. H., Zhang, X., Wei, Z., & Hol, G. W. H. (2017). Where less may be more: How the rare biosphere pulls ecosystems strings. *ISME Journal, 11*(4), 853–862. https://doi.org/10.1038/ismej.2016.174

Kalam, S., Basu, A., Ahmad, I., Sayyed, R. Z., El-Enshasy, H. A., Dailin, D. J., & Suriani, N. L. (2020). Recent understanding of soil acidobacteria and their ecological significance: A critical review. In *Frontiers in Microbiology* (Vol. 11, p. 2712). Frontiers Media S.A. https://doi.org/10.3389/fmicb.2020.580024

Khan, A. L., Asaf, S. M., Abed, R. M., Ning Chai, Y. N., Al-Rawahi, A., Mohanta, T. K., Al-Rawahi, A., Schachtman, D. P., & Al-Harrasi, A. (2020). Rhizosphere microbiome of arid land medicinal plants and extra cellular enzymes contribute to their abundance. *Microorganisms, 8*(2), 213.

Krevor, S., De Coninck, H., Gasda, S. E., Ghaleigh, N. S., de Gooyert, V., Hajibeygi, H., Juanes, R., Neufeld, J., Roberts, J. J., & Swennenhuis, F. (2023). Subsurface carbon dioxide and hydrogen storage for a sustainable energy future. *Nature Reviews Earth & Environment, 4*(2), 102–118.

Kumar, D., Saraf, M., Joshi, C. G., & Joshi, M. (2022). Host plant rhizo-microbiome interactions: Seasonal variation and microbial community structure analysis associated with Barleria prionitis. *Ecological Genetics and Genomics, 22*, 100109. https://doi.org/10.1016/j.egg.2021.100109

Kuzyakov, Y., Friedel, J. K., & Stahr, K. (2000). Review of mechanisms and quantification of priming effects. *Soil Biology and Biochemistry, 32*(11–12), 1485–1498.

Le Boulch, M., Déhais, P., Combes, S., & Pascal, G. (2019). The MACADAM database: A MetAboliC pAthways DAtabase for Microbial taxonomic groups for mining potential metabolic capacities of archaeal and bacterial taxonomic groups. *Database, 2019*(1). https://doi.org/10.1093/database/baz049

Letcher, T. M. (2021). Global warming—a complex situation. In *Climate change* (pp. 3–17). Elsevier.

Lidbury, I. D. E. A., Raguideau, S., Borsetto, C., Murphy, A. R. J., Bottrill, A., Liu, S., Stark, R., Fraser, T., Goodall, A., Jones, A., Bending, G. D., Tibbet, M., Hammond, J. P., Quince, C., Scanlan, D. J., Pandhal, J., & Wellington, E. M. H. (2022). Stimulation of distinct rhizosphere bacteria drives phosphorus and nitrogen mineralization in oilseed rape under field conditions. *MSystems, 7*(4), e0002522. https://doi.org/10.1128/msystems.00025-22

Long, S. P., Taylor, S. H., Burgess, S. J., Carmo-Silva, E., Lawson, T., De Souza, A. P., Leonelli, L., & Wang, Y. (2022). Into the shadows and back into sunlight: Photosynthesis in fluctuating light. *Annual Review of Plant Biology, 73*, 617–648.

López-Mondéjar, R., Zühlke, D., Becher, D., Riedel, K., & Baldrian, P. (2016). Cellulose and hemicellulose decomposition by forest soil bacteria proceeds by the action of structurally variable enzymatic systems. *Scientific Reports, 6*(1), 25279.

Louca, S., Parfrey, L. W., & Doebeli, M. (2016). Decoupling function and taxonomy in the global ocean microbiome. *Science, 353*(6305), 1272–1277. https://doi.org/10.1126/science.aaf4507

Matthews, M. L. (2023). Engineering photosynthesis, nature's carbon capture machine. *PLoS Biology, 21*(7), e3002183.

Mavrodi, O. V., McWilliams, J. R., Peter, J. O., Berim, A., Hassan, K. A., Elbourne, L. D. H., LeTourneau, M. K., Gang, D. R., Paulsen, I. T., Weller, D. M., et al. (2021). Root exudates alter the expression of diverse metabolic, transport, regulatory, and stress response genes in rhizosphere Pseudomonas. *Frontiers in Microbiology, 12*, 698.

McDaniel, M. D., Saha, D., Dumont, M. G., Hernández, M., & Adams, M. A. (2019). The effect of land-use change on soil CH_4 and N_2O fluxes: A global meta-analysis. *Ecosystems, 22*(6), 1424–1443.

Meadows-smith, M., Wigley, P., & Turner, S. (2017). Integrated plant breeding methods for complementary pairings of plants and microbial consortia. *Google Patents*.

Meena, M., Yadav, G., Sonigra, P., Nagda, A., Mehta, T., Swapnil, P., Harish, Marwal, A., & Kumar, S. (2023). Multifarious responses of forest soil microbial community toward climate change. *Microbial Ecology, 86*(1), 49–74.

Mukhtar, H., Wunderlich, R. F., Muzaffar, A., Ansari, A., Shipin, O. V., Cao, T. N.-D., & Lin, Y.-P. (2023). Soil microbiome feedback to climate change and options for mitigation. *Science of the Total Environment*, 163412.

Mutio, J. M., Kebeney, S., Njoroge, R., Churu, H., Ng'etich, W., Mugaa, D., Alkamoi, B., & Wamalwa, F. (2023). Effect of land rehabilitation measures on soil organic carbon fractions in semi-arid environment. *Frontiers in Sustainable Food Systems, 7*, 1095865.

Nacke, H., Fischer, C., Thürmer, A., Meinicke, P., & Daniel, R. (2014). Land use type significantly affects microbial gene transcription in soil. *Microbial Ecology, 67*, 919–930.

Naylor, D., Sadler, N., Bhattacharjee, A., Graham, E. B., Anderton, C. R., McClure, R., Lipton, M., Hofmockel, K. S., & Jansson, J. K. (2020). Soil microbiomes under climate change and implications for carbon cycling. *Annual Review of Environment and Resources, 45*(1), 29–59.

Nobori, T., Mine, A., & Tsuda, K. (2018). Molecular networks in plant–pathogen holobiont. *FEBS Letters, 592*, 1937.

Norena-Caro, D., & Benton, M. G. (2018). Cyanobacteria as photoautotrophic biofactories of high-value chemicals. *Journal of CO2 Utilization, 28*, 335–366.

Pathak, H. (2023). Impact, adaptation, and mitigation of climate change in Indian agriculture. *Environmental Monitoring and Assessment, 195*(1), 52.

Pérez Castro, S., Cleland, E. E., Wagner, R., Al Sawad, R., & Lipson, D. A. (2019). Soil microbial responses to drought and exotic plants shift carbon metabolism. *The ISME Journal, 13*(7), 1776–1787.

Pessarrodona, A., Franco-Santos, R. M., Wright, L. S., Vanderklift, M. A., Howard, J., Pidgeon, E., Wernberg, T., & Filbee-Dexter, K. (2023). Carbon sequestration and climate change mitigation using macroalgae: a state of knowledge review. *Biological Reviews, 98*, 1945.

Poole, P., Ramachandran, V., & Terpolilli, J. (2018). Rhizobia: From saprophytes to endosymbionts. *Nature Reviews Microbiology, 16*(5), 291.

Preece, C., & Peñuelas, J. (2020). A return to the wild: Root exudates and food security. *Trends in Plant Science, 25*(1), 14–21.

Raza, T., Qadir, M. F., Khan, K. S., Eash, N. S., Yousuf, M., Chatterjee, S., Manzoor, R., ur Rehman, S., & Oetting, J. N. (2023). Unrevealing the potential of microbes in decomposition of organic matter and release of carbon in the ecosystem. *Journal of Environmental Management, 344*, 118529.

Reed, L., & Glick, B. R. (2023). The recent use of plant-growth-promoting bacteria to promote the growth of agricultural food crops. *Agriculture, 13*(5), 1089.

Richardson, E. J., Bacigalupe, R., Harrison, E. M., Weinert, L. A., Lycett, S., Vrieling, M., Robb, K., Hoskisson, P. A., Holden, M. T. G., & Feil, E. J. (2018). Gene exchange drives the ecological success of a multi-host bacterial pathogen. *Nature Ecology & Evolution, 1*, 1468.

Schauer, K., Rodionov, D. A., & de Reuse, H. (2008). New substrates for TonB-dependent transport: Do we only see the 'tip of the iceberg'? *Trends in Biochemical Sciences, 33*(7), 330–338.

Schimel, J., Balser, T. C., & Wallenstein, M. (2007). Microbial stress-response physiology and its implications for ecosystem function. *Ecology, 88*(6), 1386–1394.

Semenov, M. V. (2021). Metabarcoding and metagenomics in soil ecology research: achievements, challenges, and prospects. *Biology Bulletin Reviews, 11*, 40–53.

Shaw, R., & Mukherjee, S. (2022). The development of carbon capture and storage (CCS) in India: A critical review. *Carbon Capture Science & Technology, 2*, 100036.

Sinsabaugh, R. L., Antibus, R. K., & Linkins, A. E. (1991). An enzymic approach to the analysis of microbial activity during plant litter decomposition. *Agriculture, Ecosystems & Environment, 34*(1–4), 43–54.

Song, S., Liu, Y., Wang, N. R., & Haney, C. H. (2021). Mechanisms in plant--microbiome interactions: Lessons from model systems. *Current Opinion in Plant Biology, 62*, 102003.

Stirnimann, L. (2023). *Plankton dynamics of the open Southern Ocean and surrounding the (Sub) Antarctic islands*.

Sunagawa, S., Coelho, L. P., Chaffron, S., Kultima, J. R., Labadie, K., Salazar, G., Djahanschiri, B., Zeller, G., Mende, D. R., & Alberti, A. (2015). Structure and function of the global ocean microbiome. *Science, 348*(6237), 1261359.

Tatsis, E. C., & O'Connor, S. E. (2016). New developments in engineering plant metabolic pathways. *Current Opinion in Biotechnology, 42*, 126–132.

Terry, N. (1977). Photosynthesis, growth, and the role of chloride. *Plant Physiology, 60*(1), 69–75.

Theis, K. R., Dheilly, N. M., Klassen, J. L., Brucker, R. M., Baines, J. F., Bosch, T. C. G., Cryan, J. F., Gilbert, S. F., Goodnight, C. J., & Lloyd, E. A. (2016). Getting the hologenome concept right: An eco-evolutionary framework for hosts and their microbiomes. *Msystems, 1*(2), e00028–e00016.

Tolstanova, G., Akulenko, I., Serhiichuk, T., Dovbynchuk, T., & Stepanova, N. (2023). Pro-pre-and synbiotic supplementation and oxalate homeostasis in 3 PM context: Focus on microbiota oxalate-degrading activity. In *Microbiome in 3P medicine strategies: The first exploitation guide* (pp. 335–353). Springer.

van der Heijden, M. G. A., Martin, F. M., Selosse, M. A., & Sanders, I. R. (2015). Mycorrhizal ecology and evolution: The past, the present, and the future. In *New Phytologist* (Vol. 205, Issue 4, pp. 1406–1423). Blackwell Publishing Ltd. https://doi.org/10.1111/nph.13288

Vishwakarma, P., & Kumar Dubey, S. (2007). The effect of soil type and plant age on the population size of rhizospheric methanotrophs and their activities in tropical rice soils. *Journal of Basic Microbiology, 47*(4), 351–357.

Vodovnik, M., Duncan, S. H., Reid, M. D., Cantlay, L., Turner, K., Parkhill, J., Lamed, R., Yeoman, C. J., Miller, M. E. B., White, B. A., et al. (2013). Expression of cellulosome components and type IV pili within the extracellular proteome of Ruminococcus flavefaciens 007. *PLoS One, 8*(6), e65333.

Walitang, D. I., Kim, C.-G., Kim, K., Kang, Y., Kim, Y. K., & Sa, T. (2018). The influence of host genotype and salt stress on the seed endophytic community of salt-sensitive and salt-tolerant rice cultivars. *BMC Plant Biology, 18*(1), 51.

Whatley, J. M., & Whatley, F. R. (1981). Chloroplast evolution. *New Phytologist, 87*(2), 233–247.

Xu, L., Dong, Z., Chiniquy, D., Pierroz, G., Deng, S., Gao, C., Diamond, S., Simmons, T., Wipf, H. M. L., Caddell, D., Varoquaux, N., Madera, M. A., Hutmacher, R., Deutschbauer, A., Dahlberg, J. A., Guerinot, M. L., Purdom, E., Banfield, J. F., Taylor, J. W., et al. (2021). Genome-resolved metagenomics reveals role of iron metabolism in drought-induced rhizosphere microbiome dynamics. *Nature Communications, 12*(1). https://doi.org/10.1038/s41467-021-23553-7

Yu, P., He, X., Baer, M., Beirinckx, S., Tian, T., Moya, Y. A. T., Zhang, X., Deichmann, M., Frey, F. P., Bresgen, V., Li, C., Razavi, B. S., Schaaf, G., von Wirén, N., Su, Z., Bucher, M., Tsuda, K., Goormachtig, S., Chen, X., & Hochholdinger, F. (2021). Plant flavones enrich rhizosphere Oxalobacteraceae to improve maize performance under nitrogen deprivation. *Nature Plants, 7*(4), 481–499. https://doi.org/10.1038/s41477-021-00897-y

Zhang, Y., Zhao, M., Cui, Q., Fan, W., Qi, J., Chen, Y., Zhang, Y., Gao, K., Fan, J., Wang, G., et al. (2017). Processes of coastal ecosystem carbon sequestration and approaches for increasing carbon sink. *Science China Earth Sciences, 60*, 809–820.

Zhang, C., Dang, H., Azam, F., Benner, R., Legendre, L., Passow, U., Polimene, L., Robinson, C., Suttle, C. A., & Jiao, N. (2018). Evolving paradigms in biological carbon cycling in the ocean. *National Science Review, 5*(4), 481–499.

Leveraging Plant Growth Promoting Microbes with the Goal of Minimising the Impact of Climate Change on Agriculture

9

Khushbu Rathod, Pinakin Dhandhukia, and Janki N. Thakker

Abstract

Agriculture production and food security are seriously threatened by climate change. There is a need to develop effective measures for the improvement of crop production in such changing environmental conditions. Microorganisms responsible for improving yield in crops may also demonstrate tolerance against abiotic stress and induce tolerance in plants as well. By improving soil quality and assisting plants to survive under stress from the environment, plant growth-promoting bacteria (PGPB) can assist in resolving this problem. By fusing endogenous signals with environmental cues to produce hormones that stimulate plant growth, PGPB promotes sustainable agriculture and slows the exponential growth of modern agriculture. Through the root, PGPB communicates with the plant's defence mechanism and encourages the development of these enzymes. Abiotic stress responses in plants entail interactions between hormone balance, metabolism, and gene expression, which in turn regulate crop production. The consequences of climate change on agriculture are covered in this chapter, along with how bacteria support plant growth in such conditions.

Keywords

PGP · Fertility improvement · Abiotic stress · Food nutrients

K. Rathod · J. N. Thakker (✉)
Department of Biotechnology, P D Patel Institute of Applied Sciences, Charotar University of Science and Technology, Gujarat, India
e-mail: jankithakker.bt@charusat.ac.in

P. Dhandhukia
Department of Microbiology, School of Science and Technology, Vanita Vishram Women's University, Surat, Gujarat, India

M. Saraf et al. (eds.), *Climate Change and Soil Microorganisms for Environmental Sustainability*, Microorganisms for Sustainability 28,
https://doi.org/10.1007/978-981-96-3425-5_9

9.1 Introduction

One of the major problems faced by the globe now is climate change. It is characterised as notable variations in the average values of meteorological components, such as humidity and temperature (Tubiello & Fischer, 2007). One of the many human endeavours and environmental services that are put at risk from harmful anthropogenic climate change is food security. Concern exists about how the world's agricultural productivity will be affected by climate change and its variability (Malhi et al., 2021). As shown by numerous climatic models, such as erratic rainfall, heat waves, cyclones, etc., Asia is experiencing grave challenges as a result of climate change and variability. Abiotic and biotic pressures that have a deleterious impact on agricultural output to an unsustainable extent are mostly caused by global climate change, which poses a danger to production increase and sustainable agriculture (Rathod et al., 2024a). Abiotic stresses brought on by climate change, such as salinity, drought, severe temperatures, and others, have an impact on plant traits through a variety of channels and processes, eventually impeding plant growth, development, and productivity (Cao et al., 2023). Climate variability in Asia, including changes in temperature and precipitation, as well as climate-driven extremes have a severe impact on both the quantity and quality of agricultural products (Habib-ur-Rahman et al., 2022). The quality characteristics of wheat grains, particularly their protein content, sugar and starch percentages, are being negatively impacted by climate change. High temperatures and increased carbon dioxide levels boost growth features while lowering protein levels in wheat grains (Asseng et al., 2019). The cultivation of plants for a variety of uses, including the provision of food, fibre, and fuel, constitutes the fundamental and age-old practice of agriculture. It is an essential part of human civilisation, giving cultures throughout history food and resources. As a result, it's critical to develop effective strategies to safeguard crops against climate change, which could worsen malnutrition as well as increase hunger in the world.

A variety of measures have been implemented to safeguard crops from climate change. Crop diversification, early planting, harvesting rainwater, market responses like diversification of income and credit schemes, meteorological forecasting capability development, better agricultural markets, and information provision are just a few examples of the many different ways that agriculture is adapting to climate change (Gebre et al., 2023). However, each of these approaches has its own shortcomings. Therefore, it is imperative to formulate a feasible strategy for this matter. Here, plant growth-promoting bacteria can serve as a promising solution (Compant et al., 2010). Due to their capacity to enhance soil qualities and confer climate stress in plants, plant growth-promoting bacteria (PGPB) can be used to more effectively provide sustainable agriculture with the rapid expansion of contemporary agriculture (Thakker et al., 2023; Rathod et al., 2023a). In order to control the plant's defensive reaction to diverse abiotic challenges, PGPB produces a variety of hormones that promote plant growth (Saraf & Dhandhukia, 2005) and protect plants from stress (Rathod et al., 2024b). These hormones combine environmental cues with endogenous signals (Li et al., 2021). However, plant hormones are produced by microorganisms for their own survival and growth (Badrakia et al., 2021). Indole acetic acid (IAA), which is produced by bacteria, is highly beneficial to plants because it encourages plant root length (Saraf et al., 2006b), which increases nutrient absorption (Patel & Thakker, 2019). The primary purpose of microbial IAA, however, is not to lengthen roots, but rather to encourage the overproduction of lateral roots and root hairs in plants, which leads to the release of saccharides from plant cell walls. Saccharides are a source of nutrients for microorganisms and can improve the efficiency of bacteria linked with plants to colonise territories previously unoccupied (Etesami et al., 2015). Similarly, gibberellins produced by microbes promote seed germination and cell elongation in plants (Castro-Camba et al., 2022) but have a role in pathogenicity in bacteria, where they operate as virulence factors, possibly by inhibiting the production of jasmonic acid, which compromises the defence mechanism of their host (Salazar-Cerezo et al., 2018). They will therefore play a significant part in abiotic stress protection as well. This chapter addresses the significance of PGPB in reducing environmental stressors including decreased soil fertility and decreased food output and quality, typically brought on by climatic change.

9.2 Soil Fertility

The term "soil fertility" describes a soil's capacity to support agricultural plant growth, i.e., to provide a habitat for plants to produce high-quality yields over an extended period of time. It is the capability of soil to provide vital plant nutrients and water in sufficient quantities to allow plant growth. Hence, for increasing crop production the soil should be highly fertile. When soil is described as being highly fertile, it means that it should be abundant in both the micronutrients and the macronutrients that plants require for their growth and development (Brar et al., 2019). The role of such nutrients is depicted in Fig. 9.1. It is not anticipated that the pH of the soil will vary quickly as a result of the direct consequences of climate change because it is determined by the parent material, the rate of weathering, the vegetation, and the environment. However, an increase in precipitation may exacerbate leaching and cause soil acidity (Norton & Ouyang, 2019). Acidic soil decreases the availability of nutrients for plants. Aluminium (Al) and iron (Fe) combine with phosphate ions at acidic pH levels to yield less soluble molecules. When soil pH is above 7.5, the majority of other nutrients—particularly micronutrients—tend to be less readily available. The exception is molybdenum (Mo), which seems to be more accessible at moderately alkaline pH levels and less accessible at acidic pH levels (Jensen, 2010). Additionally, the efficient use of nitrogen by plants and various reactions involving nitrogen in a soil mixture can be influenced by the pH of the soil. The process of nitrification, which involves turning ammonium into nitrate, is carried out by rhizobacteria and proceeds most quickly in soils with a pH of 7 to 8. At a pH of 5, nitrification in the soil is almost negligible. Therefore, an acidic pH of soil brought on by a change in the climate can impact soil nutrients required for plants, which will have an impact on plant growth (Norton & Ouyang, 2019).

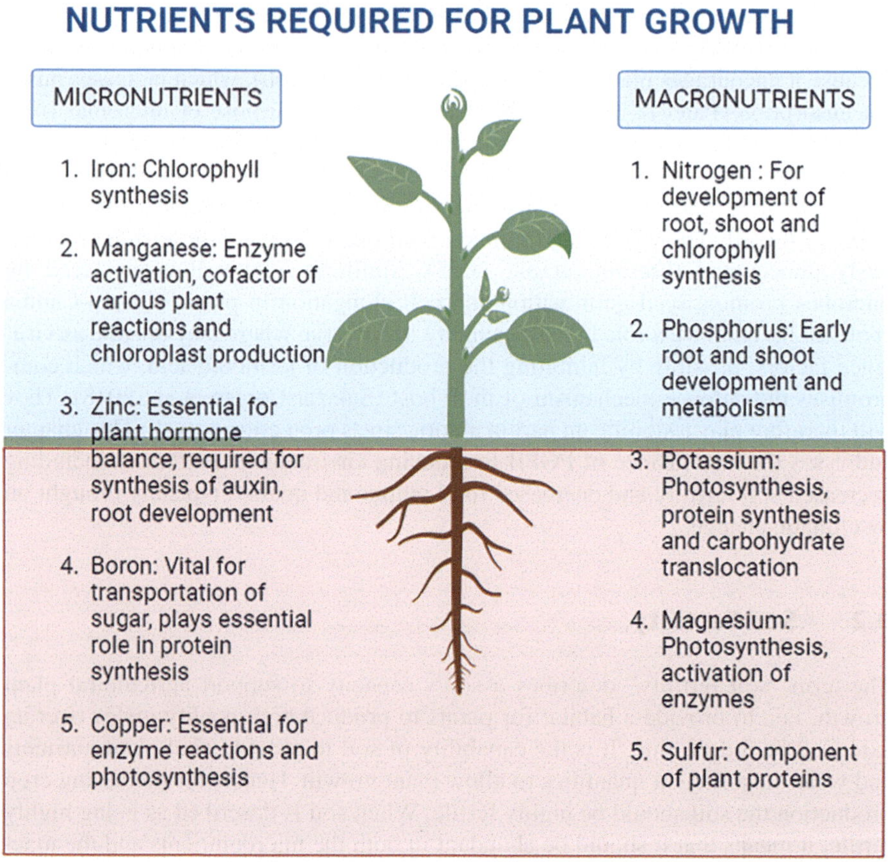

Fig. 9.1 Role of micro- and macronutrients in plant growth. (Created with BioRender)

Microbes assist plants in surviving under such acidic pH by improving soil fertility and by activating defence enzymes in plants (Goswami et al., 2019). Given adequate nutrient availability, the pH range of 5.5–7 is ideal for plant growth (Msimbira & Smith, 2020). This is also true for the majority of soil microorganisms, in part because plants in this type of range grow thoroughly and generate more root exudates, which serve as a carbon source that allows microbes to survive and multiply. While certain microorganisms can change the pH of soil by releasing compounds (acidic or basic) in their environment, most bacteria thrive in an environment with a pH of approximately neutral (Husnain et al., 2021). Organic material is broken down by microbes, which then utilise the carbon and nutrients for their own growth. Any extra nutrients (N, P, and S) that are present in the organic matter are broken down and released into the soil in forms that plants can utilise. This process is called mineralization (Gan et al., 2020). The release of nutrients during mineralization by microbes can assist the plant in proper growth as nutrients are depleted in acidic soil. Abiotic stress responses in plants are a result of interactions between various signalling molecules and hormonal balance, which control crop performance by changing levels of metabolism and gene expression (Cramer et al., 2011). More ROS are produced in the cell as a result of this alteration in metabolism. Overproduction of ROS triggers an oxidative burst in the cell, which in turn starts the oxidation of proteins and lipids, alters membrane permeability, damages DNA and RNA, and, in extreme situations, results in cell death. ROS has a direct impact on protein oxidation by oxidising the side chains of amino acids. It adversely changes protein function (Dumanović et al., 2021). The dual function of ROS makes cellular ROS equilibrium crucial for plants. They function as signalling molecules that control plant growth and development and acclimatise plants to stress, but when their concentration rises under stressful circumstances, they have negative effects on cellular metabolism. As a result, plants have a sophisticated antioxidant system that controls the equilibrium of cellular redox potential (Hasanuzzaman et al., 2021). Several enzymes, including catalase, superoxide mutase (SOD), and peroxidase (POX), are necessary for the detoxification of ROS (Menon & Thakker, 2020; Rathod et al., 2023c). In order to activate the plant's defence system, PGPB communicates with it via root. There are numerous compounds that PGPR are known to synthesise that function as signalling molecules (Jalmi & Sinha, 2022). The signal causes the plant to increase the growth of these enzymes (Rathod et al., 2023).

9.3 Decrease in Production

Crop production is also impacted by climate change. For instance, decreasing agricultural output may be a result of expected temperature rises, modifications to precipitation patterns, modifications to extreme weather events, and a decrease in water availability (Leisner, 2020). Crop production might also be impacted by higher carbon dioxide (CO_2) levels. Even though increased CO_2 can promote plant development, it also lowers the nutritional content of the majority of food crops. Most plant species, including wheat, soybeans, and rice, have lower protein and critical mineral concentrations due to rising atmospheric carbon dioxide levels. Warmer, wetter, and higher CO_2 levels are favourable to the growth of many weeds, pests, and fungi. As a result, weed control will waste more time and money. Besides CO_2 levels temperature has the most effect on plants as the temperature around a plant determines how quickly it grows and develops, and each species has a certain temperature range that is represented by a minimum, maximum, and optimal temperature (Hatfield & Prueger, 2015). It is predicted that heat waves and other high-temperature occurrences will grow more intense, more common, and continue longer than what has been seen in recent years (Arulalan et al., 2023). High temperature results in oxidative stress, DNA damage, water loss, hormonal changes, transcriptional changes and reduced photosynthetic ability causing reduction in yield of crops (Parthasarathi et al., 2022). Depending on the type of stress being experienced and the stage of development, a plant's reaction to these stressors will vary. A single stressor can have an impact on plants individually or in combination with other stressors (Singh et al., 2019). The reaction of plants to such stresses is a very intricate process that involves both physiological and biochemical alterations in the host plants. The formation of reactive oxygen species (ROS), changing relative water content, alteration in the aspect ratio of root-shoot length, leaf wilting, leaf abscission, and accumulation of free radicals that can interfere with cellular homeostasis and impair cell viability are some of these changes (Zulfiqar et al., 2020). The formation of reactive oxygen species, that may additionally alter gene regulation and enzyme activity, is one of the early signals (Patel et al., 2020). Plant hormones are yet another crucial signal. Two significant hormones that control how plants respond to various abiotic stressors are abscisic acid (ABA) and ethylene. A gaseous phytohormone, ethylene promotes the ripening of fruit, the wilting and senescence of leaves, abscission, germination, root architecture, and flowering in plants (Singh et al., 2015). Stress conditions have been shown to increase ethylene production, which ultimately leads to a reduction in the formation of root and shoot, inhibiting plant growth. Here, plant growth inhibition is prevented via the utilisation of microbial ACC deaminase to reduce plant ethylene synthesis (Jha et al., 2021). Acetyl-CoA carboxylase (ACC), a key enzyme in the ethylene synthesis pathway, is broken down by the ACC deaminase enzyme into ammonia and α-ketobutyrate, lowering the level of ethylene in the plants (Gupta & Pandey, 2019) hence, inhibiting plant growth inhibitor. Ammonia and α-ketobutyrate produced from this reaction are used by the microbe itself as a nitrogen and carbon source for its growth (Glick, 2014). Besides ethylene the other hormone which works as a growth inhibitor in abiotic stress is ABA. Plant growth is inhibited by the ABA by working against the hormones that encourage plant development (Brookbank et al., 2021). This hormone also keeps seeds dormant and works as an antagonist against gibberellin (Liu & Hou, 2018). There is still very little known about the exact microbial metabolism pathway of the phytohormone abscisic acid. However, when radioactive ABA was added to non-sterile soil, it degraded quickly into phaseic acid and dehydrophaseic acid, indicating a variety of biological transformation pathways for ABA by soil microbes. In vitro, ABA was converted to dehydrovomifoliol [1´-hydroxy-4´-α-keto-a-ionone] by a bacteria isolated from ABA-supplemented soil (and phenotypically identified as *Corynebacterium* sp.). This bacterium also has vomifoliol dehydrogenase activity (Belimov et al., 2014). PGPR *Rhodococcus* sp. P1Y was also found to metabolize ABA to dehydrovomifoliol and utilise it as a sole carbon source (Yuzikhin et al., 2021).

9.4 Decrease in Food Quality

The synthesis of minerals and proteins in crop plants is impacted by climate change, with elevated carbon dioxide being the primary cause of the majority of the observed changes. The type, severity, and length of the imparted stress, the genotype of the plant, and the stage of development all clearly influence the outcomes (Soares et al., 2019). Given that more soluble sugars result in an unbalanced C:N ratio, hastened leaf senescence, and/or constrained growth rate, elevated CO2 may cause photosynthetic modification. In the presence of high levels of carbon dioxide, a study reported that the concentration levels of protein, nitrate, magnesium, iron, and zinc decreased by 9.5%, 18.0%, 9.2%, and 16.0%, respectively (Dong et al., 2018). Another study demonstrated that higher levels of carbon dioxide have a direct impact on plants, lowering the nutritional value of food by reducing the concentrations of minerals and B vitamins by up to 30% and proteins and minerals by 5–15%, respectively (Kaplan et al., 2012). According to several research, C3 plants (such as wheat, rice, potatoes, and barley) improve photosynthesis, which can increase crop yields. However, as plants acquire more carbs and less micronutrients (such as iron and zinc), which can have a negative impact on human nutrition, the nutritional quality of those gains is reduced (Zhu et al., 2018). Because stressed ecosystems absorb less carbon during dry years, research has shown that the concentration of carbon dioxide in the atmosphere increases more quickly in drought.

The potential of mineral weathering bacteria to release macro- and micronutrients from soil minerals in croplands can boost plant growth and nutrient content as well (Patel & Thakker, 2020). With the exception of nitrogen, plant nutrients are ultimately produced by the weathering of fundamental minerals (Ribeiro et al., 2020). As a result, the subsequent settling of plants depends on the pioneer microbial activity. Thus, increased food quality will result from supplying plants with adequate soil nutrients. The yield and quality of crops are significantly improved in the presence of such nutrients as they are a part of enzymes, vitamins, and chlorophyll molecules and are essential for the formation of proteins, nucleic acids, and amino acids. It is crucial for cell division and the development of developing tissues (Carranca et al., 2018). Studies conducted in the past examining the effects of bacteria inoculation on plants raised in low-nutrient environments show a comparable rise in foliar nitrogen and phosphorus content. A study on strawberry plants treated with different PGPR strains *Staphylococcus* MFDCa-1, *Agrobacterium* A18, *Pantoea* FF1 and *Bacillus* M3 had an increase in N, P, K, Ca, Fe, Cu, and Mn foliar nutrient content (Ipek et al., 2014) and wheat plants grown under reduced fertility had an increase in N, P, K, Fe, Cu, Mn, and Zn when treated with various PGPR strains (Rana et al., 2012). In a different study, it was discovered that the *Pseudomonas* strains *P. poae* 29G9 and *P. fluorescens* 90F12-2 boosted the number of petunia flowers and plant biomass under drought and low-nutrient environments. When cultivated in low-nutrient environments, plants treated with bacteria had higher leaf nutrient content than untreated plants (Nordstedt et al., 2020).

9.5 Conclusion

PGPR have been regarded as the primary components of rhizosphere engineering due to their capacity to enhance plant growth in the face of biotic and abiotic stress. Numerous PGPR strains have been identified, characterised, and employed during the last two to three decades to aid in the normal and stressed growth and development of a number of different plants. With the drastic change of climate and its adverse effect on agricultural production and quality, PGPR can serve as a potential solution to this problem. However, more in-depth knowledge and examination of this scenario are required for developing stress-tolerant biofertilizers in future.

References

Arulalan, T., AchutaRao, K., & Sagar, A. D. (2023). Climate science to inform adaptation policy: Heat waves over India in the 1.5°C and 2°C warmer worlds. *Climatic Change, 176*(5). https://doi.org/10.1007/s10584-023-03527-y

Asseng, S., Martre, P., Maiorano, A., Rötter, R. P., O'Leary, G. J., Fitzgerald, G. J., Girousse, C., Motzo, R., Giunta, F., Babar, M. A., Reynolds, M. P., Kheir, A. M. S., Thorburn, P. J., Waha, K., Ruane, A. C., Aggarwal, P. K., Ahmed, M., Balkovič, J., Basso, B., et al. (2019). Climate change impact and adaptation for wheat protein. *Global Change Biology, 25*(1), 155–173. https://doi.org/10.1111/gcb.14481

Badrakia, J., Patel, K. B., Dhandhukia, P., & Thakker, J. N. (2021). Mycoparasitic Pseudomonas spp. against infection of Fusarium chlamydosporum pathogen in soyabean (Glycine max) plant. *Archives of Phytopathology and Plant Protection, 54*(19–20), 2160–2170. https://doi.org/10.1080/03235408.2021.1925433

Belimov, A. A., Dodd, I. C., Safronova, V. I., Dumova, V. A., Shaposhnikov, A. I., Ladatko, A. G., & Davies, W. J. (2014). Abscisic acid metabolizing rhizobacteria decrease ABA concentrations in planta and alter plant growth. *Plant Physiology and Biochemistry, 74*, 84–91. https://doi.org/10.1016/j.plaphy.2013.10.032

Brar, P. S., Kaushal, R., & Bhardwaj, G. (2019). A review on beneficial effects of PGPR and noble liquid manures in enhancing soil fertility and sustainability. *International Journal of Current Microbiology and Applied Sciences, 8*(04), 409–415. https://doi.org/10.20546/ijcmas.2019.804.045

Brookbank, B. P., Patel, J., Gazzarrini, S., & Nambara, E. (2021). Role of basal aba in plant growth and development. *Genes, 12*(12) MDPI. https://doi.org/10.3390/genes12121936

Cao, H., Ding, R., Kang, S., Du, T., Tong, L., Zhang, Y., Chen, J., & Shukla, M. (2023). Drought, salt, and combined stresses in plants: Effects, tolerance mechanisms, and strategies. In D. Sparks (Eds.), *Advances in agronomy* (Vol. 178). Academic Press.

Carranca, C., Brunetto, G., & Tagliavini, M. (2018). Nitrogen nutrition of fruit trees to reconcile productivity and environmental concerns. *Plants, 7*(1). https://doi.org/10.3390/plants7010004

Castro-Camba, R., Sánchez, C., Vidal, N., & Vielba, J. M. (2022). Plant development and crop yield: The role of gibberellins. *Plants, 11*(19) MDPI. https://doi.org/10.3390/plants11192650

Compant, S., Van Der Heijden, M. G. A., & Sessitsch, A. (2010). Climate change effects on beneficial plant-microorganism interactions. *FEMS Microbiology Ecology, 73*(2), 197–214. https://doi.org/10.1111/j.1574-6941.2010.00900.x

Cramer, G. R., Urano, K., Delrot, S., Pezzotti, M., & Shinozaki, K. (2011). Effects of abiotic stress on plants: A systems biology perspective. *BMC Plant Biology, 11*, 163. https://doi.org/10.1186/1471-2229-11-163

Dong, J., Gruda, N., Lam, S. K., Li, X., & Duan, Z. (2018). Effects of elevated CO2 on nutritional quality of vegetables: A review. *Frontiers in Plant Science, 9*. Frontiers Media S.A. https://doi.org/10.3389/fpls.2018.00924

Dumanović, J., Nepovimova, E., Natić, M., Kuča, K., & Jaćević, V. (2021). The significance of reactive oxygen species and antioxidant defense system in plants: A concise overview. *Frontiers in Plant Science, 11*. Frontiers Media S.A. https://doi.org/10.3389/fpls.2020.552969

Etesami, H., Alikhani, H. A., & Hosseini, H. M. (2015). Indole-3-acetic acid (IAA) production trait, a useful screening to select endophytic and rhizosphere competent bacteria for rice growth promoting agents. *MethodsX, 2*, 72–78. https://doi.org/10.1016/j.mex.2015.02.008

Gan, H. Y., Schöning, I., Schall, P., Ammer, C., & Schrumpf, M. (2020). Soil organic matter mineralization as driven by nutrient stoichiometry in soils under differently managed forest stands. *Frontiers in Forests and Global Change, 3*. https://doi.org/10.3389/ffgc.2020.00099

Gebre, G. G., Amekawa, Y., Fikadu, A. A., & Rahut, D. B. (2023). Farmers' use of climate change adaptation strategies and their impacts on food security in Kenya. *Climate Risk Management, 40*, 100495. https://doi.org/10.1016/j.crm.2023.100495

Glick, B. R. (2014). Bacteria with ACC deaminase can promote plant growth and help to feed the world. *Microbiological Research, 169*(1), 30–39. https://doi.org/10.1016/j.micres.2013.09.009

Goswami, D., Dhandhukia, P., & Thakker, J. N. (2019). Microbial enzymes: The types and roles in the bio-control of fungal phytopathogens. In S. Abdel-Aziz, N. Garg, A. Aeron, et al. (Eds.), *Microbial catalysts* (Vol. 1). Nova Science.

Gupta, S., & Pandey, S. (2019). Unravelling the biochemistry and genetics of ACC deaminase-An enzyme alleviating the biotic and abiotic stress in plants. *Plant Gene, 18*, 100175. https://doi.org/10.1016/j.plgene.2019.100175

Habib-ur-Rahman, M., Ahmad, A., Raza, A., Hasnain, M. U., Alharby, H. F., Alzahrani, Y. M., Bamagoos, A. A., Hakeem, K. R., Ahmad, S., Nasim, W., Ali, S., Mansour, F., & El Sabagh, A. (2022). Impact of climate change on agricultural production; Issues, challenges, and opportunities in Asia. *Frontiers in Plant Science, 13*. https://doi.org/10.3389/fpls.2022.925548

Hasanuzzaman, M., Parvin, K., Bardhan, K., Nahar, K., Anee, T. I., Masud, A. A. C., & Fotopoulos, V. (2021). Biostimulants for the regulation of reactive oxygen species metabolism in plants under abiotic stress. *Cells, 10*(10) MDPI. https://doi.org/10.3390/cells10102537

Hatfield, J. L., & Prueger, J. H. (2015). Temperature extremes: Effect on plant growth and development. *Weather and Climate Extremes, 10*, 4–10. https://doi.org/10.1016/j.wace.2015.08.001

Husnain, A., Hussain, I., Bakar Ijaz, A., & Zafar, A. (2021). Influence of soil Ph and microbes on mineral solubility and plant nutrition: A review author's details. *International Journal of Agriculture and Biological Sciences*, 71–81. https://www.researchgate.net/publication/350040537

Ipek, M., Pirlak, L., Esitken, A., Figen Dönmez, M., Turan, M., & Sahin, F. (2014). Plant Growth-Promoting Rhizobacteria (PGPR) increase yield, growth and nutrition of strawberry under high-calcareous soil conditions. *Journal of Plant Nutrition, 37*(7), 990–1001. https://doi.org/10.1080/01904167.2014.881857

Jalmi, S. K., & Sinha, A. K. (2022). Ambiguities of PGPR-induced plant signaling and stress management. In *Frontiers in microbiology* (Vol. 13). https://doi.org/10.3389/fmicb.2022.899563

Jensen, T. (2010). *Soil pH and the availability of plant nutrients*. Plant Nutrition Today.

Jha, C. K., Sharma, P., Shukla, A., Parmar, P., Patel, R., Goswami, D., & Saraf, M. (2021). Microbial enzyme, 1-aminocyclopropane-1-carboxylic acid (ACC) deaminase: An elixir for plant under stress. In *Physiological and molecular plant pathology* (Vol. 115, p. 101664). Academic Press. https://doi.org/10.1016/j.pmpp.2021.101664

Kaplan, F., Zhao, W., Richards, J. T., Wheeler, R. M., Guy, C. L., & Levine, L. H. (2012). Transcriptional and metabolic insights into the differential physiological responses of arabidopsis to optimal and supraoptimal atmospheric co2. *PLoS One, 7*(8), e43583. https://doi.org/10.1371/journal.pone.0043583

Leisner, C. P. (2020). Review: Climate change impacts on food security- focus on perennial cropping systems and nutritional value. *Plant Science, 293*. Elsevier Ireland Ltd. https://doi.org/10.1016/j.plantsci.2020.110412

Li, N., Euring, D., Cha, J. Y., Lin, Z., Lu, M., Huang, L. J., & Kim, W. Y. (2021). Plant hormone-mediated regulation of heat tolerance in response to global climate change. *Frontiers in Plant Science, 11*. Frontiers Media S.A. https://doi.org/10.3389/fpls.2020.627969

Liu, X., & Hou, X. (2018). Antagonistic regulation of ABA and GA in metabolism and signaling pathways. *Frontiers in Plant Science, 9*. Frontiers Media S.A. https://doi.org/10.3389/fpls.2018.00251

Malhi, G. S., Kaur, M., & Kaushik, P. (2021). Impact of climate change on agriculture and its mitigation strategies: A review. *Sustainability (Switzerland), 13*(3), 1–21. MDPI. https://doi.org/10.3390/su13031318

Menon, S., & Thakker, J. (2020). Assessment of Amaranthus viridis L. Leaves on Growth, Antifungal and Antioxidant Activity of Cicer arientinum L. *International Journal of Pharma Medicine and Biological Sciences, 9*(2), 70–74. https://doi.org/10.18178/ijpmbs.9.2.70-74

Msimbira, L. A., & Smith, D. L. (2020). The roles of plant growth promoting microbes in enhancing plant tolerance to acidity and alkalinity stresses. *Frontiers in Sustainable Food Systems, 4*. Frontiers Media S.A. https://doi.org/10.3389/fsufs.2020.00106

Nordstedt, N. P., Chapin, L. J., Taylor, C. G., & Jones, M. L. (2020). Identification of Pseudomonas Spp. that increase ornamental crop quality during abiotic stress. *Frontiers in Plant Science, 10*. https://doi.org/10.3389/fpls.2019.01754

Norton, J., & Ouyang, Y. (2019). Controls and adaptive management of nitrification in agricultural soils. *Frontiers in Microbiology, 10*(Aug) Frontiers Media S.A. https://doi.org/10.3389/fmicb.2019.01931

Parthasarathi, T., Firdous, S., Mariya David, E., Lesharadevi, K., & Djanaguiraman, M. (2022). Effects of high temperature on crops. In *Advances in plant defense mechanisms*. IntechOpen. https://doi.org/10.5772/intechopen.105945

Patel, K. B., & Thakker, J. N. (2019). Growth promotion and biocontrol activity of Nocardiopsis dassonvillei strain YM12: an isolate from coastal agricultural land of Khambhat. *Vegetos, 32*(4), 571–582. https://doi.org/10.1007/s42535-019-00064-x

Patel, K. B., & Thakker, J. N. (2020). Deliberating plant growth promoting and mineral-weathering proficiency of streptomyces nanhaiensis strain YM4 for nutritional benefit of millet crop (Pennisetum glaucum). *Journal of Microbiology, Biotechnology and Food Sciences, 9*(4), 721–726. https://doi.org/10.15414/jmbfs.2020.9.4.721-726

Patel, P. P., Patel, K. B., Dhandhukia, P. C., & Thakker, J. N. (2020). Elicitation of plant defense against fusarium Oxysporum f.sp. Ciceris in chickpea plant using marine Micrococcus Sp. *Journal of Microbiology, Biotechnology and Food Sciences, 10*(3), 361–365. https://doi.org/10.15414/jmbfs.2020.10.3.361-365

Rana, A., Saharan, B., Nain, L., Prasanna, R., & Shivay, Y. S. (2012). Enhancing micronutrient uptake and yield of wheat through bacterial PGPR consortia. *Soil Science and Plant Nutrition, 58*(5), 573–582. https://doi.org/10.1080/00380768.2012.716750

Rathod, K., Rana, S., Dhandukia, P., & Thakker, J. N. (2023a). Investigating marine Bacillus as an effective growth promoter for chickpea. *Journal of Genetic Engineering and Biotechnology, 21*(1), 137. https://doi.org/10.1186/s43141-023-00608-4

Rathod, K., Rana, S., Dhandhukia, P., & Thakker, J. N. (2023b). Marine Bacillus as a potent biocontrol agent against Fusarium oxysporum f.sp. ciceris. *Plant Stress, 10*, 100289. https://doi.org/10.1016/j.stress.2023.100289

Rathod, K., Rana, S., Dhandhukia, P., & Thakker, J. N. (2023c). Marine Bacillus subtilis as an effective biocontrol agent against Fusarium oxysporum f. sp. ciceris. *European Journal of Plant Pathology, 167*, 759. https://doi.org/10.1007/s10658-023-02720-0

Rathod, K., Dhandhukia, P., & Thakker, J. N. (2024a). Chickpea crop management: an endeavor of biocontrol methods. In *Bio-inoculants in horticultural crops* (pp. 269–280). Elsevier. https://doi.org/10.1016/b978-0-323-96005-2.00002-7

Rathod, K., Rana, S., Dhandhukia, P., & Thakker, J. N. (2024b). From sea to soil: Marine Bacillus subtilis enhancing chickpea production through in vitro and in vivo plant growth promoting traits. *Brazilian Journal of Microbiology, 55*, 823. https://doi.org/10.1007/s42770-023-01238-1

Ribeiro, I. D. A., Volpiano, C. G., Vargas, L. K., Granada, C. E., Lisboa, B. B., & Passaglia, L. M. P. (2020). Use of mineral weathering bacteria to enhance nutrient availability in crops: A review. In *Frontiers in plant science* (Vol. 11). Frontiers Media S.A). https://doi.org/10.3389/fpls.2020.590774

Salazar-Cerezo, S., Martínez-Montiel, N., García-Sánchez, J., Pérez-y-Terrón, R., & Martínez-Contreras, R. D. (2018). Gibberellin biosynthesis and metabolism: A convergent route for plants, fungi and bacteria. *Microbiological Research, 208*, 85–98). Elsevier GmbH. https://doi.org/10.1016/j.micres.2018.01.010

Saraf, M., & Dhandhukia, P. (2005). Response of Sinorhizobium meliloti to high salt concentration and effect of added osmotica. *Journal of Microbial World, 7*(2), 250–257.

Saraf, M., Dhiman, S., & Dhandhukia, P. (2006b). Increase in plant growth potential using some new isolates of PGPR in different combinations in Trigonell phoenum. *Journal of Phytological Research, 19*(1), 63–69.

Singh, R. P., Shelke, G. M., Kumar, A., & Jha, P. N. (2015). Biochemistry and genetics of ACC deaminase: A weapon to "stress ethylene" produced in plants. *Frontiers in Microbiology, 6*(Sep) Frontiers Media S.A. https://doi.org/10.3389/fmicb.2015.00937

Singh, D. P., Gupta, V. K., & Prabha, R. (2019). Microbial interventions in agriculture and environment: Volume 2: Rhizosphere, microbiome and agro-ecology. In *Microbial interventions in agriculture and environment: Volume 2: Rhizosphere, microbiome and agro-ecology*. Springer Singapore. https://doi.org/10.1007/978-981-13-8383-0

Soares, J. C., Santos, C. S., Carvalho, S. M. P., Pintado, M. M., & Vasconcelos, M. W. (2019). Preserving the nutritional quality of crop plants under a changing climate: importance and strategies. In *Plant and soil* (Vol. 443, p. 1–2). Springer International Publishing. https://doi.org/10.1007/s11104-019-04229-0

Thakker, J. N., Badrakia, J., Patel, K., Makwana, U., Parmar, K., & Dhandhukia, P. (2023). Potential of a marine Pseudomonas aeruginosa strain OG101 to combat Fusarium oxysporum associated wilt in legume crops. *Archives of Phytopathology and Plant Protection, 56*(4), 284–294. https://doi.org/10.1080/03235408.2023.2183800

Tubiello, F. N., & Fischer, G. (2007). Reducing climate change impacts on agriculture: Global and regional effects of mitigation, 2000–2080. *Technological Forecasting and Social Change, 74*(7), 1030–1056. https://doi.org/10.1016/j.techfore.2006.05.027

Yuzikhin, O. S., Gogoleva, N. E., Shaposhnikov, A. I., Konnova, T. A., Osipova, E. V., Syrova, D. S., Ermakova, E. A., Shevchenko, V. P., Nagaev, I. Y., Shevchenko, K. V., Myasoedov, N. F., Safronova, V. I., Shavarda, A. L., Nizhnikov, A. A., Belimov, A. A., & Gogolev, Y. V. (2021). Rhizosphere bacterium rhodococcus sp. P1y metabolizes ab-scisic acid to form dehydrovomifoliol. *Biomolecules, 11*(3), 1–16. https://doi.org/10.3390/biom11030345

Zhu, C., Kobayashi, K., Loladze, I., Zhu, J., Jiang, Q., Xu, X., Liu, G., Seneweera, S., Ebi, K. L., Drewnowski, A., Fukagawa, N. K., & Ziska, L. H. (2018). *E C O L O G Y Carbon dioxide (CO2) levels this century will alter the protein, micronutrients, and vitamin content of rice grains with potential health consequences for the poorest rice-dependent countries*. https://www.science.org.

Zulfiqar, F., Akram, N. A., & Ashraf, M. (2020). Osmoprotection in plants under abiotic stresses: new insights into a classical phenomenon. *Planta, 251*(1) Springer, 3. https://doi.org/10.1007/s00425-019-03293-1

Plant Responses to Drought Stress and Interactions with the Root Microbiome for Sustainable Agriculture

10

Nitin Baliyan, Sandhya Dhiman, Kamal Prasad,
Sudhakaran Rajasekaran, and Dinesh Kumar Maheshwari

Abstract

With global warming, drought stress is the most important abiotic factor limiting plant growth and productivity worldwide. Water scarcity is increasing due to low rainfall frequency and high frequency of dry spells. Drought stress disrupts soil fertility, reduces soil microbial populations (SMPs), fails in some microbial activities and nutrient cycling, changes physiology and morphology, and produces excessive reactive oxygen species (ROS), resulting in reduced crop yields and economic losses. Different soil and root microbial communities play an important role in plant adaptation to water deprivation. Rhizosphere and endophytic plant growth-promoting bacteria (PGPR) and fungi in response to drought, as these communities could survive under water-deficient conditions through which they can benefit plants. These beneficial microbial communities (BMC) signal drought tolerance by producing volatile compounds (VCs), phytohormones such as auxins, gibberellins, salicylic acids, cytokinin, abscisic acid, exopolysaccharides (EPS), low molecular weight osmolytes, proline, proline, other

N. Baliyan (✉)
Research and Development, Xenesis, Absolute Biologicals, Gurugram, Haryana, India

Department of Botany and Microbiology, Gurukula Kangri University, Haridwar, Uttarakhand, India

S. Dhiman
Department of Oncology, Wayne State University, Detroit, MI, USA

K. Prasad · S. Rajasekaran
Research and Development, Xenesis, Absolute Biologicals, Gurugram, Haryana, India

D. K. Maheshwari (✉)
Department of Botany and Microbiology, Gurukula Kangri University, Haridwar, Uttarakhand, India

M. Saraf et al. (eds.), *Climate Change and Soil Microorganisms for Environmental Sustainability*, Microorganisms for Sustainability 28,
https://doi.org/10.1007/978-981-96-3425-5_10

amino acids, organic acids, citrulline, polyamines, and some enzymes act as antioxidants and reduce the harmful effects of dehydration. In this review, we discuss an overview of drought stress in plants, its impact on plant resistance mechanisms (PRM), and how to manage them, including the role of soil microorganisms (SMs) in coping with drought stress (DS).

Keywords

Abiotic stress · Rhizosphere · Diversity · Enzyme activities · Root colonization

10.1 Introduction

An uneven balance between soil surface evaporation and plant transpiration combined with low soil water content causes drought stress in soil, which affects 45% of all arable land, while the global aridity stress area increased from 17% to 27% over the last 30 years (Mansoor et al., 2022). This condition occurs when the water content material or humidity inside the soil and air is significantly low along with the excessive atmospheric temperature. There are four types of droughts that have been defined by the climatological community including (1) meteorological drought due to lack of precipitation, (2) hydrological drought that occurs when low water supply becomes evident, especially in streams and reservoirs, (3) agricultural drought or ecological drought that causes plant stress from a combination of evaporation and low soil moisture, and (4) socioeconomic drought that relates the supply and demand of various commodities to drought. According to the International Rescue Committee (IRC), the drought has affected more than 40 million people in East Africa as well as Somalia, Kenya, and Ethiopia. Nearly 12% of India's land area, or about one-sixth of its entire population, is at risk for drought; the most vulnerable areas are those with annual rainfall levels of up to 60 cm. Numerous issues, including instability, a severe drought, and an exponential rise in food costs, have pushed millions of people in these regions to the verge of starvation. About 80–90% of the water in a plant's fresh biomass is used in several physiological activities, including plant growth, development, and metabolism. A lack of rainfall, salt, extremes in temperature, and a high level of light intensity are just a few of the factors that contribute to drought stress and negatively affect a plant's ability to produce biomass, quality, and energy (Seleiman et al., 2021). Plants experience changes in their physiological, morphological, biochemical, and molecular features as a result of drought stress, and multidimensional stress. Plants activate their drought response mechanisms, such as morphological and structural changes, expression of drought-resistant genes, synthesis of hormones, and production of osmotic regulating chemicals, to combat drought stress (dos Santos et al., 2022).

Long-term drought disrupts the microbial community in the soil, which changes the shape of plant roots and the nutrients and nutritional content of root exudates. Actinobacteria and other gram-positive bacteria (GPB) thrive better under drought stress than gram-negative bacteria (GNB), which are more prevalent in the rhizosphere (Breitkreuz et al., 2021).

The production of polysaccharides, which enhance soil structure and water holding capacity, and the synthesis of deaminase, indoleacetic acid (IAA), and proline (Pro), which convince plants to tolerate drought stress, and improved water circulation through fungal mycelia are just a few of the mechanisms used by soil microbes to increase plant resistance to drought stress (Dhiman et al., 2019; Baliyan et al., 2021). Surprisingly, adding organic matter to soil can mitigate the negative effects that drought has on microorganisms. Additionally, drought-associated phytohormones like abscisic acid (ABA), jasmonic acid (JA), and 1-aminocyclopropane-1-carboxylic acid (ACC) can adapt microbial activity and soil microbial community structure; on the other hand, adaptation to sustained drought amends the responses of soil microbial communities to these hormones (Passos et al., 2014; Kumar et al., 2021; Rosa et al., 2023). Drought tolerance using nanoparticles, superabsorbent hydrogels, and biochar, employment of plant growth-promoting rhizobacteria (PGPR) and film farming are just a few of the methods that researchers are continuously working to develop (Dheeman et al., 2017; Zhang et al., 2020a, Zhang et al., 2020b; Baliyan et al., 2022a, Baliyan et al., 2022b; Rosa et al., 2023).

10.2 Impact of Drought Stress on Plant Growth

Due to drought stress circumstances, plants experience severe losses in a variety of processes. Explain in Table 10.1 the various molecular, biochemical, physiological, morphological, and ecological traits and processes that result in turgor loss, disordered enzyme activity, and reduced energy transfer from photosynthesis. These factors adversely affect on cell multiplication, elongation, and specialization (Osakabe et al., 2014; Ayub et al., 2021).

Plants that experience drought have slower cell division and growth rates, slower germination rates, smaller leaves, disrupted stomatal responses, and lower chlorophyll levels. However, plants may outsmart a lack of water by lowering water loss by shutting stomata, decreasing leaf surface area, etc., or boosting water intake by deep roots (Kozlowski & Pallardy, 2002; Sayer et al., 2021). Lack of water significantly inhibits the growth of roots, shoots, and leaves, which is followed by compact growth and plant development. In addition to these alterations, plants engage in a variety of metabolic activities, from photosynthesis to antioxidant and solute buildup as components of water shortage resistance. According to various findings, dryness reduced plant metabolism and production by decreasing chlorophyll content, net photosynthetic rate, maximum quantum yield, stomatal conductance, and transpiration (Osakabe et al., 2014; Zhang et al., 2020a, Zhang et al., 2020b). In times of drought, stomatal and nonstomatal variables may be useful in reducing the photosynthetic rate (Yang et al., 2021). Water deprivation encourages the production of ABA, which closes stomata, lowering intercellular CO_2 concentrations and impeding photosynthesis. Plants have a propensity to adapt to various challenges, and they do so by changing the expression of certain genes via intricate transcriptional networks in response to drought stress (Singh & Laxmi, 2015). In plants, drought induces the production of ROS (superoxide radical, hydrogen peroxide,

Table 10.1 Morphological, anatomical, physiological, biochemical, and molecular responses of plants to drought stress

Type of impact	Effect of drought stress on plant	Reference
Morphological and anatomical	High growth in the root system and decrease in plant height	Yang et al. (2021) and Adiguzel et al. (2023)
	Reduction in seedling characteristics	Qaderi et al. (2012) and Liu et al. (2023)
	Reduced size, area, and number of leaves	Jabbari et al. (2013) and Sayer et al. (2021)
	More thick leaves with reduced number of stomata and wax formation surface of leaves	Sayer et al. (2021) and Shim et al. (2023)
	Increase mesophyll tissue and number of cell layers, but decrease the volume the intercellular space	Chartzoulakis et al. (2002) and Hewedy et al. (2020)
	Reduced the number of flowers and nodulation	Furlan et al. (2012), Hanaka et al. (2021), and Lumactud et al. (2023)
Physiological and biochemical	Decrease of photosynthetic rate and prevents transpiration water loss	Ma et al. (2015) and Yang et al. (2021)
	Synthesis of ROS and Increase in proline and ABA content	Hewedy et al. (2020) and Wang et al. (2023)
	Oxidation of carbohydrates, proteins, nucleic acids, lipids, as well as the destruction of the cell membrane	Sabra et al. (2012) and Khalvandi et al. (2021)
	Lipid peroxidation and accumulation of inorganic ions, sugar and H_2O_2	Tarazona et al. (2015) and Bao et al. (2020)
Transcriptomic and metabolomic	Expression of genes drought-tolerant (DT) and drought-susceptible (DS) sesame genotypes	You et al. (2019) and Chen et al. (2022)
	Encoding for chaperones, including heat shock proteins (HSP), DNA J proteins, aquaporins, orthologs of ABFs (ABA-Responsive-Element binding factors), and DREB2s (Dehydration-Responsive Element-Binding Proteins 2), glutaredoxins (GRXs), glutathione S-transferase (GST).	Wu et al. (2017), Zhang et al. (2020a) and Zhang et al. (2020 b)

hydroxyl radical, and singlet oxygen), which in turn causes oxidation. As a result of lipid peroxidation, chlorophyll disintegration, membrane instability, and ion leakage, ROS formation in plants is detrimental to several physiological and metabolic activities, including photosynthesis and antioxidant defense system (Hewedy et al., 2020).

The negative effects of drought stress on plants include ROS-mediated cell damage, elevated cellular temperature and viscosity of cellular contents, altered protein interaction, aggregation, and denaturation. To counter the negative impacts of drought stress on crops, plant biotechnologists have developed innovative cultivars with high drought tolerance using a variety of molecular approaches, including transcriptomics, proteomics, and metabolomics (Adiguzel et al., 2023). The expression of a number of genes that are sensitive to drought is reportedly regulated by plants in response to water stress scenarios via a complex transcriptome network (You et al., 2019). Several physiological, cellular, and molecular processes, including the up- or down-regulation of the expression of various genes responsible for osmolyte accumulation, increased content of the enzymatic and enzymatic antioxidant system, decreased rate of transpiration, growth of the shoot, and tillering, are in charge of the expansion of drought tolerance in plants (Zhang et al., 2020a, Zhang et al., 2020b; Jiang et al., 2021).

10.3 Root System Architecture and Its Role in Nutrient Uptake

For a plant to be productive, its roots must carry out a variety of functions, including nutrient and water absorption, symbiotic interactions with other bacteria in the rhizosphere, soil anchoring, and storage organ function (SOF). Root system architecture (RSA) is the spatial arrangement of roots in the soil profile and is largely influenced by branching, angle, length, and thickness (Pandey & Bennett, 2019). Some of the characteristics that distinguish RSA include root-to-shoot ratio, rooting depth, root growth angle, root length density, root diameter, root surface area, root volume, root tip frequency, root distribution, and root hair development (Liu et al., 2023). An array of interactions between a root and its environment are influenced by the organization and structure of the root, from the cellular to the plant. The stele is situated inside the root and is surrounded by concentric layers of epidermal, cortical, and endodermal tissues, which are composed of the xylem, phloem, and pericycle (Sidhu et al., 2023).

The root apical meristem (RAM) creates the main stem cell pool (SCP), which is where other cell types derive from. This RAM also contains the quiescent centre (QC), a group of cells that infrequently divide and act as a cue to the cells around it to arrange and protect the initial stem cells (Timilsina et al., 2019). Genetical, physiological, and environmental factors can all affect the root system's length, growth angles off the main axis, lateral branching, and lifetime (Adiguzel et al., 2023).

The spatial organization and distribution of the roots in the soil establish the root system architecture (RSA), which effectively regulates how rapidly plants can absorb nutrients and water from the soil (Fitter, 2002). For nutrition absorption (NA), the root's apex, also its most active phase is essential. The life of the developed root hairs of the root tip is usually only 15 days. To replace the dead root hairs and relocate the roots in the soil, they can attract nutrients and the elongation zone creates new root hairs (Yang et al., 2021). Ion pumps, channels, and transporters are the three primary categories of transport proteins that lodge nutrient transport systems (NTSs) in the plasma membrane of plants (Reid, 2014; Mitani-Ueno & Ma,

2021). Transporting nutrients from the bulk soil to the root surface involves mass flow and diffusion systems. The mass flow, or convective movement of nutrients dissolved in solution from the bulk of soil towards the root, is driven by the water potential gradient created by roots absorbing water from soils. On the other hand, diffusion is the random movement of nutrient ions based on the idea of thermal movement when there is an ion concentration differential between the soil and the root (Nye, 1980; Zhao et al., 2020; Adiguzel et al., 2023).

10.4 Physiological and Molecular Mechanisms Underlying Root Responses to Soil Drought

Roots have developed into the first organ capable of detecting changes in soil moisture on the morphological, anatomical, and molecular scales (De Micco & Aronne, 2012; Saenz Rodriguez & Cassab, 2021). The normal growth and development of crops are caused by the coordination of morphological functions between the ground and subsurface components. In response to drought stress, plants grew their roots and shoots more slowly, which resulted in an increase in the root-to-shoot ratio (Yamaguchi et al., 2010). The capacity to alter RSA characteristics in response to water stress has been proven by roots. According to Lee et al. (2016), drought conditions have an effect on root length, diameter, surface area, tissue density, biomass, and branch count. In contrast, roots have mechanisms to maintain functions and grow towards a more mesic environment to support water intake in situations when the soil water supply is constrained and overall development is hampered (Dietrich, 2018). In order to acquire potential water resources (PWRs) from a deeper soil layer to ensure water supply and improve crop adaptation to drought stress, the root system can therefore continue to grow and form a relatively well-developed root system under drought conditions. This is true even though water-deficit stress (WDS) often negatively affects root biomass, root length, root number, etc. The primary root, the first root tissue to emerge from the seed and contact with the environment, can maintain growth relative to shoot growth by maintaining cell elongation in the apical regions of several crops, including maize, soybean, and cotton, under conditions with relatively low water content (Yamaguchi et al., 2010; Kang et al., 2022). Some of the key mechanisms for primary root maintenance include osmolyte accumulation (OA), hormone regulation (HR), cell-wall remodeling, and specialized control of antioxidant systems (Voothuluru et al., 2020). However, further investigation is required to pinpoint the particular molecular mechanisms governing this process in crops. The bulk of the root systems and other components, lateral roots make up the root system (Ranjan et al., 2022). Because they establish contact with the soil through a network of branching lateral roots (LRs), LRs are essential for water intake. According to Xiong et al. (2006), lateral root growth (LRG) is commonly limited when there is water scarcity. However, in water-scarce situations, it is possible to promote the development of a certain root system composition, which will help the development of LRs (Dowd et al., 2020).

Root hydrotropism, LRG, and deeper rooting are three essential aspects of root development and the formation of root systems that have lately attracted the attention of molecular approaches. Mizu KUSSEI1 (MIZ1) has been investigated in Arabidopsis and shown to influence root hydrotropism by encoding a protein accompanying the cytoplasmic side of the endoplasmic reticulum membrane (Moriwaki et al., 2013). Hydrotropism, which controls the angle of roots, is dependent on MIZU KUSSEI1 (MIZ1) (Shkolnik et al., 2018). It influences Ca^{2+} transport associated with hydrotropism through interactions with the endoplasmic reticulum Ca^{2+}ATPase (AtECA1) (Resentini et al., 2021). MIZ1 has been discovered to function in conjunction with ABA signaling and be related to root gravitropism (Akita & Miyazawa, 2022). The rice roots were used to clone and characterize the MIZ1 gene in plants, OsMIZ1. In the roots of rice lines resistant to drought, OsMIZ1 expression levels were markedly increased (Kang et al., 2022). Another important gene for controlling root growth is AUXIN RESPONSE FACTOR 7 (ARF7), which primarily controls the development of lateral roots through hydropatterning, the process of root branching into locations with more water content (Orosa-Puente et al., 2018). In the area phase of the root, ARF7 is SUMOylated (modified by a small ubiquitin-like modifier (SUMO)) and suppressed by indole-3-acetic acid (Zhang et al., 2023). In crops, DEEPER ROOTING 1 (DRO1), a quantitative trait locus affecting root development angle, has been found and primarily researched in rice for its function in developing deeper roots under drought circumstances (Kitomi et al., 2020). Although little is known about the molecular mechanisms behind these impacts, several more genes and proteins have also been discovered that influence the formation or growth of roots in situations that are prone to drought.

The plant-specific transcription factors NAM, ATAF1/2, and CUC2 (NACs) were found to regulate root growth in transgenic Arabidopsis lines by increasing the length and number of LRs, and some NAC family members were specifically expressed in soybean roots under water-deficit conditions (Yuan et al., 2020). In rice, it has been demonstrated that RePRP interacts with the cytoskeleton to regulate root growth in response to water-deficit stress (Hsiao et al., 2020). RePRP is an inherently disordered protein that functions as a flexible stress modulator. The vacuolar-type H^+ pyrophosphatase gene (ZmVPP1), which imparts drought-inducible expression in drought-tolerant maize genotypes, is overexpressed in maize, and this results in enhanced tolerance in a transgenic line, which is most likely due to increased root growth (Liu et al., 2022). Earlier, Lum et al. (2014) claim that proline, typically soluble sugars and carbohydrates are frequently found in improved osmolytes. Additionally, Altuntas and Terzi (2020) studies have shown that roots of crops like soybean, maize, and wheat that were experiencing water stress had higher levels of transcripts for several crucial genes involved in proline and sucrose metabolism, including the sucrose synthase gene (SUS) and the pyrroline-5-carboxylate reductase gene (P5CS). Expansins, a well-known class of cell wall remodeling agents, are known to have an impact on the mechanical properties of the cell wall by generating local mechanical alterations in the cell wall and rupturing noncovalent bonds in cell wall polymers. In particular, it has been found that EXPANSIN A1 functions as a cell wall modulator in Arabidopsis during the beginning of lateral roots (Ramakrishna et al., 2019). The antioxidative abilities of these enzymes and metabolites have been thoroughly demonstrated in crop roots under drought conditions. Ascorbate peroxidase (APX), catalase (CAT), glutathione reductase (GR), dehydroascorbate reductase (DHAR), and superoxide dismutase (SOD) have been found to have increased abundances in the growth zone of maize primary roots under severe water-deficit stress by transcriptomics and proteomics studies (Zhu et al., 2007; Khanna & Semwal, 2011; Zhang et al., 2020a, Zhang et al., 2020b).

10.5 Nutrient Availability and Uptake Under Soil Drought Conditions

Even in agricultural regions, environmental stressors like drought may cause nutrient shortages since each nutrient's capacity to be absorbed relies on the physicochemical properties of the soil (Nawaz et al., 2020). Earlier, Kheradmand et al. (2014) claimed that when water availability decreases, nutrients in the soil become less accessible, causing a decrease in nutrient concentration in plant tissues. How nutrients are taken by the roots and transported to the shoots is greatly impacted by water scarcity conditions. It has been demonstrated that drought stress causes plants' N contents to rise and P contents to decrease and K has little effect (Waraich et al., 2011). A good nitrogen treatment can affect the growth of cotton roots under conditions of water stress by increasing the root-to-shoot ratio, root density, lipid peroxidation, and antioxidative enzyme activity (Luo et al., 2015; Iqbal et al., 2020). Additionally, it has been shown that ammonium (NH_4^+), as opposed to nitrate, helps rice seedlings endure drought to a larger extent. With NO_3 therapy, drought-induced oxidative damage in roots was considerably worse than with NH_4^+ treatment due to the accumulation of ROS (Cao et al., 2019; Yuan-Yuan et al., 2021).

The root system reduced the respiratory burden of root growth (RG) and changed root hair formation to enhance phosphorus uptake for the phosphorus-acquiring genotypes. However, they also seemed to limit water intake and spatial competition (Lynch & Ho, 2005; Kang et al., 2022). These genotypes' root architecture patterns enhanced topsoil foraging and phosphate uptake. Potassium is an essential cation element that most plants require to operate properly. It enhances water absorption, preserves aquaporin functionality, increases cell membrane integrity, and affects osmotic adjustment favorably, all of which may support RG (Wang et al., 2013). Researchers have discovered that during droughts, the high-affinity potassium transporter gene OsHAK1 is preferentially expressed in rice roots, enhancing potassium transport and absorption as well as crop productivity (Chen et al., 2017). Additionally, it has been shown that plants' levels of calcium decrease during drought stress (Hosseini et al., 2019). To make sure that plants received the proper quantity of mineral nutrients, the resilience of membranes in the roots was essential. Under drought stress conditions, roots are ineffective at absorbing nutrients from the soil because of inadequate root activity and slow water diffusion rates (Noori, 2023). Water shortage has several effects on plants, including stomatal closure, reduced transpiration, and reduced nutrient transport from the root to the upper parts of the plant (Hu & Schmidhalter, 2005; Xiong et al., 2021). Therefore, when there is a water deficit, there is a decrease in the availability of soil nutrients and a decrease in the consistency of how plants transfer nutrients.

10.6 Interactions Between Root Systems and Soil Microorganisms Under Drought Stress

In the rhizosphere, crop roots, soil, and microbes interact strongly. According to Berg et al. (2014), plants have intimate relationships with the bacterial populations that dwell in and around their roots, including the rhizosphere (soils that are near enough to be impacted by the release of root exudate). Changes in plant exudate profiles brought on by drought can alter the makeup and activity of the adjacent soil microbiome. This may result in further modifications to the soil geochemistry, which in turn may alter the magnitude and direction of soil community transitions (Liu et al., 2021; Prasad & Yadav, 2023). Recent studies have found that the duration of a drought significantly affects the makeup and activity of the microbiomes connected to plant roots (Waqar et al., 2022; Fadiji et al., 2023). Rhizosphere symbiotic bacteria (RSB) are useful because they improve the host plant's ability to absorb nutrients and water through their roots (Dheeman et al., 2020). Here, the focus is mostly on the function of rhizosphere microorganisms, which include bacteria, rhizobia, PGPRs, and arbuscular mycorrhizal fungus (AMF), in improving the drought resistance of roots and reducing RSA in crops (Prasad, 2022).

Earlier, Agami et al. (2016) proposed this association in their seminal work on the enhancement of rhizobacterial plant drought stress tolerance. Engineering stable links between plants and a desired microbiome is an enticing objective for crop development through stress tolerance since a plant's relative fitness and health increase following the recruitment of bacteria with PGP activities (Samain et al., 2022). Under drought stress, PGP microbes, which include members of the genera *Arthrobacter, Azospirillum, Azotobacter, Bacillus, Burkholderia, Enterobacter, Klebsiella, Lysinibacillus, Paenibacillus, Pseudomonas, Serratia*, and *Streptomyces*, are important for PG and SH (Passos et al., 2014; Murali et al., 2021; Waqar et al., 2022; Shankar & Prasad, 2023). PGP microbes help crops withstand drought conditions through a variety of mechanisms, including the secretion of phytohormones, exopolysaccharides, and antioxidants; solubilization of essential micro and macronutrients; accumulation of osmolytes; and induction of stress-responsive genes, as shown in Table 10.2.

The enzyme 1-aminocyclopropane-1-carboxylate (ACC) deaminase may be garnering the most attention, although several PGP characteristics are likely to contribute to drought resistance. In times of drought, ACC deaminase activity prevents the plant hormone ethylene from reaching inhibitory levels, promoting normal RG, and delaying senescence (Gamalero & Glick, 2015; Lotfi et al., 2022). Another phytohormone that PGPB are known for synthesizing is the auxin analog indole-3-acetic acid (IAA), which may encourage shoot and root growth in addition to other plant developmental processes (Dhiman et al., 2020; Baliyan et al., 2021). According to

Table 10.2 Recent studies on PGPRs plant responses to drought stress

PGPRs	Plants	PGP activity	Effect on crop	Reference
Bacillus megaterium (MU2) and *Bacillus licheniformis* (MU8)	Wheat	Solubilization of phosphorus, potassium, and exopolysaccharide production	Increased the germination index 46%, fresh weight 192% and dry weight 226%	Rashid et al. (2021)
Cellulomonas pakistanensis (NCCP11) and *Sphingobacterium pakistanensis* (NCCP246)	Vicia faba L.	IAA production and phosphate solubilization	Increased 28.04% seed Germination and 36.36% dry weight roots	Nafees et al. (2021)
Bacillus amyloliquefaciens (MMR04), *Bacillus subtilis* (MMR18) and *Stenotrophomonas maltophilia* (MMR36)]	Pennisetum glaucum	Production of ACC deaminase and antioxidants	Enhancement in seed Germination 91.75%	Murali et al. (2021)
Pseudomonas azotoformans FAP5	Wheat	Biofilm-forming, IAA production, solubilized tricalcium phosphate and ACC deaminase	Improvement in root growth and photosynthetic pigment efficiency	Ansari et al. (2021)
Bacillus velezensis and *Bacillus amyloliquefaciens*	Persian walnut	Produce siderophore, hydrogen cyanide (HCN), Indole-3-acetic acid (IAA) and gibberellic acid (GA_3).	Increase biomass in fruit trees	Lotfi et al. (2022)
Planomicrobium chinense (MF616408) and *Pseudomonas putida* (KX574857)	Soybean	Increase plant antioxidant enzymes (superoxide dismutase (SOD) and catalase (CAT))	Increase in plant biomass	Waqar et al. (2022)
Bacillus amyloliquefaciens QST713	Alfalfa (*Medicago sativa* L.)	Promoted the content of leaf relative water content, the accumulation of chlorophyll content and the activities of antioxidant enzymes	Enhanced the photosynthetic capacity of alfalfa seedlings	Han et al. (2022)
Paenibacillus sp. strain B2 (PB2) and *Arthrobacter* spp. strain AA	Wheat	Induced systemic resistance	Increased dry biomass and root length	Samain et al. (2022)
Bacillus cereus L90	Walnut (Juglans regia L.)	Produces high levels of cytokinins and indoleacetic acid (IAA)	Seedlings increased by 28.65% in the control	Liu et al. (2023)

(continued)

Table 10.2 (continued)

PGPRs	Plants	PGP activity	Effect on crop	Reference
Pseudomonas putida-RA	Chickpea	Responsive Car-miR166 of chickpea in drought mitigation with phytohormonal crosstalk in transgenic Arabidopsis lines	Highest germination rate in T2 (90%) and highest root length was observed in drought treated inoculated T1 lines (29%)	Yadav et al. (2023)
Pseudomonas fluorescens SBW25 and *Pseudomonas putida* KT2440	Barley (Hordeum vulgare L.)	IAA, siderophore production and phosphate solubilization	Increase 94% seed germination, 3.18 g plant biomass, and photosynthetic activity	Zaib et al. (2023)
Enterobacter cloacae BHUAS1, *Bacillus cereus* BHUAS2, *Bacillus megaterium* BHUIESDAS3	Wheat (*Triticum aestivum* L.)	Improvement in enzymatic activities of several antioxidant enzymes such as guaiacol peroxidase (POD), catalase (CAT), and ascorbate peroxidase (APX)	Improvement in plant height, root length, biomass, chlorophyll and carotenoid content, membrane stability index (MSI), leaf relative water content (RWC), total soluble sugar, total phenol, proline, and total soluble protein	Shankar and Prasad (2023)
Paraburkholderia phytofirmans and *Bacillus* sp.	Soybean	Exopolysaccharides, and enzyme 1-aminocyclopropane-1-carboxylate (ACC) deaminase	14% grain yield and 20% enzymatic antioxidant activities	Nawaz et al. (2023)

studies by Samain et al. (2022) on better plant performance, PGPB has been proven to promote fine root formation, increase root surface area overall, and decrease stress volatile emission. According to Gontia-Mishra et al. (2016) and Zaib et al. (2023), they may also play a role in the cycling of nutrients under drought, including siderophore production, phosphorus solubilization, and diazotrophic activity. Because early flowering is a drought-resistance strategy that plants can evolve, senescence and even earlier seed set times have been connected to them (Martignago et al., 2020).

It has been demonstrated in some cases, even if the exact mechanism by which PGPB functions to promote drought resistance is difficult to explain. According to

Costa et al. (2018), PGPB generated extracellular matrices to maintain a moist root environment and promote root-adhering soil and stability. Similar results were found by Altaf and Khan (2020), who discovered that a mutant of *Paenibacillus polymyxa* lacking a Sfptype 40-phosphopantetheinyl transferase had increased biofilm production. When there are Arabidopsis roots that are under stress from dryness, *Bacillus megaterium* BOFC15 secretes polyamine spermidine. Spermidine increases photosynthesis and root system architecture by scavenging ROS, upregulating ABA production and response genes, and scavenging ROS (Zhou et al., 2016). In a manner like this, *Pseudomonas chloroaphis* O6 will generate 2R-3R-butanediol, a substance involved in the SA signaling pathway that results in stomatal closure in Arabidopsis (Nam et al., 2021). Modifying plant gene expression is a typical strategy employed by PGPB to provide plants with drought resilience. The overexpression of genes associated with drought response, such as DREB1B-like or ERD15, which have a wide number of downstream targets, has frequently been used to demonstrate this in a broad fashion (Gagne-Bourque et al., 2016). It should be emphasized, however, that there is significant crosstalk among signaling pathways linked to biotic and abiotic stress responses, which suggests mechanisms for integrating overall plant stress responses, despite the expression of specific genes related to drought stress in some situations.

10.7 Techniques for Studying Plant Root Systems and Nutrient Availability Under Soil Drought Conditions

Variables such as soil temperature, moisture, nutrients, and pH have an impact on the root system architecture (RSA), which has a considerable impact on crop growth and yield. Numerous root traits, such as gravitropism, morphological plasticity, and root tip diameter, enable plants to adapt to various environmental situations and respond to those conditions (Wasaya et al., 2018). The ability of different root types to absorb water as a survival strategy in dry conditions can be compared using their distinctive characteristics, such as primary root length, length and number of lateral roots, crown root number, cortical cell file number, and cell size (Ranjan et al., 2022).

The cornerstone of image-based phenotyping of plant roots is nondestructive (where possible) optical investigations of plant characteristics. Root phenotyping is essential for crop breeding despite being a very difficult method in field settings (Wasaya et al., 2018). In root phenotyping methods, imaging and image processing are typically utilized with some amount of automation. Image analysis techniques are now available in many software programs, such as WinRhizo, EZ-Rhizo, optima's analysis software Image, IJ Rhizo, root trace, and root system analyzer (Arsenault et al., 1995; Abramoff et al., 2004; Lobet et al., 2011; Leitner et al., 2014; Moussa et al., 2021). Image processing gets more challenging when employing soil as a growth medium. Soil-filled rhizotrons and columns are used to study soil-root interaction in a more realistic setting because it is difficult or even impossible to produce soil compaction or drying effects in soil-free devices. For this, soil or another growth substrate is placed inside columns or rhizotrons (Wasaya et al., 2018). Then, a transparent wall or a soil column is used for root assessment while employing X-ray-based computed tomography (CT) to observe the 3-D root architecture (Grift et al., 2011; Li et al., 2021).

Without having to dig around or remove the roots, it might be difficult to visually evaluate the roots in the field when using markers for selection. The research on root density, depth, and angle may be done using the soil core method and standard excavation method, which are recognized as the best phenotyping methods for the field (Shao et al., 2021). Soil cores are acquired using excavation techniques in order to estimate the weights or densities of vertical root lengths. In field investigations, mini-rhizotron systems are also utilized to evaluate the surroundings of the root soil. These systems consist of Plexiglas tubes with a small camera or scanner implanted in the soil (Xu et al., 2022). Mini-rhizotrons can be used to monitor a small number of genotypes. The examination of RSA may also be done indirectly using a variety of techniques, such as root-pulling resistance (Yang et al., 2021) or leaf abscisic acid (ABA) concentration (Wang et al., 2021).

Canopy temperature is one of the crucial physiological factors connected to plant water status and a variety of metabolic processes, such as photosynthesis, respiration, and transpiration (Rebetzke et al., 2013). When agricultural plants are exposed to stress conditions, notably droughts, canopy temperature depression (CTD) is another useful technique for phenotyping them in the field (Fukai & Mitchell, 2022). Canopy temperature and CTD are recognized as accurate predictors of yield and transpiration rate during drought conditions (Lepekhov, 2022). Several physiological processes, including stomatal conductance, grain production, and leaf water potential, are all linked with CTD, which is the difference between air temperature and CT, under stressful conditions (Sofi et al., 2019). The water balance of plants is a clear sign of how agricultural plants react to drought. This is related to the plant's transpiration mechanism, which is the main driver of changes in leaf temperature and is inversely proportional to stomatal conductance (Sobejano-Paz et al., 2020). CTD has been used in breeding programs to remove genotypes that are stress-tolerant or susceptible, particularly in regions that are prone to drought. When there is not enough water for transpiration, the stomata shut, which increases the canopy's temperature. It is possible to pick many plants based on their availability of CO_2 and ability to absorb water using thermal imaging to detect this change in stomatal conductance and leaf transpiration in response to water stress (Brewer et al., 2022).

Finding quantitative trait loci (QTL) linked to drought tolerance via marker-assisted selection is one of the promising ways (Yu et al., 2019). There have been several attempts to identify the genes and QTL that are in charge of root traits. Numerous researchers have looked at the connection between QTL and characteristics that improve the ability of the roots of agricultural plants to forage (White, 2019; Fatemi et al., 2022; Panda et al., 2021). The molecular mapping of root attributes in wheat indicated the existence of several QTLs for different root features, including root number, length, seminal root number and angle, total root biomass, root system depth, lateral root number and length, and root surface area. Numerous QTLs have been identified in significant grains and legumes, such as wheat, rice, maize, barley, pearl millet, sorghum, chickpea, and soybean, which are adapted to drought conditions (Wasaya et al., 2018; Singh et al., 2022). Finding genes or molecular markers associated with root architecture and root development is beneficial for breeding and molecular selection to improve root attributes. In a manner similar to this, Liu and Qin (2021) revealed that a large QTL for leaf ABA concentration was advantageous for root structure and grain yield in maize.

10.8 Implications for Sustainable Agriculture and Future Research Directions

First, data from conventional experimental models that only change the amount of water supplied to plants under controlled conditions and under drought conditions may lead to inaccurate conclusions about microbiota response. Root organisms to external stress due to drought. Future research should focus on the association between gene expression, protein regulation, metabolite accumulation, and nutrient fertilization with growth characteristics and root phenotype of plants under poor growing conditions. Water creates a network that includes reactions in various aspects (e.g., plant hormone crosstalk, water), nutrient interactions, and microbial-soil feedback. This network can be used as a guide to essential traits that can be enhanced by selective breeding to improve root tolerance to water stress. Drought tolerance of major root crops (such as maize, rice, wheat, soybean, and cotton) is expected to increase soon through the adoption of integrated technologies. New genetic resources and genotypes that are more resilient to future water shortages and climate change need to be created using advanced breeding processes. Our understanding of the physiology of plant stress will be enhanced through studies exploring how plant roots respond to drought stress in the rhizosphere. This study will also provide recommendations on livestock and agronomic practices to produce high-yielding, drought-tolerant crops. Mathematical models of agroecosystems can also facilitate the design of experiments by considering several factors involved and providing more accurate feedback on specific responses generated due to the perturbation of water deficiency parameters. Research on root microbiota under drought conditions has been significantly hampered by the method used to explain how soil moisture affects root-associated bacterial populations. Future trials should consider the developmental stage and context of the study to explain any differences from previous studies. A topic that needs further investigation in the context of microbial communities during drought is the specific cause of the growth of different bacterial species under water-limited conditions.

10.9 Conclusion

Plant growth and development are negatively affected by nature's wrath in the form of water shortages. Water shortages are severe in agricultural areas that rely heavily on rainwater, endangering both humans and the environment. PGP bacteria can be effectively introduced to increase drought tolerance and water retention. It is possible to perform genomic studies, such as identifying, cloning, and functionalizing bacterial genes that adapt to stress and provide resistance to drought stress. Systems biology techniques can be used in studies to test how stress-adapted microorganisms change their metabolic profiles in the absence of water. The most important step is to decipher the complex genetic and metabolic networks that govern the interactions between plants and microorganisms in difficult contexts.

References

Abramoff, M. D., Magelhaes, P. J., & Ram, S. J. (2004). Image processing with ImageJ. *Biophotonics International, 11*, 36–42.

Adiguzel, P., Nyirahabimana, F., Shimira, F., et al. (2023). Applied biotechnological approaches for reducing yield gap in melon grown under saline and drought stresses: An overview. *Journal of Soil Science and Plant Nutrition, 23*(1), 139–151.

Agami, R. A., Medani, R. A., Abd El-Mola, I. A., et al. (2016). Exogenous application with plant growth promoting rhizobacteria (PGPR) or proline induces stress tolerance in basil plants (*Ocimum basilicum* L.) exposed to water stress. *International Journal of Environmental & Agriculture Research, 2*(5), 78.

Akita, K., & Miyazawa, Y. (2022). The mechanosensitive Ca^{2+} channel, OSCA1. 1, modulates root hydrotropic bending in Arabidopsis thaliana. *Environmental and Experimental Botany, 197*, 104825.

Altaf, M. M., & Khan, M. S. A. (2020). Plant growth promoting rhizobacteria: Biofilm and role in sustainable agriculture. In *Microbial biofilms* (pp. 161–174). CRC Press.

Altuntas, C., & Terzi, R. (2020). Dithiothreitol and PEG induced combined stress may affect the expressions of ABA aldehyde oxidase, sucrose synthase and proline metabolic genes in maize seedlings. *Phyton, 89*(3), 487.

Ansari, F. A., Jabeen, M., Ahmad, I., et al. (2021). *Pseudomonas azotoformans* FAP5, a novel biofilm-forming PGPR strain, alleviates drought stress in wheat plant. *International journal of Environmental Science and Technology, 18*, 3855–3870.

Arsenault, J. L., Pouleur, S., Messier, C., et al. (1995). WinRhizo, a root measuring system with a unique overlap correction method. *HortScience, 30*, 906.

Ayub, M., Ashraf, M. Y., & Kausar, A. (2021). Growth and physio-biochemical responses of maize (*Zea mays* L.) to drought and heat stresses. *Plant Biosystems, 155*(3), 535–542.

Baliyan, N., Dhiman, S., Dheeman, S., et al. (2021). Optimization of indole-3-acetic acid using response surface methodology and its effect on vegetative growth of chickpea. *Rhizosphere, 17*, 100321.

Baliyan, N., Qureshi, K. A., Jaremko, M., et al. (2022a). Bioformulation containing cohorts of *Ensifer adhaerens* MSN12 and *Bacillus cereus* MEN8 for the nutrient enhancement of *Cicer arietinum* L. *Plants, 11*(22), 3123.

Baliyan, N., Dhiman, S., Dheeman, S., et al. (2022b). Bacteriophage cocktails as antibacterial agents in crop protection. *Environmental Sustainability, 5*(3), 305–311.

Bao, G., Tang, W., An, Q., et al. (2020). Physiological effects of the combined stresses of freezing-thawing, acid precipitation and deicing salt on alfalfa seedlings. *BMC Plant Biology, 20*, 204.

Berg, G., Grube, M., Schloter, M., et al. (2014). Unraveling the plant microbiome: Looking back and future perspectives. *Frontiers in Microbiology, 5*, 10–28.

Breitkreuz, C., Herzig, L., Buscot, F. M., et al. (2021). Interactions between soil properties, agricultural management and cultivar type drive structural and functional adaptations of the wheat rhizosphere microbiome to drought. *Environmental Microbiology, 23*, 5866–5882.

Brewer, K., Clulow, A., Sibanda, M., et al. (2022). Estimation of maize foliar temperature and stomatal conductance as indicators of water stress based on optical and thermal imagery acquired using an unmanned aerial vehicle (UAV) platform. *Drones, 6*(7), 169.

Cao, X., Zhu, C., Zhong, C., et al. (2019). Nitric oxide synthase-mediated early nitric oxide burst alleviates water stress-induced oxidative damage in ammonium-supplied rice roots. *BMC Plant Biology, 19*, 108.

Chartzoulakis, K., Patakas, A., Kofidis, G., et al. (2002). Water stress affects leaf anatomy, gas exchange, water relations and growth of two avocado cultivars. *Scientia Horticulturae, 95*, 39–50.

Chen, G., Liu, C., Gao, Z., et al. (2017). OsHAK1, a high-affinity potassium transporter, positively regulates responses to drought stress in rice. *Frontiers in Plant Science, 8*, 1885.

Chen, L., Shen, Y., Yang, W., et al. (2022). Comparative metabolic study of two contrasting Chinese cabbage genotypes under mild and severe drought stress. *International Journal of Molecular Sciences, 23*(11), 5947.

Costa, O. Y., Raaijmakers, J. M., Kuramae, E. E., et al. (2018). Microbial extracellular polymeric substances: Ecological function and impact on soil aggregation. *Frontiers in Microbiology, 9*, 1636.

De Micco, V., & Aronne, G. (2012). Morpho-anatomical traits for plant adaptation to drought. In V. De Micco (Ed.), *Plant responses to drought stress: From morphological to molecular features* (pp. 37–61). Springer Berlin Heidelberg.

Dheeman, S., Maheshwari, D. K., Baliyan, N., et al. (2017). Bacterial endophytes for ecological intensification of agriculture. In D. K. Maheshwari (Ed.), *Endophytes: Biology and biotechnology* (Vol. 1, pp. 193–231). Springer International Publishing.

Dheeman, S., Baliyan, N., Dubey, R. C., et al. (2020). Combined effects of rhizo-competitive rhizosphere and non-rhizosphere *Bacillus* in plant growth promotion and yield improvement of *Eleusine coracana* (Ragi). *Canadian Journal of Microbiology, 16*(2), 111–124.

Dhiman, S., Dubey, R. C., Baliyan, N., et al. (2019). Application of potassium-solubilising *Proteus mirabilis* MG738216 inhabiting cattle dung in improving nutrient use efficiency of *Foeniculum vulgare* Mill. *Environmental Sustainability, 2*, 401–409.

Dhiman, S., Baliyan, N., & Maheshwari, D. K. (2020). Buffalo dung-inhabiting bacteria enhance the nutrient enrichment of soil and proximate contents of *Foeniculum vulgare* Mill. *Archives of Microbiology, 202*, 2461–2470.

Dietrich, D. (2018). Hydrotropism: How roots search for water. *Journal of Experimental Botany, 69*, 2759–2771.

Dos Santos, T. B., Ribas, A. F., de Souza, S. G. H., et al. (2022). Physiological responses to drought, salinity, and heat stress in plants: A review. *Stresses, 2*(1), 113–135.

Dowd, T. G., Braun, D. M., Sharp, R. E., et al. (2020). Maize lateral root developmental plasticity induced by mild water stress. II: Genotypespecific spatio-temporal effects on determinate development. *Plant, Cell & Environment, 43*, 2409–2427.

Fadiji, A. E., Yadav, A. N., Santoyo, G., et al. (2023). Understanding the plant-microbe interactions in environments exposed to abiotic stresses: An overview. *Microbiological Research*, 127368.

Fatemi, F., Kianersi, F., Pour-Aboughadareh, A., et al. (2022). Overview of identified genomic regions associated with various agronomic and physiological traits in barley under abiotic stresses. *Applied Sciences, 12*(10), 5189.

Fitter, A. (2002). Characteristics and functions of root systems. In A. Fitter (Ed.), *Plant roots* (pp. 49–78). CRC Press.

Fukai, S., & Mitchell, J. (2022). Role of canopy temperature depression in rice. *Crop and Environment, 7*, 25–42.

Furlan, A., Llanes, A., Luna, V., et al. (2012). Physiological and biochemical responses to drought stress and subsequent rehydration in the symbiotic association peanut-*Bradyrhizobium* sp. *ISRN Agronomy, 2012*, 890–895.

Gagne-Bourque, F., Bertrand, A., Claessens, A., et al. (2016). Alleviation of drought stress and metabolic changes in timothy (*Phleum pratense* L.) colonized with *Bacillus subtilis* B26. *Frontiers in Plant Science, 7*, 584.

Gamalero, E., & Glick, B. R. (2015). Bacterial modulation of plant ethylene levels. *Plant Physiology, 169*(1), 13–22.

Gontia-Mishra, I., Sapre, S., Sharma, A., et al. (2016). Amelioration of drought tolerance in wheat by the interaction of plant growth-promoting rhizobacteria. *Plant Biology, 18*(6), 992–1000.

Grift, T. E., Novais, J., Bohn, M., et al. (2011). High-throughput phenotyping technology for maize roots. *Biosystems Engineering, 110*, 40–48.

Han, L., Zhang, M., Du, L., et al. (2022). Effects of *Bacillus amyloliquefaciens* QST713 on photosynthesis and antioxidant characteristics of Alfalfa (*Medicago sativa* L.) under drought stress. *Agronomy, 12*(9), 2177.

Hanaka, A., Ozimek, E., Reszczynska, E., et al. (2021). Plant tolerance to drought stress in the presence of supporting bacteria and fungi: An efficient strategy in horticulture. *Horticulturae, 7*, 390.

Hewedy, O. A., Abdel Lateif, K. S., Seleiman, M. F., et al. (2020). Phylogenetic diversity of *Trichoderma* strains and their antagonistic potential against soil-borne pathogens under stress conditions. *Biology, 9*, 189.

Hosseini, S. A., Rethore, E., Pluchon, S., et al. (2019). Calcium application enhances drought stress tolerance in sugar beet and promotes plant biomass and beetroot sucrose concentration. *International Journal of Molecular Sciences, 20*(15), 3777.

Hsiao, A. S., Wang, K., Ho, T. H. D., et al. (2020). An intrinsically disordered protein interacts with the cytoskeleton for adaptive root growth under stress. *Plant Physiology, 183*, 570–587.

Hu, Y., & Schmidhalter, U. (2005). Drought and salinity: A comparison of their effects on mineral nutrition of plants. *Journal of Plant Nutrition and Soil Science, 168*(4), 541–549.

Iqbal, A., Dong, Q., Wang, X., et al. (2020). High nitrogen enhance drought tolerance in cotton through antioxidant enzymatic activities, nitrogen metabolism and osmotic adjustment. *Plants, 9*(2), 178.

Jabbari, H., Akbari, G. A., Sima, N. A. K. K., et al. (2013). Relationships between seedling establishment and soil moisture content for winter and spring rapeseed genotypes. *Industrial Crops and Products, 49*, 177–187.

Jiang, C., Li, X., Zou, J., et al. (2021). Comparative transcriptome analysis of genes involved in the drought stress response of two peanut (*Arachis hypogaea* L.) varieties. *BMC Plant Biology, 21*, 64.

Kang, J., Peng, Y., Xu, W., et al. (2022). Crop root responses to drought stress: Molecular mechanisms, nutrient regulations, and interactions with microorganisms in the rhizosphere. *International Journal of Molecular Sciences, 23*(16), 9310.

Khalvandi, M., Siosemardeh, A., Roohi, E., et al. (2021). Salicylic acid alleviated the effect of drought stress on photosynthetic characteristics and leaf protein pattern in winter wheat. *Heliyon, 7*(1), 55–67.

Khanna, R., & Semwal, V. K. (2011). Superoxide dismutase and ascorbate peroxidase are constitutively more thermotolerant than other antioxidant enzymes in Chenopodium album. *Physiology and Molecular Biology of Plants, 17*, 339–346.

Kheradmand, M. A., Fahraji, S. S., Fatahi, E., et al. (2014). Effect of water stress on oil yield and some characteristics of *Brassica napus*. *International Research Journal of Applied and Basic Sciences, 8*(9), 1447–1453.

Kitomi, Y., Hanzawa, E., Kuya, N., et al. (2020). Root angle modifications by the DRO1 homolog improve rice yields in saline paddy fields. *Proceedings of the National Academy of Sciences, 117*(35), 21242–21250.

Kozlowski, T. T., & Pallardy, S. G. (2002). Acclimation and adaptive responses of woody plants to environmental stresses. *The Botanical Review, 68*(2), 270–334.

Kumar, S., Dheeman, S., Dubey, R. C., et al. (2021). Cyclic siloxane biosurfactant-producing *Bacillus cereus* BS14 biocontrols charcoal rot pathogen *Macrophomina phaseolina* and induces growth promotion in *Vigna mungo* L. *Archives of Microbiology, 203*, 5043–5054.

Lee, D. K., Chung, P. J., Jeong, J. S., et al. (2016). The rice OsNAC6 transcription factor orchestrates multiple molecular mechanisms involving root structural adaptions and nicotianamine biosynthesis for drought tolerance. *Plant Biotechnology Journal, 15*, 754–764.

Leitner, D., Felderer, B., Vontobel, P., et al. (2014). Recovering root system traits using image analysis exemplified by two-dimensional neutron radiography images of lupine. *Plant Physiology, 164*, 24–35.

Lepekhov, S. B. (2022). Canopy temperature depression for drought-and heat stress tolerance in wheat breeding. *Journal of Genetics and Breeding, 26*(2), 196.

Li, X., Wang, S., Chen, P., et al. (2021). 3-D inspection method for industrial product assembly based on single X-ray projections. *IEEE Transactions on Instrumentation and Measurement, 70*, 1–14.

Liu, S., & Qin, F. (2021). Genetic dissection of maize drought tolerance for trait improvement. *Molecular Breeding, 41*, 1–13.

Liu, Y., Cheng, J., & Liu, X. (2021). Tracking the changes of wetland soil bacterial community and metabolic potentials under drought and flooding conditions in experimental microcosms. *Journal of Soils and Sediments, 21*, 2404–2417.

Liu, S., Liu, X., & Zhang, X. (2022). Co-expression of ZmVPP1 with ZmNAC111 confers robust drought resistance in maize. *Genes, 14*(1), 8.

Liu, F., Ma, H., Liu, B., et al. (2023). Effects of plant growth-promoting rhizobacteria on the physioecological characteristics and growth of walnut seedlings under drought stress. *Agronomy, 13*(2), 290.

Lobet, G., Pages, L., & Draye, X. (2011). A novel image-analysis toolbox enabling quantitative analysis of root system architecture. *Plant Physiology, 157*, 29–39.

Lotfi, N., Soleimani, A., & Çakmakçı, R. (2022). Characterization of plant growth-promoting rhizobacteria (PGPR) in Persian walnut associated with drought stress tolerance. *Scientific Reports, 12*(1), 12725.

Lum, M. S., Hanafi, M. M., & Rafii, Y. M. (2014). Effect of drought stress on growth, proline and antioxidant enzyme activities of upland rice. *Journal of Animal & Plant Sciences, 24*, 1487–1493.

Lumactud, R. A., Dollete, D., Liyanage, D. K., et al. (2023). The effect of drought stress on nodulation, plant growth, and nitrogen fixation in soybean during early plant growth. *Journal of Agronomy and Crop Science, 209*(3), 345–354.

Luo, H. H., Tao, X. P., & Hu, Y. Y. (2015). Response of cotton root growth and yield to root restriction under various water and nitrogen regimes. *Journal of Plant Nutrition and Soil Science, 178*, 384–392.

Lynch, J. P., & Ho, M. D. (2005). Rhizoeconomics: Carbon costs of phosphorus acquisition. *Plant and Soil, 269*, 45–56.

Ma, P., Bai, T. H., & Ma, F. W. (2015). Effects of progressive drought on photosynthesis and partitioning of absorbed light in apple trees. *Journal of Integrative Agriculture, 14*, 681–690.

Mansoor, S., Khan, T., & Farooq, I. (2022). Drought and global hunger: Biotechnological interventions in sustainability and management. *Planta, 256*(5), 97.

Martignago, D., Rico-Medina, A., Blasco-Escamez, D., et al. (2020). Drought resistance by engineering plant tissue-specific responses. *Frontiers in Plant Science, 10*, 1676.

Mitani-Ueno, N., & Ma, J. F. (2021). Linking transport system of silicon with its accumulation in different plant species. *Soil Science & Plant Nutrition, 67*(1), 10–17.

Moriwaki, T., Miyazawa, Y., Kobayashi, A., et al. (2013). Molecular mechanisms of hydrotropism in seedling roots of *Arabidopsis thaliana* (Brassicaceae). *American Journal of Botany, 100*(1), 25–34.

Moussa, A. A., Mandozai, A., Jin, Y., et al. (2021). Genome-wide association screening and verification of potential genes associated with root architectural traits in maize (*Zea mays* L.) at multiple seedling stages. *BMC Genomics, 22*, 1–19.

Murali, M., Singh, S. B., Gowtham, H. G., et al. (2021). Induction of drought tolerance in *Pennisetum glaucum* by ACC deaminase producing PGPR-*Bacillus amyloliquefaciens* through Antioxidant defense system. *Microbiological Research, 253*, 126891.

Nafees, M., Ullah, S., Ahmed, I., et al. (2021). Morphological and elemental evaluation of biochar through analytical techniques and its combined effect along with plant growth promoting rhizobacteria on *Vicia faba* L. under induced drought stress. *Microscopy Research and Technique, 84*(12), 2947–2959.

Nam, H. S., Anderson, A. J., & Kim, Y. C. (2021). Effects of volatiles from *Pseudomonas chlororaphis* O6 on the plasticity of growth in *Arabidopsis thaliana*. *Trends in Agriculture & Life Sciences, 59*, 1–13.

Nawaz, F., Shehzad, M. A., & Majeed, S. (2020). *Role of mineral nutrition in improving drought and salinity tolerance in field crops* (pp. 129–147). Springer.

Nawaz, F., Rafeeq, R., Majeed, S., et al. (2023). Biochar amendment in combination with endophytic bacteria stimulates photosynthetic activity and antioxidant enzymes to improve soybean yield under drought stress. *Journal of Soil Science and Plant Nutrition, 23*(1), 746–760.

Noori, M. S. (2023). Wheat production under drought stress conditions: Physiological and biochemical responses and mitigation options. *Journal of Stress Physiology & Biochemistry, 19*(2), 69–81.

Nye, P. H. (1980). Diffusion of ions and uncharged solutes in soils and soil clays. *Advances in Agronomy, 31*, 225–272.

Orosa-Puente, B., Leftley, N., Von Wangenheim, D., et al. (2018). Root branching toward water involves posttranslational modification of transcription factor ARF7. *Science, 362*, 1407–1410.

Osakabe, Y., Osakabe, K., Shinozaki, K., et al. (2014). Response of plants to water stress. *Frontiers in Plant Science, 5*, 1–7.

Panda, S., Majhi, P. K., Anandan, A., et al. (2021). Proofing direct-seeded rice with better root plasticity and architecture. *International Journal of Molecular Sciences, 22*(11), 6058.

Pandey, B. K., & Bennett, M. J. (2019). A new angle on how roots acclimate to sporadic rainfall. *Cell, 178*(2), 269–271.

Passos, J. F. M. D., Costa, P. B. D., Costa, M. D., et al. (2014). Cultivable bacteria isolated from apple trees cultivated under different crop systems: Diversity and antagonistic activity against *Colletotrichum gloeosporioides*. *Genetics and Molecular Biology, 37*, 560–572.

Prasad, K. (2022). *Potential impression of arbuscular mycorrhizal fungi on agricultural growth, productivity, and environment toward global sustainable development for green technology* (pp. 111–136). Springer.

Prasad, K., & Yadav, K. (2023). Global restoration of contaminated agricultural soils through the utilization of potential microbes for the future generation. In *Integrative strategies for bioremediation of environmental contaminants* (pp. 85–97). Academic Press.

Qaderi, M. M., Kurepin, L. V., & Reid, D. M. (2012). Effects of temperature and watering regime on growth gas exchange and abscisic acid content of canola (*Brassica napus* L.) seedlings. *Environmental and Experimental Botany, 75*, 107–113.

Ramakrishna, P., Duarte, P. R., & Rance, G. A. (2019). EXPANSIN A1-mediated radial swelling of pericycle cells positions anticlinal cell divisions during lateral root initiation. *Proceedings of the National Academy of Sciences, 116*, 8597–8602.

Ranjan, A., Sinha, R., Singla-Pareek, S. L., et al. (2022). Shaping the root system architecture in plants for adaptation to drought stress. *Physiologia Plantarum, 174*(2), 13651.

Rashid, U., Yasmin, H., Hassan, M. N., et al. (2021). Drought-tolerant *Bacillus megaterium* isolated from semi-arid conditions induces systemic tolerance of wheat under drought conditions. *Plant Cell Reports*, 1–21.

Rebetzke, G. J., Rattey, A. R., Farquhar, G. D., et al. (2013). Genomic regions for canopy temperature and their genetic association with stomatal conductance and grain yield in wheat. *Functional Plant Biology, 40*, 14–33.

Reid, R. (2014). Understanding the boron transport network in plants. *Plant and Soil, 385*, 1–13.

Resentini, F., Ruberti, C., & Grenzi, M. (2021). The signatures of organellar calcium. *Plant Physiology, 187*(4), 1985–2004.

Rosa, A. P., Dias, T., Mouazen, A. M., et al. (2023). Finding optimal microorganisms to increase crop productivity and sustainability under drought-a structured reflection. *Journal of Plant Interactions, 18*(1), 2178680.

Sabra, A., Daayf, F., & Renault, S. (2012). Differential physiological and biochemical responses of three Echinacea species to salinity stress. *Scientia Horticulturae, 135*, 23–31.

Saenz Rodriguez, M. N., & Cassab, G. I. (2021). Primary root and mesocotyl elongation in maize seedlings: Two organs with antagonistic growth below the soil surface. *Plants, 10*(7), 1274.

Samain, E., Ernenwein, C., Aussenac, T., et al. (2022). Effective and durable systemic wheat-induced resistance by a plant-growth-promoting rhizobacteria consortium of *Paenibacillus* sp. strain B2 and *Arthrobacter* spp. strain AA against *Zymoseptoria tritici* and drought stress. *Physiological and Molecular Plant Pathology, 119*, 101830.

Sayer, E. J., Crawford, J. A., Edgerley, J., et al. (2021). Adaptation to chronic drought modifies soil microbial community responses to phytohormones. *Communications Biology, 4*, 516.

Seleiman, M. F., Al-Suhaibani, N., Ali, N., et al. (2021). Drought stress impacts on plants and different approaches to alleviate its adverse effects. *Plants, 10*(2), 259.

Shankar, A., & Prasad, V. (2023). Potential of desiccation-tolerant plant growth-promoting rhizobacteria in growth augmentation of wheat (*Triticum aestivum* L.) under drought stress. *Frontiers in Microbiology, 14*, 1017167.

Shao, M. R., Jiang, N., Li, M., et al. (2021). Complementary phenotyping of maize root architecture by root pulling force and X-ray computed tomography. *BioRxiv, 3*, 226–241.

Shim, Y., Seong, G., Choi, Y., et al. (2023). Suppression of cuticular wax biosynthesis mediated by rice LOV KELCH REPEAT PROTEIN 2 supports a negative role in drought stress tolerance. *Plant, Cell & Environment, 46*(5), 1504–1520.

Shkolnik, D., Nuriel, R., Bonza, M. C., et al. (2018). MIZ1 regulates ECA1 to generate a slow, long-distance phloemtransmitted Ca^{2+} signal essential for root water tracking in Arabidopsis. *Proceedings of the National Academy of Sciences, 115*, 8031–8036.

Sidhu, J. S., Ajmera, I., Arya, S., et al. (2023). Rootslice-A novel functional-structural model for root anatomical phenotypes. *Plant, Cell & Environment, 46*(5), 1671–1690.

Singh, D., & Laxmi, A. (2015). Transcriptional regulation of drought response: A tortuous network of transcriptional factors. *Frontiers in Plant Science, 6*, 895.

Singh, G., Kaur, N., Khanna, R., et al. (2022). 2Gs and plant architecture: Breaking grain yield ceiling through breeding approaches for next wave of revolution in rice (*Oryza sativa* L.). *Critical Reviews in Biotechnology*, 1–24.

Sobejano-Paz, V., Mikkelsen, T. N., Baum, A., et al. (2020). Hyperspectral and thermal sensing of stomatal conductance, transpiration, and photosynthesis for soybean and maize under drought. *Remote Sensing, 12*(19), 3182.

Sofi, P. A., Ara, A., Gull, M., et al. (2019). Canopy temperature depression as an effective physiological trait for drought screening. In G. Ondrasek (Ed.), *Drought: Detection solutions* (pp. 77–93). IntechOpen.

Tarazona, P., Feussner, K., Feussner, I., et al. (2015). An enhanced plant lipidomics method based on multiplexed liquid chromatography-mass spectrometry reveals additional insights into cold- and drought-induced membrane remodeling. *The Plant Journal, 84*, 621–633.

Timilsina, R., Kim, J. H., Nam, H. G., et al. (2019). Temporal changes in cell division rate and genotoxic stress tolerance in quiescent center cells of Arabidopsis primary root apical meristem. *Scientific Reports, 9*(1), 3599.

Voothuluru, P., Makela, P., Zhu, J., et al. (2020). Apoplastic hydrogen peroxide in the growth zone of the maize primary root. Increased levels differentially modulate root elongation under well-watered and water-stressed conditions. *Frontiers in Plant Science, 11*, 392.

Wang, M., Zheng, Q., Shen, Q., et al. (2013). The critical role of potassium in plant stress response. *International Journal of Molecular Sciences, 14*, 7370–7390.

Wang, X., Li, Q., Xie, J., et al. (2021). Abscisic acid and jasmonic acid are involved in drought priming-induced tolerance to drought in wheat. *Crop Journal, 9*(1), 120–132.

Wang, H., Li, N., Li, H., et al. (2023). Overexpression of NtGCN2 improves drought tolerance in tobacco by regulating proline accumulation, ROS scavenging ability, and stomatal closure. *Plant Physiology and Biochemistry, 198*, 107665.

Waqar, A., Bano, A., Ajmal, M., et al. (2022). Effects of PGPR bioinoculants, hydrogel and biochar on growth and physiology of soybean under drought stress. *Communications in Soil Science and Plant Analysis, 7*, 826–847.

Waraich, E. A., Ahmad, R., & Ashraf, M. Y. (2011). Role of mineral nutrition in alleviation of drought stress in plants. *Australian Journal of Crop Science, 5*(6), 764–777.

Wasaya, A., Zhang, X., & Fang, Q. (2018). Root phenotyping for drought tolerance: A review. *Agronomy, 8*(11), 241.

White, P. J. (2019). Root traits benefitting crop production in environments with limited water and nutrient availability. *Annals of Botany, 124*(6), 883–890.

Wu, Q., Hu, Y., Sprague, S. A., et al. (2017). Expression of a monothiol glutaredoxin, AtGRXS17, in tomato (*Solanum lycopersicum*) enhances drought tolerance. *Biochemical and Biophysical Research Communications, 491*, 1034–1039.

Xiong, L., Wang, R. G., Mao, G., et al. (2006). Identification of drought tolerance determinants by genetic analysis of root response to drought stress and abscisic acid. *Plant Physiology, 142*, 1065–1074.

Xiong, R., Liu, S., Considine, M. J., et al. (2021). Root system architecture, physiological and transcriptional traits of soybean (*Glycine max* L.) in response to water deficit: A review. *Physiologia Plantarum, 172*(2), 405–418.

Xu, W., Yu, G., Cui, Y., et al. (2022). PRMI: A dataset of minirhizotron images for diverse plant root study. *ArXiv, 08002*.

Yadav, A., Kumar, S., Verma, R., et al. (2023). Overexpression of PGPR responsive chickpea miRNA166 targeting ATHB15 for drought stress mitigation. *Plant Cell, Tissue and Organ Culture, 154*, 1–18.

Yamaguchi, M., Valliyodan, B., Zhang, J., et al. (2010). Regulation of growth response to water stress in the soybean primary root. I. Proteomic analysis reveals region-specific regulation of phenylpropanoid metabolism and control of free iron in the elongation zone. *Plant, Cell & Environment, 33*, 223–243.

Yang, X., Lu, M., Wang, Y., et al. (2021). Response mechanism of plants to drought stress. *Horticulturae, 7*, 50.

You, J., Zhang, Y., Liu, A., et al. (2019). Transcriptomic and metabolomic profiling of drought-tolerant and susceptible sesame genotypes in response to drought stress. *BMC Plant Biology, 19*, 267.

Yu, X., Brown, J. M., Graham, S. E., et al. (2019). Detection of quantitative trait loci associated with drought tolerance in St. Augustinegrass. *PloS One, 14*(10), 0224620.

Yuan, C., Li, C., Lu, X., et al. (2020). Comprehensive genomic characterization of NAC transcription factor family and their response to salt and drought stress in peanut. *BMC Plant Biology, 20*, 1–21.

Yuan-Yuan, Z., Jun-Ping, T., Jun, C., et al. (2021). Effects of different NH_4^+/NO_3^- ratios on the photosynthetic and physiology responses of blueberry (*Vaccinium* spp.) seedlings growth. *Journal of Plant Nutrition, 44*(6), 854–864.

Zaib, S., Zubair, A., Abbas, S., et al. (2023). Plant Growth-Promoting Rhizobacteria (PGPR) reduce adverse effects of salinity and drought stresses by regulating nutritional profile of Barley. *Applied and Environmental Soil Science, 2023*, 1–12.

Zhang, G., Ba, Q., Chen, S., et al. (2020a). Exogenous application of glycine betaine alleviates cadmium toxicity in super black waxy maize by improving photosynthesis, the antioxidant system and glutathione-ascorbic acid cycle metabolites. *Cereal Research Communications, 48*, 449–458.

Zhang, Q., Liu, H., Wu, X., et al. (2020b). Identification of drought tolerant mechanisms in a drought-tolerant maize mutant based on physiological, biochemical and transcriptomic analyses. *BMC Plant Biology, 20*, 315.

Zhang, C., Yang, Y., Yu, Z., et al. (2023). SUMO E3 ligase AtMMS21-dependent SUMOylation of AUXIN/INDOLE-3-ACETIC ACID 17 regulates auxin signaling. *Plant Physiology, 191*(3), 1871–1883.

Zhao, B., Ma, W., Zhang, P., et al. (2020). Permeation and diffusion of nutrient ions in poly (vinyl alcohol) hydrogel membrane. *Chemical Papers, 74*, 3913–3923.

Zhou, C., Ma, Z., Zhu, L., et al. (2016). Rhizobacterial strain *Bacillus megaterium* BOFC15 induces cellular polyamine changes that improve plant growth and drought resistance. *International Journal of Molecular Sciences, 17*(6), 976.

Zhu, J., Alvarez, S., Marsh, E. L., et al. (2007). Cell wall proteome in the maize primary root elongation zone. II. Region-specific changes in water soluble and lightly ionically bound proteins under water deficit. *Plant Physiology, 145*, 1533–1548.

Rhizosphere Dynamics: The Microbial Influence on Root Architecture

11

Rohit Patel, Sugandha Asthana, and Neeru Sood

Abstract

The rhizosphere, the soil region immediately surrounding plant roots, is a dynamic interface where intricate interactions between root architecture and microbial communities occur. Root architecture, which refers to the spatial configuration and structural organization of a plant's root system, is crucial for nutrient and water uptake, anchorage, and overall plant growth. Various root architectures, such as taproot and fibrous systems, and branching patterns like herringbone, enable plants to adapt to different environmental conditions and efficiently explore soil resources. Microbial communities in the rhizosphere, particularly plant growth-promoting rhizobacteria (PGPR), significantly influence root development by producing phytohormones and other bioactive compounds that modify root branching and root hair development, enhancing nutrient acquisition and plant growth. Conversely, plant roots release exudates, including sugars, amino acids, and organic acids, which serve as signals and nutrients, shaping the composition and activity of microbial communities in the rhizosphere. These microbial interactions can further modify root architecture, promoting adaptive responses to environmental stresses and enhancing plant resilience. This reciprocal relationship between root architecture and the rhizosphere microbiome underlines the complexity of plant-soil-microbe interactions and their combined role in determining plant health, productivity, and adaptation.

R. Patel
Department of Microbiology & Biotechnology, University School of Sciences, Gujarat University, Ahmedabad, Gujarat, India

S. Asthana · N. Sood (✉)
Department of Biotechnology, Birla Institute of Technology & Science, Pilani, Dubai Campus, Dubai, UAE
e-mail: sood@dubai.bits-pilani.ac.in

M. Saraf et al. (eds.), *Climate Change and Soil Microorganisms for Environmental Sustainability*, Microorganisms for Sustainability 28,
https://doi.org/10.1007/978-981-96-3425-5_11

Keywords

Rhizosphere · Root architecture · Microbiome · Plant-microbe interaction · Biofilms

11.1 Introduction

The term "rhizosphere," the most dynamic part of the soil and root interface, first given by Hiltner in 1904, describes the area of soil where all the microbial and geochemical activities take place under the effect of proper root (Berg & Smalla, 2009). Rhizosphere holds central importance for plant growth and development because all the biochemical processes such as nutrient cycling, various biosignaling pathways, nutrient/gaseous exchange, microbes-derived carbon sequestration, and various ecological processes are controlled by the rhizosphere. Multiple biotic and abiotic factors are responsible for affecting the structural and functional diversity of microbial communities in the rhizosphere (Berg & Smalla, 2009). Researchers commonly use the conceptual term "rhizosphere" to describe the volume of soil influenced by roots. They may also extend the term to describe root-soil interfaces, which can be further separated into ecto-rhizosphere (rhizosphere soil), rhizoplane (root surface), and endo-rhizosphere (inner root) (Oburger & Schmidt, 2016). Rhizosphere is the small area of root and soil interface that plays a crucial role in obtaining carbon input (Hartmann et al., 2009) in the form of sugars, amino acids, mucilage, border cells, and other organic substances (Nguyen, 2003). Rhizosphere secretions can promote plant growth by increasing nutrient availability, influencing plant hormonal signaling, and repelling or outcompeting pathogenic microbial strains (McNear Jr., 2013).

Rhizosphere has two main regulators of the local environment: plant roots and microbiome. The spatial configuration and structural organization of a plant's root system, also known as root architecture, plays a critical role in nutrient and water uptake, anchorage, and overall plant growth and development. There are various types of root architectures, such as taproot systems, which are characterized by a single, dominant root that grows vertically downward, and fibrous root systems, which consist of numerous fine roots that spread out horizontally. Additionally, the root system can be further classified based on branching patterns, such as herringbone, dichotomous, and more complex structures. The importance of root architecture lies in its ability to adapt to different environmental conditions, enhancing plant's capacity to explore soil resources efficiently (Fitter, 2002). The complexity and plasticity of root architecture allow plants to optimize nutrient and water acquisition, interact with symbiotic organisms, and respond to abiotic and biotic stresses, which are crucial for survival and productivity (Smith & de Smet, 2012). Therefore, understanding root architecture is vital for improving agricultural practices and crop yields by selecting or engineering plants with optimal root traits suited for specific environments. Furthermore, the microbes living in the rhizosphere are also of interest as they might be able to change the characteristics of plant roots and consequently modify the plant root architecture.

Microbes in the rhizosphere, also known as plant growth-promoting rhizobacteria (PGPR), play a crucial role in influencing plant health, growth, and development. PGPRs can modify root system architecture by producing phytohormones that enhance lateral root branching and root hair development, ultimately improving plant nutrient acquisition and growth (Vacheron et al., 2013). Additionally, plants can also influence the chemical environment of the rhizosphere through the secretion of root exudates, which act as signals to attract specific beneficial microbes or deter pathogens, thereby shaping the microbial community structure and function in this zone (Hartmann et al., 2009). Overall, the microbial community and plant roots are in close contact and frequently modulate each other's behaviors within the rhizosphere which may result in the modification of root architecture, in turn affecting plant growth, resilience, and productivity. Understanding these interactions provides valuable insights into sustainable agricultural practices and helps in developing crops with enhanced stress tolerance and nutrient efficiency.

11.2 Microbial Populations in the Rhizospheres

The microbial communities within the rhizosphere exhibit considerable diversity, influenced by a multitude of factors. Different plant species, and even different genotypes within the same species, can significantly impact microbial populations through the secretion of root exudates. In the case of different cultivars of barley namely, landrace (Chevallier) and modern (NFC Tipple) barley (*Hordeum vulgare*) cultivars, specific microbial recruitment at the species level has also been reported (Pacheco-Moreno et al., 2024). Furthermore, the metagenomics analysis of *Triticum aestivum*, *Brassica napus*, *Medicago truncatula,* and *Arabidopsis thaliana* provides additional evidence of the selective capabilities of host plants over rhizospheric microbiome using root exudates as an effective tool (Achouak et al., 2019). Such selective capabilities can also contribute to the reduction of alpha diversity in rhizospheric soil compared to bulk soil (Kumar et al., 2022). The quality, quantity, and composition of root exudates can attract or repel specific microbial groups, providing essential nutrients and serving as chemical signals for microbial activity modulation (Pacheco-Moreno et al., 2024). At different stages of *A. thaliana* growth, the quantity and quality of root exudates show changes that are reflected in the microbial community structures (M. Zhao et al., 2021).

The physical and chemical properties of soil, such as texture, structure, pH, organic matter content, and nutrient availability, are fundamental in shaping microbial community structure (Naveed et al., 2016) (Fig. 11.1). The soil microbiome exhibits substantial variability with depth, microbial activity generally decreases with depth, leading to shifts in the functional potential of the microbial community, particularly in carbon cycling. Some bacteria such as *Chloroflexi* and *Firmicutes* are commonly found in deeper soils, while *Cyanobacteria, Planctomycetes,* and *Proteobacteria* often decline with depth (Naylor et al., 2022). Variations in soil type

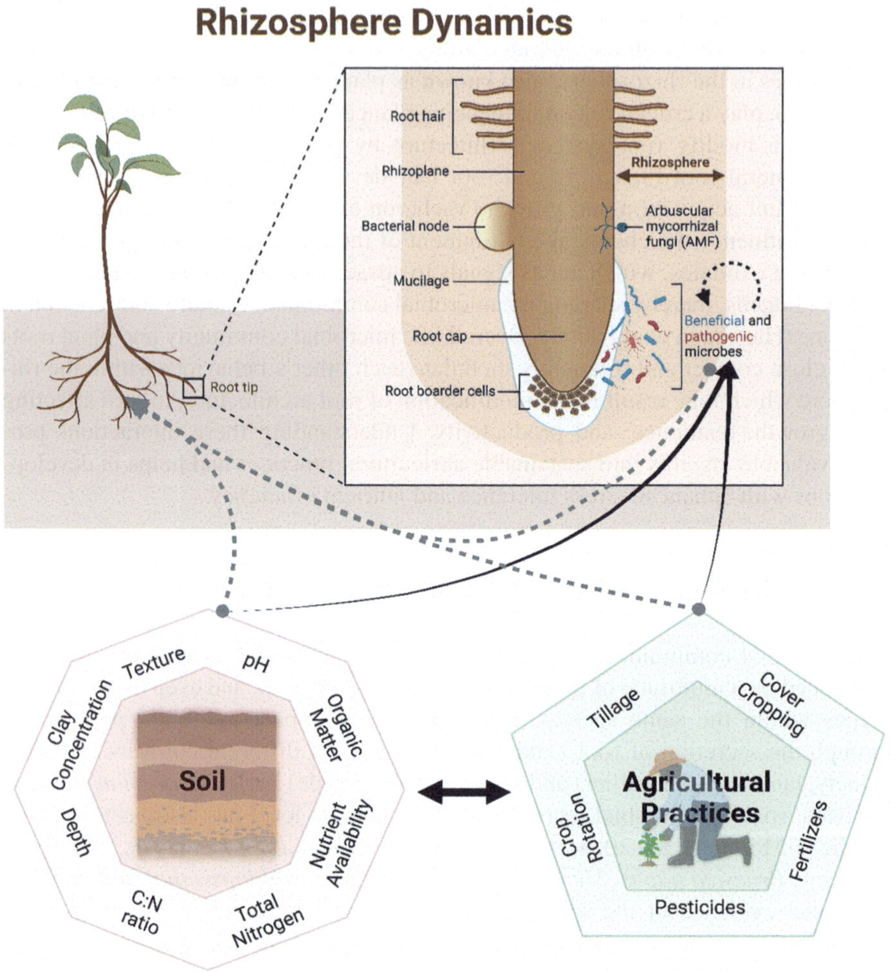

Fig. 11.1 Dynamic interactions among plant roots, soil properties, agricultural practices, and the microbiome that shape the rhizosphere

create distinct microenvironments that support different microbial communities. Clay-rich soils support diverse bacterial communities due to high nutrient content and moisture retention (Biesgen et al., 2020). In contrast, sandy soils, with larger particles and lower nutrients, host different bacterial taxa (Hemkemeyer et al., 2018). Desert soils with leguminous plants also exhibit significant differences in bacterial structures across rhizocompartments, influenced by harsh conditions and specific plant-soil interactions (Zhou et al., 2020).

Agricultural practices, including crop rotation, tillage, cover cropping, and the application of fertilizers and pesticides, significantly influence rhizosphere microbial diversity (Omotayo et al., 2021). No-till farming, for example, can enhance microbial diversity by preserving soil structure and organic matter content, whereas conventional tillage may disrupt microbial habitats and reduce diversity (Bziuk et al., 2021). Crop rotation has been found to enhance microbial diversity and soil health by providing varied organic inputs and disrupting pest and pathogen cycles (Fadiji et al., 2021). Cover cropping contributes to increased microbial biomass and activity, thereby improving soil health and nutrient cycling. The application of organic fertilizers can enhance microbial diversity and activity, whereas chemical fertilizers may have mixed effects, sometimes reducing microbial diversity by favoring certain microbial groups over others (Omotayo et al., 2021). Intra- and interspecific interactions among soil microbes, such as competition, predation, and symbiosis, play vital roles in determining microbial community dynamics, dictating the survival, dominance, and functional roles of different microbial species (Hortal et al., 2017). Predatory interactions within microbial communities are crucial for regulating community structure and nutrient cycling, with nematode predation significantly affecting bacterial community dynamics and enhancing bacterial diversity (Jiang et al., 2017). Hence, the microbiome of soil remains highly dynamic as it is at any given time being influenced by a number of factors as mentioned above.

11.3 Mechanisms of Microbial Influence on Root Architecture

Microbes can influence root architecture through a variety of mechanisms, such as through the production of phytohormones and signaling molecules, modulating nutrient availability and formation of biofilms, and mycorrhizae.

11.3.1 Phytohormones

Phytohormones play a crucial role in the development of root architecture by regulating various physiological and morphological processes essential for plant growth and adaptation. Auxins, for instance, are known to promote lateral root initiation and elongation, thereby enhancing the root's ability to explore and exploit soil resources (Egamberdieva et al., 2017). Cytokinins, on the other hand, regulate cell division and differentiation, which are critical for root apical meristem activity and root branching (Orozco-Mosqueda et al., 2023). Gibberellins and abscisic acid modulate root growth in response to environmental stresses, thus contributing to the plant's stress tolerance mechanisms. Alongside, microbes within the rhizosphere can produce various phytohormones that directly affect root architecture (Lu et al., 2022). Together phytohormones and root-associated microbial communities further fine-tune root development, optimizing nutrient uptake and improving plant resilience to adverse conditions. Larger root systems provide more extensive habitats for microbes, which in turn produce more phytohormones, thereby enhancing root growth and development (Khan et al., 2020; Sharma et al., 2021).

Microbes commonly manipulate root architecture to facilitate interactions with plants, a strategy employed by both beneficial plant growth-promoting rhizobacteria (PGPR) and detrimental phytopathogens. For instance, rhizobacteria can produce auxins, cytokinins, and other phytohormones that integrate into the plant's endogenous hormone pool, altering root growth patterns and architecture (Hussain et al., 2020). Some microbes independently modulate phytohormone pathways, influencing root architecture regulation through ethylene signaling, separate from auxin pathways. This demonstrates the intricate control microbes have over root development (Gonin et al., 2023). Further exploration of plant-microbe interactions has identified many microbes that alter the development of roots for their benefit. One such instance occurs with the plant pathogenic bacterium *Ralstonia solanacearum* which has been shown to alter root development in *Arabidopsis thaliana* by promoting auxin accumulation, which increases root hair formation and induces earlier xylem differentiation, highlighting the bacterium's ability to manipulate host root developmental pathways both dependently and independently of auxin (Xue et al., 2022). Beneficial microbes like *Pseudomonas fluorescens* GM30 and the fungal species *Serendipita indica* promotes lateral root proliferation in *Arabidopsis* through the modulation of host auxin signaling and transport components, emphasizing the importance of auxin signaling in microbe-induced root development (Matthiadis et al., 2020). The ectomycorrhizal fungus *Laccaria bicolor* enhances root colonization and development by modulating host sensitivity to phytohormones such as jasmonate, salicylate, gibberellin, and ethylene, facilitating both root architecture modifications and symbiotic interactions (Basso et al., 2020). Various rhizobacteria also influence root growth by producing cytokinins and indole-3-acetic acid (IAA), which alter gene expression in the root apical meristem, demonstrating their capacity to impact root architecture through phytohormone regulation (Hussain et al., 2020). These studies collectively illustrate the diverse mechanisms through which microbes interact with phytohormone signaling pathways to shape root system architecture.

11.3.2 Biofilms

The most important trait of rhizospheric microorganisms, which is responsible for the adherence on the plant root surface and prevents them from being detached, is known as biofilm formation. It is caused by various natural processes occurring in the soil (Davies & Whitbread, 1989). The microbial biofilm matrix comprises numerous polysaccharides, known as EPS (exopolysaccharides), which is 50%–90% of the total organic carbon of the biofilms (Sutherland, 2001). Biofilm formation is the ultimate response of the accumulation and assembly of various communities of microbial cells bounded by EPS, which includes metal ions, proteins, DNA, lipids, and other humic substances (Teschler et al., 2015). Apart from enhancing soil fertility, biofilms play a significant role in the colonization of roots of plants and facilitate the proliferation of microbial communities in the desired niche (Velmourougane et al., 2017). This is also responsible for maintaining the resilience power of plants against environmental stresses such as temperature, pH, pathogens, and salinity with growth-promoting traits improving nutrient and water availability (Yujia Li et al., 2024). It has been reported that biofilm formation can kill or reduce the penetrance of *F. graminearum* by forming a root-hair endophyte stack (Mousa et al., 2016). Some reports have shown the efficacy of biofilms of *Anabaena* in maintaining root rot of *Gossypium* and diseases in *Abelmoschus esculentus* (Babu et al., 2015; Bharti et al., 2017; Manjunath et al., 2016; Prasanna et al., 2013).

Biofilm formation around the root surface can enhance the population of other plant growth-promoting microbes, ultimately promoting greater metabolic activity among living cells in the soil (Yujia Li et al., 2024) (Fig. 11.2). Some important plant growth-promoting hormones such as gibberellic acid and IAA released by these plant growth-promoting bacteria's (PGPBs) stimulates the development of lateral root extension, root elongation, and stem tissue (Chuks Kenneth, 2017; Yaxley et al., 2001; Gupta et al., 2017). Strong biofilm-forming PGPBs unveiled better root colonization with better survival rates (Altaf & Ahmad, 2017) with enhanced root length, and root numbers (Ansari et al., 2021). Biofilm production by multiple species of bacteria can enhance the soil viscosity and formation of larger soil aggregates which helps bacteria to access more complex nutrient and mineral sources (Xiong et al., 2021). Therefore, it is now considered that biofilm-forming microbial communities are very effective in colonizing the root, improving the structure of the root by various secretions resulting in plant growth–promotion activity.

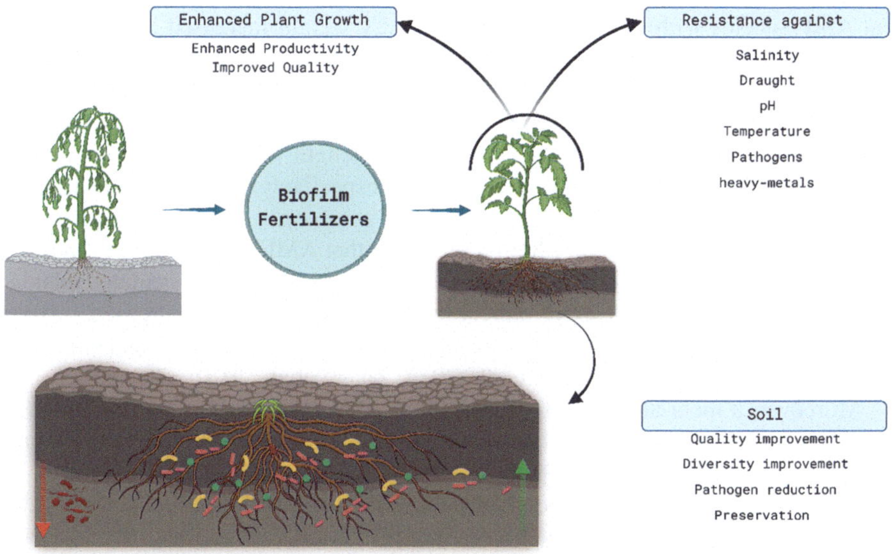

Fig. 11.2 Microbial biofilms and their impact on plant and soil characteristics

11.3.3 Mycorrhizae

Plants can change the root structure such as root fineness or specific root length (root length per gram root weight), root-to-shoot ratio, root hair length, and root number under nutrient-limiting conditions. Mycorrhizal symbiosis is another alternative to such changes. It is observed that plants with high degrees of dependence on symbiosis have coarser root systems, less plasticity in root/shoot ratio, and fewer root hairs in low-fertility soils (Hetrick, 1991).

Mycorrhizal biofilms are responsible for increasing the assimilative area of plants inside the soil by forming a mycelial network. They also facilitate enzymatic mobilization of nutrients, chiefly nitrogen, and phosphorus. Furthermore, they may also mitigate water stress, block heavy metal uptake, and enable symbiotic interactions with other microorganisms (Harding, 2017). Mycorrhizal network around the plant species gives a lot of insight for developing changes related to root morphology. For example- In tea plants, arbuscular mycorrhizal fungi (AMF) remarkably improved total root length and volume, while decreasing root hair length and number (Shao et al., 2018). Root colonization is also highly dependent on the variety of AMF species. They are involved in increasing the nutrition absorption ability, thereby increasing root colonization (C. Li et al., 2019). AMF hyphae could penetrate the soil around the roots of the plant so that the plant gets a larger volume of soil, resulting in a higher uptake of nutrients from the soil, ultimately responsible for more root colonization (Huang et al., 2020).

Murugesan Chandrasekaran concluded in his research, that after treating different AMF species with walnut seedlings, root dry weight increased by 49%. Root morphological study showed that root lengths, volume, and surface area increased by 37, 65, and 31%, under drought stress. In the same study, the researcher concluded that for root traits, AMF-inoculated plants at a moderate level of drought stress showed that the mean effect size of root length increased by 43% and at severe and mild levels of stress by 32 and 23%, respectively, in root length augmentation (Chandrasekaran, 2022).

Furthermore, Quiroga et al. (2020) confirmed that AMF colonization influences root dry weight in all drought stress levels, compared to non-inoculated plants. Thus, it will be helpful for extra-radical hyphae persistence and increased surface area of the mycelia network of the mycorrhiza for root biomass changes (Liang et al., 2021; Wu et al., 2017; Yang et al., 2014; Zhang et al., 2020).

Moreover, a meta-analysis of the effects of AMF on plant growth and nutrient uptake revealed that AMF could increase plant nitrogen (N), phosphorus (P), and K uptake by 22.1, 36.3, and 18.5%, respectively (Chandrasekaran, 2020). It is important to note that the enhanced nutrient uptake by AMF is associated with the regulation of root elongation, lateral root hair formation, root surface area, and root volume expansion (Chandrasekaran, 2020). These studies have confirmed that mycorrhizal species provide an excellent ability for nutrient uptake through multiple mechanisms.

11.3.4 Unconventional Signaling Molecules

Microbes possess the remarkable ability to produce a wide range of molecules, including secondary metabolites, which play crucial roles in various ecological interactions and plant growth processes. Secondary metabolites, unlike primary metabolites that are essential for basic cellular functions, are often involved in complex processes such as defense mechanisms, signaling, and symbiotic relationships (Pang et al., 2021). These compounds include antibiotics, pigments, toxins, and volatile organic compounds (VOCs), which can influence plant health and development (Khare et al., 2020).

Lipochitooligosaccharides (LCOs), or Nod factors, produced by nitrogen-fixing bacteria, are critical for establishing symbiotic relationships between legumes and rhizobia, inducing root hair curling, cortical cell division, and nodule formation to enhance nitrogen fixation and root system architecture. LCOs are signaling molecules consisting of an N-acetylglucosamine backbone with fatty acyl chains, playing key roles in plant-root symbioses. These molecules trigger symbiotic signaling pathways that facilitate the colonization of plant roots by beneficial microbes (Poinsot & Couderc, 2017). Recent studies have shown that the ectomycorrhizal fungus *Laccaria bicolor* produces LCOs that activate nuclear calcium spiking in its host, *Populus*, and use the common symbiosis pathway to enhance root colonization and lateral root development (Cope et al., 2019). Additionally, distinct genetic bases for root responses to LCOs from different microbial origins have been identified in *Medicago truncatula*, indicating that both Nod-LCOs and Fung-LCOs stimulate root branching but through different molecular mechanisms (Bonhomme et al., 2021). These studies highlight the pivotal role of LCOs in modulating root architecture and promoting beneficial plant-microbe interactions.

Volatile organic compounds (VOCs), emitted by soil bacteria and fungi, are one group of low-molecular-weight compounds that play a significant role in modulating root growth patterns, promoting root hair formation, and enhancing root branching. For instance, *Erwinia amylovora* emits VOCs such as 1-nonanol and 1-dodecanol, which modulate root growth and oxidative stress responses in *Arabidopsis thaliana* (Parmagnani et al., 2023). Another VOC, dimethyl disulfide (DMDS), produced by *rhizobacteria* and fungi, has been shown to affect root system architecture in *Arabidopsis* by modulating auxin signaling pathways, leading to changes in lateral root formation and overall root growth (Tyagi et al., 2019). Additionally, VOCs from *Bacillus subtilis* have been reported to significantly increase plant biomass and modify root architecture in soybeans, promoting greater root length, diameter, surface area, and volume, thereby enhancing nutrient uptake and plant growth (Bavaresco et al., 2020).

Quorum sensing (QS) molecules are small chemical signals that enable bacteria to communicate and coordinate their behavior based on population density. These molecules, such as N-acyl homoserine lactones (AHLs), play crucial roles in regulating various physiological processes, including biofilm formation, virulence, and symbiotic interactions. AHLs, in particular, have been found to influence plant root architecture by promoting root elongation and lateral root formation through interactions with plant hormone pathways (Hartmann, 2020). Moreover, AHLs produced by beneficial rhizobacteria can enhance root growth and resistance to pathogens via defense response induction, contributing to improved plant health and productivity (Ortiz-Castro & López-Bucio, 2019; Shrestha et al., 2020).

Siderophores are iron-chelating compounds produced by various microorganisms, including bacteria, fungi, and plants, to facilitate the uptake of iron under iron-limited conditions (Crowley, 2000). Siderophores produced by the arbuscular mycorrhizal fungus *Gigaspora margarita* were found to accumulate in the roots of *Tagetes patula nana*, enhancing iron mobilization and promoting root development (Haselwandter et al., 2020). Furthermore, the presence of hydroxamate-type siderophores in *Pinus sylvestris* roots demonstrated a positive effect on root hair development and overall root architecture, suggesting their potential in enhancing plant growth under iron-limited conditions (Mucha et al., 2019). These unconventional signaling molecules underscore the intricate and diverse interactions between plants and their associated microbiomes, which are crucial for optimizing root architecture and improving plant resilience and productivity.

11.4 Impact of Environmental Factors on Rhizosphere Dynamics

Multiple physical and chemical factors in the local region of the rhizosphere can have profound effects in creating changes at the structural level of roots. Soil moisture and structure, pH, temperature, water availability, and redox potential are some of the crucial factors, on which rhizosphere dynamics relies.

11.4.1 Soil Type

The rhizosphere is the close association of soil and the outer root zone, which is the place for various chemicals, substances, and living organisms. Microbial activity is found highest in the nearest area of the root, which makes the rhizosphere a hothouse for numerous microbial activities (de Graaff et al., 2010) with distinct hydraulic properties (Carminati et al., 2016a). Soil microorganisms produce various enzymes that help in maintaining the nutrient cycle, total plant growth, organic matter decomposition, and biocontrol activity (Bastida et al., 2016). It has been shown in some studies that soil type has a remarkable effect on rhizosphere competence and biocontrol activity of bacterial colonies along with their effect on the indigenous soil microbial community (Schreiter et al., 2014). One researcher evaluated the disease suppression effect of three different types of soil. Bacterial inoculants *P. jessenii* RU47 and *S. plymuthica* 3Re4–18 were used without and with *R. solani* inoculation which was responsible for bottom rot disease. The disease severity index was reported highest in diluvial soil, followed by alluvial loam and loess loam soil. The influence of soil types had a surprisingly low effect on rhizosphere competence and biocontrol activity compared to reducing the severity of the bottom rot disease (Schreiter et al., 2014).

Fine root anatomy gives an important understanding of the relationship between root function and soil environment. It is reported that the fine roots of lower root order were significantly affected by various macro and micro molecules inside the soil. Among them, the cortical tissue of first-order fine roots was positively correlated with potassium and phosphorus but negatively correlated with nitrogen while second- and third-order fine roots were positively correlated with soil total potassium and negatively correlated with nitrogen and phosphorus, in the case of root system of tropical forests of Hainan Island, China (Hogan et al., 2020). This proves that all the macro and micro molecules can increase or decrease the root growth depending upon the concentration of these minerals inside the soil. The local soil environment is responsible for the changes in fine root anatomical structures (Hogan et al., 2020). Soil bulk density is also likely to be an important driver for positive variation (Eissenstat et al., 2015), on the fine root diameter (Clark et al., 2003). Plants with higher fine root diameters have greater root elongation performance in denser soils (Freschet & Roumet, 2017), whereas reduced diameter can be found in less dense soils (Eissenstat et al., 2015). This is because lower mechanical impedance is generated by smaller soil bulk density, which promotes root dispersal (Cao et al., 2021). Additionally, regarding soil nutrients (nitrogen, phosphorus, potassium) and the anatomical structure of fine roots, a lack of nutrients in the soil usually leads to root elongation and structural development (Hogan et al., 2020). These studies show that changes in the soil environment are responsible for the variations in the distribution and morphological structure of the fine root system.

One of the important factors nowadays that significantly affects soil quality is land degradation. It refers to a substantial waning in soil quality and fertility, which can take various forms, including loss of nutrients, acidification, soil erosion, desertification, compaction, excessive salinity, and chemical pollution (Wang et al., 2023), which is ultimately responsible for the huge decrease in crop yield. The result is soil compaction. Soil compaction did not affect the number of seminal and seminal-adventitious roots but did reduce their length. Additionally, compaction decreased both the number and total length of nodal roots. This study was reported in the case of triticale and maize by Grzesiak et al. (2013). In compacted soil, deep roots are almost exclusively found growing in pre-existing pore spaces, underscoring the importance of macropores for root growth (Islam & Hallett, 2021; Valentine et al., 2012; White & Kirkegaard, 2010). A decrease in macroporosity was also expected with increasing bulk density of soil (Colombi et al., 2017) The increase in macropores in structured soil could enhance oxygen diffusion, microbial activities, and biochemical reactions, thereby benefiting lateral roots (Bauke et al., 2017; Horn & Smucker, 2005; Giuliani et al., 2024).

The primary supplier of carbon is plant roots. Soil pores could have phenomenal effects on allocating belowground carbon. Lucas et al. (2023) identified root-soil contact for the fate of belowground carbon (C) and plant-derived carbon by combining X-ray and computed tomography with 14C plant labeling. Greater contact of the root with the surrounding soil having small pores results in strong rhizodeposition but in areas of high microbial activity. This is responsible for greater C inputs for many soil pores (Lucas et al., 2023). Different soil texture and compaction treatments were also used to check the quantified changes in soil structure for three plant species (pea, tomato, and wheat) and significant increases in porosity at the immediate root surface were found in the case of clay loam and loamy sand soils (Helliwell et al., 2019). All these findings strongly support the effect of different soil types on the remodeling of root morphology.

11.4.2 Water Availability

Water condition in the soil is crucial for maintaining the resilience of plants. The root is the plant's first part that emerges from the seed, comes in contact with the soil, and is affected by water deficit conditions (Bishopp & Lynch, 2015; Dinneny, 2019; Gewin, 2010). During water deficit (WD) conditions, root exudates play a significant role in multiple functions such as attracting symbiotic mycorrhiza (Pace et al., 1999), chemicals for root growth for easier penetration in bulk soil, maintaining the soil structure for stabilization of particles and formation of rhizosheaths with increase in water conductivity during drying conditions (Moriwaki et al., 2013; Oldroyd & Leyser, 2020).

Root exudates could affect plant growth positively by attracting symbionts, resisting parasites or competitors, and negatively by using phytotoxins and allelopathy (Durand et al., 2016). Previous work has focused on root exudates' capability to develop biotic relations between plant and microbial communities (Xu et al., 2019). It is also reported that multiple morphological alterations can be created by root exudates such as root diameter, root length, root surface area, biomass, no. of root branches, and tissue density under drought-type conditions (Durand et al., 2016; Fitter, 2002; Lee et al., 2017). There are multiple mechanisms by which a plant can increase its survival rate even with less water availability such as the production of osmolytes, regulation of various phytohormones and antioxidants, and modification of cell wall (Kang et al., 2022; Voothuluru et al., 2020; Voothuluru et al., 2016).

It is proposed that amino acids profiling by root exudates can be changed in the response to WD conditions which is linked to a variety of physiological responses of the plants (Bobille et al., 2019). Under water stress, certain amino acids play a key role in enhancing plant stress tolerance by acting as osmoprotectants, precursors for important metabolites, signaling molecules, or by aiding in the detoxification of ROS (reactive oxygen species) (Okumoto et al., 2016; Wienkoop et al., 2009). Two of the most prevalent osmolytes accumulated by plant cells in response to osmotic stress are GABA (Gama amino butyric acid) and Proline, both originating from the glutamine/glutamate family (Okumoto et al., 2016; Vendruscolo et al., 2007).

The release and diffusion of root exudates and the shaping of biochemical profiles in the rhizosphere are crucially controlled by root hairs and soil water content. Adaptation of rhizosphere properties under drought conditions is made essential by root hairs to maintain plant nutrition when the continuous mass flow of water transporting nutrients to the roots is interrupted (Zhang et al., 2023). This shows that the availability of water plays a very crucial part in maintaining the root structures and functions.

11.4.3 Temperature and pH

11.4.3.1 Temperature

Rhizosphere processes are strongly dependent on each other and highly dynamic in nature. Key factors including pH changes, temperature, and soil water content in the proximity of the rhizosphere are responsible for changes in a variety of metabolic processes such as root exudation, respiration, and nutrient supply. Temperature can affect plant fine roots directly and indirectly by changing root morphology (Sorensen et al., 2008). For example- increasing the diameter of fine roots and producing more numbers of absorbing roots in plants is responsible for a higher intake of water and nutrients at low temperatures (Zadworny et al., 2016), to improve the adaptability of fine roots to environmental change. It is also worth noting that fine root growth also increases significantly at high temperatures (Malhotra et al., 2020). It is reported that negative effects on soil bacterial community can be exerted by low temperatures by reducing the accretion of carbon and nitrogen (Zhang et al., 2022). Total growth, respiration of soil microorganisms, and enzyme dynamics can be limited by changing temperatures inside the soil (Steinweg et al., 2012; Zhong et al., 2021).

Plant growth, development, soil fertility, nutrient uptake ability, and resistance from competitive species can be directly influenced by plant-microbe interactions (Eisenhauer, 2012). Microbes produce a variety of signaling molecules to affect the development of plant roots and alter their architecture during pH change and temperature (Yucong Li et al., 2024). On the contrary, The root system of plants can secrete various microbially beneficial vitamins, enzymes, plant growth regulators, and amino acids (Chamkhi et al., 2022; Mommer et al., 2016), which in turn have an impact on the species, number and distribution of rhizosphere microorganisms with temperature change (Vives-Peris et al., 2020). The effect of temperature rise on the rhizosphere environment of soybean roots has been examined by researchers to clarify the rhizosphere crosstalk between roots and soil in response to warming temperatures in the context of global climate change (Zhao et al., 2023). One study reported that β-glucosidase secretion increased due to higher root enzyme activity with rising temperature in rhizosphere soil in comparison to non-rhizosphere soil (Zhao et al., 2023). Rhizosphere microorganisms affect plant growth by influencing the decomposition of soil organic carbon (Zhao et al., 2023). Increasing a little change in soil temperature was found to affect various chemical reactions and ion compositions in the soil solution, leading to increased soil nutrient availability and microbial biomass, and consequently enhancing soil enzyme activity.

11.4.3.2 pH

An ideal pH is an essential key factor for maintaining proper plant growth and development because it is required for the appropriate absorption of water, uptake of suitable nutrients, sustaining the physiological activities of PGPRs, fertilizer use efficiency, etc. Soil acidification has been expanding in many areas of the world due to industrial activities (Wang et al., 2020). It has been proposed that high pH levels of alkaline soil can bind nutrients inside the soil and make them less available to plants which ultimately leads to poor plant growth and development. In addition, poor soil structure and decreased water-holding capacity can be seen in alkaline soil due to the lack of organic matter (López-Bucio et al., 2000). Microbial activity also gets affected by soil pH. Shifts in the plant-associated microbial communities inhabiting the rhizosphere and root endosphere are predominantly associated with soil pH (Lopes et al., 2021).

In general, soil acidification increased root biomass while reducing root decomposition (Wang et al., 2020). One study evaluated the effect of soil on root biomass which was observed highest in the alkaline Langhorne Creek soil and lowest in the acidic Mount Bold soil (Wang et al., 2012). Lopes et al. (2021) reported a positive correlation between *Actinobacteria* and soil pH ranging from 3.5 to 9.0 (Lauber et al., 2009), but highly alkaline soils (pH > 9.0) can decrease the relative abundance of *actinibacteria.* Other proteobacterial families commonly known to interact with plants like *Bradyrhizobiaceae* and *Sphingomonadaceae* decreased in relative abundance in alkaline soils, probably inhibited by the high soil pH (Lindemann et al., 2011; Luo et al., 2019). Along with soil pH, plant species were identified as one of the most important factors affecting the structure of microbial communities in the root endosphere (Lopes et al., 2021). This study was the extension of the report which assessed the combined effect of acidic soil, pH, and plant diversity/vegetation type on the bulk soil microbiome (Jiabao Li et al. 2018; Ren et al., 2018; Yuan et al., 2016). A higher degree of association was found in microbiomes of alkaline soils compared to neutral soils.

In summary, we can conclude that root exudation, moisture, soil type, water availability, nutrient uptake by roots, root biomass, dry weight of roots, and presence of competitive plant species are affected by soil temperature and pH. These findings can be useful for future management of crop rhizosphere communities.

11.5 Applications and Implications for Agriculture

Dynamics of changing the root structure is one of the important phenomena through which we can make improved crop varieties with better climatic resilience, increased nutrient uptake, and enhanced biocontrol activity for global food security & sustainability. It can be treated as interdisciplinary thought for reducing global challenges now.

11.5.1 Root Architecture in Agriculture

Changing root architecture is a fundamentally important strategy to enhance nutrient acquisition, especially under nutrient-deficient conditions. Multiple studies have been published on the genetic, molecular, and physiological regulation of root architecture to increase plant nutrient efficiency. Many genes have been identified in *Arabidopsis* (Młodzińska et al., 2015), maize (Lin et al., 2013), rice (Wu & Wang, 2008), and soybean (Guo et al., 2011), which are responsible for changing root architecture, ultimately increasing the nutrient uptake efficiency. Additionally, regulators including transcription factors, (Dai et al., 2012) proteins (Araya et al., 2014), and miRNAs (Vidal et al., 2010) have also been exhibited to participate in regulatory networks linking root architecture in response to nutrient deficiency. Concerning genetic research in this area, quantitative trait loci (QTL) analysis is a powerful approach for understanding genetic variation, control of genetically compiled root architecture, and transport traits in response to nutrient starvation. Many QTLs for root architecture to N/P deficiency stress have been characterized in maize (Li et al., 2015), rice (Li et al., 2009), bean/common bean (Cichy et al., 2009; Liao et al., 2004), and soybean (Liang et al., 2010), suggesting that these putative QTLs can be used in marker-assisted breeding and facilitate the genetic improvement for N/P efficiency (Li et al., 2016).

Nitrate is considered the most mobile mineral nutrient ion due to which its absorption is a little bit difficult but enhanced root elongation with deeper root systems has been observed to absorb nitrate in staple crops such as rice, wheat, and maize (Ju et al., 2015; Rasmussen et al., 2015; Wang et al., 2002; Yu et al., 2015). Few changes in root structure such as fewer crown roots and deeper roots were more effective for N absorption in N-deprived soil in the case of maize (Saengwilai et al., 2014). Strong and deep lateral roots have been proposed for efficient N attainment in maize (Mi et al., 2010) and were further confirmed by field trials (Yu et al., 2015).

It is well documented that Auxin plays a very crucial role in root development and its architecture, especially in response to N availability (Li et al., 2016), This finding is supported by an auxin biosynthetic gene, TAR2 which is found to be involved in remodeling of root structure under low N-condition in *Arabidopsis* (Ma et al., 2014). Multiple signaling pathways are linked to soil N availability and plant N status to changes in root architecture (Forde, 2014; Walch-Liu et al., 2005; Zhang et al., 2007), examples include the plasma membrane proton pump, AHA2, CLE peptides, and a CLV1 leucine-rich repeat receptor-like kinase in *Arabidopsis*, which are involved in the regulation of root growth response to N availability (Araya et al., 2014; Młodzińska et al., 2015).

Recent progress has been made in understanding the molecular and physiological mechanisms underlying the alterations in root architecture modeling for better nutrient acquisition. However, there is much more to be discovered before being able to translate these findings to improve nutrient efficiency in crops via breeding for the development of new root traits.

11.5.2 Bioproducts in Agriculture

Sustainable agriculture can be achieved with the help of plant growth-promoting rhizobacteria (PGPR). They are very important in the improvement of plant growth through different processes such as phosphate solubilization, biological nitrogen fixation, siderophore production, and various phytohormone synthesis (Riaz et al., 2020). These microbes have the capacity to increase the soil nutrient status via expansion of root surface area. The improved surface area makes plants absorb more biochemicals, phytohormones, and macro/micronutrients. Studies show that phytohormones like IAA and gibberellins produced by PGPRs in rhizospheric soil are responsible for enhancing the total no of root tips and root surface area in some herbaceous plant species (Han et al., 2005). Morphological changes in root and shoot such as an increase in root surface area, root initiation, cell division, cell enlargement, and increased growth of lateral and adventitious roots have been confirmed by the use of PGPR-expressed cytokinins (Riaz et al., 2020).

The most extensively studied species of rhizomicrobes are *Pseudomonas* and *Bacillus* (Prasad et al., 2019) and reflected as two very significant genera in disease management attributable to their effective antibiotic mechanism (Dominguez-Nuñez et al., 2016; Jayaprakashvel & Mathivanan, 2011). The biocontrol effects of *Bacillus* spp. are mainly due to the biosynthesis of cyclic lipo-peptides from the iturin, fengycin, and surfactin families, which are involved in creating direct effects by destabilizing cell membranes of phytopathogens (Bonmatin et al., 2012; Han et al., 2015; Zhang et al., 2013) and indirect effects by inducing defense responses in the host plant (Jourdan et al., 2009; Rabbee Muhammad Fazle et al., 2019). A very recent report suggests that *Bacillus swezeyi* strain, obtained from two halophytes (*Anabasis truncate* and *Anabasis eriopoda)*, has a wide range of activities such as enzymatic, biocontrol, and nutrient utilization especially different types of carbon sources, under high salt stress conditions (Gao et al., 2024).

The rhizosphere directly impacts the microbial matter to promote the richness of the population. Species of the genus *Pseudomonas* have outstanding root-colonizing abilities, catabolic adaptability, and capacity to produce a wide range of antifungal metabolites such as DAPG and phenazine, which makes them an attractive biocontrol agent. Due to biocontrol and growth-promoting capacity for control of pathogens with diverse mechanisms (Shanmugaiah et al., 2010), fluorescent *pseudomonades* (FPs) have been attracting much attention worldwide in recent days (Varatharaju et al., 2020).

The role of QS (Quorum sensing) in biofilm formation and the release of antifungal compounds, such as phenazine by *Pseudomonas*, has been well-documented for wheat rhizospheres also. In addition, surface chemistry on roots, and the production of surfactin by *Bacillus* were found to influence biofilm formation and root colonization (Altaf & Ahmad, 2017). PGPRs are the most attractive and safe options as biofertilizers and biopesticides to maintain our soil fertility and reduce the impact of excessive harmful chemicals in the near future. It is the urgent need of the hour to cut down or minimize the use of chemical fertilizers to make our environment more safe, productive, and sustainable (Riaz et al., 2020).

11.6 Future Directions and Research Needs

Despite significant findings in understanding the various subtopics related to root morphological alterations under different environmental circumstances, there is still scope for further research to unveil the remaining gaps. As we know the importance of macropores in the diffusion, transportation of various gases, chemicals, and nutrients, and providing the least resistance for roots (Giuliani et al., 2024) in finding their path for better diffusion, further studies should also consider the analysis of pore connectivity and tortuosity.

To drive the root growth, soil pores play a significant role but how they affect the carbon input inside the soil, is still a big question. The mechanism of locating pore networks by roots is also unclear. More studies can be done to understand the interaction between different structures of roots and soil (Gao et al., 2016). It is known that root hairs and soil water content can control the biochemical properties of the rhizosphere and diffusion of root exudates. Further studies about root hairs can be explored to investigate their roles in minimizing the harmful effects of drought on microbial activity (Carminati et al., 2016a).

We can optimize root structural changes and plant health by making engineered microbial communities. There is still more potential for engineering microbial communities to optimize root architecture and plant health. More emphasis could be placed on the role and function of carbon- and nitrogen-fixing bacteria in the rhizosphere soil material cycle of some species such as *M. sieversii* (Zhang et al., 2022). Additionally, more clarity is needed to know the effect of soil pH of the root endosphere compartment on various fungal and bacterial species (Lopes et al., 2021). Because of the naturally hidden condition of the roots, their architectural study should be more to unveil the specific interactions between soil and the rhizosphere (Finér et al., 2011).

References

Achouak, W., Abrouk, D., Guyonnet, J., Barakat, M., Ortet, P., Simon, L., et al. (2019). Plant hosts control microbial denitrification activity. *FEMS Microbiology Ecology, 95*(3), fiz021.

Altaf, M. M., & Ahmad, I. (2017). In vitro and In vivo biofilm formation by Azotobacter Isolates and its relevance to rhizosphere colonization. *Rhizosphere, 3*, 138–142.

Ansari, F. A., Jabeen, M., & Ahmad, I. (2021). Pseudomonas Azotoformans FAP5, a novel biofilm-forming PGPR strain, alleviates drought stress in wheat plant. *International Journal of Environmental Science and Technology, 18*(12), 3855–3870.

Araya, T., von Wirén, N., & Takahashi, H. (2014). CLE peptides regulate lateral root development in response to nitrogen nutritional status of plants. *Plant Signaling & Behavior, 9*(7), 2029–2034.

Babu, S., et al. (2015). Evaluating microbe-plant interactions and varietal differences for enhancing biocontrol efficacy in root rot disease challenged cotton crop. *European Journal of Plant Pathology, 142*(2), 345–362.

Basso, V., et al. (2020). An ectomycorrhizal fungus alters sensitivity to Jasmonate, Salicylate, Gibberellin, and Ethylene in host roots. *Plant Cell and Environment, 43*(4), 1047–1068.

Bastida, F., et al. (2016). The active microbial diversity drives ecosystem multifunctionality and is physiologically related to carbon availability in Mediterranean semi-arid soils. *Molecular ecology, 25*(18), 4660–4673.

Bauke, S. L., et al. (2017). Macropore effects on phosphorus acquisition by wheat roots – A Rhizotron study. *Plant and Soil, 416*(1–2), 67–82.

Bavaresco, L. G., et al. (2020). Bacillus subtilis can modulate the growth and root architecture in soybean through volatile organic compounds. *Theoretical and Experimental Plant Physiology, 32*(2), 99–108.

Berg, G., & Smalla, K. (2009). Plant species and soil type cooperatively shape the structure and function of microbial communities in the rhizosphere. *FEMS Microbiology Ecology, 68*(1), 1–13.

Bharti, A., Velmourougane, K., & Prasanna, R. (2017). Phototrophic biofilms: Diversity, ecology and applications. *Journal of Applied Phycology, 29*(6), 2729–2744.

Biesgen, D., Frindte, K., Maarastawi, S., & Knief, C. (2020). Clay content modulates differences in bacterial community structure in soil aggregates of different size. *Geoderma, 376*(114), 544.

Bishopp, A., & Lynch, J. P. (2015). The hidden half of crop yields. *Nature Plants, 1*(8), 1–2.

Bobille, H., et al. (2019). Effect of water availability on changes in root amino acids and associated rhizosphere on root exudation of amino acids in Pisum Sativum L. *Phytochemistry, 161*, 75–85.

Bonhomme, M., et al. (2021). Distinct genetic basis for root responses to Lipo-Chitooligosaccharide signal molecules from different microbial origins. *Journal of Experimental Botany, 72*(10), 3821–3834.

Bonmatin, J.-M., Laprevote, O., & Peypoux, F. (2012). Diversity among microbial cyclic Lipopeptides: Iturins and Surfactins. Activity-structure relationships to design new bioactive agents. *Combinatorial Chemistry & High Throughput Screening, 6*(6), 541–556.

Bziuk, N., et al. (2021). Tillage shapes the soil and rhizosphere microbiome of barley-but not its susceptibility towards Blumeria Graminis f. Sp. Hordei. *FEMS Microbiology Ecology, 97*(3), fiab018.

Cao, Y., Yangui, S., Huang, G., & Li, Y. (2021). Soil texture modulates the intensity of fertile islands through affecting the distribution of shrub fine roots. *Journal of Arid Environments, 189*, 104503.

Carminati, A., et al. (2016a). Biophysical rhizosphere processes affecting root water uptake. *Annals of Botany, 118*(4), 561–571.

Chamkhi, I., et al. (2022). Is the rhizosphere a source of applicable multi-beneficial microorganisms for plant enhancement? *Saudi Journal of Biological Sciences, 29*(2), 1246–1259.

Chandrasekaran, M. (2020). A meta-analytical approach on Arbuscular Mycorrhizal fungi inoculation efficiency on plant growth and nutrient uptake. *Agriculture (Switzerland), 10*(9), 1–12.

Chandrasekaran, M. (2022). Arbuscular Mycorrhizal Fungi mediated alleviation of drought stress via non-enzymatic antioxidants: A meta-analysis. *Plants, 11*(19), 2448.

Chuks Kenneth, O. (2017). Plant growth Promoting Rhizobacteria (PGPR): A bioprotectant bioinoculant for sustainable agrobiology. A review. *International Journal of Advanced Research in Biological Sciences (IJARBS), 4*(5), 123–142.

Cichy, K. A., et al. (2009). QTL analysis of root architecture traits and low phosphorus tolerance in an Andean bean population. *Crop Science, 49*(1), 59–68.

Clark, L. J., Whalley, W. R., & Barraclough, P. B. (2003). How do roots penetrate strong soil? *Plant and Soil, 255*(1), 93–104.

Colombi, T., Braun, S., Keller, T., & Walter, A. (2017). Artificial Macropores attract crop roots and enhance plant productivity on compacted soils. *Science of the Total Environment, 574*, 1283–1293.

Cope, K. R., et al. (2019). The ectomycorrhizal fungus Laccaria Bicolor produces Lipochitooligosaccharides and uses the common symbiosis pathway to colonize populus roots. *Plant Cell, 31*(10), 2386–2410.

Crowley, D. (2000). Function of siderophores in the plant rhizosphere. In *The rhizosphere* (pp. 239–278). CRC Press.

Dai, X., Wang, Y., Yang, A., & Zhang, W. H. (2012). OsMYB2P-1, an R2R3 MYB transcription factor, is involved in the regulation of phosphate-starvation responses and root architecture in rice. *Plant Physiology, 159*(1), 169–183.

Davies, K. G., & Whitbread, R. (1989). Factors affecting the colonisation of a root system by fluorescent pseudomonads: The effects of water, temperature and soil microflora. *Plant and Soil, 116*(2), 247–256.

Dinneny, J. R. (2019). Developmental responses to water and salinity in root systems. *Annual Review of Cell and Developmental Biology, 35*, 239–257.

Dominguez-Nuñez, J. A., Benito, B., Berrocal-Lobo, M., & Albanesi, A. (2016). Mycorrhizal fungi: Role in the solubilization of potassium BT – Potassium solubilizing microorganisms for sustainable agriculture. In *Potassium solubilizing microorganisms for sustainable agriculture* (Vol. 77–98). Springer. https://doi.org/10.1007/978-81-322-2776-2_6

Durand, M. Ë., et al. (2016). Water deficit enhances C Export to the roots in Arabidopsis Thaliana plants with contribution of sucrose transporters in both shoot and roots. *Plant Physiology, 170*(3), 1460–1479.

Egamberdieva, D., Wirth, S. J., Alqarawi, A. A., AbdAllah, E. F., & Hashem, A. (2017). Phytohormones and beneficial microbes: Essential components for plants to balance stress and fitness. *Frontiers in Microbiology, 8*, 2104.

Eissenstat, D. M., et al. (2015). Linking root traits to nutrient foraging in arbuscular mycorrhizal trees in a temperate forest. *New Phytologist, 208*(1), 114–124.

Eisenhauer, N. (2012). Aboveground-belowground interactions as a source of complementarity effects in biodiversity experiments. *Plant and Soil, 351*(1–2), 1–22.

Fadiji, A. E., Kanu, J. O., & Babalola, O. O. (2021). Metagenomic profiling of rhizosphere microbial community structure and diversity associated with maize plant as affected by cropping systems. *International Microbiology, 24*(3), 325–335.

Fazle, R. M., et al. (2019). Bacillus Velezensis: A valuable member of bioactive molecules within plant microbiomes. *Molecules, 24*(1046), 2–13.

Finér, L., Ohashi, M., Noguchi, K., & Hirano, Y. (2011). Factors causing variation in fine root biomass in forest ecosystems. *Forest Ecology and Management, 261*(2), 265–277.

Fitter, A. (2002). Characteristics and functions of root systems. In *Plant roots* (pp. 49–78). CRC Press.

Forde, B. G. (2014). Nitrogen signalling pathways shaping root system architecture: An update. *Current Opinion in Plant Biology, 21*, 30–36.

Freschet, G. T., & Roumet, C. (2017). Sampling roots to capture plant and soil functions. *Functional Ecology, 31*(8), 1506–1518.

Gewin, V. (2010). Food: An underground revolution. *Nature, 466*(7306), 552–553.

Giuliani, L. M., Hallett, P. D., & Loades, K. W. (2024). Effects of soil structure complexity to root growth of plants with contrasting root architecture. *Soil and Tillage Research, 238*(106), 023.

Gonin, M., Salas-González, I., Gopaulchan, D., Frene, J. P., Roden, S., Van de Poel, B., et al. (2023). Plant microbiota controls an alternative root branching regulatory mechanism in plants. *Proceedings of the National Academy of Sciences, 120*(15), e2301054120.

De Graaff, M. A., Classen, A. T., Castro, H. F., & Schadt, C. W. (2010). Labile soil carbon inputs mediate the soil microbial community composition and plant residue decomposition rates. *New Phytologist, 188*(4), 1055–1064.

Grzesiak, S., et al. (2013). Changes in root system structure, leaf water potential and gas exchange of maize and triticale seedlings affected by soil compaction. *Environmental and Experimental Botany, 88*, 2–10.

Gao, W.,Hodgkinson, L., Jin, K., Watts, C. W., Ashton, R. W., Shen, J. & Ren, T et al. (2016). Deep roots and soil structure. *Plant, Cell & Environment, 39*(8), 1662–1668.

Guo, W., et al. (2011). A soybean β-Expansin Gene GmEXPB2 intrinsically involved in root system architecture responses to abiotic stresses. *Plant Journal, 66*(3), 541–552.

Gupta, G., Snehi, S. K., & Singh, V. (2017). Role of PGPR in biofilm formations and its importance in plant health. *Biofilms in Plant and Soil Health, 27*, 27–42.

Han, J., et al. (2005). Characterization of a novel plant growth-promoting bacteria strain Delftia Tsuruhatensis HR4 Both as a Diazotroph and a potential biocontrol agent against various plant pathogens. *Systematic and Applied Microbiology, 28*(1), 66–76.

Han, Q., et al. (2015). The bacterial Lipopeptide Iturins induce Verticillium Dahliae cell death by affecting fungal signalling pathways and mediate plant defence responses involved in pathogen-associated molecular pattern-triggered immunity. *Environmental Microbiology, 17*(4), 1166–1188.

Harding, K. T. (2017). *Sclerocactus wetlandicus: Habitat characterization, seed germination and Mycorrhizal analysis*. Utah State University.

Hartmann, A. (2020). Quorum sensing N-acyl-homoserine lactone signal molecules of plant beneficial Gram-negative rhizobacteria support plant growth and resistance to pathogens. *Rhizosphere, 16*(100), 258.

Hartmann, A., Schmid, M., van Tuinen, D., & Berg, G. (2009). Plant-driven selection of microbes. *Plant and Soil, 321*(1–2), 235–257.

Haselwandter, K., Haas, H., Häninger, G., & Winkelmann, G. (2020). Siderophores in plant root tissue: Tagetes Patula Nana colonized by the Arbuscular Mycorrhizal Fungus Gigaspora Margarita. *BioMetals, 33*(2–3), 137–146.

Helliwell, J. R., et al. (2019). The role of plant species and soil condition in the structural development of the rhizosphere. *Plant Cell and Environment, 42*(6), 1974–1986.

Hemkemeyer, M., Dohrmann, A. B., Christensen, B. T., & Tebbe, C. C. (2018). Bacterial preferences for specific soil particle size fractions revealed by community analyses. *Frontiers in Microbiology, 9*, 149.

Hetrick, B. A. D. (1991). Mycorrhizas and root architecture. *Experientia, 47*(4), 355–362.

Hogan, J. A., Valverde-Barrantes, O. J., Ding, Q., Xu, H., & Baraloto, C. (2020). Morphological variation of fine root systems and leaves in primary and secondary tropical forests of Hainan Island, China. *Annals of Forest Science, 77*, 1–21.

Horn, R., & Smucker, A. (2005). Structure formation and its consequences for gas and water transport in unsaturated arable and forest soils. *Soil and Tillage Research, 82*(1), 5–14.

Hortal, S., Lozano, Y. M., Bastida, F., Armas, C., Moreno, J. L., Garcia, C., & Pugnaire, F. I. (2017). Plant-plant competition outcomes are modulated by plant effects on the soil bacterial community. *Scientific Reports, 7*(1), 17756.

Huang, G. M., et al. (2020). Mycorrhizal roles in plant growth, gas exchange, root morphology, and nutrient uptake of walnuts. *Plant, Soil and Environment, 66*(6), 295–302.

Hussain, A., Husna, I. U., & Naseem, M. (2020). Plant-associated microbes alter root growth by modulating root apical meristem. *Methods in Molecular Biology, 2094*, 49–58.

Islam, M. D. D., Price, A. H., & Hallett, P. D. (2021). Contrasting ability of deep and shallow rooting rice genotypes to grow through plough pans containing simulated biopores and cracks. *Plant and Soil, 467*(1–2), 515–530.

Jayaprakashvel, M., & Mathivanan, N. (2011). Management of plant diseases by microbial metabolites. In *Bacteria in agrobiology: Plant nutrient management* (pp. 237–265).

Jiang, Y., et al. (2017). Nematode grazing promotes bacterial community dynamics in soil at the aggregate level. *ISME Journal, 11*(12), 2705–2717.

Jourdan, E., et al. (2009). Insights into the defense-related events occurring in plant cells following perception of Surfactin-Type Lipopeptide from Bacillus subtilis. *Molecular Plant-Microbe Interactions, 22*(4), 456–468.

Ju, C., et al. (2015). Root and shoot traits for rice varieties with higher grain yield and higher nitrogen use efficiency at lower nitrogen rates application. *Field Crops Research, 175*, 47–55.

Kang, J., Voothuluru, P., Hoyos-Miernyk, E., Alexander, D., Oliver, M. J., & Sharp, R. E. (2022). Antioxidant metabolism underlies different metabolic strategies for primary root growth maintenance under water stress in cotton and maize. *Antioxidants, 11*(5), 820.

Khan, N., Bano, A., Ali, S., & Babar, M. A. (2020). Crosstalk amongst Phytohormones from Planta and PGPR under biotic and abiotic stresses. *Plant Growth Regulation, 90*(2), 189–203.

Khare, S., et al. (2020). Plant secondary metabolites synthesis and their regulations under biotic and abiotic constraints. *Journal of Plant Biology, 63*(3), 203–216.

Kumar, U., Saqib, H. S. A., Islam, W., Prashant, P., Patel, N., Chen, W., et al. (2022). Landscape composition and soil physical–chemical properties drive the assemblages of bacteria and fungi in conventional vegetable fields. *Microorganisms, 10*(6), 1202.

Lauber, C. L., Hamady, M., Knight, R., & Fierer, N. (2009). Pyrosequencing-based assessment of soil PH as a predictor of soil bacterial community structure at the continental scale. *Applied and Environmental Microbiology, 75*(15), 5111–5120.

Lee, D. K., et al. (2017). The rice OsNAC6 transcription factor orchestrates multiple molecular mechanisms involving root structural adaptions and Nicotianamine biosynthesis for drought tolerance. *Plant Biotechnology Journal, 15*(6), 754–764.

Lei, G., Ma, J-B. Huang, Y., Muhammad, M.. Lian, H-T., Shurigin, V., Egamberdieva, D., Li W-J, & Li, L. (2024). Insight into endophytic microbial diversity in two halophytes and plant beneficial attributes of Bacillus swezeyi. *Frontiers in Microbiology, 15*, 1447755.

Li, J., et al. (2009). Root and shoot traits responses to phosphorus deficiency and QTL analysis at seedling stage using introgression lines of rice. *Journal of Genetics and Genomics, 36*(3), 173–183.

Li, P., et al. (2015). A genetic relationship between nitrogen use efficiency and seedling root traits in maize as revealed by QTL analysis. *Journal of Experimental Botany, 66*(11), 3175–3188.

Li, X., Zeng, R., & Liao, H. (2016). Improving crop nutrient efficiency through root architecture modifications. *Journal of Integrative Plant Biology, 58*(3), 193–202.

Li, C., Zhou, J., Wang, X., & Liao, H. (2019). A purple acid phosphatase, GmPAP33, participates in Arbuscule degeneration during Arbuscular Mycorrhizal symbiosis in Soybean. *Plant Cell and Environment, 42*(6), 2015–2027.

Li, Y., Narayanan, M., Shi, X., Chen, X., Li, Z., & Ma, Y. (2024). Biofilms formation in plant growth-promoting bacteria for alleviating agro-environmental stress. *Science of the Total Environment, 907*, 167774. https://doi.org/10.1016/j.scitotenv.2023.167774

Li, Y., Chen, Y., Fu, Y., Shao, J., Liu, Y., Xuan, W., Xu, G., Zhang, R. (2024). Signal communication during microbial modulation of root system architecture. *Journal of Experimental Botany, 75*(2), 526–537. https://doi.org/10.1093/jxb/erad263

Li JiaBao, L. J., Shen ZeHao, S. Z., Li ChaoNan, L. C., Kou YongPing, K. Y., Wang YanSu, W. Y., Tu Bo, T. B., & Li XiangZhen, L. X. (2018). *Stair-step pattern of soil bacterial diversity mainly driven by pH and vegetation types along the elevational gradients of Gongga Mountain*. China.

Liang, Q., et al. (2010). QTL analysis of root traits as related to phosphorus efficiency in Soybean. *Annals of Botany, 106*(1), 223–234.

Liang, S. M., et al. (2021). Physiological responses of Mycorrhizal symbiosis to drought stress in white clover. *Notulae Botanicae Horti Agrobotanici Cluj-Napoca, 49*(1), 1–10.

Liao, H., et al. (2004). Genetic mapping of basal root gravitropism and phosphorus acquisition efficiency in common bean. *Functional Plant Biology, 31*(10), 959–970.

Lin, H. J., et al. (2013). Transcriptional responses of maize seedling root to phosphorus starvation. *Molecular Biology Reports, 40*(9), 5359–5379.

Lindemann, A., et al. (2011). Isovaleryl-Homoserine Lactone, an unusual branched-chain quorum-sensing signal from the soybean symbiont Bradyrhizobium Japonicum. *Proceedings of the National Academy of Sciences of the United States of America, 108*(40), 16765–16770.

López-Bucio, J., et al. (2000). Agriculture for marginal lands: Transgenic plants towards the third millennium. *Developments in Plant Genetics and Breeding, 5*(C), 159–165.

Lu, Y., et al. (2022). Roots and microbiome jointly drive the distributions of 17 Phytohormones in the plant soil continuum in a Phytohormone-specific manner. *Plant and Soil, 470*(1–2), 153–165.

Lucas, M., et al. (2023). The soil pore structure encountered by roots affects plant-derived carbon inputs and fate. *New Phytologist, 240*(2), 515–528.

Luo, Y., Wang, F., Huang, Y., Zhou, M., Gao, J., Yan, T., et al. (2019). Sphingomonas sp. Cra20 increases plant growth rate and alters rhizosphere microbial community structure of Arabidopsis thaliana under drought stress. *Frontiers in Microbiology, 10*, 1221.

Lucas, D. L., Jingjie, H., & Daniel, P, (2021). Schachtman, Alkaline soil pH affects bulk soil, rhizosphere and root endosphere microbiomes of plants growing in a Sandhills ecosystem. *FEMS Microbiology Ecology, 97*(4), fiab028. https://doi.org/10.1093/femsec/fiab028

Ma, W., et al. (2014). Auxin biosynthetic gene TAR2 is involved in low nitrogen-mediated reprogramming of root architecture in arabidopsis. *Plant Journal, 78*(1), 70–79.

Malhotra, A., et al. (2020). Peatland warming strongly increases fine-root growth. *Proceedings of the National Academy of Sciences of the United States of America, 117*(30), 17627–17634.

Manjunath, M., et al. (2016). Beneficial cyanobacteria and eubacteria synergistically enhance bioavailability of soil nutrients and yield of okra. *Heliyon, 2*(2), e00066.

Matthiadis, A., Sukumar, P., DeLeon, A., Pelletier, D., Labbé, J. L., Tuskan, G. A., & Kalluri, U. C. (2020). Host auxin machinery is important in mediating root architecture changes induced by Pseudomonas fluorescens GM30 and Serendipita indica. *BioRxiv*, 2020–2008.

McNear, D. H., Jr. (2013). The rhizosphere – Roots, soil and everything in between. *Nature Education Knowledge, 4*(3), 1.

Mi, G. H., et al. (2010). Ideotype root architecture for efficient nitrogen acquisition by maize in intensive cropping systems. *Science China Life Sciences, 53*(12), 1369–1373.

Młodzińska, E., Kłobus, G., Christensen, M. D., & Fuglsang, A. T. (2015). The plasma membrane H+-ATPase AHA2 contributes to the root architecture in response to different nitrogen supply. *Physiologia Plantarum, 154*(2), 270–282.

Mommer, L., Kirkegaard, J., & van Ruijven, J. (2016). Root-root interactions: Towards a rhizosphere framework. *Trends in Plant Science, 21*(3), 209–217.

Moriwaki, T., Miyazawa, Y., Kobayashi, A., & Takahashi, H. (2013). Molecular mechanisms of hydrotropism in seedling roots of arabidopsis thaliana (Brassicaceae). *American Journal of Botany, 100*(1), 25–34.

Mousa, W. K., Shearer, C., Limay-Rios, V., Ettinger, C. L., Eisen, J. A., & Raizada, M. N. (2016). Root-hair endophyte stacking in finger millet creates a physicochemical barrier to trap the fungal pathogen Fusarium graminearum. *Nature Microbiology, 1*(12), 1–12.

Mucha, J., Pawłowski, T. A., Klupczyńska, E. A., Guzicka, M., & Zadworny, M. (2019). The effect of Hydroxamic Siderophores structure on acetylation of histone H3 and alpha Tubulin in Pinus sylvestris root cells. *International Journal of Molecular Sciences, 20*(23), 6099.

Naveed, M., et al. (2016). Spatial variability of microbial richness and diversity and relationships with soil organic carbon, texture and structure across an agricultural field. *Applied Soil Ecology, 103*, 44–55.

Naylor, D., McClure, R., & Jansson, J. (2022). Trends in microbial community composition and function by soil depth. *Microorganisms, 10*(3), 540.

Nguyen, C. (2003). Rhizodeposition of organic C by plants: Mechanisms and controls. *Agronomie, 23*(5–6), 375–396.

Oburger, E., & Schmidt, H. (2016). New methods to unravel rhizosphere processes. *Trends in Plant Science, 21*(3), 243–255.

Okumoto, S., Funck, D., Trovato, M., & Forlani, G. (2016). Amino acids of the glutamate family: Functions beyond primary metabolism. *Frontiers in Plant Science, 7*, 318.

Oldroyd, G. E., & Leyser, O. (2020). A plant's diet, surviving in a variable nutrient environment. *Science, 368*(6486), eaba0196.

Omotayo, O. P., Igiehon, O. N., & Babalola, O. O. (2021). Metagenomic study of the community structure and functional potentials in maize rhizosphere microbiome: elucidation of mechanisms behind the improvement in plants under normal and stress conditions. *Sustainability, 13*(14), 8079.

Orozco-Mosqueda, M. D. C., Santoyo, G., & Glick, B. R. (2023). Recent advances in the bacterial phytohormone modulation of plant growth. *Plants, 12*(3), 606.

Ortiz-Castro, R., & López-Bucio, J. (2019). Review: Phytostimulation and root architectural responses to quorum-sensing signals and related molecules from Rhizobacteria. *Plant Science, 284*, 135–142.

Pace, P. F., et al. (1999). Drought-induced changes in shoot and root growth of young cotton plants. *Journal of Cotton Science, 3*(4), 183–187.

Pacheco-Moreno, A., Bollmann-Giolai, A., Chandra, G., Brett, P., Davies, J., Thornton, O., et al. (2024). The genotype of barley cultivars influences multiple aspects of their associated microbiota via differential root exudate secretion. *Plos Biology, 22*(4), e3002232.

Pang, Z., Chen, J., Wang, T., Gao, C., Li, Z., Guo, L., et al. (2021). Linking plant secondary metabolites and plant microbiomes: A review. *Frontiers in Plant Science, 12*(621), 276.

Parmagnani, A. S., Kanchiswamy, C. N., Paponov, I. A., Bossi, S., Malnoy, M., & Maffei, M. E. (2023). Bacterial volatiles (mVOC) emitted by the Phytopathogen Erwinia amylovora promote Arabidopsis thaliana growth and oxidative stress. *Antioxidants, 12*(3), 600.

Poinsot, V., & Couderc, F. (2017). *Formation of lipochitin oligosaccharide signaling molecules. Biogenesis of Fatty Acids, Lipids and Membranes* (pp. 1–23). Springer International Publishing.

Prasad, M., Srinivasan, R., Chaudhary, M., Choudhary, M., & Jat, L. K. (2019). Plant growth promoting rhizobacteria (PGPR) for sustainable agriculture: perspectives and challenges. In *PGPR amelioration in sustainable agriculture* (pp. 129–157).

Prasanna, R., et al. (2013). Cyanobacteria mediated plant growth promotion and bioprotection against Fusarium Wilt in Tomato. *European Journal of Plant Pathology, 136*(2), 337–353.

Quiroga, G., Erice, G., Aroca, R., Delgado-Huertas, A., & Ruiz-Lozano, J. M. (2020). Elucidating the possible involvement of maize aquaporins and arbuscular mycorrhizal symbiosis in the plant ammonium and urea transport under drought stress conditions. *Plants, 9*(2), 148.

Rasmussen, I. S., Dresbøll, D. B., & Thorup-Kristensen, K. (2015). Winter wheat cultivars and nitrogen (N) fertilization-effects on root growth, N uptake efficiency and N use efficiency. *European Journal of Agronomy, 68*, 38–49.

Ren, B., Hu, Y., Chen, B., Zhang, Y., Thiele, J., Shi, R., et al. (2018). Soil pH and plant diversity shape soil bacterial community structure in the active layer across the latitudinal gradients in continuous permafrost region of Northeastern China. *Scientific Reports, 8*(1), 5619.

Riaz, U., et al. (2020). Plant growth-promoting Rhizobacteria (PGPR) as Biofertilizers and Biopesticides. In *Microbiota and biofertilizers: A sustainable continuum for plant and soil health* (pp. 181–196).

Saengwilai, P., Tian, X., & Lynch, J. P. (2014). Low crown root number enhances nitrogen acquisition from low-nitrogen soils in maize. *Plant Physiology, 166*(2), 581–589.

Schreiter, S., Sandmann, M., Smalla, K., & Grosch, R. (2014). Soil type dependent rhizosphere competence and biocontrol of two bacterial inoculant strains and their effects on the rhizosphere microbial community of field-grown lettuce. *Plos one, 9*(8), e103726.

Shanmugaiah, V., Mathivanan, N., & Varghese, B. (2010). Purification, crystal structure and antimicrobial activity of Phenazine-1-Carboxamide produced by a growth-promoting biocontrol bacterium, pseudomonas aeruginosa MML2212. *Journal of Applied Microbiology, 108*(2), 703–711.

Shao, Y. D., et al. (2018). Mycorrhiza-induced changes in root growth and nutrient absorption of tea plants. *Plant, Soil and Environment, 64*(6), 283–289.

Sharma, M., Singh, D., Saksena, H. B., Sharma, M., Tiwari, A., Awasthi, P., et al. (2021). Understanding the intricate web of phytohormone signalling in modulating root system architecture. *International Journal of Molecular Sciences, 22*(11), 5508.

Shrestha, A., Grimm, M., Ojiro, I., Krumwiede, J., & Schikora, A. (2020). Impact of quorum sensing molecules on plant growth and immune system. *Frontiers in Microbiology, 11*, 1545.

Smith, S., & de Smet, I. (2012). Root system architecture: Insights from arabidopsis and cereal crops. *Philosophical Transactions of the Royal Society B: Biological Sciences, 367*(1595), 1441–1452.

Sorensen, P. L., et al. (2008). Plant and microbial uptake and allocation of organic and inorganic nitrogen related to plant growth forms and soil conditions at two subarctic tundra sites in Sweden. *Arctic, Antarctic, and Alpine Research, 40*(1), 171–180.

Steinweg, J. M., Dukes, J. S., & Wallenstein, M. D. (2012). Modeling the effects of temperature and moisture on soil enzyme activity: Linking laboratory assays to continuous field data. *Soil Biology and Biochemistry, 55*, 85–92.

Sutherland, I. W. (2001). Biofilm exopolysaccharides: A strong and sticky framework. *Microbiology, 147*(1), 3–9.

Teschler, J. K., et al. (2015). Living in the matrix: Assembly and control of vibrio Cholerae biofilms. *Nature Reviews Microbiology, 13*(5), 255–268.

Tyagi, S., Kim, K., Cho, M., & Lee, K. J. (2019). Volatile dimethyl disulfide affects root system architecture of arabidopsis via modulation of canonical auxin signaling pathways. *Environmental Sustainability,* 2(2), 211–216.

Vacheron, J., Desbrosses, G., Bouffaud, M. L., Touraine, B., Moënne-Loccoz, Y., Muller, D., et al. (2013). Plant growth-promoting rhizobacteria and root system functioning. *Frontiers in plant science, 4*, 356.

Valentine, T. A., et al. (2012). Soil strength and macropore volume limit root elongation rates in many UK agricultural soils. *Annals of Botany, 110*(2), 259–270.

Varatharaju, G., et al. (2020). Biocontrol properties and functional characterization of rice rhizobacterium pseudomonas Sp. VsMKU4036. *Journal of Pure and Applied Microbiology, 14*(2), 1545–1556.

Velmourougane, K., Prasanna, R., & Saxena, A. K. (2017). Agriculturally important microbial biofilms: Present status and future prospects. *Journal of Basic Microbiology, 57*(7), 548–573.

Vendruscolo, E. C. G., et al. (2007). Stress-induced synthesis of proline confers tolerance to water deficit in transgenic wheat. *Journal of Plant Physiology, 164*(10), 1367–1376.

Vidal, E. A., et al. (2010). Nitrate-responsive MiR393/AFB3 regulatory module controls root system architecture in Arabidopsis Thaliana. *Proceedings of the National Academy of Sciences of the United States of America, 107*(9), 4477–4482.

Vives-Peris, V., de Ollas, C., Gómez-Cadenas, A., & Pérez-Clemente, R. M. (2020). Root exudates: From plant to rhizosphere and beyond. *Plant Cell Reports, 39*(1), 3–17.

Voothuluru, Priyamvada, Jeffrey C. Anderson, Robert E. Sharp, and Scott C. Peck. 2016. "Plasma membrane proteomics in the maize primary root growth zone: Novel insights into root growth adaptation to water stress." *Plant Cell and Environment* 39(9): 2043–54.

Voothuluru, P., Mäkelä, P., Zhu, J., Yamaguchi, M., Cho, I. J., Oliver, M. J., et al. (2020). Apoplastic hydrogen peroxide in the growth zone of the maize primary root. Increased levels differentially modulate root elongation under well-watered and water-stressed conditions. *Frontiers in Plant Science, 11*, 392.

Walch-Liu, P., Filleur, S., Gan, Y., & Forde, B. G. (2005). Signaling mechanisms integrating root and shoot responses to changes in the nitrogen supply. *Photosynthesis Research, 83*(2), 239–250.

Wang, X. B., Ping, W., Bin, H., & Chen, Q. S. (2002). Effects of nitrate on the growth of lateral root and nitrogen absorption in rice. *Acta Botanica Sinica, 44*(6), 678–683.

Wang, Y., Zhang, F., & Marschner, P. (2012). Soil PH is the main factor influencing growth and rhizosphere properties of wheat following different pre-crops. *Plant and Soil, 360*(1–2), 271–286.

Wang, P., Guo, J., Xu, X., Yan, X., Zhang, K., Qiu, Y., et al. (2020). Soil acidification alters root morphology, increases root biomass but reduces root decomposition in an alpine grassland. *Environmental Pollution, 265*(115), 016.

Wang, J., et al. (2023). Remote sensing of soil degradation: Progress and perspective. *International Soil and Water Conservation Research, 11*(3), 429–454.

White, R. G., & Kirkegaard, J. A. (2010). The distribution and abundance of wheat roots in a dense, structured subsoil – Implications for water uptake. *Plant, Cell and Environment, 33*(2), 133–148.

Wienkoop, S., et al. (2009). Carbon metabolism and Bacteroid functioning are involved in the regulation of nitrogen fixation in Medicago Truncatula under drought and recovery. *Molecular Plant-Microbe Interactions, 22*(12), 1565–1576.

Wu, P., & Wang, X. M. (2008). Role of OsPHR2 on phosphorus homeostasis and root hairs development in rice (Oryza Sativa L.). *Plant Signaling and Behavior, 3*(9), 674–675.

Wu, H. H., Zou, Y. N., Rahman, M. M., Ni, Q. D., & Wu, Q. S. (2017). Mycorrhizas alter sucrose and proline metabolism in trifoliate orange exposed to drought stress. *Scientific Reports, 7*(1), 42389.

Xiong, X., et al. (2021). Keystone species determine the 'selection mechanism' of multispecies biofilms for bacteria from soil aggregates. *Science of the Total Environment, 773*, 1–17.

Xu, C., et al. (2019). Increasing impacts of extreme droughts on vegetation productivity under climate change. *Nature Climate Change, 9*(12), 948–953.

Xue, H., et al. (2022). Ralstonia Solanacearum alters root auxin-dependent and independent manners. *bioRxiv, 75*, 297–306. https://doi.org/10.1101/2022.06.22.497157

Yang, Y., et al. (2014). Arbuscular Mycorrhizal fungi alter fractal dimension characteristics of Robinia Pseudoacacia L. seedlings through regulating plant growth, leaf water status, photosynthesis, and nutrient concentration under drought stress. *Journal of Plant Growth Regulation, 33*(3), 612–625.

Yaxley, J. R., Ross, J. J., Sherriff, L. J., & Reid, J. B. (2001). Gibberellin Biosynthesis mutations and root development in pea. *Plant Physiology, 125*(2), 627–633.

Yu, P., Li, X., White, P. J., & Li, C. (2015). A large and deep root system underlies high nitrogen-use efficiency in maize production. *PLoS ONE, 10*(5), e0126293.

Yuan, X., et al. (2016). Plant community and soil chemistry responses to long-term nitrogen inputs drive changes in alpine bacterial communities. *Ecology, 97*(6), 1543–1554.

Zadworny, M., et al. (2016). Scots pine fine roots adjust along a 2000-Km latitudinal climatic gradient. *The New phytologist, 212*(2), 389–399.

Zhang, H., Rong, H., & Pilbeam, D. (2007). Signalling mechanisms underlying the morphological responses of the root system to nitrogen in Arabidopsis Thaliana. *Journal of Experimental Botany, 58*(9), 2329–2338.

Zhang, B., et al. (2013). New insights into membrane-active action in plasma membrane of fungal hyphae by the Lipopeptide Antibiotic Bacillomycin L. *Biochimica et Biophysica Acta – Biomembranes, 1828*(9), 2230–2237.

Zhang, F., Zou, Y. N., Wu, Q. S., & Kuča, K. (2020). Arbuscular mycorrhizas modulate root polyamine metabolism to enhance drought tolerance of trifoliate orange. *Environmental and Experimental Botany, 171*, 103926.

Zhang, H., Phillip, F. O., Wu, L., Zhao, F., Yu, S., & Yu, K. (2022). Effects of temperature and nitrogen application on carbon and nitrogen accumulation and bacterial community composition in apple rhizosphere soil. *Frontiers in Plant Science, 13*(859), 395.

Zhang, X., et al. (2023). The spatial distribution of rhizosphere microbial activities under drought: Water availability is more important than root-hair-controlled exudation. *New Phytologist, 237*(3), 780–792.

Zhao, M., et al. (2021). Root exudates drive soil-microbe-nutrient feedbacks in response to plant growth. *Plant Cell and Environment, 44*(2), 613–628.

Zhao, W., Wen, M., Zhao, C., Zhang, S., Dou, R., Liang, X., et al. (2023). Warm temperature increments strengthen the crosstalk between roots and soil in the rhizosphere of soybean seedlings. *Plants, 12*(24), 4135.

Zhong, C., et al. (2021). Organic nitrogen sources promote Andrographolide biosynthesis by reducing nitrogen metabolism and increasing carbon accumulation in Andrographis Paniculata. *Plant Physiology and Biochemistry, 164*, 82–91.

Zhou, Z., Yu, M., Ding, G., Gao, G., & He, Y. (2020). Diversity and structural differences of bacterial microbial communities in rhizocompartments of desert leguminous plants. *PLoS One, 15*(12), e0241057.

Mechanistic Insights of Plant-Microbe Interactions in Mitigation of Salinity Stress

12

Drashti Patel, Dimple Pardhi, Tejalba Rathod, Anjali Solanki, Vikram H. Raval, Rakeshkumar R. Panchal, Rushikesh Joshi, and Kiransinh N. Rajput

Abstract

Salinity significantly hampers crop production by disrupting water and nutrient uptake in plants, which leads to reduced yields and diminished soil fertility. Plant growth-promoting bacteria (PGPB) can be used as a long-term strategy for reducing the negative effects of salinity by producing phytohormones, ACC deaminase, osmolytes, ROS scavenging enzymes, exopolysaccharides, and salicylic acid. Phytohormones play a crucial role in this stress response by controlling several aspects of plant growth and adaptability. Abscisic acid activates stress-responsive genes and induces stomatal closure to reduce water loss. Auxins promote root initiation and elongation while cytokinins maintain the balance between root and shoot growth by cell division. ACC deaminase-producing PGPB mitigates inhibitory effects of ethylene under salinity stress, while osmolytes and ROS scavenging maintain cellular homeostasis. Exopolysaccharides protect roots from dehydration, and salicylic acid enhances salt tolerance by

D. Patel
Department of Microbiology and Biotechnology, School of Sciences, Gujarat University, Ahmedabad, Gujarat, India

Department of Biochemistry and Forensic Sciences, School of Sciences, Gujarat University, Ahmedabad, Gujarat, India
e-mail: drashti.patel@gujaratuniversity.ac.in

D. Pardhi · T. Rathod · A. Solanki · V. H. Raval · R. R. Panchal · K. N. Rajput (✉)
Department of Microbiology and Biotechnology, School of Sciences, Gujarat University, Ahmedabad, Gujarat, India
e-mail: rajputkn@gujaratuniversity.ac.in

R. Joshi
Department of Biochemistry and Forensic Sciences, School of Sciences, Gujarat University, Ahmedabad, Gujarat, India

M. Saraf et al. (eds.), *Climate Change and Soil Microorganisms for Environmental Sustainability*, Microorganisms for Sustainability 28,
https://doi.org/10.1007/978-981-96-3425-5_12

improving ion balance, boosting antioxidative responses, and activating the defense pathways. The chapter highlights the potential of PGPB as a promising approach to improve crop resilience in saline conditions by highlighting key biochemical and physiological mechanisms. The necessity of combining PGPB-mediated approaches with traditional farming methods is an effective way of achieving long-term salinity mitigation and ensuring sustainable agriculture.

Keywords

Salinity stress · Plant growth-promoting bacteria (PGPB) · ACC deaminase · Osmolytes · Reactive oxygen species · Sustainable agriculture

12.1 Introduction

India will need approximately 311 million tons of food grains to feed an estimated population of 1.43 billion people by 2030. This demand is projected to rise to 350 million tons by 2050 when the population is expected to reach around 1.8 billion. To ensure food security, efforts must focus on both expanding agricultural land and increasing crop productivity. Currently, about 6.74 million hectares of land in India are salt-affected. Estimates indicate that 10% of additional land becomes salinized each year, and by 2050, around 50% of the arable land could be affected by salinity (Kumar & Sharma, 2020).

The excessive use of chemical fertilizers and accumulation of salt resulting in saline soils appears a significant hazard to both environmental health and agricultural output. Chemical fertilizers can cause the accumulation of sodium, chloride, and other soluble salts by adding salts directly to the soil and disrupting the natural nutritional balance. The salts build up in the root zone over time and cause nutritional imbalances, ion toxicity, and osmotic stress, all of which hinder plant growth (Shahid et al., 2020). Additionally, improper irrigation practices and poor drainage can exacerbate salt accumulation, as water evaporates and leaves behind concentrated salts (Mohamed, 2017).

Salt stress in plants exhibits a variety of symptoms, including stunted growth, delayed germination, and reduced overall development. Leaves may show burn-like symptoms, necrosis at the edges and tips, chlorosis due to nutrient imbalances, and premature drop. Plants may show discoloration, spotting, and leaf curling or rolling (Majeed & Muhammad, 2019). Plants often wilt despite adequate soil moisture because high salt concentrations hinder water uptake, mimicking drought stress. Root systems become less extensive and may exhibit tip necrosis, further impairing water and nutrient absorption (Bartels & Sunkar, 2005). Reproductive processes are also affected, leading to flower drop, reduced fruit set, and poor fruit quality (Pengfei et al., 2017). Physiologically, salt stress causes the accumulation of toxic ions, altered hormone levels, and oxidative stress due to increased reactive oxygen species (Nawaz et al., 2010).

Plant growth-promoting bacteria (PGPB) increase plant growth and resistance through a variety of pathways, offering an environmentally friendly approach to salinity stress in agriculture. Under saline conditions, beneficial microbes produce phytohormones like auxins, cytokinins, and gibberellins that promote root growth and enhance plant vigor. It produces osmoprotectants like proline and glycine betaine, which help in osmoregulation, as well as improve nutrient availability by solubilizing important minerals. PGPB generates ACC deaminase activity to lower ethylene levels. Furthermore, it has the ability to create systemic resistance, which strengthens the salinity tolerance of the plants. Utilizing PGPB improves soil health by increasing microbial diversity and activity while lowering the demand for chemical pesticides and fertilizers, thereby supporting environmental conservation. The naturally occurring increase in crop yield even in salinized environments supports sustainable agricultural production and food security. In order to integrate PGPB into traditional farming practices and advance sustainable agriculture, it is imperative to perform extensive field trials, educate farmers, and produce tailored PGPB inoculants in order to maximize advantages.

12.2 Plant Growth-Promoting Bacteria (PGPB)

Plant growth-promoting bacteria (PGPB) are a diverse group of beneficial microorganisms that significantly enhance plant growth and health through various mechanisms. These bacteria are found in several natural environments, particularly in the soil, where they interact closely with plant roots. The most common habitat for PGPB is the rhizosphere, the narrow region of soil directly influenced by root secretions and associated soil microorganisms. This nutrient-rich environment attracts and supports diverse microbial communities, creating a dynamic interface where beneficial interactions between plants and bacteria can flourish (Kour et al., 2019). Some PGPBs also colonize the internal tissues of plants as endophytes, living symbiotically within plant tissues without causing harm.

PGPB are broadly categorized into two main types based on their habitat and interaction with plants: intracellular PGPB (iPGPB) and extracellular PGPB (ePGPB). Intracellular PGPB, also referred to as endophytic bacteria, inhabit the internal tissues of plants, including roots, stems, and leaves. These bacteria establish a symbiotic or associative relationship with their host plants, providing multiple benefits. It enhances nutrient uptake by facilitating the absorption of essential elements like nitrogen, phosphorus, and potassium. They also produce phytohormones such as indole-3-acetic acid (IAA), gibberellins, and cytokinins, which regulate growth processes, improve development, and promote overall plant health. Moreover, iPGPB can alleviate stress by producing ACC deaminase, which reduces ethylene levels and mitigates stress responses (Saleem et al., 2024). They also play a role in disease resistance by inducing systemic resistance and enhancing the ability of plants to resist pathogens. Examples of iPGPB include *Rhizobium* species, which form nitrogen-fixing nodules in leguminous plants, and *Frankia* species, which perform a similar function in non-leguminous plants (Azeem et al., 2021).

Extracellular PGPB resides outside plant tissues, primarily in the rhizosphere or on the root surface. These bacteria interact with the plant roots and the surrounding soil to promote growth and health through various mechanisms (Mohammadipanah & Zamanzadeh, 2019). They can solubilize nutrients such as phosphate, potassium, and calcium, making them more available to plants. Certain ePGPB, such as *Azospirillum* and *Azotobacter*, are capable of fixing atmospheric nitrogen, enriching the soil with this essential nutrient. It also produces siderophores, which bind and transport iron from the soil to the plant, enhancing iron availability and reducing competition with pathogens (Pérez-Álvarez et al., 2022). Additionally, ePGPB can suppress soil-borne pathogens by producing antibiotics and other antimicrobial compounds, which help protect plants from diseases (Kolomiiets et al., 2024). Key genera of ePGPB include *Bacillus*, *Pseudomonas*, *Azospirillum*, and *Azotobacter* (Abdelaal et al., 2021).

PGPB appears in various forms, each playing a unique role in enhancing plant growth and health through diverse mechanisms. Free-living bacteria, such as species of *Pseudomonas*, *Bacillus*, and *Azospirillum*, reside independently in the soil and rhizosphere. They contribute significantly to plant growth by solubilizing essential nutrients like phosphate, producing growth-stimulating phytohormones, and protecting plants from pathogens through the production of antibiotics and other antimicrobial substances (Saeed et al., 2021). Symbiotic bacteria, such as those in the *Rhizobium* and *Frankia* genera, establish mutually beneficial relationships with specific plants (Martin et al., 2017). For instance, *Rhizobium* species form root nodules on leguminous plants where they fix atmospheric nitrogen into ammonia, a form of nitrogen that plants can readily absorb and utilize (Raza et al., 2020). Similarly, *Frankia* species form nodules on the roots of non-leguminous plants like alder trees, also contributing to nitrogen fixation (Muhammad et al., 2019).

Endophytic bacteria, including species of *Bacillus*, *Enterobacter*, and *Pseudomonas*, live inside plant tissues without causing any harm. These bacteria can colonize roots, stems, and leaves, providing diverse benefits such as improved nutrient uptake, enhanced resistance to diseases, and overall growth promotion (Afzal et al., 2019). Rhizobacteria, such as *Pseudomonas*, *Bacillus*, and *Azotobacter*, specifically colonize the rhizosphere, the soil region directly influenced by root exudates. These bacteria are crucial for nutrient cycling, disease suppression, and enhancing plant growth through multiple mechanisms, including the production of phytohormones and biocontrol agents.

Phyllosphere bacteria, such as *Methylobacterium* and *Pantoea*, colonize the above-ground parts of plants, including leaves, stems, flowers, and fruits, and play a crucial yet often overlooked role in plant health and growth. These diverse microbial communities, adapted to the harsh conditions of the leaf surface, contribute to plant growth by producing phytohormones like indole-3-acetic acid (IAA), fixing atmospheric nitrogen, and solubilizing nutrients. Additionally, phyllosphere bacteria protect plants from foliar pathogens through the production of antimicrobial compounds and by inducing systemic resistance (Dong et al., 2019). Their ability to form biofilms and produce quorum-sensing molecules further enhances their protective functions. Beyond their impact on individual plants, phyllosphere bacteria participate in broader environmental processes, such as the degradation of pollutants and contribution to global carbon and nitrogen cycles (Lukman et al., 2024). Actinobacteria, particularly those in the *Streptomyces* genus, are known for their ability to produce a wide range of antibiotics and other secondary metabolites (Riahi et al., 2019). These bacteria play a pivotal role in suppressing soil-borne pathogens, enhancing nutrient availability, and promoting overall plant health.

12.3 Mechanisms of Plant Growth Promotion

Plant growth-promoting bacteria (PGPB) promote plant development, providing a comprehensive strategy for enhancing plant resilience and health through various direct and indirect mechanisms. Direct mechanisms include nitrogen fixation, where bacteria convert atmospheric nitrogen into ammonia, a form that plants can readily use for growth. These bacteria also solubilize phosphate, making it available to plants by converting insoluble phosphate compounds in the soil into soluble forms. PGPB synthesizes phytohormones like indole-3-acetic acid (IAA), gibberellins, and cytokinins, which stimulate various aspects of plant growth and development. The synthesis of siderophores molecules that bind and transfer iron to plants improves iron uptake and restricts pathogen access to it. Some PGPBs also possess ACC deaminase activity, which breaks down ACC, the precursor of the plant hormone ethylene. By reducing ethylene levels, these bacteria help alleviate stress in plants, promoting better growth under adverse conditions (Ghosh et al., 2018).

PGPB supports plant health including biocontrol of pathogens by indirect mechanism. These bacteria also produce antibiotics, lytic enzymes, and secondary metabolites that inhibit the growth of harmful pathogens. PGPB can also induce systemic resistance in plants, enhancing their immune system and enabling them to better resist attacks by pathogens and pests. Furthermore, some PGPBs produce exopolysaccharides that improve soil structure, enhancing water retention, root penetration, and overall soil health (Premachandra et al., 2016). Figure 12.1 shows direct and indirect mechanisms induced by PGPB on plants.

PGPBs have an effect on the overall health of plants by assisting in enhanced nutrient absorption by host plants, protecting against phytopathogenic bacteria, and promoting tolerance to a range of abiotic stresses (Majeed et al., 2018). Numerous PGPB strains have been shown to increase crop yields, exhibit biocontrol, enhance foliar disease resistance, promote nodulation in legumes, and enhance seedling emergence (Gamalero & Glick, 2019).

12.4 Salinity Stress Induced Pathway By PGPB

Salinity stress in plants triggers an intricate network of physiological and biochemical pathways aimed at mitigating its harmful effects, with key components including phytohormones, ACC deaminase enzymes, osmolytes, ROS scavenging pathways, and exopolysaccharides (EPS) (Table 12.1 and Fig. 12.2).

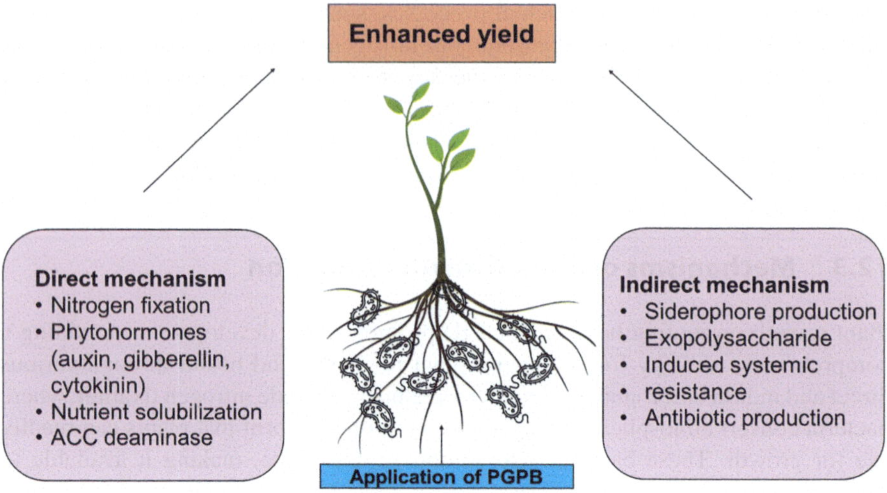

Fig. 12.1 Direct and indirect mechanisms of PGPB

Table 12.1 Compounds produced by PGPB in salt stress

Compounds	Types
Stress alleviating enzymes	ACC deaminase
Osmolytes/ Osmoprotectants	Proline, glycine betaine, soluble sugars, and other amino acids (aspartic acid, alanine, serine, threonine)
ROS scavenging enzymes	Glutathione reductase (GR), catalase (CAT), Ascorbate peroxidase (APX), superoxide dismutase (SOD),
Polysaccharides	Exopolysaccharides (EPS)
Phytohormones	IAA, GA, ethylene, ABA, cytokinin

Phytohormones play a crucial role in regulating plant response. Abscisic acid (ABA) increases in response to salinity stress, which causes stomatal closure to minimize water loss and the activation of stress-responsive genes. Auxins facilitate the growth and development of roots, which improves the capacity of plants to penetrate the soil for nutrients and water. Cytokinins help balance growth between roots and shoots. The negative effects of ethylene, which normally limits development under stress, are mitigated by the ACC deaminase activity of PGPB. Salicylic acid (SA) and jasmonic acid (JA) also contribute by modulating plant defense mechanisms. Table 12.2 shows various types of salt-tolerant PGPBs that have been reported.

12.4.1 Phytohormones

Under salinity stress, the synergistic effects of auxins, gibberellic acid (GA), and cytokinins are crucial for maintaining plant growth and productivity. Auxins

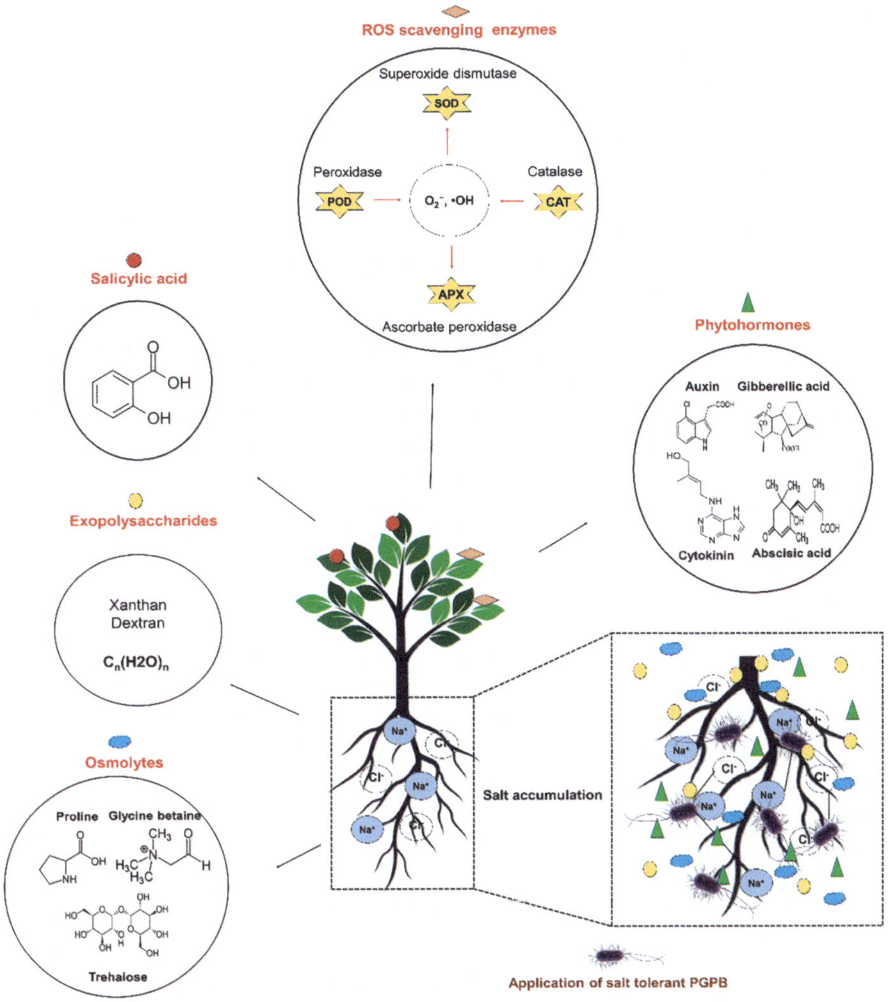

Fig. 12.2 PGPB induces various mechanisms in the presence of salinity stress

stimulate root initiation and elongation, enhancing nutrient and water uptake, while gibberellic acid promotes stem elongation and shoot development, counteracting growth stunting caused by salinity. Cytokinins support root branching and delay leaf senescence, preserving chlorophyll levels and photosynthetic capacity (Kaya et al., 2009). Together, these hormones optimize root and shoot growth, improve stress tolerance, and ensure a balanced hormonal environment, which collectively helps plants adapt to saline conditions and maintain overall health and productivity.

Table 12.2 Salt-tolerant PGPBs

Crop	Salt tolerant PGPB	Citation
Wheat	*Bacillus mojavensis*	Ghazala et al. (2023)
Tomato	*Bacillus aryabhattai H19-1, Bacillus mesonae H20-5*	Yoo et al. (2019)
	Staphylococcus sciuri ET101	Taj et al. (2024)
Rice	*Bacillus siamensis*	Oubaha et al. (2024)
	Bacillus altitudinis NKA32	Bharti et al. (2024)
	Brevibacillus parabrevis BRAM_Y3	Roy et al. (2024)
Barley	*Bacillus pumilus*	Mahmoud et al. (2024)
Pomegranate	*Pseudomonas fluorescens*	Ahmadi et al. (2024)
Arabidopsis	*Bacillus altitudinis NKA32*	Peng et al. (2024)
	Pseudomonas koreensis S4T10	Kalleku et al. (2024)
Turtleweed	*Brevibacterium EB3*	Sierra-García et al. (2024)
Maize	*Microbacterium azadirachtae CNUC13*	Luo et al. (2024)
	Azospirillum brasilense	da Cunha et al. (2024)
Wheat	*Priestia aryabhattai ADJ6* *Bacillus cereus ADJ1*	Dhar et al. (2024)
	Variovorax sp. *P1R9*	Acuña et al. (2024)
Mungbean	*Pseudomonas fluorescens, Enterococcus hirae*	Kumawat et al. (2024)
	Citrobacter sp.	Chakraborty and Mondal (2024)
Chickpea	*Mesorhizobium*	Ben Gaied et al. (2024)
Sunflower	*Azospirillum brasilense, Pseudomonas fluorescens*	Ullah et al. (2024)
Pigeon pea	*Leclercia adecarboxylata*	Saeed et al. (2024)
Canola	*Rahnella victoriana B38, Bacillus paramycoides N38*	Saeed et al. (2024)
Carrot	*Pseudescherichia liriopis*	Kim et al. (2024)
Yellow pea	*Ensifer alkalisoli YIC4027*	Dong et al. (2024)
Onion	*Sphingobacterium multivorum, Rhizobium tropici*	Khumairah et al. (2024)

12.4.1.1 Auxin

Auxins are a class of phytohormones that play a critical role in plant growth and development, particularly in response to environmental stresses such as salinity. The interaction between plants and microbes significantly influences auxin dynamics, enhancing plant resilience to salinity stress (Etesami & Glick, 2024). Auxins, particularly indole-3-acetic acid (IAA), are central to various plant growth processes, including cell elongation, root initiation, and branching. Under normal conditions, auxins are synthesized in the shoot apical meristem and transported to other parts of the plant, where they regulate growth and developmental processes (Teale et al., 2006). Salinity stress disrupts plant hormone balance, including auxin distribution and signaling, leading to adverse effects on plant growth such as reduced root length and shoot biomass. High salinity can inhibit root growth, making it difficult for plants to access water and nutrients, exacerbating stress conditions (Fahad et al., 2015).

Certain PGPB strains can produce IAA or influence the plant's own auxin production. Microbial production of IAA is often stimulated by the presence of root exudates (Tsavkelova et al., 2007). By producing IAA, PGPB can directly enhance root growth and development, improving the ability of plants to explore the soil for water and nutrients under saline conditions. It can influence the auxin transport mechanisms of plants, promoting a more effective distribution of auxin within the plant (Gao et al., 2022). This can lead to enhanced root architecture, including increased lateral root formation and root hair development. Improved root systems enable plants to better cope with salinity stress by accessing deeper soil layers and more nutrients. Salinity stress can lead to the accumulation of stress hormones like ethylene, which inhibit auxin signaling pathways. PGPB with ACC deaminase activity can lower ethylene levels, thereby reducing its inhibitory effect on auxin signaling. PGPB helps maintain auxin signaling pathways, ensuring continuous root growth and development even under stress conditions by modulating ethylene levels (Glick, 2014). The combined effects of microbial auxin production improved auxin transport and reduced ethylene inhibition resulting in enhanced root growth. It not only improves water and nutrient uptake but also increases the overall stress tolerance of the plant. A robust root system is crucial for maintaining plant health and productivity under saline conditions.

12.4.1.2 Abscisic Acid

Abscisic acid (ABA) producing plant growth-promoting bacteria (PGPB) plays a crucial role in enhancing plant resilience to salinity stress through several integrated mechanisms. ABA produced by these bacteria helps in regulating stomatal closure, which reduces water loss through transpiration and improves water use efficiency, essential for maintaining hydration under saline conditions (Gupta et al., 2022). This hormone also aids in osmotic adjustment by inducing the accumulation of osmoprotectants such as proline, glycine betaine, and soluble sugars, which protect cells from osmotic stress caused by high salinity (Sharma et al., 2019). Additionally, ABA regulates ion transport by influencing the expression and activity of ion transporters, maintaining ion homeostasis, and ensuring the balance of sodium (Na^+) and potassium (K^+) ions within plant cells, thus enhancing cellular stability (Osakabe et al., 2014). Moreover, ABA acts as a signaling molecule that activates stress-responsive genes involved in antioxidant defense, osmo-protection, and overall stress signaling pathways. This activation enhances the intrinsic defense mechanisms of plants, enabling them to cope more effectively with salinity stress (Chen et al., 2020). ABA also boosts antioxidant defenses by upregulating the production of crucial antioxidant enzymes as well as increasing the synthesis of antioxidant compounds like ascorbate and glutathione (Hasanuzzaman et al., 2013). These enhanced antioxidant defenses play a significant role in mitigating oxidative stress and protecting cellular components from reactive oxygen species (ROS)-induced damage.

ABA promotes root growth and development, leading to a more extensive and robust root system. This improved root architecture enhances the ability of plants to absorb water and nutrients from the soil, which is particularly beneficial under saline conditions where resource availability is often limited. Additionally, ABA influences the production of root exudates that attract beneficial microbes, further supporting plant health and stress resilience (Liu et al., 2021). The combined effects of these mechanisms significantly improve plant health, growth, and productivity under saline conditions, underscoring the vital role of ABA-producing PGPB in managing and mitigating the impacts of salinity stress on plants.

12.4.1.3 Gibberellic Acid

Gibberellic acid (GA) is a crucial phytohormone that regulates a wide range of plant growth processes, including seed germination, stem elongation, flowering, and fruit development (Gupta & Chakrabarty, 2013). Under salinity stress, the production and action of GA can be disrupted, leading to negative impacts on plant growth and productivity. PGPB can produce GA which provides significant benefits to plants facing salinity stress. GA promotes the breakdown of seed dormancy and activates the enzymes needed for seed germination. It induces the synthesis of hydrolytic enzymes such as amylase, which mobilize stored nutrients for the growing seedling (Tuan et al., 2019). It also stimulates cell elongation and division, leading to increased stem height and overall plant growth. It regulates the timing of flowering and affects fruit set and development, influencing crop yields and quality (Castro-Camba et al., 2022).

High salt concentrations can inhibit the enzymes involved in GA biosynthesis, leading to lower GA levels. This reduction affects growth processes like seed germination and stem elongation. Salinity stress can interfere with GA signaling pathways, including the regulation of GA receptors and transcription factors. This interference can impair the ability of plants to respond to GA, resulting in stunted growth and reduced vigor. PGPB strains have the capability to produce GA or stimulate its production in plants. These microbes synthesize GA from precursors available in root exudates or through direct interaction with plant tissues (Sindhu et al., 2017). PGPB directly promotes seed germination, root growth, and shoot elongation, helping plants better cope with the constraints imposed by salinity by increasing GA levels. Increased GA production helps counteract the growth-limiting effects of salinity, supporting normal growth processes and improving overall plant health.

12.4.1.4 Cytokinin

Cytokinins are vital plant hormones that regulate various growth and developmental processes, including cell division, leaf senescence, and root-shoot balance. Under salinity stress, the balance and effectiveness of cytokinins can be disrupted, affecting plant health and productivity (Bielach et al., 2017). Cytokinins promote cell division in the root and shoot meristems, driving growth and the development of new tissues. They are crucial for the formation of lateral roots and shoots, enhancing plant structural complexity and resource acquisition. It delays the aging of leaves by inhibiting chlorophyll degradation and protein turnover. It can extend the photosynthetic activity of leaves, contributing to improved nutrient and energy availability for the plant. It also helps balance growth between roots and shoots. They influence root branching and elongation, ensuring that plants maintain a functional root system while promoting shoot growth (Kieber & Schaller, 2014).

High salt concentrations can interfere with cytokinin biosynthesis, reducing the levels of this hormone in plant tissues. This reduction impairs cell division and elongation, leading to stunted growth and reduced plant vigor. It can also affect the transport of cytokinins within the plant, limiting their distribution to key growth sites. This impaired transport can disrupt normal growth patterns and affect overall plant development. Salinity stress can interfere with the signaling pathways of cytokinins, affecting their ability to regulate gene expression and physiological processes. Disruption in cytokinin signaling can accelerate leaf senescence, reduce root and shoot growth, and impair stress response mechanisms (Yu et al., 2021).

Certain PGPB strains produce cytokinins or cytokinin-like substances, such as zeatin, which can be absorbed by plant roots (Cassán et al., 2014). This microbial cytokinin production can directly enhance cell division and growth. PGPB may also stimulate the cytokinin production of plants by altering root exudate profiles or by secreting compounds that influence plant hormone biosynthesis (Khan et al., 2016). PGPB can affect the biosynthesis pathways of cytokinins in plants, either by directly providing cytokinin precursors or by influencing the internal hormone synthesis mechanisms of plants. PGPB helps alleviate the growth-inhibiting effects of salinity stress, supporting normal development and improving overall plant health by boosting cytokinin levels (Sevim et al., 2023). It can enhance the expression of genes involved in cytokinin perception and signaling. This can include increasing the activity of cytokinin receptors or enhancing downstream signal transduction pathways (Heyl & Schmülling, 2003). Improved cytokinin signaling helps maintain proper root and shoot growth, delays leaf senescence, and optimizes stress response mechanisms. Extended leaf lifespan and improved growth contribute to enhanced photosynthetic capacity and overall plant productivity, even under saline conditions (Gu et al., 2018).

Understanding and leveraging these interactions can be key to improving crop resilience and sustainability in saline conditions.

12.4.2 ACC Deaminase

ACC deaminase (1-aminocyclopropane-1-carboxylate deaminase) is an enzyme produced by specific plant growth-promoting rhizobacteria that plays a pivotal role in mitigating the negative effects of salinity stress on plants. Its primary function is to break down 1-aminocyclopropane-1-carboxylate (ACC), a precursor in the biosynthesis of ethylene, a plant hormone that can exacerbate stress responses when present in elevated levels (Gamalero et al., 2023). In plants, ACC is converted into ethylene, a hormone that regulates various physiological processes, including stress responses.

Under stress conditions such as high salinity, ethylene levels often increase, leading to adverse effects such as reduced root elongation, accelerated leaf senescence, and stunted overall growth (Ilangumaran & Smith, 2017). ACC deaminase catalyzes the deamination of ACC, converting it into α-ketobutyrate and ammonia. This reaction effectively reduces the amount of ACC available for ethylene production, thereby lowering ethylene levels in the plant (Singh et al., 2015). The ethylene signaling pathway is shown in Fig. 12.3. High ethylene levels inhibit root growth and elongation, limiting the ability of plants to access water and nutrients. By reducing ethylene levels, ACC deaminase alleviates this inhibition, promoting better root growth and enhancing the plant's ability to forage for water and nutrients even under saline conditions. Ethylene also affects shoot growth and leaf longevity. Lower ethylene levels, due to ACC deaminase activity, help in maintaining shoot elongation and delaying leaf senescence, which contributes to improved overall plant vigor and productivity (Singh et al., 2022).

Plants exhibit better growth performance, characterized by increased biomass, healthier foliage, and improved physiological functions with reduced ethylene levels. This improved strategy is crucial in saline environments where high salt levels can otherwise lead to significant growth reductions. Ethylene is associated with increased production of reactive oxygen species (ROS), which cause oxidative damage to plant cells. By lowering ethylene levels, ACC deaminase helps reduce oxidative stress and protects the plant from ROS-induced damage (Chandwani & Amaresan, 2022). Specific PGPB strains possess ACC deaminase activity and can be applied to plant roots. These beneficial microbes produce ACC deaminase in the rhizosphere, where they break down ACC and reduce ethylene levels. The interaction between PGPB with ACC deaminase and plants can create a more favorable growing environment (Naing et al., 2021). The microbes help the plant to overcome the detrimental effects of salinity, leading to enhanced plant health and productivity.

Utilizing PGPB strains with ACC deaminase activity offers a sustainable approach to managing salinity stress in crops. These microbial treatments can be integrated into agricultural practices to improve crop resilience, reduce the need for chemical inputs, and enhance overall soil health. By enhancing stress tolerance and growth performance, ACC deaminase-producing PGPB can help maintain and increase crop yields in saline soils, contributing to more resilient and productive agricultural systems.

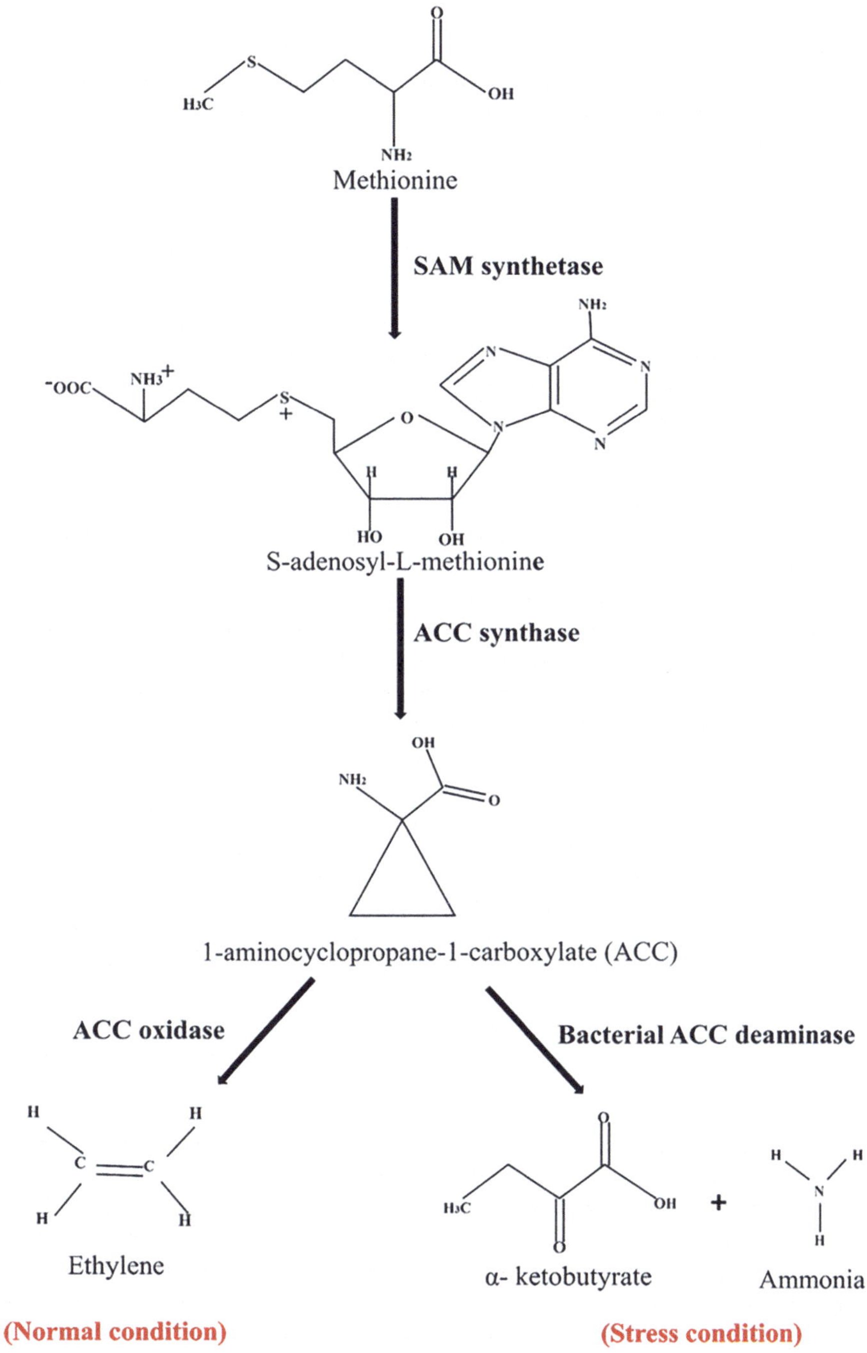

Fig. 12.3 ACC deaminase role in ethylene signaling

12.4.3 Osmolytes

Osmolytes are small, low-molecular-weight molecules that help plants cope with salinity stress by stabilizing proteins, and cellular structures, and maintaining cellular osmotic balance (Khan et al., 2023). Under saline conditions, osmolytes play a crucial role in mitigating the detrimental effects of high salt concentrations. Osmolytes help regulate the osmotic balance within plant cells by accumulating in response to high salt concentrations. This accumulation counteracts the osmotic pressure exerted by external salts, maintaining cellular turgor pressure and preventing dehydration. It also helps cells retain water, reducing the risk of dehydration and maintaining cell function even under saline conditions by maintaining osmotic balance (Nahar et al., 2016).

Osmolytes stabilize protein structures by preferentially excluding denaturing agents like high salt concentrations. They help maintain protein function and prevent denaturation, ensuring that crucial enzymatic and structural proteins remain active (Rabbani & Choi, 2018). Osmolytes support the activity of key enzymes involved in stress responses and metabolic processes, contributing to overall plant resilience by stabilizing proteins. It protects cellular membranes and organelles from damage caused by osmotic stress. They help maintain membrane integrity and function, which is crucial for cell survival under high salt conditions (Choudhary et al., 2023). Some osmolytes, such as certain amino acids and sugars, have antioxidant properties that help mitigate oxidative stress induced by high salinity which scavenge reactive oxygen species (ROS) and reduce oxidative damage to cellular components (Ejaz et al., 2020). Plants regulate the synthesis of osmolytes through gene expression changes. Stress-responsive genes are activated to produce these compounds in response to high salt concentrations. The production and accumulation of osmolytes involve adjustments in metabolic pathways, ensuring that plants can produce and manage these molecules effectively under stress conditions (Jogawat, 2019). Compatible solutes are critical for plant adaptation to high salinity environments, where they help manage osmotic stress, protect cellular components, and maintain overall plant health.

12.4.3.1 Proline

Proline accumulates in plant cells in response to salinity and other stress conditions. Its role in osmoregulation involves increasing the internal osmotic pressure to counterbalance the external osmotic stress caused by high salt concentrations (Heuer, 2010). This adjustment helps cells maintain turgor pressure, which is essential for normal cell function and structure. It stabilizes proteins by forming hydrogen bonds with their structures. This action prevents the unfolding and aggregation of proteins that can occur under stressful conditions. By stabilizing protein structures, proline ensures that enzymes and other proteins maintain their functional conformations, which is crucial for metabolic processes (Wijayasinghe et al., 2017). Proline also protects cellular membranes from damage caused by osmotic stress. It helps maintain membrane fluidity and integrity, which is important for proper cell function. By stabilizing the lipid bilayer of membranes, it prevents leakage and maintains the function of membrane-bound proteins and receptors (Pathak et al., 2020).

12.4.3.2 Glycine Betaine

Glycine betaine is a potent osmoprotectant that stabilizes cells under high salinity by balancing osmotic pressure. It accumulates in the cytoplasm and prevents the loss of water from cells, thereby maintaining turgor pressure and cellular function. It protects proteins from denaturation and aggregation by stabilizing their three-dimensional structures (Hashem & Mohamed, 2020). This stabilization is achieved through its interaction with protein molecules, which prevents changes in protein conformation that can lead to loss of function. It helps protect cellular membranes and organelles from damage. It stabilizes membrane structures and prevents the disruption of organelles like mitochondria and chloroplasts, which are crucial for cellular energy production and photosynthesis, respectively. Glycine betaine enhances overall stress tolerance by maintaining cellular homeostasis (Chen & Murata, 2011). It supports various stress-related physiological processes, including stress-induced gene expression and repair mechanisms.

12.4.3.3 Trehalose

Trehalose acts as a protective agent for cellular proteins and membranes. It helps stabilize protein structures by forming a protective matrix around proteins, which reduces the likelihood of denaturation. For membranes, trehalose maintains fluidity and prevents damage caused by osmotic stress. It contributes to osmotic balance by modulating the internal osmotic pressure of cells. It helps to retain water and maintain turgor pressure, which is critical for cell survival under high salinity conditions (Jain & Roy, 2009). Trehalose has antioxidant properties that help reduce oxidative stress by scavenging reactive oxygen species (ROS). This reduction in oxidative stress helps prevent damage to cellular components and supports overall plant health (Sadak et al., 2019).

12.4.3.4 Sucrose

Sucrose plays a key role in balancing internal osmotic pressure. It accumulates in response to high salt concentrations and helps to adjust the osmotic potential within the plant cells, thus maintaining water balance and turgor pressure. It serves as an energy source for various metabolic processes that are crucial during stress. It also protects cellular components, including proteins and membranes, from damage due to its ability to stabilize these structures under osmotic stress (Poonam et al., 2016).

12.4.4 ROS Scavenging Pathway

Reactive oxygen species (ROS), such as superoxide anions (O_2^-), hydrogen peroxide (H_2O_2), and hydroxyl radicals (•OH), are highly reactive molecules that accumulate in plants under salinity stress. Elevated ROS levels lead to oxidative stress, causing damage to proteins, lipids, nucleic acids, and cellular membranes. PGPB can alleviate oxidative stress through a variety of mechanisms that enhance the plant's ability to scavenge ROS and maintain cellular integrity (Sagar et al., 2023).

12.4.4.1 ROS Scavenging Enzymes

PGPB enhances plant resilience to salinity stress by improving the ability of plants to scavenge reactive oxygen species (ROS) through various mechanisms (Mishra et al., 2021). These include increasing the production of antioxidative enzymes like superoxide dismutase (SOD), catalase (CAT), and peroxidase (POD), boosting antioxidant defense systems such as the ascorbate-glutathione cycle, and promoting the synthesis of antioxidative compounds like flavonoids, phenolics, and siderophores (Khanna et al., 2018). Additionally, PGPB can modulate gene expression to enhance stress responses, improve root architecture for better water and nutrient uptake, and produce volatile organic compounds (VOCs) that further support plant stress tolerance. PGPB helps mitigate oxidative damage, support plant health, and improve productivity under saline conditions by enhancing the ROS scavenging pathway.

High salt concentrations lead to an increase in ROS production due to disruptions in cellular processes, including electron transport chains and metabolic pathways. This oxidative burst can overwhelm the natural antioxidant defense of plants (Møller, 2001). ROS can cause lipid peroxidation, protein denaturation, and DNA damage. Lipid peroxidation leads to membrane disruption and loss of cellular compartmentalization, protein denaturation affects enzyme function and structural proteins, and DNA damage impairs genetic integrity and cellular function (Blokhina et al., 2003).

PGPB can stimulate the production of SOD, an enzyme that catalyzes the dismutation of superoxide radicals (O_2^-) into hydrogen peroxide (H_2O_2) and oxygen. This reaction reduces the harmful effects of superoxide radicals and makes hydrogen peroxide, a less reactive molecule, easier to manage (Sharma et al., 2022). PGPB can enhance the synthesis of CAT, an enzyme that decomposes hydrogen peroxide into water and oxygen. This step is crucial in reducing hydrogen peroxide levels and preventing its accumulation, which could otherwise lead to further oxidative damage (Vocciante et al., 2022). Peroxidases (POD) are involved in breaking down hydrogen peroxide and other peroxides (Teng et al., 2016). PGPB can boost POD activity, which helps in detoxifying peroxides and reducing oxidative stress. This enzyme also plays a role in the biosynthesis of lignin and other secondary metabolites that contribute to plant defense (Pandey et al., 2017).

PGPB can stimulate the activity of key enzymes in the ascorbate-glutathione cycle, such as ascorbate peroxidase (APX) and glutathione reductase (GR) (Nivetha et al., 2021). APX catalyzes the reduction of H_2O_2 using ascorbate, while GR regenerates reduced ascorbate from its oxidized form. PGPB helps maintain low ROS levels and protect cells from oxidative stress, supporting overall plant health and growth by promoting the ascorbate-glutathione cycle. Ascorbate acts as a crucial antioxidant by directly neutralizing hydrogen peroxide (H_2O_2) in water. This reduces the potential for oxidative damage caused by H_2O_2, which is a more stable and less reactive ROS compared to others (Orabi & Abou-Hussein, 2019). Glutathione, a tripeptide composed of glutamate, cysteine, and glycine, plays a dual role as both a direct antioxidant and a regenerating agent. It helps in converting oxidized ascorbate back to its reduced form, thus maintaining a pool of active ascorbate. Additionally, glutathione itself detoxifies ROS through conjugation with reactive molecules and acts as a substrate for various antioxidant enzymes (Narayanankutty et al., 2019).

12.4.4.2 Flavonoids

Flavonoids are a diverse group of polyphenolic compounds known for their strong antioxidant properties. They neutralize ROS, stabilize cell membranes, and protect cellular components from oxidative damage (Gupta et al., 2016). Flavonoids also have anti-inflammatory and stress-mitigating effects. Phenolics, including phenolic acids and tannins, are potent antioxidants that scavenge ROS and chelate metal ions that catalyze ROS formation. It also contributes to the stabilization of cellular structures and modulation of stress responses (Šamec et al., 2021). PGPB can stimulate the biosynthesis of flavonoids and phenolic compounds by influencing plant metabolic pathways which increases in secondary metabolites and helps in neutralizing ROS and enhancing stress tolerance. Elevated levels of flavonoids and phenolics due to PGPB activity contribute to reduced oxidative stress, improved cellular stability, and enhanced plant resilience to salinity stress (Sharma et al., 2023).

12.4.4.3 Siderophores

Siderophores are specialized organic molecules produced by certain PGPBs. These molecules play a crucial role in managing oxidative stress in plants, particularly under challenging conditions such as salinity stress. Siderophores are synthesized and secreted by PGPB to bind iron (Fe^{+3}) in the soil or rhizosphere. These molecules have a high affinity for iron, which they sequester from the surrounding environment (Scavino & Pedraza, 2013). The binding of iron by siderophores forms a stable siderophore-iron complex, which is then transported into the bacterial cells or made available to the plant. Iron is a crucial element in plant and microbial metabolism, acting as a cofactor for many enzymes (Sah & Singh, 2015). However, iron also plays a role in generating ROS through reactions such as the Fenton reaction, where iron catalyzes the conversion of hydrogen peroxide (H_2O_2) into highly reactive hydroxyl radicals (•OH) (Remucal & Sedlak, 2011). The Fenton reaction is a process where iron catalyzes the formation of hydroxyl radicals from hydrogen peroxide. These hydroxyl radicals are highly reactive and can cause severe oxidative damage to cellular components. By chelating and reducing the availability of free iron, siderophores inhibit the Fenton reaction and other iron-dependent ROS-generating processes. This reduction in free iron limits the production of harmful ROS, thus helping to prevent oxidative damage (Chen et al., 2014). By mitigating ROS formation, siderophores help in alleviating oxidative stress in plants which contributes to reducing damage to cellular structures, proteins, and nucleic acids, which in turn supports plant health and growth under saline conditions (Desoky et al., 2020). The reduction of ROS through siderophore activity can trigger systemic defense responses in plants. Plants may enhance their own antioxidant systems in response to lower oxidative stress, leading to a more robust defense mechanism against salinity and other stress factors (Ali et al., 2022). The presence of PGPB-producing siderophores often creates a beneficial microbial community in the rhizosphere. This community can work synergistically to support plant health, further enhancing stress tolerance and reducing oxidative damage.

12.4.4.4 Volatile Organic Compounds (VOCs)

Volatile organic compounds (VOCs) released by PGPB play a crucial role in enhancing plant resilience to salinity stress by inducing systemic resistance and boosting antioxidant defenses (Chatterjee & Niinemets, 2022). VOCs act as signaling molecules that trigger the production of key antioxidant enzymes, such as superoxide dismutase (SOD), catalase (CAT), and ascorbate peroxidase (APX), as well as antioxidants like ascorbate and glutathione (Rafedzi et al., 2023). This activation improves the capacity of plants to scavenge reactive oxygen species (ROS) and reduces oxidative damage, thereby enhancing stress tolerance and overall plant health. Additionally, VOCs contribute to beneficial plant-microbe interactions, influencing microbial community dynamics in the rhizosphere and facilitating better stress adaptation (Velmourougane et al., 2017). This comprehensive effect of VOCs supports improved growth and productivity under saline conditions, highlighting their significant role in managing oxidative stress and enhancing plant resilience.

12.4.5 Exopolysaccharides (EPS)

EPS produced by PGPB are high-molecular-weight carbohydrates secreted into the extracellular environment. These compounds play multifaceted roles in helping plants cope with salinity stress through several mechanisms (Paul et al., 2024). PGPB secrete EPS into the rhizosphere, where it accumulates around plant roots which helps to maintain osmotic balance within plant cells (Saberi Riseh et al., 2021). EPS molecules have a high water-binding capacity, which helps in retaining water in the root zone. This assists in maintaining cell turgor and prevents dehydration under saline conditions, where osmotic stress can cause water loss from plant cells (Morcillo & Manzanera, 2021). EPS ensures that plant cells retain water even when soil salinity increases by stabilizing cellular hydration levels. This is crucial for sustaining metabolic activities and preventing cellular damage.

EPS forms a mucilaginous layer around plant roots, acting as a physical barrier that limits the direct exposure of roots to high salt concentrations. EPS can bind to sodium ions (Na^+), reducing their availability to plant roots and thereby decreasing salt uptake. The protective barrier minimizes the detrimental effects of salt ions on plant root cells, reducing ion toxicity and maintaining better root function (Liu et al., 2023). The gel-like nature of EPS enhances the ability of soil to retain water, particularly around the root zone. This increased water retention helps maintain a consistent moisture level in the rhizosphere, crucial for plant survival under saline conditions. Plants benefit from a more stable supply of water, which is essential for their growth and stress tolerance, especially when external water availability is limited due to high soil salinity (Jha et al., 2022). EPS binds soil particles together, promoting the formation of soil aggregates. This improves soil structure, increases porosity, and enhances aeration. Better soil structure facilitates root penetration and expansion, allowing for more efficient water and nutrient uptake (Hartmann & Six, 2023). A well-structured soil environment supports a robust root system, which is essential for plant stability and nutrient absorption under stress conditions.

EPS creates a favorable microenvironment for beneficial microbes in the rhizosphere, fostering a diverse and active microbial community. The presence of beneficial microbes can enhance nutrient availability and promote plant growth by fixing nitrogen, solubilizing phosphorus, and producing growth-promoting substances. Enhanced microbial activity and diversity contribute to overall plant health and resilience, as these microbes can assist in nutrient uptake and provide additional protection against pathogens and environmental stresses (Naik et al., 2019). EPS helps reduce the osmotic stress-induced production of ROS by maintaining hydration and stabilizing cellular structures. EPS can also indirectly promote the production of antioxidant compounds by the plant, further mitigating oxidative damage (Liu et al., 2022). Reduced ROS levels help protect cellular components from oxidative damage, improving overall plant health and stress tolerance. These multifaceted roles contribute to better hydration stability, reduced salt toxicity, enhanced root growth, and overall improved plant health and productivity under saline conditions. EPS thus plays a critical role in enabling plants to thrive in challenging environments caused by high salinity.

12.4.6 Salicylic Acid (SA)

Salicylic acid (SA) is a key plant hormone that plays a fundamental role in the plant response to salinity stress, which is caused by high concentrations of soluble salts in the soil. SA helps plants cope with the stress of saline stress through a number of processes. Firstly, SA aids in osmotic adjustment by promoting the synthesis of osmoprotectants like proline, which helps maintain cellular turgor and stability under osmotic stress. This allows plants to better handle the osmotic imbalance caused by excess salts (Das et al., 2024). Additionally, SA regulates ion homeostasis by influencing the expression of ion transporters and channels, which helps to reduce the accumulation of toxic sodium ions (Na^+) and enhance the uptake of essential potassium ions (K^+) (Jayakannan et al., 2015).

SA also plays a vital role in mitigating oxidative stress, which is exacerbated by high salinity. It enhances the antioxidant defense system by upregulating the production of key antioxidant enzymes such as superoxide dismutase (SOD), catalase (CAT), and peroxidase (POD). These enzymes scavenge reactive oxygen species (ROS) generated under stress, thereby reducing oxidative damage and helping the plant to survive and recover from the adverse effects of salinity (Herrera-Vásquez et al., 2015).

SA influences the expression of stress-responsive genes that encode proteins and enzymes involved in the plant defense mechanisms. This regulation helps plants activate a range of responses tailored to cope with salinity stress (Singh & Gautam,

2013). SA also interacts with other plant hormones, including abscisic acid (ABA) and jasmonic acid (JA), to coordinate a comprehensive stress response, ensuring that various signaling pathways are integrated effectively to optimize defense strategies (Yang et al., 2019).

PGPB can significantly enhance the salinity stress response of plants by modulating SA pathways. Certain PGPBs, such as those from the genera *Pseudomonas* and *Bacillus*, can stimulate higher production of SA in plants (Olanrewaju et al., 2017). This increased SA level amplifies the ability of plants to manage osmotic stress, maintain ion balance, and boost antioxidant defenses. By promoting these SA-mediated processes, PGPB helps plants to better withstand the adverse effects of high salinity, leading to improved resilience and recovery. This interaction between PGPB and SA underscores the potential for using beneficial bacteria in sustainable agriculture to mitigate soil salinity impacts and enhance crop production.

In addition to its role in SAR, SA influences various growth and development processes. High levels of SA can modulate plant growth by affecting cell division and elongation, which can be a trade-off between defense and growth. For example, while SA enhances defense, it can also inhibit growth under certain conditions, thus requiring a balance between optimizing defense responses and maintaining overall plant vigor (Züst & Agrawal, 2017).

12.5 Conclusion

Plant-microbe interactions are fundamental to sustainable agriculture and environmental conservation. These relationships (symbiotic, commensal, or pathogenic) enhance plant health, improve soil quality, and help mitigate environmental stresses. Beneficial microbes, such as plant growth-promoting bacteria (PGPB), boost plant resilience to salinity stress through mechanisms like phytohormone production, osmolyte accumulation, ROS scavenging, salicylic acid synthesis, and ACC deaminase activity. These processes support plant growth, protect against stress, and ensure continued agricultural productivity.

Integrating these natural interactions with sustainable practices and leveraging biotechnology can enhance agricultural resilience. With research, policy support, and public awareness, we can harness plant-microbe partnerships to create a more sustainable and resilient agricultural system, crucial for both environmental health and human well-being.

References

Abdelaal, K., AlKahtani, M., Attia, K., Hafez, Y., Király, L., & Künstler, A. (2021). The role of plant growth-promoting bacteria in alleviating the adverse effects of drought on plants. *O Biologico, 10*(6), 520.

Acuña, J. J., Rilling, J. I., Inostroza, N. G., Zhang, Q., Wick, L. Y., Sessitsch, A., & Jorquera, M. A. (2024). *Variovorax sp. strain* P1R9 applied individually or as part of bacterial consortia enhances wheat germination under salt stress conditions. *Scientific Reports, 14*(1), 2070.

Afzal, I., Shinwari, Z. K., Sikandar, S., & Shahzad, S. (2019). Plant beneficial endophytic bacteria: Mechanisms, diversity, host range and genetic determinants. *Microbiological Research, 221*, 36–49.

Ahmadi, S. R., Seifi, E., Varasteh, F., & Akbarpour, V. (2024). Mitigating salinity stress in pomegranate: Effects of *Pseudomonas fluorescens* and *Glomus mosseae* on stress responses of red angel and wonderful cultivars. *Scientia Horticulturae, 330*, 113036.

Ali, B., Wang, X., Saleem, M. H., Sumaira, H. A., Afridi, M. S., Khan, S., Ullah, I., Amaral Júnior, A. T., Alatawi, A., & Ali, S. (2022). PGPR-mediated salt tolerance in maize by modulating plant physiology, antioxidant defense, compatible solutes accumulation and bio-surfactant producing genes. *Plants, 11*(3), 345.

Azeem, F., Rashid, F., Shahzadi, M., Abbas, A., Batool, R., Nadeem, H., Moosa, A., Siddique, M. H., Hussain, S., & Ali, M. A. (2021). Endophytes as guardians of plants against diseases. In *Recent developments in microbial technologies* (pp. 221–242). Springer.

Bartels, D., & Sunkar, R. (2005). Drought and salt tolerance in plants. *Critical Reviews in Plant Sciences, 24*(1), 23–58.

Ben Gaied, R., Sbissi, I., Tarhouni, M., & Brígido, C. (2024). Bacterial endophytes from legumes native to arid environments are promising tools to improve Mesorhizobium–chickpea symbiosis under salinity. *O Biologico, 13*(2), 96.

Bharti, C., Fatima, T., Mishra, P., Verma, P., Bhattacharya, A., Alaylar, B., & Arora, N. K. (2024). Salt-tolerant endophytic *Bacillus altitudinis* NKA32 with ACC deaminase activity modulates physiochemical mechanisms in rice for adaptation in saline ecosystem. *Environmental Sustainability, 14*, 1–9.

Bielach, A., Hrtyan, M., & Tognetti, V. B. (2017). Plants under stress: Involvement of auxin and cytokinin. *International Journal of Molecular Sciences, 18*(7), 1427.

Blokhina, O., Virolainen, E., & Fagerstedt, K. V. (2003). Antioxidants, oxidative damage and oxygen deprivation stress: A review. *Annals of Botany, 91*(2), 179–194.

Cassán, F., Vanderleyden, J., & Spaepen, S. (2014). Physiological and agronomical aspects of phytohormone production by model plant-growth-promoting rhizobacteria (PGPR) belonging to the genus *Azospirillum*. *Journal of Plant Growth Regulation, 33*, 440–459.

Castro-Camba, R., Sánchez, C., Vidal, N., & Vielba, J. M. (2022). Plant development and crop yield: The role of gibberellins. *Plants, 11*(19), 2650.

Chakraborty, S., & Mondal, S. (2024). Halotolerant *Citrobacter sp.* remediates salinity stress and promotes the growth of *Vigna radiata (L)* by secreting extracellular polymeric substances (EPS) and biofilm formation: A novel active cell for microbial desalination cell (MDC). *International Microbiology, 27*(1), 291–301.

Chandwani, S., & Amaresan, N. (2022). Role of ACC deaminase producing bacteria for abiotic stress management and sustainable agriculture production. *Environmental Science and Pollution Research, 29*(16), 22843–22859.

Chatterjee, P., & Niinemets, Ü. (2022). Improving plant stress resistance by growth-promoting bacteria and evaluating the improvements by volatile emissions. *Plant and Soil, 476*(1), 403–419.

Chen, T. H., & Murata, N. (2011). Glycinebetaine protects plants against abiotic stress: Mechanisms and biotechnological applications. *Plant, Cell & Environment, 34*(1), 1–20.

Chen, L. H., Yang, S. L., & Chung, K. R. (2014). Resistance to oxidative stress via regulating siderophore-mediated iron acquisition by the citrus fungal pathogen Alternaria alternata. *Microbiology, 160*(5), 970–979.

Chen, H., Lai, L., Li, L., Liu, L., Jakada, B. H., Huang, Y., He, Q., Chai, M., Niu, X., & Qin, Y. (2020). AcoMYB4, an *Ananas comosus L.* MYB transcription factor, functions in osmotic stress through negative regulation of ABA signaling. *International Journal of Molecular Sciences, 21*(16), 5727.

Choudhary, S., Wani, K. I., Naeem, M., Khan, M. M., & Aftab, T. (2023). Cellular responses, osmotic adjustments, and role of osmolytes in providing salt stress resilience in higher plants: Polyamines and nitric oxide crosstalk. *Journal of Plant Growth Regulation, 42*(2), 539–553.

da Cunha, E. T., Pedrolo, A. M., & Arisi, A. C. (2024). Thermal and salt stress effects on the survival of plant growth-promoting bacteria *Azospirillum brasilense* in inoculants for maize cultivation. *Journal of Science and Food Agriculture, 104*(9), 5360–5367.

Das, A. K., Ghosh, P. K., Nihad, S. A., Sultana, S., Keya, S. S., Rahman, M. A., Ghosh, T. K., Akter, M., Hasan, M., Salma, U., & Hasan, M. M. (2024). Salicylic acid priming improves cotton seedling heat tolerance through photosynthetic pigment preservation, enhanced antioxidant activity, and Osmoprotectant levels. *Plants, 13*(12), 1639.

Desoky, E. S., Saad, A. M., El-Saadony, M. T., Merwad, A. R., & Rady, M. M. (2020). Plant growth-promoting rhizobacteria: Potential improvement in antioxidant defense system and suppression of oxidative stress for alleviating salinity stress in *Triticum aestivum (L.)* plants. *Biocatalysis and Agricultural Biotechnology, 30*, 101878.

Dhar, S. K., Kaur, J., Singh, G. B., Chauhan, A., Tamang, J., Lakhara, N., Asyakina, L., Atuchin, V., Mudgal, G., & Abdi, G. (2024). Novel *Bacillus* and *Prestia* isolates from Dwarf century plant enhance crop yield and salinity tolerance. *Scientific Reports, 14*(1), 14645.

Dong, C. J., Wang, L. L., Li, Q., & Shang, Q. M. (2019). Bacterial communities in the rhizosphere, phyllosphere and endosphere of tomato plants. *PLoS One, 14*(11), e0223847.

Dong, X., Li, Z., Wang, Q., Xie, Z., Li, Y., & Luo, Y. (2024). Enhancing the growth performance of *Sesbania cannabina* using Ensifer alkalisoli and biochar under salt stress. *Rhizosphere, 30*, 100888.

Ejaz, S., Fahad, S., Anjum, M. A., Nawaz, A., Naz, S., Hussain, S., & Ahmad, S. (2020). Role of osmolytes in the mechanisms of antioxidant defense of plants. In *Sustainable agriculture reviews* (Vol. 39, pp. 95–117). Springer.

Etesami, H., & Glick, B. R. (2024). Bacterial indole-3-acetic acid: A key regulator for plant growth, plant-microbe interactions, and agricultural adaptive resilience. *Microbiological Research, 281*, 127602.

Fahad, S., Hussain, S., Matloob, A., Khan, F. A., Khaliq, A., Saud, S., Hassan, S., Shan, D., Khan, F., Ullah, N., & Faiq, M. (2015). Phytohormones and plant responses to salinity stress: A review. *Plant Growth Regulation, 75*, 391–404.

Gamalero, E., & Glick, B. R. (2019). Plant growth-promoting bacteria in agricultural and stressed soils. In *Modern soil microbiology* (pp. 361–380). CRC Press.

Gamalero, E., Lingua, G., & Glick, B. R. (2023). Ethylene, ACC, and the plant growth-promoting enzyme ACC deaminase. *O Biologico, 12*(8), 1043.

Gao, Y., Zou, H., Wang, B., & Yuan, F. (2022). Progress and applications of plant growth-promoting bacteria in salt tolerance of crops. *International Journal of Molecular Sciences, 23*(13), 7036.

Ghazala, I., Chiab, N., Saidi, M. N., & Gargouri-Bouzid, R. (2023). The plant growth-promoting bacteria strain *Bacillus mojavensis* I4 enhanced salt stress tolerance in durum wheat. *Current Microbiology, 80*(5), 178.

Ghosh, P. K., De, T. K., & Maiti, T. K. (2018). Role of ACC deaminase as a stress ameliorating enzyme of plant growth-promoting rhizobacteria useful in stress agriculture: A review. In *Role of rhizospheric microbes in soil: Volume 1: Stress management and agricultural sustainability* (pp. 57–106). Springer.

Glick, B. R. (2014). Bacteria with ACC deaminase can promote plant growth and help to feed the world. *Microbiological Research, 169*(1), 30–39.

Gu, J., Li, Z., Mao, Y., Struik, P. C., Zhang, H., Liu, L., Wang, Z., & Yang, J. (2018). Roles of nitrogen and cytokinin signals in root and shoot communications in maximizing of plant productivity and their agronomic applications. *Plant Science, 274*, 320–331.

Gupta, R., & Chakrabarty, S. K. (2013). Gibberellic acid in plant: Still a mystery unresolved. *Plant Signaling & Behavior, 8*(9), e25504.

Gupta, J., Gupta, A., & Gupta, A. K. (2016). Flavonoids: Its working mechanism and various protective roles. *International Journal of Chemical Studies, 4*(4), 190–198.

Gupta, A., Mishra, R., Rai, S., Bano, A., Pathak, N., Fujita, M., Kumar, M., & Hasanuzzaman, M. (2022). Mechanistic insights of plant growth promoting bacteria mediated drought and salt stress tolerance in plants for sustainable agriculture. *International Journal of Molecular Sciences, 23*(7), 3741.

Hartmann, M., & Six, J. (2023). Soil structure and microbiome functions in agroecosystems. *Nature Reviews Earth & Environment, 4*(1), 4–18.

Hasanuzzaman, M., Nahar, K., Gill, S. S., & Fujita, M. (2013). Drought stress responses in plants, oxidative stress, and antioxidant defense. In *Climate change and plant abiotic stress tolerance* (pp. 209–250).

Hashem, H. A., & Mohamed, A. H. (2020). Strategies for drought tolerance in xerophytes. In *Plant ecophysiology and adaptation under climate change: Mechanisms and perspectives I: General consequences and plant responses* (pp. 269–293). Springer Nature.

Herrera-Vásquez, A., Salinas, P., & Holuigue, L. (2015). Salicylic acid and reactive oxygen species interplay in the transcriptional control of defense genes expression. *Frontiers in Plant Science, 6*, 171.

Heuer, B. (2010). Role of proline in plant response to drought and salinity. In *Handbook of plant and crop stress* (Vol. 3, pp. 213–238).

Heyl, A., & Schmülling, T. (2003). Cytokinin signal perception and transduction. *Current Opinion in Plant Biology, 6*(5), 480–488.

Ilangumaran, G., & Smith, D. L. (2017). Plant growth promoting rhizobacteria in amelioration of salinity stress: A systems biology perspective. *Frontiers in Plant Science, 8*, 1768.

Jain, N. K., & Roy, I. (2009). Effect of trehalose on protein structure. *Protein Science, 18*(1), 24–36.

Jayakannan, M., Bose, J., Babourina, O., Rengel, Z., & Shabala, S. (2015). Salicylic acid in plant salinity stress signalling and tolerance. *Plant Growth Regulation, 76*, 25–40.

Jha, S., Singh, A. A., & Thakur, N. (2022). Role of exopolysaccharide and biofilms in microorganisms for alleviating salt stress. In *Mitigation of plant abiotic stress by microorganisms* (pp. 205–230). Academic Press.

Jogawat, A. (2019). Osmolytes and their role in abiotic stress tolerance in plants. In *Molecular plant abiotic stress: Biology and biotechnology* (Vol. 12, pp. 91–104). John Wiley & Sons.

Kalleku, J. N., Ihsan, S., Al-Azzawi, T. N., Khan, M., Hussain, A., Chebitok, F., Das, A. K., Moon, Y. S., Mun, B. G., Lee, I. J., & Ali, S. (2024). Halotolerant *Pseudomonas koreensis* S4T10 mitigate salt and drought stress in Arabidopsis thaliana. *Physiologia Plantarum, 176*(2), e14258.

Kaya, C., Tuna, A. L., & Yokaş, I. (2009). The role of plant hormones in plants under salinity stress. In *Salinity and water stress: Improving crop efficiency* (pp. 45–50). Springer.

Khan, M. M., Khatun, A., & Islam, M. T. (2016). Promotion of plant growth by phytohormone producing bacteria. In *Microbes in action* (pp. 1–43). Nova Science Publishers.

Khan, S., Siraj, S., Shahid, M., Haque, M. M., & Islam, A. (2023). Osmolytes: Wonder molecules to combat protein misfolding against stress conditions. *International Journal of Biological Macromolecules, 234*, 123662.

Khanna, K., Kohli, S. K., Bali, S., Kaur, P., Saini, P., Bakshi, P., Ohri, P., Mir, B. A., & Bhardwaj, R. (2018). Role of micro-organisms in modulating antioxidant defence in plants exposed to metal toxicity. In *Plants under metal and metalloid stress: Responses, tolerance and remediation* (pp. 303–335). Springer.

Khumairah, F. H., Simarmata, R., Erdayani, E., Khairina, Y., Christita, M., Widowati, T., & Lekatompessy, S. J. (2024) Allium Cepa salinity tolerance induced by plant growth promoting bacteria through alteration of rhizosphere microbiome diversity. Available at SSRN 4725037.

Kieber, J. J., & Schaller, G. E. (2014). Cytokinins. *The Arabidopsis Book/American Society of Plant Biologists, 12*, e0168.

Kim, I., Woo, H., Chhetri, G., Park, S., & Seo, T. (2024). A novel exopolysaccharide-producing bacterium, *Pseudescherichia liriopis* sp. nov. isolated from Liriope platyphylla, enhances the growth of *Daucus carota subsp. sativus* under drought and salinity stress. *Frontiers in Plant Science, 15*, 1417639.

Kolomiiets, Y., Butsenko, L., Yemets, A., & Blume, Y. (2024). The use of PGPB-based bioformulations to control bacterial diseases of vegetable crops in Ukraine. *The Open Agriculture Journal, 18*(1). https://doi.org/10.2174/0118743315283724231220104524

Kour, D., Rana, K. L., Yadav, N., Yadav, A. N., Kumar, A., Meena, V. S., Singh, B., Chauhan, V. S., Dhaliwal, H. S., & Saxena, A. K. (2019). Rhizospheric microbiomes: Biodiversity, mechanisms of plant growth promotion, and biotechnological applications for sustainable agriculture. In *Plant growth promoting rhizobacteria for agricultural sustainability: From theory to practices* (pp. 19–65). Springer.

Kumar, P., & Sharma, P. K. (2020). Soil salinity and food security in India. *Frontiers in Sustainable Food Systems, 4*, 533781.

Kumawat, K. C., Sharma, P., Sirari, A., Sharma, B., Kumawat, G., Nair, R. M., & Bindumadhava, H. (2024). Co-existence of halo-tolerant *Pseudomonas fluorescens* and *Enterococcus hirae* with multifunctional growth promoting traits to ameliorate salinity stress in *Vigna radiata*. *Chemosphere, 349*, 140953.

Liu, H. Q., Lu, X. B., Li, Z. H., Tian, C. Y., & Song, J. (2021). The role of root-associated microbes in growth stimulation of plants under saline conditions. *Land Degradation & Development, 32*(13), 3471–3486.

Liu, X., Chai, J., Zhang, Y., Zhang, C., Lei, Y., Li, Q., & Yao, T. (2022). Halotolerant rhizobacteria mitigate the effects of salinity stress on maize growth by secreting exopolysaccharides. *Environmental and Experimental Botany, 204*, 105098.

Liu, X., Yao, T., Chai, J., & Han, J. (2023). Adsorption of sodium ions by exopolysaccharides from *Pseudomonas simiae* MHR6 and its improvement of Na+/K+ homeostasis in maize under salt stress. *Journal of Agricultural and Food Chemistry, 71*(50), 19949–19957.

Lukman, G., Waturangi, D. E., Julyantoro, P. G., & Papuangan, N. (2024). Phyllosphere bacteria with antiquorum sensing and antibiofilm activities against fish pathogenic bacteria. *BMC Research Notes, 17*(1), 5.

Luo, H., Win, C. S., Lee, D. H., He, L., & Yu, J. M. (2024). *Microbacterium azadirachtae* CNUC13 enhances salt tolerance in maize by modulating osmotic and oxidative stress. *O Biologico, 13*(4), 244.

Mahmoud, O. M., Hidri, R., Abdelly, C., & Debez, A. (2024). *Bacillus pumilus* isolated from sabkha rhizosphere ameliorates the behavior of the facultative halophyte Hordeum marinum when salt-challenged by improving nutrient uptake and soil health-related traits. *Plant Stress, 11*, 100383.

Majeed, A., & Muhammad, Z. (2019). Salinity: A major agricultural problem—Causes, impacts on crop productivity and management strategies. In *Plant abiotic stress tolerance: Agronomic, molecular and biotechnological approaches* (pp. 83–99). Springer International Publishing.

Majeed, A., Muhammad, Z., & Ahmad, H. (2018). Plant growth promoting bacteria: Role in soil improvement, abiotic and biotic stress management of crops. *Plant Cell Reports, 37*(12), 1599–1609.

Martin, F. M., Uroz, S., & Barker, D. G. (2017). Ancestral alliances: Plant mutualistic symbioses with fungi and bacteria. *Science, 356*(6340), eaad4501.

Mishra, P., Mishra, J., & Arora, N. K. (2021). Plant growth promoting bacteria for combating salinity stress in plants–recent developments and prospects: A review. *Microbiological Research, 252*, 126861.

Mohamed, N. N. (2017). Management of salt-affected soils in the Nile Delta. In *The Nile Delta* (pp. 265–295). Springer.

Mohammadipanah, F., & Zamanzadeh, M. (2019). Bacterial mechanisms promoting the tolerance to drought stress in plants. In *Secondary metabolites of plant growth promoting rhizomicroorganisms: Discovery and applications* (pp. 185–224). Springer.

Møller, I. M. (2001). Plant mitochondria and oxidative stress: Electron transport, NADPH turnover, and metabolism of reactive oxygen species. *Annual Review of Plant Biology, 52*(1), 561–591.

Morcillo, R. J., & Manzanera, M. (2021). The effects of plant-associated bacterial exopolysaccharides on plant abiotic stress tolerance. *Metabolites, 11*(6), 337.

Muhammad, S., Abbas, Z., Bashir, S., Qadir, G., & Khan, S. (2019). Nitrogen fixation with non-leguminous plants. *Specialty Journal of Agricultural Sciences, 5*(3), 35–42.

Nahar, K., Hasanuzzaman, M., & Fujita, M. (2016). Roles of osmolytes in plant adaptation to drought and salinity. In *Osmolytes and plants acclimation to changing environment: Emerging omics technologies* (pp. 37–68). Springer.

Naik, K., Mishra, S., Srichandan, H., Singh, P. K., & Sarangi, P. K. (2019). Plant growth promoting microbes: Potential link to sustainable agriculture and environment. *Biocatalysis and Agricultural Biotechnology, 21*, 101326.

Naing, A. H., Maung, T. T., & Kim, C. K. (2021). The ACC deaminase-producing plant growth-promoting bacteria: Influences of bacterial strains and ACC deaminase activities in plant tolerance to abiotic stress. *Physiologia Plantarum, 173*(4), 1992–2012.

Narayanankutty, A., Job, J. T., & Narayanankutty, V. (2019). Glutathione, an antioxidant tripeptide: Dual roles in carcinogenesis and chemoprevention. *Current Protein & Peptide Science, 20*(9), 907–917.

Nawaz, K., Hussain, K., Majeed, A., Khan, F., Afghan, S., & Ali, K. (2010). Fatality of salt stress to plants: Morphological, physiological and biochemical aspects. *African Journal of Biotechnology, 9*(34).

Nivetha, N., Lavanya, A. K., Vikram, K. V., Asha, A. D., Sruthi, K. S., Bandeppa, S., Annapurna, K., & Paul, S. (2021). PGPR-mediated regulation of antioxidants: Prospects for abiotic stress management in plants. In *Antioxidants in plant-microbe interaction* (pp. 471–497).

Olanrewaju, O. S., Glick, B. R., & Babalola, O. O. (2017). Mechanisms of action of plant growth promoting bacteria. *World Journal of Microbiology and Biotechnology, 33*, 1–6.

Orabi, S. A., & Abou-Hussein, S. D. (2019). Antioxidant defense mechanisms enhance oxidative stress tolerance in plants: A review. *Current Science International, 8*(3), 565–576.

Osakabe, Y., Yamaguchi-Shinozaki, K., Shinozaki, K., & Tran, L. S. (2014). ABA control of plant macroelement membrane transport systems in response to water deficit and high salinity. *The New Phytologist, 202*(1), 35–49.

Oubaha, B., Rathore, R. S., Bagri, J., Singhal, N. K., Mazumdar, K., Rishi, V., Pareek, A., & Singla-Pareek, S. L. (2024). *Bacillus siamensis* strain BW enhances rice growth and salinity tolerance through redox equilibrium and hormone modulation. *Current Plant Biology, 37*, 100321.

Pandey, V. P., Awasthi, M., Singh, S., Tiwari, S., & Dwivedi, U. N. (2017). A comprehensive review on function and application of plant peroxidases. *Biochemistry and Analytical Biochemistry, 6*(1), 308.

Pathak, J., Ahmed, H., Kumari, N., Pandey, A., & Rajneesh, S. R. P. (2020). Role of calcium and potassium in amelioration of environmental stress in plants. In *Protective chemical agents in the amelioration of plant abiotic stress: Biochemical and molecular perspectives* (pp. 535–562). Wiley Blackwell.

Paul, S., Parvez, S. S., Goswami, A., & Banik, A. (2024). Exopolysaccharides from agriculturally important microorganisms: Conferring soil nutrient status and plant health. *International Journal of Biological Macromolecules, 262*, 129954.

Peng, Y., Jiang, L., Jeon, D., Cho, D., Kim, Y., Kim, C. Y., Lee, J. H., & Lee, J. (2024). *Pseudoxanthomonas sp.* JBR18, a halotolerant endophytic bacterium, improves the salt tolerance of Arabidopsis seedlings. *Plant Physiology and Biochemistry, 207*, 108415.

Pengfei, Z., Yanyan, D., Masateru, S., Natsumi, M., & Kengo, I. (2017). Interactions of salinity stress and flower thinning on tomato growth, yield, and water use efficiency. *Communications in Soil Science and Plant Analysis, 48*(22), 2601–2611.

Pérez-Álvarez, S., Ardisana, E. F., Magallanes-Tapia, M. A., Bonilla, C. M., Garcia, C. U., González, M. M., Flores-Cordova, M. A., & Guerrero-Morales, S. (2022). Microorganisms used as growth regulators in modern agriculture. In *Beneficial microorganisms in agriculture* (pp. 37–84). Springer Nature Singapore.

Poonam, B. R., Handa, N., Kaur, H., Rattan, A., Bali, S., Gautam, V., Sharma, A., Ohri, P., Thukral, A. K., & Sirhindi, G. (2016). Sugar signalling in plants: A novel mechanism for drought stress management. In *Water stress and crop plants: A sustainable approach* (Vol. 1, pp. 287–302). John Wiley & Sons.

Premachandra, D., Hudek, L., & Brau, L. (2016). Bacterial modes of action for enhancing of plant growth. *Journal of Biotechnology and Biomaterials, 6*(3), 1.

Rabbani, G., & Choi, I. (2018). Roles of osmolytes in protein folding and aggregation in cells and their biotechnological applications. *International Journal of Biological Macromolecules, 109*, 483–491.

Rafedzi, E. A., Bharudin, I., Kamaruddin, S., Diba, F., & Murad, A. M. (2023). Volatile Organic Compounds (VOCs) emission and antioxidant activities of rice treated with a consortium of plant growth-promoting rhizobacteria under drought conditions. *Universal Journal of Agricultural Research, 11*(5), 916–928.

Raza, A., Zahra, N., Hafeez, M. B., Ahmad, M., Iqbal, S., Shaukat, K., & Ahmad, G. (2020). Nitrogen fixation of legumes: Biology and physiology. In *The plant family Fabaceae: Biology and physiological responses to environmental stresses* (pp. 43–74). Springer.

Remucal, C. K., & Sedlak, D. L. (2011). The role of iron coordination in the production of reactive oxidants from ferrous iron oxidation by oxygen and hydrogen peroxide. In *Aquatic redox chemistry* (pp. 177–197). American Chemical Society.

Riahi, K., Hosni, K., Raies, A., & Oliveira, R. (2019). Unique secondary metabolites of a *Streptomyces* strain isolated from extreme salty wetland show antioxidant and antibacterial activities. *Journal of Applied Microbiology, 127*(6), 1727–1740.

Roy, B., Maitra, D., Bhattacharya, A., Mondal, A., Pal, N., Nandy, A., Bakshi, B., Ghosh, J., & Mitra, A. K. (2024). Alleviation of abiotic stress in *Oryza sativa* by the application of novel polyextremophilic plant growth promoting *Bacillus*. *Biocatalysis and Agricultural Biotechnology, 60*, 103272.

Saberi Riseh, R., Ebrahimi-Zarandi, M., Gholizadeh Vazvani, M., & Skorik, Y. A. (2021). Reducing drought stress in plants by encapsulating plant growth-promoting bacteria with polysaccharides. *International Journal of Molecular Sciences, 22*(23), 12979.

Sadak, M. S., El-Bassiouny, H. M., & Dawood, M. G. (2019). Role of trehalose on antioxidant defense system and some osmolytes of quinoa plants under water deficit. *Bulletin of the National Research Centre, 43*(1), 1.

Saeed, Q., Xiukang, W., Haider, F. U., Kučerik, J., Mumtaz, M. Z., Holatko, J., Naseem, M., Kintl, A., Ejaz, M., Naveed, M., & Brtnicky, M. (2021). Rhizosphere bacteria in plant growth promotion, biocontrol, and bioremediation of contaminated sites: A comprehensive review of effects and mechanisms. *International Journal of Molecular Sciences, 22*(19), 10529.

Saeed, E. E., Sudalaimuthuasari, N., Purayil, F. T., Rafi, M., Kundu, B., Mishra, A. K., Al-Maskari, R. S., Abdelfattah, A. M., Eldin, A. K., George, S., & Procter, M. (2024). The genome, pangenome, and physiological analysis of Leclercia adecarboxylata (kcgeb_e1), a plant growth-promoting bacterium. *Discover Applied Sciences, 6*(3), 76.

Sagar, L., Singh, S., Sharma, A., Maitra, S., Attri, M., Sahoo, R. K., Ghasil, B. P., Shankar, T., Gaikwad, D. J., Sairam, M., & Sahoo, U. (2023). Role of soil microbes against abiotic stresses induced oxidative stresses in plants. In *Microbial symbionts and plant health: Trends and applications for changing climate* (pp. 149–177). Springer.

Sah, S., & Singh, R. (2015). Siderophore: Structural and functional characterisation–a comprehensive review. *Agriculture (Pol'nohospodárstvo), 61*(3), 97–114.

Saleem, S., Mushtaq, N. U., Shah, W. H., Rasool, A., & Rehman, R. U. (2024). Plant growth-promoting bacteria in food security and plant disease management. In *Biocontrol agents for improved agriculture* (pp. 275–292). Academic Press.

Šamec, D., Karalija, E., Šola, I., Vujčić Bok, V., & Salopek-Sondi, B. (2021). The role of polyphenols in abiotic stress response: The influence of molecular structure. *Plants, 10*(1), 118.

Scavino, A. F., & Pedraza, R. O. (2013). The role of siderophores in plant growth-promoting bacteria. In *Bacteria in agrobiology: Crop productivity* (pp. 265–285). Springer.

Sevim, G., Ozdemir-Kocak, F., & Unal, D. (2023). Hormonal signaling molecules triggered by plant growth-promoting bacteria. In *Phytohormones and stress responsive secondary metabolites* (pp. 187–196). Academic Press.

Shahid, M. A., Sarkhosh, A., Khan, N., Balal, R. M., Ali, S., Rossi, L., Gómez, C., Mattson, N., Nasim, W., & Garcia-Sanchez, F. (2020). Insights into the physiological and biochemical impacts of salt stress on plant growth and development. *Agronomy, 10*(7), 938.

Sharma, A., Shahzad, B., Kumar, V., Kohli, S. K., Sidhu, G. P., Bali, A. S., Handa, N., Kapoor, D., Bhardwaj, R., & Zheng, B. (2019). Phytohormones regulate accumulation of osmolytes under abiotic stress. *Biomolecules, 9*(7), 285.

Sharma, P., Chouhan, R., Bakshi, P., Gandhi, S. G., Kaur, R., Sharma, A., & Bhardwaj, R. (2022). Amelioration of chromium-induced oxidative stress by combined treatment of selected plant-growth-promoting rhizobacteria and earthworms via modulating the expression of genes related to reactive oxygen species metabolism in *Brassica juncea. Frontiers in Microbiology, 13*, 802512.

Sharma, P., Bakshi, P., Kaur, R., Sharma, A., Bhardwaj, R., El-Sheikh, M. A., Tyagi, A., & Ahmad, P. (2023). Inoculation of plant-growth-promoting rhizobacteria and earthworms in the rhizosphere reinstates photosynthetic attributes and secondary metabolites in *Brassica juncea L.* under chromium toxicity. *Plant and Soil, 483*(1), 573–587.

Sierra-García, I. N., Ferreira, M. J., Torres-Ballesteros, A., Louvado, A., Gomes, N., & Cunha, A. (2024). *Brevibacterium* EB3 inoculation enhances rhizobacterial community interactions leading to improved growth of *Salicornia europaea. Applied Soil Ecology, 196*, 105306.

Sindhu, S. S., Sehrawat, A., Sharma, R., Dahiya, A., & Khandelwal, A. (2017). Belowground microbial crosstalk and rhizosphere biology. In *Plant-microbe interactions in agro-ecological perspectives: Volume 2: Microbial interactions and agro-ecological impacts* (pp. 695–752). Springer.

Singh, P. K., & Gautam, S. (2013). Role of salicylic acid on physiological and biochemical mechanism of salinity stress tolerance in plants. *Acta Physiologiae Plantarum, 35*, 2345–2353.

Singh, R. P., Shelke, G. M., Kumar, A., & Jha, P. N. (2015). Biochemistry and genetics of ACC deaminase: A weapon to "stress ethylene" produced in plants. *Frontiers in Microbiology, 6*, 937.

Singh, R. P., Ma, Y., & Shadan, A. (2022). Perspective of ACC-deaminase producing bacteria in stress agriculture. *Journal of Biotechnology, 352*, 36–46.

Taj, Z., Bakka, K., & Challabathula, D. (2024). Halotolerant PGPB *Staphylococcus sciuri* ET101 protects photosynthesis through activation of redox dissipation pathways in *Lycopersicon esculentum. Plant Physiology and Biochemistry, 208*, 108482.

Teale, W. D., Paponov, I. A., & Palme, K. (2006). Auxin in action: Signalling, transport and the control of plant growth and development. *Nature Reviews Molecular Cell Biology, 7*(11), 847–859.

Teng, K., Xiao, G. Z., Guo, W. E., Yuan, J. B., Li, J., Chao, Y. H., & Han, L. B. (2016). Expression of an alfalfa (*Medicago sativa L.*) peroxidase gene in transgenic *Arabidopsis thaliana* enhances resistance to NaCl and H_2O_2. *Genetics and Molecular Research, 15*. https://doi.org/10.4238/gmr.15028002

Tsavkelova, E. A., Cherdyntseva, T. A., Klimova, S. Y., Shestakov, A. I., Botina, S. G., & Netrusov, A. I. (2007). Orchid-associated bacteria produce indole-3-acetic acid, promote seed germination, and increase their microbial yield in response to exogenous auxin. *Archives of Microbiology, 188*, 655–664.

Tuan, P. A., Sun, M., Nguyen, T. N., Park, S., & Ayele, B. T. (2019). Molecular mechanisms of seed germination. In *Sprouted grains* (pp. 1–24). Woodhead Publishing.

Ullah, S., Ikram, M., Sarfaraz, S., Ul Haq, I., Khan, A., Murad, Z., & Munsif, F. (2024). Influence of Plant Growth Promoting Rhizobacteria (PGPR) on the growth and yield of sunflower (*Helianthus annus L.*) under salt stress. *Journal of Crop Health, 76*, 1–4.

Velmourougane, K., Saxena, G., & Prasanna, R. (2017). Plant-microbe interactions in the rhizosphere: Mechanisms and their ecological benefits. In *Plant-microbe interactions in agro-ecological perspectives: Volume 2: Microbial interactions and agro-ecological impacts* (pp. 193–219). Springer.

Vocciante, M., Grifoni, M., Fusini, D., Petruzzelli, G., & Franchi, E. (2022). The role of plant growth-promoting rhizobacteria (PGPR) in mitigating plant's environmental stresses. *Applied Sciences, 12*(3), 1231.

Wijayasinghe, Y. S., Tyagi, A., & Poddar, N. K. (2017). Regulation of cell volume by osmolytes. In *Cellular osmolytes: From chaperoning protein folding to clinical perspectives* (pp. 195–228). Springer.

Yang, J., Duan, G., Li, C., Liu, L., Han, G., Zhang, Y., & Wang, C. (2019). The crosstalks between jasmonic acid and other plant hormone signaling highlight the involvement of jasmonic acid as a core component in plant response to biotic and abiotic stresses. *Frontiers in Plant Science, 10*, 1349.

Yoo, S. J., Weon, H. Y., Song, J., & Sang, M. K. (2019). Induced tolerance to salinity stress by halotolerant bacteria *Bacillus aryabhattai* H19-1 and *B. mesonae* H205 in tomato plants. *Journal of Microbiology and Biotechnology, 29*(7), 1124–1136.

Yu, Y., Li, Y., Yan, Z., & Duan, X. (2021). The role of cytokinins in plant under salt stress. *Journal of Plant Growth Regulation, 41*, 1–3.

Züst, T., & Agrawal, A. A. (2017). Trade-offs between plant growth and defense against insect herbivory: An emerging mechanistic synthesis. *Annual Review of Plant Biology, 68*(1), 513–534.

Part III

Future Perspectives on Soil Bacterial Research

The Hidden Players: Soil Microbes in Ecosystems Sustainability

13

Salam Suresh Singh, Emilynruwaka Lamare,
Balu Ananda Chopade, Rajdeep Chanda,
Ningthoujam Ranjana Devi, Keshav Kumar Upadhyay,
Khoisnam Vramari Devi, and Shri Kant Tripathi

Abstract

Soil microbes constitute an array of microscopic life forms (e.g. bacteria, fungi, archaea) present in the soil, which constitute the dynamic and essential components of terrestrial ecosystems. These microbes perform various essential functions (cycling of carbon and nutrients, fixation of nitrogen, transformation, and maintenance of soil structure and stability) that lead to ecosystem sustainability worldwide. Increasing industrialisation and agricultural intensification have altered the composition and ability of soil microbes, ultimately reducing the quality of soil to serve society. The dumping of waste into the environment can lead to alterations in soil microbial communities, disrupting their function and ecosystem services, and affecting the intricate relationship between plants and microbes. However, in the last couple of decades, studies have focused on this hidden player and attempted to understand their multifaceted nature to sustain the ecosystem. In recent years, microbes with specific functions have been used to improve soil fertility, sequester carbon, and decontaminate problem soils to achieve sustainability goals. This chapter highlights the multifaceted nature of soil microbes, emphasising their silent yet transformative functions in sustaining the intricate tapestry of life on the Earth and fostering a sustainable future.

S. S. Singh · E. Lamare · R. Chanda · N. R. Devi · K. K. Upadhyay · K. V. Devi
S. K. Tripathi (✉)
Department of Forestry, School of Earth Science and Natural Resources Management, Mizoram University, Aizawl, Mizoram, India

B. A. Chopade
Dr. Babasaheb Ambedkar Marathwada University, Aurangabad, India

M. Saraf et al. (eds.), *Climate Change and Soil Microorganisms for Environmental Sustainability*, Microorganisms for Sustainability 28,
https://doi.org/10.1007/978-981-96-3425-5_13

Keywords

Nutrient cycling · Bioremediation · Biotransformation · Climate change · Carbon sequestration

13.1 Introduction

Soil is a dynamic and intricate ecosystem containing an unseen workforce of micro- and mesofauna (e.g. fungi, bacteria, archaea, and other minuscule life forms), which together play an indispensable role in the global sustainability of ecosystems through various functions (Gupta et al., 2017). These microorganisms guard important ecological processes and play a significant role in plant health, soil structure, nutrient cycling, and ecosystem functions, and maintain the equilibrium of the ecosystem. Therefore, the invisible yet monumental impact on the intricate web of life beneath our feet underscores the significance of understanding and appreciating the role of microbes in fostering ecosystem sustainability (Yadav et al., 2018). A sustainable ecosystem relies on a delicate balance of essential nutrients (i.e. nitrogen and phosphorus) in the soil. The depletion of nutrients in the soil owing to plant uptake is compensated by the regular supply of these nutrients through the process of organic matter decomposition and nutrient release (Glick, 2018). The role of microbes in ecosystem sustainability is important in regulating the decomposition of dead plants and animals and maintaining a regular supply of nutrients in the soil (Doube & Gupta, 1997) (Fig. 13.1).

In addition to nutrient cycling, soil microbes also significantly influence the physical structure of the soil (Gasperi-Mago & Troeh, 1979), play a pivotal role in shaping plant health and resilience, and contribute to carbon sequestration by incorporating carbon-rich compounds into stable organic matter. In addition, they act as natural clean-up crews for the remediation of pollutants (e.g. toxic metals, pesticides, nutrients, and industrial waste) by transforming them into non-toxic substances or reducing the concentration of pollutants by bringing them to a considerable limit. Thus, the invisible realm, governed by soil microbes, is a linchpin for ecosystem sustainability. This chapter focuses on the role of soil microbes in ecosystem sustainability.

13.2 Role of Microbes in Nutrient Cycling

Nutrients are essential elements for plant growth, development, and survival. Plants require large amounts of nutrients, such as nitrogen (N), phosphorus (P), carbon (C), oxygen (O), calcium (Ca), potassium (K), and magnesium (Mg), as well as small amounts of micronutrients, such as boron(B), iron (Fe), manganese (Mn), zinc (Zn), copper (Cu), and chlorine (Cl), where each element is characterised by a unique biogeochemical cycle (Pandey, 2018). Plants acquire nutrients from the soil via various metabolic processes. Upon uptake, plants utilise nutrients for physiological functions, contributing to their growth, development, and overall health.

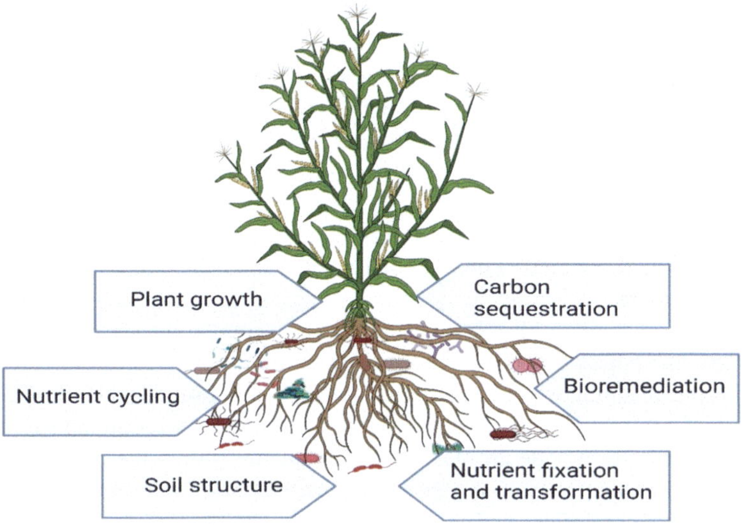

Fig. 13.1 Role of microbes in ecosystem sustainability

Subsequently, plants naturally return these nutrients to the soil through processes such as litterfall, pruning, or root senescence, which occur in both managed and unmanaged ecosystems. These plant fragments are broken down into simpler substances and release nutrients which were previously sequestered within the plant's biomass by intricate activity by a diverse array of soil microorganisms, making them easily available for the plants (Koshila Ravi et al., 2019). In these nutrient cycling processes, a diverse array of microbes plays vital roles in the absorption, transformation, and release of nutrients in different components.

The term nutrient cycling refers to the ongoing movement and exchange of nutrients within and between various abiotic and biotic components of ecosystems, including mineral weathering, soil biota activity, and other changes in the biosphere, lithosphere, and hydrosphere (Jordan, 1985). The cycling of nutrients is controlled by various abiotic factors (climate, site, topography, and parent material factors) and biotic (microbes) factors. The nutrient cycling model consists of a soil-plant-animal system separated into different pools with inputs into the system that come from fertilisers, rain, dust, or organic materials and outputs that result from erosion, leaching, plant harvest, denitrification, volatilisation of N, and burning (Nair et al., 1995). The nutrient cycle encompasses various processes, each focusing on specific elements and their movements in ecosystems. The elements that pass through an ecosystem's components are not lost; rather, they are recycled or gathered in locations known as reservoirs (or 'unsinks'), where they can be stored for a very long time. Depending on these reservoirs, ecosystems constitute different nutrient cycles, such as gaseous (nitrogen, carbon, hydrogen, and oxygen cycles), sedimentary (phosphorus and sulfur cycles), and hydrological (water cycles) cycles. Microorganisms play an indispensable role in shaping and maintaining nutrients in various ecosystems in a sustainable manner. Numerous ecological processes linked to soil production, nutrient cycling, and other ecosystem attributes and services are significantly influenced by the actions and interactions of microorganisms (Cardinale et al., 2012). Therefore, understanding the intricate relationship between microbial communities and their functioning and management is important for enhancing the provision of ecosystem services, because many complex soil processes are driven by soil microorganisms. (Nemergut et al., 2013; Pereira e et al., 2012F; Wagg et al., 2019). Microorganisms act as agents for recycling various nutrients in an eco-friendly manner, allowing sustainable life on Earth and balancing essential elements in the environment.

The significance of soil has been escalating at a concerning pace because of the rapidly growing population. This is because the availability of fertile soils for future demand is diminishing owing to increased urbanisation, biofuel production, and the extraction of natural resources. Soils serve as the foundation for all land-based ecosystems and harbour a wide array of macro-organisms within the rhizosphere. The vitality of soil emanates from living organisms, including microorganisms, which encompass genes, species, and communities. Soil microorganisms are excellent indicators of soil health (Nielsen & Winding, 2002). These rhizospheric microorganisms play critical roles in the function of soil-plant systems by regulating nutrient cycling reactions necessary to preserve soil quality and assist in the formation and maintenance of soil structure in a sustainable way (Meena et al., 2015a; Raghavendra et al., 2016; Zahedi, 2016; Rawat et al., 2016; Yasin et al., 2016; Teotia et al., 2016).

Microbes rapidly adapt to several environmental conditions as they respond quickly to changes. Changes in microbial populations and activity may therefore serve as a good signal of changes in soil health, since this adaptation may make it possible for microbial tests to be discriminatory in soil health evaluation (Kennedy & Papendick, 1995; Pankhurst & Hawke, 1995; Singh, 2015). Combinations of sand, silt, and clay constitute soils that offer a variety of microhabitats that are home to a wide range of organisms. These organisms work together to change the environment between the soil particles. Most functional features, such as the breakdown of plant litter, the process of food webs, and closed nutrient cycles, are the consequence of microbial communities interacting intimately with one another rather than with a single organism. These microorganisms play a crucial role in both nitrogen and carbon cycles, facilitating essential processes for the environment and life on Earth.

In nature, nitrogen is recycled through a process known as the nitrogen cycle. Microbes in the soil or plant roots convert nitrogen gas into ammonia, nitrates, and nitrites, which can then be taken up by plants. However, plant organic matter can either be converted to dissolved organic carbon (DOC) molecules or broken down into carbon dioxide by soil bacteria. These microbial activities have far-reaching consequences and influence nutrient availability, soil fertility, greenhouse gas concentrations, and the overall ecosystem health. Microbes maintain the equilibrium of these essential biogeochemical cycles.

13.3 Role of Microbes in Nutrient Fixation and Transformation

Nutrient fixation is crucial for ecosystem function through various natural and biological mechanisms (Yousuf et al., 2022). This ensures that essential nutrients are made available to plants and other organisms, forming the basis of the food chain and supporting biodiversity, which contributes to the stability, health, and sustainability of the natural environment. Nutrient fixation is essential for soil health; for instance, nitrogen-fixing bacteria convert atmospheric nitrogen into forms that plants can absorb, thereby reducing the need for synthetic fertilisers. Nutrient fixation can be influenced by various factors, including natural and human-induced elements, such as climate, pH, moisture, soil texture, microbial activity, nutrient availability, and plant-microbe interactions (Yang et al., 2022; Stahl & Olson, 2021). Understanding and managing nutrient fixation is critical for sustainable agriculture and environmental conservation.

The atmosphere contains more than 78% nitrogen in the form of N_2 and serves as the main nitrogen reservoir. In terrestrial environments, nitrogen is predominantly present as a component of soil organic matter, with litter and soil inorganic nitrogen comprising most of its distribution. Additionally, nitrogen can be found in organic compounds, such as amino acids, amines, and amides, as well as in inorganic compounds, such as nitrate, nitrite, ammonia, hydrazine, nitrous oxide, and nitrogen dioxide. The importance of nitrogen lies in the fact that, next to carbon, it predominantly regulates biological activity within the soil. The cycling of nitrogen on a global scale is defined by the movement of nitrogen between land, water, and atmosphere. Variations in soil pH, nutrient availability, temperature, water conditions, and other factors affect nitrogen content in various ecosystems. In addition to this abiotic factor, soil microorganisms also affect the nitrogen cycle in different ecosystems, and the amount of nitrogen in the soil affects the growth, survival, and metabolic actions of various microbial communities. Understanding the nature of nitrogen-fixing bacteria and other microbes and their intricate relationship is essential for maintaining a sustainable ecosystem (Werner & Newton, 2005).

Microorganisms play a vital role in the nitrogen cycle through various processes such as nitrogen fixation, ammonification, nitrification, and denitrification (Rana et al., 2019). Nitrogen atoms form a strong triple bond, making nitrogen gas an extremely stable chemical bond which requires a significant amount of energy to break. Diazotrophic microorganisms possess a distinctive ability to convert atmospheric nitrogen into a biologically accessible form, which is facilitated by the activity of nitrogenase enzymes. Diazotrophic microorganisms are primarily found in major taxonomic groups, such as cyanobacteria and chlorobi (green sulfur bacteria), as well as in protobacterial families, such as Azotobacteraceae, *Rhizobia*, and *Azotobacter Frankia*, which play a crucial role in atmospheric nitrogen fixation. The process of converting atmospheric nitrogen into biologically available nitrogen is referred to as nitrogen fixation. According to a study by Yeager et al. (2005), mixed conifer soil is predominantly inhabited by nitrogen-fixing bacteria such as *Beijinckia derxii* ssp., *Venezuelae*, *Frankia* sp., *Paenibacillus* sp., *and Clostridium pasteurianum*, which are capable of fixing nitrogen in the environment. Although symbiotic diazotrophs are known to fix approximately 100 times more nitrogen than non-symbiotic diazotrophs (Cleveland et al., 1999), free-living microbes also play an important role in contributing nitrogen to soil ecosystems. For instance, free-living diazotrophic microorganisms form associations with ectomycorrhiza that colonise *Pseudotsuga menziesii* (Douglas fir) roots (Li & Hung, 1987) and enhance the establishment of both mycorrhizae and conifer seedlings (Cracknell & Lousier, 1988; Paul et al., 2007). However, it has been observed that forest soils exhibit a higher abundance of nitrogen fixation genes than agricultural soils (Morales et al., 2010). Future studies should prioritise comparing the abundance of nitrogen-fixing bacteria to fungal symbiosis and litter decomposition, as these factors are key to understanding how nitrogen fixers influence carbon dynamics in the soil. This is essential because (i) fixed nitrogen can affect interconnected nitrogen and carbon processes, such as mycorrhizal symbiosis and decomposition rates (Larson et al., 1978); (ii) nitrogen fixation in soil can enhance carbon availability (Johnson & Curtis, 2001); and (iii) soil carbon is closely linked to the abundance of nitrogen-fixing microorganisms.

Nitrification involves two steps: conversion of ammonia to nitrite and then to nitrate. The oxidation of ammonia to nitrite is conducted by chemo-lithotrophic ammonia oxidisers such as *Nitrosospira, Nitrosomonas, and Nitrosococcus* (Koops & Möller, 1992); nitrite oxidisers are required in this process. The former can be further categorised into two groups: ammonia-oxidising bacteria (AOB) and archaea (AOA). The activities of these bacteria, whether influenced by natural or anthropogenic factors, can periodically alleviate the constraints on the growth and activity of ammonia oxidisers. However, evidence suggests that Archaea may be the predominant contributor to ammonia oxidation in some soils. According to Bru et al. (2011), archaea were found to be 10–400 times more abundant than ammonia-oxidising bacteria across 77 sampled sites. However, their abundance was either lower or similar to that of ammonia-oxidising bacteria at nine specific sites. Petersen et al. (2012) observed that ammonia-oxidising bacterial genes were 8–18 times more prevalent than archaeal genes in both coniferous and deciduous forest soils. This suggests a spatial variability in the relative prevalence of these microbial groups.

Denitrification is a crucial process that closes the nitrogen cycle by converting nitrate back into atmospheric nitrogen (Rana et al., 2019). Similar to immobilisation or assimilation, denitrification is a reduction process; however, nitrogen is not directly utilised for cellular synthesis. It is characterised by the full or partial dissimilative reduction of nitrate by microorganisms into dinitrogen gas, which acts as the primary route for emissions of nitrous oxide from the soil (Colliver & Stephenson, 2000; Kowalchuk & Stephen, 2001; Shaw et al., 2006). Denitrifying microorganisms are commonly found in soil, accounting for 0.5% and 5% of the total bacterial population (Henry et al., 2006; Demanèche et al., 2009; Bru et al., 2011), including heterotrophic bacteria (e.g. *Paracoccus denitrificans*, *Pseudomonas aeruginosa, Pseudomonas stutzeri*) (Carlson & Ingraham, 1983), and autotrophic bacteria (e.g. *Thiobacillus denitrificans*) (Baalsrud & Baalsrud, 1954). It also occurs in fungi such as Ascomycota (*Fusarium solani*), *Cylindrocarpon tonkinense*, *Fusarium oxysporum*, *Gibberella fujiuroii*, and Basidiomycota (*Trichosporon cutaneum*) (Bollag & Tung, 1972; Shoun et al., 1992). In grassland ecosystems, Laughlin and Stevens (2002) found that fungal denitrification contributes up to 89% of nitrous oxide emissions. The rates and end-products of this process are affected by soil pH, oxygen levels, organic carbon content, temperature, and moisture (Knowles, 1982; Tiedje, 1988; Brumme & Beese, 1992; Saad & Conrad, 1993; Thomsen et al., 1994; Bergaust et al., 2008, 2010). Although this process produces nitrogen gas, it can also produce nitrous oxide, which is a potent greenhouse gas. This process is vital for maintaining ecosystem health, mitigating nitrogen pollution, and contributing to the regulation of global nitrogen cycles; however, it is essential to monitor and manage nitrous oxide to minimise it.

In a nutshell, nitrogen being the element most often limits productivity in natural ecosystems, and the assimilation of inorganic nitrogen into plants and soil microbial biomass is indispensable for preserving soil fertility and ensuring productivity in terrestrial ecosystems. To maintain such valuable elements, microorganisms hold great significance for actively contributing to soil fertility through their intricate involvement in nitrogen cycles, thus maintaining a sustainable ecosystem (Delgado & Follett, 2002; Dominy & Haynes, 2002; Kätterer & Andrén, 2001; McLean & Parkinson, 1997).

13.4 Role of Microbes in Soil Structure and Stability

Soil serves as a natural home for countless hidden players in the Earth's 'microorganisms' and living organisms encompassing a wide array of genera and species (Bugmann, 1996). The quantity, type, and function of these microorganisms differ from soil to soil and are influenced by many factors such as soil organic matter, texture, pH, moisture, temperature, aeration, and other conditions (Kätterer & Andrén, 2001; Peuke, 2000; Ruark & Zarnoch, 1992; Smith et al., 2000; Subler & Kirsch, 1998). Among the different microorganisms, bacteria are the most abundant in the soil, followed by actinomycetes, fungi, and algae. These organisms are particularly prolific in the uppermost soil layer, where they thrive owing to the abundant supply of organic nutrients in the form of decomposed plants and other waste materials. A diverse range of major soil microbes exist in the soil, where the C:N ratio for bacteria is 5:1, 6:1 for actinomycetes, and 10:1 for fungi (Hernández-Hernández & López-Hernández, 2002; Ruark & Zarnoch, 1992; Subler & Kirsch, 1998). Microbes perform the vital task of converting complex organic compounds such as proteins, carbohydrates, and cellulose into simpler forms, which plants can readily utilise for their growth and development. Aga presence of various microorganisms in the soil creates health and higher productivity, and their activity in the soil is a potential indicator of soil quality.

Soil formation is a gradual process that has unfolded over millions of years and encompasses physicochemical weathering and biological processes (Chen et al., 2023). Microbes play a crucial role in the formation of soil aggregates and soil stability, which, in turn, contribute to soil fertility and productivity. Reciprocally, the soil structure also affects the diversity, growth, and activity of organisms. Aga soil structure can be defined by its shape, stability, and resiliency (Kay, 1990). The structural form pertains to the current arrangement of solids and voids, whereas stability denotes the capacity of the soil to maintain this arrangement when subjected to stress. Resiliency refers to the ability of a soil to regain its structural form or stability after the removal of applied stress, and the presence of various microbial communities that have high resilience. In soil, solid particles typically exist in the form of clusters known as aggregates, pods, or clods. Soil structural stability is frequently assessed based on its ability to withstand various stresses, including those caused by rapid wetting and mechanical compression. Tisdall and Oades (1982) introduced an aggregation model for Australian grassland soils, which seems to be applicable to soils where organic matter primarily serves as a stabilising agent for aggregates. Few studies have focused on the role of specific rhizosphere microorganisms in soil aggregation. For instance, Gouzou et al. (1993) reported that the inoculation of wheat roots with a rhizosphere strain of *Bacillus polymyxa* resulted in an increase in the amount of soil adhering to the roots. Bowen and Rovira (1991) also reported that microbial extracellular polysaccharides are commonly found in the rhizosphere of plants and can serve as binding agents, potentially acting as cementing materials for soil particles.

Microorganisms play a vital role in various processes such as the decomposition of organic matter, formation of mycorrhizal associations with plants, and secretion of extracellular substances (Tunlid et al., 2016; Ekblad et al., 2013; Maillard et al., 2023). These activities can influence the soil structure, water retention, pore size distribution, and nutrient availability, ultimately affecting the overall health and fertility of the soil. Strong et al. (1998) reported that the volume of mesopores (0.6–10.0 m) increased by 67% because of substrate addition, which accompanied the growth of microbes at the expense of both smaller and larger pores. Aspiras et al. (1971) discovered that when various microbial species were exposed to an extra carbon source, they generated distinct chemical and physical binding agents. Employing various extractants and oxidants, these researchers showed that *Agrobacterium radiobacter*, *A. tumefaciens*, and *Rhizobium trifoli* caused binding by polysaccharides when introduced into sterilised silt loam and silty clay loam aggregates. Gasperi-Mago and Troeh (1979) examined how the addition of microbes to soil affects soil stability during simulated rainfall. Sterile and non-sterile soil (<3 mm) were sprayed with nutrient broth containing *Penicillium* sp., *Pseudomonas aeruginosa*, and *Streptomyces* sp., which enhanced soil stability within 15 days with *Penicillium* sp. proving to be the most effective.

Fungi produce hyphae that create networks, bind soil particles, and form stable aggregates. Harris et al. (1966a, b) isolated ten fungal species, and several bacterial spore formers were investigated to check their effects on soil aggregates. It was found that fungi showed varying abilities to influence aggregation, which depend on the presence of sucrose. The formation of stable aggregates >2 mm in size coincided with the visible presence of macroscopic mycelia. In another laboratory experiment, Lynch (1981) revealed that introducing 7-day cultures of *Azotobacter chroococcum*, *Lipomyces starkeyi*, and *Pseudomonas* sp. into silt loam soil led to an immediate enhancement in soil stability, which was directly proportional to the number of cells and biomass added.

The relationships among soil microorganisms, soil structure, and stability changes are complex and highly significant (Yergeau et al., 2007). Microbial products located within soil aggregates modify the soil's physical environment. To boost soil stability and structure, decomposition of plant residues by soil organisms, especially fungi and *actinomycetes*, is highly effective. Beneficial organisms decompose plant and animal residues, release plant nutrients, and generate substances that enhance the soil's structural stability. Deterioration of the soil structure can inhibit these beneficial groups while promoting the growth of microbial groups that make nutrients less available.

13.5 Role of Microbes in Sustainable Agriculture

The global population has been increasing exponentially over the past few centuries and is projected to reach approximately 9 billion by 2050. Therefore, there is a need to increase food production corresponding to an increasing population to ensure food sustenance (Kumar & Dubey, 2020). Conventional agricultural practices have significantly increased production, with intense reliance on the application of chemical fertilisers and pesticides, without compromising ecosystem health and services, resulting in detrimental consequences for diverse ecosystems (Tal, 2018). However, the symbiotic relationship between plants and microbes offers immense promise and acts as a cornerstone for sustainable agriculture, without disturbing ecosystem health and services.

Microbes, such as bacteria, fungi, and archaea, play a critical role in forming an essential agricultural environment that influences soil health, plant growth, and crop productivity. Their intricate association with plants and the environment enhances soil physicochemical properties, such as nutrient cycling, soil health, and the soil microbiome (Srivastava et al., 2023). Some microbial communities, such as plant growth-promoting bacteria (PGPB) and mycorrhizal fungi, help boost plant health and productivity. Studies have also reported that their interaction is not only for nutrient acquisition and productivity but also helps to suppress soil-borne diseases and increase the resilience of the plant, thus reducing the reliance on pesticide applications (Koshila Ravi et al., 2019).

13.5.1 Plant Growth Promotion

Plant growth-promoting microbes (PGPM) are microscopic organisms which help boost the growth and resilience of crops in various stressful environments. These microscopic organisms are secret weapons of nature that promote the growth and health of crops and help reduce reliance on synthetic fertilisers (Srivastava et al., 2023).

However, the function of PGPM is vast; that is, some PGPMs produce hormones and boost crop growth, whereas others help in unlocking the nutrients from the soil. For example, some PGPMs produce various enzymes such as gibberlins, auxins, and cytokinins, which help in stem growth and stimulate the growth of roots, shoots, flowers, and fruits (Srivastava et al., 2023). However, some PGPMs produce metabolites which help unlock the essential nutrients present in the soil. Nutrients, such as phosphorus and iron, are often locked in the soil, making them unavailable to plants. However, some microbes produce enzymes that specialise in iron delivery. The production of siderophores helps bind iron, making it easily available to plants (Kumar & Verma, 2019). In addition, some PGPMs help in coping with stressful environments and toxic metals. By harnessing the power of these microscopic allies, we can cultivate crops that are more resilient to environmental challenges and less reliant on synthetic input.

13.5.2 Soil Health Enhancement

Microbes are architects of soil health and are the main characteristics behind nutrient cycling, organic matter decomposition, and shaping soil formation. These tireless unsung heroes (microbes) break down complex organic materials, transform them into simple substances, and release essential nutrients, such as nitrogen, phosphorus, and potassium, which can be easily absorbed by plants (Shah et al., 2021). Microbes such as rhizobia have symbiotic relationships with legumes which help to fix atmospheric nitrogen and convert it into an available form, reducing reliance on synthetic fertilisers (Das et al., 2022). Moreover, various soil microbes release various enzymes and metabolites which further help improve the soil structure, soil water-holding capacity, and aeration, creating a conducive environment for root growth and overall plant health (Takoutsing et al., 2016).

13.5.3 Biocontrol Agents

In addition to enhancing crop growth and productivity, some microbes are eco-friendly for pest control. However, we often rely on the application of pesticides to protect crops, which results in the destruction of soil physicochemical properties and microorganisms. The application of such pesticides further contaminates food and water and disturbs the delicate balance of ecosystems.

Some microbes act as natural agents to control or fight plant disease-causing organisms. Some microscopic organisms, such as *Bacillus subtilis (bacteria)* and Trichoderma (fungi), battle against disease-causing organisms and suppress the growth of pathogens or bring them to an inactive state. However, some microbes compete for nutrients and space with harmful microbes and stress them. Similarly, other microbes also produce enzymes and act as vaccines for their growth and suppress the growth of disease-causing organisms (Kumar & Verma, 2019). By harnessing the power of this microscopic natural agent, we can reduce our reliance on synthetic pesticides and others and achieve sustainable and eco-friendly crop production.

13.5.4 Bioremediation

In addition to their pivotal role in enhancing the growth and productivity of crops, microbes serve as natural janitors for removing various contaminants from the environment. In this process, various microorganisms can break down harmful substances and convert them into simpler forms; this process is known as bioremediation (Basit et al., 2021).

In the realm of agriculture, various toxic substances are present in herbicides, pesticides, and other agrochemicals which can leach into the soil and water sources and act as agents for destroying soil health and fertility (Singh et al., 2024a; Singh et al., 2024b). However, microorganisms, such as bacteria and fungi, are capable of consuming or detoxifying various toxic compounds present in numerous herbicides, pesticides, and agrochemicals and converting or detoxifying them into less harmful substances (Basit et al., 2021; Rashid et al., 2019). Thus, their conscientiousness, which breaks down or detoxifies various toxic substances, helps mitigate ecological damage. Harnessing this natural process of remediation helps to reduce not only cost-effectiveness and manpower but also promotes the restoration of soil health and fertility. Microbes play an indispensable role in maintaining consciousness, productivity, and sustainability in agriculture.

13.5.5 Biofertilisers

Like humans, plants also contain essential nutrients, such as nitrogen, phosphorus, potassium, and other micronutrients for growth and yield. However, conventional agricultural practices often rely on the application of synthetic fertilisers to obtain nutrients and achieve more efficient productivity without compromising ecosystem health and services (Reddy et al., 2020).

Biofertilisers are also known as living fertilisers and are composed of beneficial microbes. Harnessing this microbe can enhance the availability of plant nutrients in win-win solutions (Reddy et al., 2020). For instance, microbes, such as nitrogen-fixing bacteria (Rhizobia and Azobacter), have a remarkable potential to convert atmospheric nitrogen into nitrate and ammonia which can be utilised by plants. Similarly, microbes such as phosphate-solubilising bacteria (Pseudomonas and Bacillus species) have the potential to convert inorganic phosphorus in the soil into a readily usable form (Rawat et al., 2020). In harmony with nature, this microscopic biofertiliser plays a pivotal role in converting various nutrients into usable forms, leading to an increase in healthier crops, yield, and soil health in a sustainable and eco-friendly manner (Singh et al., 2022). With the help of microscopic natural helpers, we can reduce our reliance on the application of synthetic chemical fertilisers, which contribute to soil pollution and degradation.

13.5.6 Disease Suppression

Some microbes act as bodyguards of soil health by producing antibiotics and antifungal compounds which help inhibit the growth of pathogens, thus playing a crucial role in suppressing soil-borne plant pathogens (Galani et al., 2024). In addition, some microbes act as keen competitors with pathogens for the resource, making it harder for them to thrive on the resource and induce systemic resistance in plants, making them less susceptible to diseases. For instance, Pseudomonas fluorescens releases a powerful antibiotic known as phenazine, which acts as a shield against the sprouting of various soil-borne diseases (Riaz et al., 2021). Thus, other microorganisms help reduce reliance on synthetic pesticide application which causes various harmful effects on ecosystem health and services.

13.5.7 Stress Tolerance

Some microbes release exopolysaccharide molecules into the rhizosphere environment. The release of such molecules helps retain soil moisture under abiotic stress (drought, salinity, and extreme temperatures), thus enhancing plant tolerance to abiotic stresses. Similarly, other microbes help in the synthesis of osmolytes such as proline and glycine betaine, which protect plant cells from osmotic stress (Khan et al., 2023). Under stress, plants produce some chemicals known as reactive oxygen species (ROS) that are highly reactive and harmful to plants. However, certain microbes produce antioxidants which can neutralise ROS, protect cells from further cellular damage, and act as a shield against stress (Jagadesh et al., 2024).

13.5.8 Microbiome Engineering

With advances in knowledge and technologies, scientists have obtained a deep understanding of the intricate relationship between plant-microbe interactions. The introduction of genomics and microbiome research has revealed a new frontier in sustainable agriculture through microbiome engineering. The introduction of such techniques helps to promote plant growth, productivity, resistance to abiotic and biotic stress, and defence against pests and diseases (Chandel et al., 2024).

In this process, a carefully selected group of beneficial microbes is utilised to create a customised microbial consortium and introduced into the soil or plant rhizosphere. This customised microscopic microbiome acts as a specialised team to synergistically boost plant health and productivity. The introduction of such microbes helps increase the availability of nutrients in the rhizosphere, reducing reliance on synthetic fertilisers, pesticides, and other agrochemicals (Omotayo, 2022; Khan et al., 2023). In addition to boosting agricultural productivity, the introduction of microbiome consortia helps in disease suppression and stress tolerance and holds immense potential for developing sustainable agricultural practices.

In summary, microbes are indispensable partners in sustainable agriculture and offer various benefits such as soil health enhancement, bioremediation, biocontrol, and plant growth promotion. The interactions between microbe-plant and microbe-environment interactions are key to developing sustainable agricultural practices that can feed a growing population without compromising ecosystem health and services. Harnessing the power of microbes through biofertilisers, biocontrol agents, and microbiome engineering represents a paradigm shift towards a more eco-friendly and resilient agricultural system. By fostering a deeper understanding of plant-microbe interactions, we can unlock the full potential of these microscopic partners to cultivate a sustainable future for agriculture.

13.6 Role of Microbes in Carbon Sequestration

The continuously increasing carbon dioxide (CO_2) concentration is the main cause of global warming and a major concern worldwide. Starting from 280 ppm in the pre-industrial area, the present CO_2 concentration exceeds 422 ppm (https://climate.nasa.gov/vital-signs/carbon-dioxide/) and is predicted to reach 600–800 ppm by the end of the twentieth century (Knohl & Veldkamp, 2011). To sustain human life on Earth, it is very important to minimise or control the rapid release of CO_2 into the atmosphere by controlling the emission of CO_2 or sequestering the CO_2 from the atmosphere. Among the different types of C sequestration, biotic sequestration accounts for eco-friendly and sustainable processes. Again, among biotic sequestration, microbial carbon sequestration is one of the biotic types and plays a vital role in capturing carbon dioxide from the atmosphere and store in the soil for a longer period as compared with other sequestration techniques (Grover et al., 2015). This chapter focused on the role of microbes in carbon sequestration.

Microbes play a significant role in carbon sequestration and account for long-term storage of atmospheric carbon (Gougoulias et al., 2014). The roles of microbes in capturing and storing carbon from the atmosphere are diverse and they occur in various ecosystems (Zhang et al., 2011). This process is crucial for mitigating climate change by reducing the concentration of greenhouse gases in the atmosphere (Ahmed et al., 2019). In soil, the total microbial biomass is composed of 0.1–2% active, dormant, and dead microorganisms accounting for 42% and 56%, respectively (Blagodatskaya & Kuzyakov, 2013), and the dead microbial biomass in soil is the largest organic carbon stock on the planet. In various ecosystems, the interaction between microbial biomass, community structure, and soil properties such as texture, porosity, and clay mineralogy determines the degree of carbon sequestration. Different microbial communities vary in their metabolic activities, which play different roles in carbon sequestration in various ecosystems. Broadly, the different strategies adopted by microbial carbon sequestration include (a) fungal and bacterial dominance for carbon sequestration, (b) mycorrhizal association, and (c) fungal metabolites (Buragohain, 2019).

In natural ecosystems, bacteria exist in a larger population than fungi but exhibit less carbon sequestration performance than fungi (Li et al., 2015). However, their small contribution around the globe to capturing carbon from the atmosphere has a significant impact on the reduction of greenhouse gases. Recent studies have shown that some bacteria can capture carbon from the atmosphere and mitigate climate change. For example, bacteria such as *Bacillus mucilaginosus* produce a metabolite containing carbonic anhydrase. This carbonic anhydrase metabolite reacts with atmospheric CO_2 to form carbonate minerals (calcium carbonate) that are deposited in the soil. Similarly, various bacterial microbiomes produce different arrays of metabolites which react with and capture atmospheric CO_2. Moreover, Xiao et al. (2015) transcribed the carbonic anhydrous-producing gene from *Bacillus mucilaginosus*, translated it into *Escherichia coli* and examined its ability to form calcium carbonate. Studies have reported that transgenic bacteria produce anhydrous carbonic acid and form calcium carbonate which helps reduce atmospheric carbon dioxide. Similarly, bacteria, such as *Pseudomonas fluorescens*, promote plant growth and productivity (Nie et al., 2015). The increase in plant productivity helps to capture atmospheric CO_2, thus mitigating global warming.

As mentioned above, the fungal population is less than the bacterial population but contributes more to the deposition of organic matter, carbon sequestration, and mineralisation (Six et al., 2006). Fungal diversity contributes to carbon sequestration through the formation of fungal biomass, by-products, and necromass. The degree of carbon sequestration by fungi depends on the type of fungal species and the nature of the environment. The formation of thread-like structures, also known as fungal hyphae, plays a vital role in carbon sequestration in fungi. Fungal hypha is a vegetative structure which spreads around the soil for nutrients and water, and fungal hyphae production can reach up to a kilogram per hectare per year (Wallander et al., 2013). Again, these hyphae exudate a specialised liquid from the tips known as glomalin, which contains mainly carbon. Wu et al. (2014a, b) reported that this glomalin helps attach the soil particles to the organic matter and form clumps. Indeed, the formation of clumps is versatile in nature, which allows the movement of roots, air, and water but resists external erosion. Once glomalin is formed, it captures carbon and remains protected for a longer period. Glomalin is insoluble in water (Pal & Pandey, 2014), which precedes carbon leaching. Moreover, this fungal hypha is made up of carbon, and the more fungal hyphae are produced, the more carbon is sequestered in the soil. After the death of these hyphae, it converts the fungal necromass, which is a very important source of carbon storage in the soil and food for living organisms. This fungal necromass contains high amounts of chitin and nitrogen, which facilitates decomposition within a short period. However, in the cell wall of the mycelium, melanin is present which resists decomposition, and the concentration of melanized cells differs among fungal species, leading to long-term sequestration of carbon in the soil (Fernandez & Koide, 2012, 2014; Drigo et al., 2012; Koide et al., 2014). Thus, some fungal species sequester carbon from the atmosphere to mitigate climate change.

13.7 Role of Microbes in Bioremediation

Microbes serve as covert agents in ecological systems, contributing not only to nutrient cycling, plant growth promotion, and food provisioning but also clandestinely participating in the remediation of pollutants across diverse ecosystems. Microbial communities engage in bioremediation through intricate biochemical processes and metabolic pathways Bioremediation is a cutting-edge and eco-friendly technique that employs natural biological processes to detoxify, clean, or reduce the concentration of pollutants in various ecosystems (Meitei & Prasad, 2021; Huidrom et al., 2024). Among the different natural biological approaches, microbial bioremediation is eco-friendly, cost-effective, and sustainable in nature (Srivastava et al., 2014). Microbes such as bacteria, fungi, and algae play a crucial role in breaking down or transforming various contaminants into less harmful substances. Owing to huge varieties of microbes and versatility in nature, microbial bioremediation can be applied to a wide range of pollutants, including heavy metals, hydrocarbons, nutrients, pesticides, and industrial wastes (Hopkin, 1998; Nanda et al., 2019) . So far, a microbial diversity of only 5% has been reported in bioremediation applications, and the remaining 95% has yet to be explored (Riccardi et al., 2005). Bioremediation is a multiphasic, heterogeneous process that depends on various biotic and abiotic factors.

Microbial bioremediation is characterised by the lack of a singular uniform mechanism; rather, it encompasses a diverse array of microbial activities and interactions that collectively contribute to the degradation, transformation, and sequestration of various environmental pollutants (Verma & Kuila, 2019). The multifaceted nature of bioremediation arises from the intricate biochemical pathways, metabolic processes, and ecological dynamics exhibited by different microorganisms under distinct environmental conditions (Tayang & Songachan, 2021). The diversity of mechanisms underscores the complexity inherent in the microbial response to pollutants, necessitating a nuanced understanding of microbial ecology and biotechnological applications for effective environmental management. For example, bacteria (*Streptomyces rimosus, Cronobacter muytjensii, and Staphylococcus aureus*) have a specialised protective layer in their cell walls, peptidoglycan, which absorbs pollutants in their cell walls without showing any effects (Vasudevan et al., 2001). The density of peptidoglycan in the microbial cell wall determines the adsorption capability of pollutants. For example, the peptidoglycan layer present in gram-positive bacteria is thicker than that in gram-negative bacteria, and teichoic acid, amino acid, and meso-diaminopimelic acids mainly comprise the peptidoglycan of gram-positive bacteria, which offers a positive site for binding various toxic pollutants from the surrounding environment, resulting in greater adsorption of pollutants (Hoyle & Beveridge, 1983). Similarly, fungal species (*Saccharomyces cerevisiae, Phanerochaeta chrysosporium, Penicillium aurantiogriseum,* and *Alternaria alternata*) exhibit good performance in the biosorption of toxic pollutants from various ecosystems (Say et al., 2001; Machado et al., 2010; Bahobil et al., 2017). The fungal mycelium is composed of negatively charged particles such as amino groups, carboxyls, phosphate, and sulfate, which offer a huge advantage for the biosorption of pollutants (Say et al., 2001). In addition, many photosynthetic algae, such as *Spirulina platensis, Chlorella vulgaris, Oscillatoria* sp., and *Sargassam* sp., are capable of adsorbing various toxic pollutants from the surrounding environment (Leong & Chang, 2020; Riskuwa-Shehu et al., 2019; Sayadi et al., 2019). These green photosynthetic algae contain various cell-polymeric substances (peptides and polysaccharides). This polymeric substance is composed of alginate, cellulose, organic proteins, lipids, and many functional groups, such as carboxyl, amino, hydroxyl, phosphate, imidazole, thiol, and sulfonate, in their cell walls. The presence of such compounds in the cell wall is a good site for the binding of various toxic pollutants (Priatni et al., 2018; Pradhan et al., 2019). Thus, microbes adsorb toxic pollutants from the surrounding environment and sequester them into their cells, resulting in reduced environmental pollutant concentrations. This toxic metal remediation process is known as microbial biosorption. In addition, some microbes can transfer pollutants into the cytoplasm or intracellular space using protein transporters. Microorganisms, such as *Pseudomonas putida* 62 BN, accumulate toxic metals inside the intracellular and periplasmic regions without showing any effects. Furthermore, the authors concluded that this microbe has great potential for the accumulation of toxic metals (Rani & Goel, 2009). This remediation process is known as microbial bioaccumulation (Table 13.1).

In the biotransformation process, toxic pollutant structures are generally modified and transformed into non-toxic or polar substances. Microbes such as *Acinetobacter* sp. and *Micrococcus* sp. interact with toxic arsenic (As III), oxidise it, and convert it into non-toxic arsenic after reducing its solubility (Nagvenkar & Ramaiah, 2010). Similarly, *the Bacillus* sp. SFC 500-1E is used for the transformation of Cr (VI) (toxic copper) into non-toxic copper (Cr III) without causing any harm to microbes (Thatoi et al., 2014). Such microbes have high growth rates, surface areas, and metabolic activities, resulting in the ability to maintain sterile conditions that allow high-rate biotransformation of toxic metals into non-toxic metals without producing any secondary pollutants. Therefore, they are widely used for the remediation of various toxic substances such as hydrocarbons, pharmaceuticals, pesticides, and toxic metals.

In summary, bioremediation through microbes (biosorption, bioaccumulation, and biotransformation) offers a promising, eco-friendly, cost-effective, and sustainable approach to mitigating environmental pollution. The diverse mechanisms employed by microorganisms highlight the potential of microbial bioremediation for addressing a range of environmental challenges. Microbial bioremediation is likely to play an important role in the future in maintaining a sustainable environment, as we need to improve our understanding of microbial ecology and biotechnological applications, along with improvements in technological advances.

Table 13.1 The role of microbes in bioremediation

SL	Type of microbes	Types of metal	Removal efficiency	Reference
1	*Acinetobacter junii* (Bacteria)	Pb (II)	1071 mg g^{-1}	Kushwaha et al. (2017)
2	*Bacillus* sp. (Bacteria)	Cu (II)	97.6%	Rani et al. (2010)
3	*Pseudomonas* sp. (Bacteria)	Cu (II)	70.4%	Azzam and Tawfik (2015)
4	*Rhodobacter capsulatus* (Bacteria)	Zn (II)	164 mg g^{-1}	Magnin et al. (2014)
5	*Staphylococcus* sp. (Bacteria)	Cd (II)	44%	Nanda et al. (2011)
6	*Aspergillus niger* (Fungi)	Pb (II)	172 mg g^{-1}	Iram et al. (2015)
7	*Cellulosimicrobium* sp. (Fungi)	Cd (II), Cu (II), Fe (II), Ni (II)	7.65%, 21.21%, 70.26%, 18.98%	Bhati et al. (2019)
8	*Termitomyces clypeatus* (Fungi)	Cr (II)	403 mg g^{-1}	Ramrakhiani et al. (2011)
9	*Penicillium* (Fungi)	Pb (II)	11.55 mg g^{-1}	Kumar et al. (2014)
10	*Trichoderma viride* (Fungi)	Pb (II)	9.14 mg g^{-1}	Kumar et al. (2014)
11	*Aspergillus oryzae*	Endosulfan	95%	Gangola et al. (2016)
12	Phlebia aurea	Heptachlor epoxide	90%	Xiao et al. (2011)
13	*Sphingobacterium* sp.	DDT	–	Fang et al. (2010)
14	Diuron	*Mortierella isabellina*	58	Tixier et al. (2000)
15	Toxaphene	*Bjerkandera* sp.	85	Romero et al. (2006)
16	Endosulfan	*Bacillus aryabhatti*	87%	Negi et al. (2014)
17	Cypermethrin	*Bacillus* sp.	81.6%	Pankaj et al. (2016)
18	Diuron, Metalaxyl, Atrazine, & Terbuthylazine	*Coriolus versicolor*	86%	Bending et al. (2002)
19	Diuron, Chlortoluron, Isoproturon, & Linuron	*Mortierella* sp. (z)	–	Badawi et al. (2009)
20	DDT, naphthalene	*Dunaliella* sp.	–	Ardal (2014)

13.8 Conclusion

In the early years of the twentieth century, human-environment interactions disrupted ecosystem services in many ways and have become a major concern worldwide. Human activities disturb ecosystem services beyond self-restoration capabilities. Understanding the diversity and nature of microbes offers a potentially dynamic and economically feasible approach to tackling this global problem. Soil contains millions of microbes and performs an intricate dance beginning with fundamental ecosystem processes, including nutrient cycling, nutrient fixation and transformation, soil texture, climate mitigation through carbon sequestration, remediation of toxic metals, and supporting the growth of plants under stressful environmental conditions and increase the agricultural productivity in a sustainable form. Understanding the multifaceted nature of soil microbes and emphasising their silent yet transformative actions in sustaining the intricate tapestry of life on Earth.

References

Ahmed, A. A. Q., Odelade, K. A., & Babalola, O. O. (2019). Microbial inoculants for improving carbon sequestration in agroecosystems to mitigate climate change. In *Handbook of climate change resilience* (pp. 1–21). Springer.

Ardal, E. (2014). Phycoremediation of pesticides using microalgae. In *Faculty of landscape architecture, horticulture and crop production science* (pp. 3–40).

Aspiras, R. B., Allen, O. N., Harris, R. F., & Chesters, G. (1971). Aggregate stabilisation by filamentous microorganisms. *Soil Science, 1*, 282–284.

Azzam, A. M., & Tawfik, A. (2015). Removal of heavy metals using bacterial bio-flocculants of Bacillus sp. and Pseudomonas sp. *Journal of Environmental Engineering and Landscape Management, 23*(4), 288–294.

Baalsrud, K., & Baalsrud, K. S. (1954). Studies on *Thiobacillus denitrificans*. *Archives of Microbiology, 20*, 34–62.

Badawi, N., Rønhede, S., Olsson, S., Kragelund, B. B., Johnsen, A. H., Jacobsen, O. S., & Aamand, J. (2009). Metabolites of the phenylurea herbicides chlorotoluron, diuron, isoproturon and linuron produced by the soil fungus Mortierella sp. *Environmental Pollution, 157*(10), 2806–2812. https://doi.org/10.1016/j.envpol.2009.04.019

Bahobil, A., Bayoumi, R. A., Atta, H. M., & El-Sehrawey, M. M. (2017). Fungal biosorption for cadmium and mercury heavy metal ions isolated from some polluted localities in KSA. *International Journal of Current Microbiology and Applied Sciences, 6*(6), 2138–2154.

Basit, A., Shah, S. T., Ullah, I., Muntha, S. T., & Mohamed, H. I. (2021). Microbe-assisted phytoremediation of environmental pollutants and energy recycling in sustainable agriculture. *Archives of Microbiology, 203*(10), 5859–5885.

Bending, G. D., Friloux, M., & Walker, A. (2002). Degradation of contrasting pesticides by white rot fungi and its relationship with ligninolytic potential. *FEMS Microbiology Letters, 212*(1), 59–63. https://doi.org/10.1016/S0378-1097(02)00710-3

Bergaust, L., Shapleigh, J., Frostegård, Å., & Bakken, L. (2008). Transcription and activities of NOx reductases in Agrobacterium tumefaciens: The influence of nitrate, nitrite and oxygen availability. *Environmental Microbiology, 10*(11), 3070–3081.

Bergaust, L., Mao, Y., Bakken, L. R., & Frostegård, A. (2010). Denitrification response patterns during the transition to anoxic respiration and posttranscriptional effects of suboptimal pH on nitrogen oxide reductase in *Paracoccus denitrificans*. *Applied and Environmental Microbiology, 76*(19), 6387–6396.

Bhati, T., Gupta, R., Yadav, N., Singh, R., Fuloria, A., Waziri, A., et al. (2019). Assessment of bioremediation potential of Cellulosimicrobium sp. for treatment of multiple heavy metals. *Microbiology and Biotechnology Letters, 47*(2), 269–277.

Bollag, J. M., & Tung, G. (1972). Nitrous oxide release by soil fungi. *Soil Biology and Biochemistry, 4*, 271–276.

Bowen, G. D., & Rovira, A. D. (1991). The rhizosphere, the hidden half of the hidden half. In *Plant roots: The hidden half* (pp. 641–669).

Bru, D., Ramette, R., Saby, N. P. A., Dequiedt, S., Ranjard, L., Jolivet, C., Arrouays, D., & Philippot, L. (2011). Determinants of the distribution of nitrogen-cycling microbial communities at the landscape scale. *The ISME Journal, 5*, 532–542.

Brumme, R., & Beese, F. (1992). Effects of liming and nitrogen fertilization on emissions of CO2 and N2O from temperate forest. *Journal of Geophysical Research, 97*, 12851–12858.

Bugmann, H. K. (1996). A simplified forest model to study species composition along climate gradients. *Ecology, 77*(7), 2055–2074.

Buragohain, P. (2019). Role of microbes on carbon sequestration. *International Journal of Microbiology Research*. ISSN, 0975-5276.

Cardinale, B. J., Duffy, J. E., Gonzalez, A., Hooper, D. U., Perrings, C., Venail, P., et al. (2012). Biodiversity loss and its impact on humanity. *Nature, 486*(7401), 59–67.

Carlson, C. A., & Ingraham, J. L. (1983). Comparison of denitrification by Pseudomonas stutzeri, Pseudomonas aeruginosa, and Paracoccus denitrificans. *Applied and Environmental Microbiology, 45*, 1247–1253.

Chandel, N., Kumar, A., & Kumar, R. (2024). Towards sustainable agriculture: Integrating agronomic practices, environmental physiology and plant nutrition. *International Journal of Plant & Soil Science, 36*(6), 492–503.

Chen, J., Li, F., Zhao, X., Wang, Y., Zhang, L., Feng, L., et al. (2023). The weathering process of carbonatite: Weathering time. *PeerJ, 11*, e15793.

Cleveland, C. C., Townsend, A. R., Schimel, D. S., Fisher, H., Howarth, R. W., Hedin, L. O., Perakis, S. S., Latty, E. F., von Fischer, J. C., Elseroad, A., & Wasson, M. F. (1999). Global patterns of terrestrial biological nitrogen (N2) fixation in natural ecosystems. *Global Biogeochemical Cycles, 13*, 623–645.

Colliver, B. B., & Stephenson, T. (2000). Production of nitrogen oxide and dinitrogen oxide by autotrophic nitrifiers. *Biotechnology Advances, 18*, 219–232.

Cracknell, P. C., & Lousier, J. D. (1988). Improving forest fertilization decision-making in British Columbia. In J. D. Lousier, H. Brix, R. Brockley, R. Carter, & V. G. Marshall (Eds.), *Improving forest fertilization decision-making in British Columbia: Proceedings of a forest fertilization workshop* (p. 293). Research Branch, Ministry of Forests.

Das, P. P., Singh, K. R., Nagpure, G., Mansoori, A., Singh, R. P., Ghazi, I. A., et al. (2022). Plant-soil-microbes: A tripartite interaction for nutrient acquisition and better plant growth for sustainable agricultural practices. *Environmental Research, 214*, 113821.

Delgado, J. A., & Follett, R. F. (2002). Carbon and nutrient cycles. *Journal of Soil and Water Conservation, 57*(6), 455–464.

Demanèche, S., Philippot, L., David, M. M., Navarro, E., Vogel, T. M., & Simonet, P. (2009). Characterization of denitrification gene clusters of soil bacteria via a metagenomic approach. *Applied and Environmental Microbiology, 75*, 534–537.

Dominy, C., & Haynes, R. (2002). Influence of agricultural land management on organic matter content, microbial activity and aggregate stability in the profiles of two Oxisols. *Biology and Fertility of Soils, 36*, 298–305.

Doube, B. M., & Gupta, V. V. S. R. (1997). Biological indicators of soil health: Synthesis. In C. E. Pankhurst, B. M. Doube, & V. V. S. R. Gupta (Eds.), *Biological indicators of soil health* (pp. 419–435). CAB International.

Drigo, B., Anderson, I. C., Kannangara, G. S. K., Cairney, J. W., & Johnson, D. (2012). Rapid incorporation of carbon from ectomycorrhizal mycelial necromass into soil fungal communities. *Soil Biology and Biochemistry, 49*, 4–10.

Ekblad, A., Wallander, H., Godbold, D. L., Cruz, C., Johnson, D., Baldrian, P., et al. (2013). The production and turnover of extramatrical mycelium of ectomycorrhizal fungi in forest soils: Role in carbon cycling. *Plant and Soil, 366*, 1–27.

Fang, H., Dong, B., Yan, H., Tang, F., & Yu, Y. (2010). Characterization of a bacterial strain capable of degrading DDT congeners and its use in bioremediation of contaminated soil. *Journal of Hazardous Materials, 184*(1–3), 281–289.

Fernandez, C. W., & Koide, R. T. (2012). The role of chitin in the decomposition of ectomycorrhizal fungal litter. *Ecology, 93*(1), 24–28.

Fernandez, C. W., & Koide, R. T. (2014). Initial melanin and nitrogen concentrations control the decomposition of ectomycorrhizal fungal litter. *Soil Biology and Biochemistry, 77*, 150–157.

Galani, B., Chauhan, J., & Gohel, S. (2024). Role of rhizospheric microbial enzymes in plant growth promotion, antagonism, and sustainable agriculture: A review. *Trends in Horticulture, 7*(2), 4902.

Gangola, S., Negi, G., Srivastava, A., & Sharma, A. (2016). Enhanced biodegradation of endosulfan by Aspergillus and Trichoderma spp. isolated from an agricultural field of tarai region of Uttarakhand. *Pesticide Research Journal, 140*(13), 223–230. www.IndianJournals.com

Gasperi-Mago, R. R., & Troeh, F. R. (1979). Microbial effects on soil erodibility. *Soil Science Society of America Journal, 43*(4), 765–768.

Glick, B. R. (2018). Soil microbes and sustainable agriculture. *Pedosphere, 28*(2), 167–169.

Gougoulias, C., Clark, J. M., & Shaw, L. J. (2014). The role of soil microbes in the global carbon cycle: Tracking the below-ground microbial processing of plant-derived carbon for manipulating carbon dynamics in agricultural systems. *Journal of the Science of Food and Agriculture, 94*(12), 2362–2371.

Gouzou, L., Burtin, G., Philippy, R., Bartoli, F., & Heulin, T. (1993). Effect of inoculation with Bacillus polymyxa on soil aggregation in the wheat rhizosphere: Preliminary examination. *Geoderma, 56*, 479–491.

Grover, M., Maheswari, M., Desai, S., Gopinath, K. A., & Venkateswarlu, B. (2015). Elevated CO2: Plant associated microorganisms and carbon sequestration. *Applied Soil Ecology, 95*, 73–85.

Gupta, A., Gupta, R., Singh, R., & L. (2017). Microbes and environment. In *Principles and applications of environmental biotechnology for a sustainable future* (pp. 43–84). Springer.

Harris, R. F., Chester, G., & Allen, O. N. (1966a). Soil aggregate stabilisation by indigenous microtlora as affected by temperature. *Soil Science Society of America Proceedings, 30*, 205–210.

Harris, R. F., Chester, G., & Allen, O. N. (1966b). Dynamics of soil aggregation. *Advances in Agronomy, 18*, 107–169.

Henry, S., Bru, D., Stres, B., Hallet, S., & Philippot, L. (2006). Quantitative detection of the nosZ gene, encoding nitrous oxide reductase, and comparison of the abundances of 16S rRNA, narG, nirK, and nosZ genes in soils. *Applied and Environmental Microbiology, 72*, 5181–5189.

Hernández-Hernández, R. M., & López-Hernández, D. (2002). Microbial biomass, mineral nitrogen and carbon content in savanna soil aggregates under conventional and no-tillage. *Soil Biology and Biochemistry, 34*(11), 1563–1570.

Hopkin, S. (1998). *Biological indicators of soil health*. In C. E. Pankhurst, B. M. Doube, & V. V. S. R. Gupta. 23× 15 cm. Pp. xii+ 451 with 70 text-figures. CAB International, 1997. Price h/b:£ 60.00, ISBN 0 85199 158 0. The New Phytologist, 139(2), 389–392.

Hoyle, B., & Beveridge, T. J. (1983). Binding of metallic ions to the outer membrane of Escherichia coli. *Applied and Environmental Microbiology, 46*(3), 749–752.

Huidrom, M., Singh, S. S., & Meitei, M. D. (2024). Role of Naga Nullah tributary as a major polluter of Loktak: A Ramsar site from Manipur, India with 3 decades in the Montreux records of the wetland's convention. *Sustainable Water Resources Management, 10*(2), 56.

Iram, S., Shabbir, R., Zafar, H., & Javaid, M. (2015). Biosorption and bioaccumulation of copper and lead by heavy metal-resistant fungal isolates. *Arabian Journal for Science and Engineering, 40*, 1867–1873.

Jagadesh, M., Dash, M., Kumari, A., Singh, S. K., Verma, K. K., Kumar, P., et al. (2024). Revealing the hidden world of soil microbes: Metagenomic insights into plant, bacteria, and fungi interactions for sustainable agriculture and ecosystem restoration. *Microbiological Research, 127764*, 127764.

Johnson, D. W., & Curtis, P. S. (2001). Effects of forest management on soil C and N storage: Meta-analysis. *Forest Ecology and Management, 140*, 227–238.

Jordan, C. F. (1985). Nutrient cycling in tropical forest ecosystems. In *Principles and their application in management and conservation*. John Wiley & Sons.

Kätterer, T., & Andrén, O. (2001). The ICBM family of analytically solved models of soil carbon, nitrogen and microbial biomass dynamics—descriptions and application examples. *Ecological Modelling, 136*(2–3), 191–207.

Kay, B. D. (1990). Rates of change of soil structure under different cropping systems. *Advances in Soil Sciences, 12*, 1–52.

Kennedy, A. C., & Papendick, R. I. (1995). Microbial characteristics of soil quality. *Journal of Soil and Water Conservation, 50*(3), 243–248.

Khan, D., Kabiraj, A., Biswas, R., Roy, R. K., & Bandopadhyay, R. (2023). Plant-microbe community dynamics in rhizosphere: Reviewing the grassroots ecology towards sustainable agriculture. *The Indian Journal of Agricultural Sciences, 93*(3), 247–257.

Knohl, A., & Veldkamp, E. (2011). Indirect feedback to rising CO2. *Nature, 475*(7355), 177–178.

Knowles, R. (1982). Denitrification. *Microbiological Reviews, 46*, 43–70.

Koide, R. T., Fernandez, C., & Malcolm, G. (2014). Determining place and process: Functional traits of ectomycorrhizal fungi that affect both community structure and ecosystem function. *New Phytologist, 201*(2), 433–439.

Koops, H. P., & Möller, U. C. (1992). The lithotrophic ammonia-oxidizing bacteria. In A. Balows, H. G. Trüper, H. Dworkin, & M.W., Schleifer, K.H. (Eds.), *The Prokaryotes* (pp. 2625–2637). Springer Verlag.

Koshila Ravi, R., Anusuya, S., Balachandar, M., & Muthukumar, T. (2019). Microbial interactions in soil formation and nutrient cycling. In *Mycorrhizosphere and pedogenesis* (pp. 363–382). Springer Nature Singapore.

Kowalchuk, G. A., & Stephen, J. R. (2001). Ammonia-oxidizing bacteria: A model for molecular microbial ecology. *Annual Review of Microbiology, 55*, 485–529.

Kumar, A., & Verma, J. P. (2019). The role of microbes to improve crop productivity and soil health. In *Ecological wisdom inspired restoration engineering* (pp. 249–265). Springer.

Kumar, R., Singh, P., Dhir, B., Sharma, A. K., & Mehta, D. (2014). Potential of some fungal and bacterial species in bioremediation of heavy metals. *Journal of Nuclear Physics, Material Sciences, Radiation and Applications, 1*(2), 213–223.

Kumar, A., & Dubey, A. (2020). Rhizosphere microbiome: Engineering bacterial competitiveness for enhancing crop production. *Journal of Advanced Research, 24*, 337–352.

Kushwaha, A., Rani, R., Kumar, S., Thomas, T., David, A. A., & Ahmed, M. (2017). A new insight to adsorption and accumulation of high lead concentration by exopolymer and whole cells of lead-resistant bacterium Acinetobacter junii L. Pb1 isolated from coal mine dump. *Environmental Science and Pollution Research, 24*, 10652–10661.

Larson, M. G., Jurgensen, M. F., & Harvey, A. E. (1978). N2-fixation associated with wood decayed by some common fungi in western Montana. *Canadian Journal of Forest Research, 8*, 341–345.

Laughlin, R. J., & Stevens, R. J. (2002). Evidence for fungal dominance of denitrification and codenitrification in a grassland soil. *Soil Science Society of America Journal, 66*(5), 1540–1548.

Leong, Y. K., & Chang, J. S. (2020). Bioremediation of heavy metals using microalgae: Recent advances and mechanisms. *Bioresource Technology, 303*, 122886.

Li, C., & Hung, L. (1987). Nitrogen-fixing (acetylene-reducing) bacteria associated with ectomycorrhizae of douglas-fir. *Plant and Soil, 98*, 425–428.

Li, N., Xu, Y. Z., Han, X. Z., He, H. B., Zhang, X. D., & Zhang, B. (2015). Fungi contribute more than bacteria to soil organic matter through necromass accumulation under different agricultural practices during the early pedogenesis of a Mollisol. *European Journal of Soil Biology, 67*, 51–58.

Lynch, J. M. (1981). Promotion and inhibition of soil aggregate stabilisation by soil microorganisms. *Journal of General Microbiology, 126*, 371–375.

Machado, M. D., Soares, E. V., & Soares, H. M. (2010). Removal of heavy metals using a brewer's yeast strain of Saccharomyces cerevisiae: Chemical speciation as a tool in the prediction and improving of treatment efficiency of real electroplating effluents. *Journal of Hazardous Materials, 180*(1–3), 347–353.

Magnin, J. P., Gondrexon, N., & Willison, J. C. (2014). Zinc biosorption by the purple non-sulfur bacterium Rhodobacter capsulatus. *Canadian Journal of Microbiology, 60*(12), 829–837.

Maillard, F., Colin, Y., Viotti, C., Buée, M., Brunner, I., Brabcová, V., et al. (2023). A cryptically diverse microbial community drives organic matter decomposition in forests. *Applied Soil Ecology, 193*, 105148.

McLean, M. A., & Parkinson, D. (1997). Soil impacts of the epigeic earthworm Dendrobaena octaedra on organic matter and microbial activity in lodgepole pine forest. *Canadian Journal of Forest Research, 27*(12), 1907–1913.

Meena, V. S., Verma, J. P., & Meena, S. K. (2015a). Towards the current scenario of nutrient use efficiency in crop species. *Journal of Cleaner Production, 102*, 556.

Meitei, M. D., & Prasad, M. N. V. (2021). Potential of Typha latifolia L. for phytofiltration of iron-contaminated waters in laboratory-scale constructed microcosm conditions. *Applied Water Science, 11*, 1–10.

Morales, S. E., Cosart, T., & Holben, W. E. (2010). Bacterial gene abundances as indicators of greenhouse gas emission in soils. *The ISME Journal, 4*, 799–808.

Nagvenkar, G. S., & Ramaiah, N. (2010). Arsenite tolerance and biotransformation potential in estuarine bacteria. *Ecotoxicology, 19*, 604–613.

Nair, P. K. R., Kang, B. T., & Kass, D. C. L. (1995). Nutrient cycling and soil-erosion control in agroforestry systems. In *Agriculture and the environment: Bridging food production and environmental protection in developing countries* (Vol. 60, pp. 117–138). American Society of Agronomy: Crop Science Society of America: Soil Science Society of America.

Nanda, M., Sharma, D., & Kumar, A. (2011). Removal of heavy metals from industrial effluent using bacteria. *International Journal of Environmental Sciences, 2*(2), 765–780.

Nanda, M., Kumar, V., & Sharma, D. K. (2019). Multimetal tolerance mechanisms in bacteria: The resistance strategies acquired by bacteria that can be exploited to 'clean-up'heavy metal contaminants from water. *Aquatic Toxicology, 212*, 1–10.

Negi, G., Srivastava, A., & Sharma, A. (2014). In situ biodegradation of endosulfan, imidacloprid, and carbendazim using indigenous bacterial cultures of agriculture fields of Uttarakhand, India. *International Journal of Bioengineering and Life Sciences, 8*(9), 973–981.

Nemergut, D. R., Schmidt, S. K., Fukami, T., O'Neill, S. P., Bilinski, T. M., Stanish, L. F., Knelman, J. E., Darcy, J. L., Lynch, R. C., Wickey, P., & Ferrenberg, S. (2013). Patterns and processes of microbial community assembly. *Microbiology and Molecular Biology Reviews, 77*, 342–356.

Nie, M., Bell, C., Wallenstein, M. D., & Pendall, E. (2015). Increased plant productivity and decreased microbial respiratory C loss by plant growth-promoting rhizobacteria under elevated CO2. *Scientific Reports, 5*(1), 9212.

Nielsen, M. N., & Winding, A. (2002). *Microorganisms as indicators of soil health*. NERI Technical Report No. 388. National Environmental Research Institute, Ministry of the Environment.

Omotayo, O. P. (2022). The role of rhizosphere microbiome in agricultural sustainability. *PONTE International Journal of Science and Research, 78*(10).

Pal, A., & Pandey, S. (2014). Role of glomalin in improving soil fertility. *International Journal of Plant and Soil Science, 3*, 112–129.

Pandey, N. (2018). Role of plant nutrients in plant growth and physiology. In *Plant nutrients and abiotic stress tolerance* (pp. 51–93). Springer Singapore.

Pankaj, S. A., Gangola, S., Khati, P., Kumar, G., & Srivastava, A. (2016). Novel pathway of cypermethrin biodegradation in a Bacillus sp. strain SG2 isolated from cypermethrin-contaminated agriculture field. *3 Biotech, 6*(1), 1–11. https://doi.org/10.1007/s13205-016-0372-3

Pankhurst, C. E., Hawke, B. G., McDonald, H. J., Kirkby, C. A., Buckerfield, J. C., Michelsen, P., ... & Doube, B. M. (1995). Evaluation of soil biological properties as potential bioindicators of soil health. Australian Journal of Experimental Agriculture, 35(7), 1015–1028.

Paul, L. R., Chapman, B. K., & Chanway, C. P. (2007). Nitrogen fixation associated with Suillus tomentosus tuberculate ectomycorrhizae on Pinus contorta var. latifolia. *Annals of Botany, 99*, 1101–1109.

Petersen, D. G., Blazewicz, S. J., Firestone, M., Herman, D. J., Turetsky, M., & Waldrop, M. (2012). Abundance of microbial genes associated with nitrogen cycling as indices of biogeochemical process rates across a vegetation gradient in Alaska. *Environmental Microbiology, 14*, 993–1008.

Peuke, A. D. (2000). The chemical composition of xylem sap in Vitis vinifera L. cv. Riesling during vegetative growth on three different Franconian vineyard soils and as influenced by nitrogen fertilizer. *American Journal of Enology and Viticulture, 51*(4), 329–339.

Pradhan, D., Shukla, L. B., Mishra, B. B., & Devi, N. (2019). Biosorption for removel of hexavalent chromium using microalgae Scenedesmus sp. *Journal of Cleaner Production, 209*, 617–616.

Priatni, S., Ratnaningrum, D., Warya, S., & Audina, E. (2018). Phycobiliproteins production and heavy metals reduction ability of Porphyridium sp. In *IOP conference series: Earth and environmental science* (Vol. 160, p. 012006). IOP Publishing.

Raghavendra, M. P., Nayaka, N. C., & Nuthan, B. R. (2016). Role of rhizosphere microflora in potassium solubilization. In V. S. Meena, B. R. Maurya, J. P. Verma, & R. S. Meena (Eds.), *Potassium solubilizing microorganisms for sustainable agriculture* (pp. 43–59). Springer.

Ramrakhiani, L., Majumder, R., & Khowala, S. (2011). Removal of hexavalent chromium by heat inactivated fungal biomass of Termitomyces clypeatus: Surface characterization and mechanism of biosorption. *Chemical Engineering Journal, 171*(3), 1060–1068.

Rana, A., Pandey, R. K., & Ramakrishnan, B. (2019). Enzymology of the nitrogen cycle and bioremediation of toxic nitrogenous compounds. In *Smart bioremediation technologies* (pp. 45–61). Academic Press.

Rani, A., & Goel, R. (2009). Strategies for crop improvement in contaminated soils using metal-tolerant bioinoculants. In *Microbial strategies for crop improvement* (pp. 85–104).

Rani, M. J., Hemambika, B., Hemapriya, J., & Kannan, V. R. (2010). Comparative assessment of heavy metal removal by immobilized and dead bacterial cells: A biosorption approach. *African Journal of Environmental Science and Technology, 4*(2).

Rashid, M. H., Kamruzzaman, M., Haque, A. N. A., & Krehenbrink, M. (2019). Soil microbes for sustainable agriculture. In *Sustainable management of soil and environment* (pp. 339–382). Springer.

Rawat, J., Sanwal, P., & Saxena, J. (2016). Potassium and its role in sustainable agriculture. In V. S. Meena, B. R. Maurya, J. P. Verma, & R. S. Meena (Eds.), *Potassium solubilizing microorganisms for sustainable agriculture* (pp. 235–253). Springer. https://doi.org/10.1007/978-81-322-2776-2_17

Rawat, P., Shankhdhar, D., & Shankhdhar, S. C. (2020). Plant growth-promoting rhizobacteria: A booster for ameliorating soil health and agriculture production. In *Soil health* (pp. 47–68). https://doi.org/10.1007/978-3-030-44364-1_3

Reddy, G. C., Goyal, R. K., Puranik, S., Waghmar, V., Vikram, K. V., & Sruthy, K. S. (2020). Biofertilizers toward sustainable agricultural development. In *Plant microbe symbiosis* (pp. 115–128). Springer.

Riaz, U., Shahzad, L., Anum, W., & Waheed, A. (2021). Favourable soil microbes for sustainable agriculture. In *Handbook of research on microbial remediation and microbial biotechnology for sustainable soil* (pp. 135–157). IGI Global.

Riccardi, C., Papacchini, M., Mansi, A., Ciervo, A., Petrucca, A., La Rosa, G., et al. (2005). Characterization of bacterial population coming from a soil contaminated by polycyclic aromatic hydrocarbons (PAHs) able to degrade pyrene in slurry phase. *Annals of Microbiology, 55*(2), 85–90.

Riskuwa-Shehu, M. L., Ismail, H. Y., & Sulaiman, M. (2019). Biosorption of heavy metals by Oscillatoria species. *Microbiology Research Journal International, 27*(6), 1–8.

Romero, M. L., Terrazas, E., Van Bavel, B., & Mattiasson, B. (2006). Degradation of toxaphene by Bjerkandera sp. strain BOL13 using waste biomass as a cosubstrate. *Applied Microbiology and Biotechnology, 71*(4), 549–554.

Ruark, G. A., & Zarnoch, S. J. (1992). Soil carbon, nitrogen, and fine root biomass sampling in a pine stand. *Soil Science Society of America Journal, 56*(6), 1945–1950.

Saad, O. A., & Conrad, R. (1993). Adaptation to temperature of nitric oxide-producing nitrate-reducing bacterial populations in soil. *Systematic and Applied Microbiology, 16*(1), 120–125.

Say, R., Denizli, A., & Arıca, M. Y. (2001). Biosorption of cadmium (II), lead (II) and copper (II) with the filamentous fungus Phanerochaete chrysosporium. *Bioresource Technology, 76*(1), 67–70.
Sayadi, M. H., Rashki, O., & Shahri, E. (2019). Application of modified Spirulina platensis and Chlorella vulgaris powder on the adsorption of heavy metals from aqueous solutions. *Journal of Environmental Chemical Engineering, 7*(3), 103169.
Shah, K. K., Tripathi, S., Tiwari, I., Shrestha, J., Modi, B., Paudel, N., & Das, B. D. (2021). Role of soil microbes in sustainable crop production and soil health: A review. *Agricultural Science and Technology, 13*(2), 109–118.
Shaw, L. J., Nicol, G. W., Smith, Z., Fear, J., Prosser, J. I., & Baggs, E. M. (2006). Nitrosopsira spp. can produce nitrous oxide via a nitrifier denitrification pathway. *Environmental Microbiology, 8*, 214–222.
Shoun, H., Kim, D. H., Uchiyama, H., & Sugiyama, J. (1992). Denitrification by fungi. *FEMS Microbiology Letters, 73*, 277–281.
Singh, J. S. (2015). Plant-microbe interactions: A viable tool for agricultural sustainability. *Applied Soil Ecology, 92*, 45–46.
Singh, D., Kumar, J., Singh, G., & Kumar, A. (2022). Use of ecofriendly fertilizers and crop residues for enhancing crop productivity and sustainable agriculture. In *Handbook of research on green technologies for sustainable management of agricultural resources* (pp. 156–179). IGI Global.
Singh, S. S., Meitei, M. D., Upadhyay, K. K., Chanda, R., Mawi, R., Singh, N. S., ... & Tripathi, S. K. (2024a). Bioremediation of toxic metals using aquatic macrophytes: Challenges and opportunities. *CLEAN–Soil, Air, Water, 52*(11), e202400273.
Singh, S. S., Chanda, R., Singh, N. S., Ramtharmawi, Devi, N. R., Devi, K. V., ... & Tripathi, S. K. (2024b). Microplastic pollution: Exploring trophic transfer pathways and ecological impacts. *Discover Environment, 2*(1), 103.
Six, J., Frey, S. D., Thiet, R. K., & Batten, K. M. (2006). Bacterial and fungal contributions to carbon sequestration in agroecosystems. *Soil Science Society of America Journal, 70*(2), 555–569.
Smith, C. K., Coyea, M. R., & Munson, A. D. (2000). Soil carbon, nitrogen, and phosphorus stocks and dynamics under disturbed black spruce forests. *Ecological Applications, 10*(3), 775–788.
Srivastava, J., Naraian, R., Kalra, S. J. S., & Chandra, H. (2014). Advances in microbial bioremediation and the factors influencing the process. *International Journal of Environmental Science and Technology, 11*, 1787–1800.
Srivastava, P., Sachan, K., Baskar, P., Saikanth, D. R. K., Lytand, W., Kumar, R. K. M., & Singh, B. V. (2023). Soil microbes expertly balancing nutrient demands and environmental preservation and ensuring the delicate stability of our ecosystems-a review. *International Journal of Plant & Soil Science, 35*(18), 989–1000.
Stahl, L. M., & Olson, J. B. (2021). Environmental abiotic and biotic factors affecting the distribution and abundance of Naegleria fowleri. *FEMS Microbiology Ecology, 97*(1), fiaa238.
Strong, D. T., Sale, P. W. G., & Helyar, K. R. (1998). The influence of the soil matrix on nitrogen mineralisation and nitrification. II. The pore system as a framework for mapping the organisation of the soil matrix. *Australian Journal of Soil Research, 36*, 855–872.
Subler, S., & Kirsch, A. S. (1998). Spring dynamics of soil carbon, nitrogen, and microbial activity in earthworm middens in a no-till cornfield. *Biology and Fertility of Soils, 26*, 243–249.
Takoutsing, B., Weber, J., Aynekulu, E., Martín, J. A. R., Shepherd, K., Sila, A., et al. (2016). Assessment of soil health indicators for sustainable production of maize in smallholder farming systems in the highlands of Cameroon. *Geoderma, 276*, 64–73.
Tayang, A., & Songachan, L. S. (2021). Microbial bioremediation of heavy metals. *Current Science, 120*(6), 00113891.
Teotia, P., Kumar, V., Kumar, M., Shrivastava, N., & Varma, A. (2016). Rhizosphere microbes: Potassium solubilization and crop productivity–present and future aspects. In *Potassium solubilizing microorganisms for sustainable agriculture* (pp. 315–325). Springer.

Thatoi, H., Das, S., Mishra, J., Rath, B. P., & Das, N. (2014). Bacterial chromate reductase, a potential enzyme for bioremediation of hexavalent chromium: A review. *Journal of Environmental Management, 146*, 383–399.

Thomsen, J. K., Geest, T., & Cox, R. P. (1994). Mass spectrometric studies of the effect of pH on the accumulation of intermediates in denitrification by *Paracoccus denitrificans. Applied and Environmental Microbiology, 60*(2), 536–541.

Tiedje, J. M. (1988). Ecology of denitrification and dissimilatory nitrate reduction to ammonium. In *Biology of anaerobic microorganisms* (pp. 179–244). John Wiley & Sons.

Tisdall, J. M., & Oades, J. M. (1982). Organic matter and water-stable aggregates. *Journal of Soil Science, 33*, 141–163.

Tixier, C., Bogaerts, P., Sancelme, M., Bonnemoy, F., Twagilimana, L., Cuer, A., Bohatier, J., & Veschambre, H. (2000). Fungal biodegradation of a phenylurea herbicide, diuron: Structure and toxicity of metabolites. *Pest Management Science, 56*(5), 455–462.

Tunlid, A., Floudas, D., Koide, R., & Rineau, F. (2016). Soil organic matter decomposition mechanisms in ectomycorrhizal fungi. *Molecular Mycorrhizal Symbiosis*, 257–275.

Vasudevan, P., Padmavathy, V., Tewari, N., & Dhingra, S. C. (2001). Biosorption of heavy metal ions. *Journal of Scientific & Industrial Research, 60*, 112–120.

Verma, S., & Kuila, A. (2019). Bioremediation of heavy metals by microbial process. *Environmental Technology & Innovation, 14*, 100369.

Wagg, C., Schlaeppi, K., Banerjee, S., Kuramae, E. E., & Van der Heijden, M. G. (2019). Fungal-bacterial diversity and microbiome complexity predict ecosystem functioning. *Nature Communications, 10*, 4841. https://doi.org/10.1038/s41467-019-12798-y

Wallander, H., Ekblad, A., Godbold, D. L., Johnson, D., Bahr, A., Baldrian, P., et al. (2013). Evaluation of methods to estimate production, biomass and turnover of ectomycorrhizal mycelium in forests soils–A review. *Soil Biology and Biochemistry, 57*, 1034–1047.

Werner, D., & Newton, W. E. (2005). *Nitrogen fixation in agriculture, forestry, ecology, and the environment*. Springer Publication.

Wu, Q. S., Cao, M. Q., Zou, Y. N., & He, X. H. (2014a). Direct and indirect effects of glomalin, mycorrhizal hyphae and roots on aggregate stability in rhizosphere of trifoliate orange. *Scientific Reports, 4*(1), 5823.

Wu, Q. S., Huang, Y. M., Li, Y., & He, X. H. (2014b). Contribution of arbuscular mycorrhizas to glomalin-related soil protein, soil organic carbon and aggregate stability in citrus rhizosphere. *International Journal of Agriculture & Biology, 16*(1).

Xiao, P., Mori, T., Kamei, I., & Kondo, R. (2011). Metabolism of organochlorine pesticide heptachlor and its metabolite heptachlor epoxide by white rot fungi, belonging to genus Phlebia. *FEMS Microbiology Letters, 314*(2), 140–146.

Xiao, L., Lian, B., Hao, J., Liu, C., & Wang, S. (2015). Effect of carbonic anhydrase on silicate weathering and carbonate formation at present day CO2 concentrations compared to primordial values. *Scientific Reports, 5*(1), 7733.

Yadav, S. K., Soni, R., & Rajput, A. S. (2018). Role of microbes in organic farming for sustainable agro ecosystem. In *Microorganisms for green revolution: Volume 2: Microbes for sustainable agro-ecosystem* (pp. 241–252). Springer.

Yang, Y., Zhang, Y., Jia, G., Wang, L., & Yu, X. (2022). Contributions of biotic and abiotic factors to soil aggregation under different thinning intensities. *Ecological Indicators, 139*, 108958.

Yasin, M., Munir, I., & Faisal, M. (2016). Can Bacillus spp. enhance K+ uptake in crop species. In *Potassium solubilizing microorganisms for sustainable agriculture* (pp. 163–170). Springer.

Yergeau, E., Kang, S., He, Z., Zhou, J., & Kowalchuk, G. A. (2007). Functional microarray analysis of nitrogen and carbon cycling genes across an Antarctic latitudinal transect. *The ISME Journal, 1*, 163–179.

Yeager, C. M., Northup, D. E., Grow, C. C., Barns, S. M., & Kuske, C. R. (2005). Changes in nitrogen-fixing and ammonia-oxidizing bacterial communities in soil of a mixed conifer forest after wildfire. *Applied and Environmental Microbiology, 71*(5), 2713–2722.

Yousuf, S., Naqash, N., & Singh, R. (2022). Nutrient cycling: An approach for environmental sustainability. In *Environmental microbiology: Advanced research and multidisciplinary applications* (Vol. 77). Bentham Science Publisher.

Zahedi, H. (2016). Growth-promoting effect of potassium-solubilizing microorganisms on some crop species. In *Potassium solubilizing microorganisms for sustainable agriculture* (pp. 31–42). Springer.

Zhang, Z., Lian, B., Hou, W., Chen, M., Li, X., & Li, Y. (2011). Bacillus mucilaginosus can capture atmospheric CO2 by carbonic anhydrase. *African Journal of Microbiology Research, 5*(2), 106–112.

Biofilms: The Unsung Heroes of Soil Health and Crop Productivity

14

Milan Dabhi, Bhavdip Anavadiya, Karan Prajapati, and Sheetal Sharma

Abstract

Biofilms, complex microbial communities embedded in an extracellular polymeric substance (EPS) matrix, play a crucial role in enhancing soil health and crop productivity. This chapter explores the multifaceted contributions of biofilms to sustainable agriculture, emphasizing their ecological and agricultural significance. Biofilms facilitate nutrient cycling, enhance water retention, suppress soil-borne pathogens, and stabilize soil structure, thereby promoting robust plant growth and resilience. Emerging technologies such as high-resolution imaging, next-generation sequencing, microfluidic devices, and synthetic biology have advanced our understanding and manipulation of biofilms. Furthermore, biofilms contribute to climate change mitigation through soil carbon sequestration and improved soil resilience. The potential of biofilms in transforming sustainable agriculture practices is immense, offering natural and effective solutions to key agricultural challenges. To fully realize this potential, interdisciplinary research, farmer education, policy support, and public-private partnerships are essential. This chapter underscores the need for continued research and practical adoption of biofilm-based technologies to achieve sustainable and resilient agricultural systems.

M. Dabhi (✉) · B. Anavadiya · K. Prajapati
Department of Microbiology and Biotechnology, University School of Sciences, Gujarat University, Ahmedabad, Gujarat, India
e-mail: milandabhi@gujaratuniversity.ac.in

S. Sharma
Construction Testing Solution, Leicester, UK

M. Saraf et al. (eds.), *Climate Change and Soil Microorganisms for Environmental Sustainability*, Microorganisms for Sustainability 28,
https://doi.org/10.1007/978-981-96-3425-5_14

Keywords

Biofilms · Soil health · Crop productivity · Sustainable agriculture · Climate change mitigation

14.1 Introduction

Biofilms are a vital yet often overlooked component of soil health and crop productivity. These complex microbial communities, enclosed in a protective matrix, play key roles in sustaining soil ecosystems and boosting agricultural yields. Biofilms interact dynamically, sharing nutrients and signals that significantly impact their environment. Beyond their biological complexity, biofilms improve soil fertility, manage water resources, combat plant pathogens, and support plant growth under stress, making them active agents in enhancing agricultural resilience and productivity.

Biofilms, the unsung heroes within the microscopic world, represent a fundamental aspect of life that intricately weaves together the fabric of various ecosystems, including those pivotal to soil health and crop productivity. These complex assemblages of microorganisms enveloped within a self-produced matrix of extracellular polymeric substances (EPS) are not just mere aggregations of cells; they are dynamic entities capable of remarkable environmental interactions and transformations (Anavadiya et al., 2023). Biofilms are multifunctional, influencing environments from aquatic systems to soil ecosystems, which are vital for agricultural productivity. This introduction explores their definition, characteristics, and significance, emphasizing their key roles in soil health and crop productivity.

A biofilm is a complex community of microorganisms, including bacteria, fungi, algae, and protozoa, embedded in a self-secreted EPS matrix composed of polysaccharides, proteins, lipids, and nucleic acids. This matrix not only provides structural support but also protects against environmental stresses, aids nutrient capture, and enables cell communication. Biofilm formation is a regulated process involving adhesion, microcolony formation, maturation, and dispersion, controlled by environmental signals and mechanisms like quorum sensing, reflecting its dynamic responsiveness to surroundings (Rather et al., 2021).

Biofilms show diverse structures and compositions, influenced by environmental conditions and microbial makeup, resulting in varied impacts. In aquatic systems, they aid nutrient cycling, and pollutant degradation, and serve as a food source. In clinical settings, they contribute to persistent infections due to their resistance to antibiotics. In industry, biofilms are beneficial in wastewater treatment but problematic in biofouling. However, their most significant role is in soil ecosystems, where they have profound implications for sustainability and life on Earth.

Soil biofilms are crucial for soil health and agricultural productivity. Located between soil particles, organic matter, and plant roots, they serve as microbial hotspots that enhance soil structure, stabilize aggregates, and improve porosity and water retention. Their microbial activity mobilizes essential nutrients, promoting nutrient cycling and reducing the need for chemical fertilizers (Kumar & Singh, 2020). Furthermore, biofilms play a critical role in protecting plants from pathogens. By forming a protective barrier around plant roots or by outcompeting harmful microorganisms, biofilms enhance plant resistance to diseases (Dewi et al., 2023).

The symbiotic relationships within biofilms also extend to plant growth-promoting rhizobacteria (PGPR), which can significantly enhance plant growth, yield, and resilience to environmental stresses. The formation of biofilms by PGPR on plant roots is a testament to the complex interplay between soil microorganisms and plants, a relationship that holds the potential for revolutionary advances in sustainable agriculture (Ajijah et al., 2023).

Studying soil biofilms is challenging due to the complexity of soil, but advances in molecular biology and imaging are revealing their crucial roles. These findings have practical applications in developing biofilm-based technologies to boost crop productivity, manage plant diseases, and reduce environmental stresses, opening new possibilities for sustainable agriculture.

14.2 Biofilms in Soil: An Overview

Soil biofilm formation enhances microbial survival, ecological interactions, and soil health by creating a protective niche. This communal structure allows microorganisms to cooperate and perform complex metabolic processes beyond the capabilities of individual cells (Flemming et al., 2016). Soil biofilms improve soil structure and fertility by binding particles into stable aggregates, enhancing porosity, water retention, and erosion resistance. They support plant roots, improve aeration, and aid in nutrient cycling and organic matter decomposition, boosting soil fertility and carbon sequestration (Nannipieri et al., 2017).

14.2.1 Composition and Formation of Soil Biofilms

Soil biofilms represent a dynamic and intricate consortium of microorganisms that include bacteria, fungi, archaea, protozoa, and algae. This microbial diversity is critical to the biofilm's functionality, resilience, and ecological impact. The bacterial component of soil biofilms is often dominated by members of the Proteobacteria, Firmicutes, Actinobacteria, and Bacteroidetes phyla (Vos et al., 2013). These bacteria perform essential functions such as nitrogen fixation, organic matter decomposition, and pathogen suppression.

Fungi, another vital constituent, contribute to biofilm formation through hyphal networks that provide structural stability and enhance nutrient acquisition. Mycorrhizal fungi, in particular, establish symbiotic relationships with plant roots, facilitating nutrient exchange and improving plant health (Smith & Read, 2010). Archaea, although less abundant than bacteria and fungi, play crucial roles in biogeochemical cycles, including methane production and ammonia oxidation, thereby influencing soil fertility and greenhouse gas emissions (Bates et al., 2011).

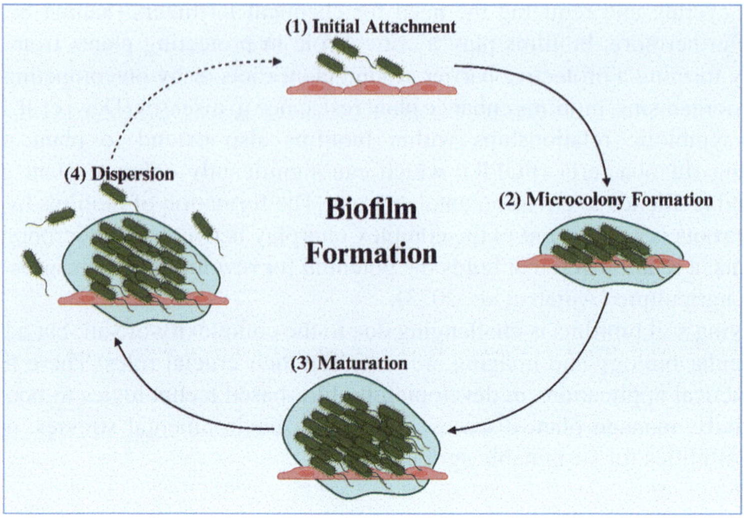

Fig. 14.1 General mechanism of biofilm formation

Protozoa and algae, though often overlooked, contribute to the microbial diversity and functionality of soil biofilms. Protozoa graze on bacteria, thereby regulating bacterial populations and stimulating nutrient cycling. Algae, particularly cyanobacteria, contribute to soil biofilms through photosynthesis, producing organic matter and oxygen, which support other microbial communities (Mager & Thomas, 2011).

14.2.1.1 Stages of Biofilm Formation

The development of soil biofilms is a highly orchestrated process involving several distinct stages: initial attachment, microcolony formation, maturation, and dispersion (Fig. 14.1).

14.2.1.1.1 Initial Attachment

The initial attachment phase is characterized by the adhesion of planktonic microorganisms to soil particles or organic matter. This attachment is initially mediated by weak, reversible physical forces such as van der Waals forces, electrostatic interactions, and hydrophobic effects. Microorganisms possess various surface structures, such as fimbriae, pili, and flagella, which facilitate this initial contact. Upon initial contact, microorganisms may strengthen their attachment through the production of adhesins and other surface proteins that bind firmly to the substrate. This transition from reversible to irreversible attachment is critical for the stability of the developing biofilm (Berne et al., 2018).

14.2.1.1.2 Microcolony Formation

Following irreversible attachment, microorganisms begin to proliferate and form microcolonies. This stage involves the production of extracellular polymeric substances (EPS), which comprise polysaccharides, proteins, nucleic acids, and lipids. The EPS matrix provides structural integrity and protects the microbial community from environmental stressors such as desiccation, predation, and antimicrobial agents (Flemming & Wingender, 2010). The microcolonies within the biofilm grow and merge, forming a complex, three-dimensional structure. This architecture includes water channels that facilitate the distribution of nutrients, oxygen, and waste products, thereby supporting microbial metabolism and growth (Stoodley et al., 2002).

14.2.1.1.3 Maturation

During maturation, the biofilm develops into a fully structured community with distinct microenvironments. The EPS matrix continues to accumulate, and the biofilm's architecture becomes more elaborate, often exhibiting mushroom-like or pillar-like structures. These structures enhance the biofilm's resilience and ability to withstand environmental fluctuations (Hall-Stoodley et al., 2004). The maturation phase is also characterized by increased microbial diversity and metabolic activity. Microorganisms within the biofilm engage in various cooperative and competitive interactions, leading to functional differentiation and specialization. This metabolic diversity allows the biofilm to exploit a wide range of substrates and adapt to changing environmental conditions (Stewart & Franklin, 2008).

14.2.1.1.4 Dispersion

The final stage of biofilm development is dispersion, where cells detach from the biofilm and return to a planktonic state. This process is critical for the colonization of new niches and the propagation of biofilm communities. Dispersion can be triggered by various environmental cues such as nutrient depletion, changes in shear forces, or the accumulation of signaling molecules (Kaplan, 2010). Dispersion involves the breakdown of the EPS matrix and the release of planktonic cells into the surrounding environment. These dispersed cells often exhibit enhanced motility and increased resistance to environmental stressors, enabling them to initiate new biofilms in favorable conditions (Boles et al., 2005).

14.2.2 Physical and Chemical Properties of Soil Biofilms

EPS is a critical component of biofilms, providing the structural framework that binds microbial cells together and anchors them to surfaces. The EPS matrix is a complex mixture of polysaccharides, proteins, nucleic acids, and lipids, each contributing to the physical properties of the biofilm. The polysaccharides in the EPS matrix often include glucans, fructans, and mannans, which form a viscous and adhesive network that holds the biofilm together (Flemming & Wingender, 2010).

Proteins within the EPS matrix serve various functions, including enzymatic activities that facilitate nutrient acquisition and environmental adaptation. Structural proteins contribute to the mechanical stability of the biofilm, allowing it to withstand shear forces and physical disturbances. Nucleic acids, particularly extracellular DNA (eDNA), play a crucial role in maintaining the biofilm's integrity by acting as a scaffold that stabilizes the matrix (Whitchurch et al., 2014).

Lipids in the EPS matrix, although less abundant, can influence the biofilm's hydrophobicity and contribute to the formation of microenvironments within the biofilm. The EPS matrix's composition can vary significantly depending on the microbial community and environmental conditions, reflecting the adaptive strategies of the biofilm (Flemming et al., 2016).

The pH within a biofilm can vary spatially, creating microenvironments that favor different microbial processes. For instance, the production of organic acids by certain bacteria can lower the pH locally, promoting the solubilization of minerals and the availability of essential nutrients like phosphorus and iron. Conversely, some microorganisms can generate alkaline conditions that precipitate certain metals, affecting their bioavailability (Stewart & Franklin, 2008).

The redox potential within biofilms is another critical factor that influences microbial metabolism. Biofilms often exhibit redox gradients, with more oxidized conditions at the surface and reduced conditions deeper within the biofilm. This gradient allows for the coexistence of aerobic and anaerobic microorganisms, enabling diverse metabolic pathways such as nitrification, denitrification, sulfate reduction, and methanogenesis. These redox processes are crucial for nutrient cycling and the maintenance of soil fertility (Flemming et al., 2016).

The physical and chemical properties of soil biofilms significantly influence soil structure and fertility. The EPS matrix acts as a binding agent that aggregates soil particles, enhancing soil structure and stability. These aggregates improve soil porosity, which facilitates water infiltration, root penetration, and gas exchange. Enhanced soil structure also reduces erosion and compaction, promoting a healthier soil environment (Chenu & Cosentino, 2011).

Biofilms contribute to soil fertility through their involvement in organic matter decomposition and nutrient cycling. Microorganisms within biofilms decompose complex organic materials into simpler compounds, releasing nutrients that are essential for plant growth. This process not only recycles nutrients but also contributes to the formation of humus, a stable form of organic matter that enhances soil fertility and water-holding capacity (Nannipieri et al., 2017).

In addition, the metabolic activities of biofilms play a crucial role in nutrient transformations. For example, nitrogen-fixing bacteria within biofilms convert atmospheric nitrogen into ammonia, which can be utilized by plants. Similarly, biofilms participate in phosphorus solubilization, making this critical nutrient more available to plants. These activities ensure a continuous supply of essential nutrients, supporting sustainable agricultural practices and promoting plant health (Richardson et al., 2009).

14.2.3 The Role of Biofilms in Soil Structure and Fertility

14.2.3.1 Contribution to Soil Aggregation and Stability

Soil biofilms play a crucial role in enhancing soil structure through the formation of soil aggregates. Aggregation is the process where soil particles bind together to form larger, more stable structures. This process is fundamental to maintaining soil health, as it influences water infiltration, root penetration, and resistance to erosion.

The EPS produced by biofilms acts as a glue that binds soil particles together. This binding action not only stabilizes soil aggregates but also improves the soil's physical properties. The EPS matrix consists of polysaccharides, proteins, lipids, and nucleic acids, which contribute to the adhesion of soil particles (Flemming & Wingender, 2010). The formation of stable aggregates enhances soil porosity, which in turn improves aeration and water retention. These physical improvements are vital for supporting plant growth and microbial activity.

In addition to enhancing physical stability, biofilms also contribute to the mechanical strength of soil aggregates. The EPS matrix protects soil particles from external stresses such as water flow and mechanical disturbances, thereby reducing the likelihood of soil erosion. By stabilizing soil structure, biofilms play a critical role in preventing soil degradation and maintaining long-term soil health (Chenu & Cosentino, 2011).

14.2.3.2 Enhancement of Soil Fertility Through Organic Matter Decomposition

Biofilms are integral to the decomposition of organic matter in soil, a process that is essential for nutrient cycling and soil fertility. The microorganisms within biofilms, including bacteria and fungi, secrete a variety of enzymes that break down complex organic compounds into simpler molecules. This decomposition process releases nutrients such as nitrogen, phosphorus, and sulfur, which are essential for plant growth (Nannipieri et al., 2017).

During decomposition, microorganisms within biofilms convert organic matter into humus, a stable form of organic matter that enhances soil fertility. Humus improves soil structure, increases water-holding capacity, and provides a reservoir of nutrients that plants can access over time. The presence of humus also enhances the cation exchange capacity (CEC) of soil, which improves its ability to retain and supply essential nutrients to plants (Lehmann & Kleber, 2015).

The metabolic activities of biofilms also contribute to the mineralization of nutrients. For instance, nitrogen-fixing bacteria within biofilms convert atmospheric nitrogen into ammonia, which can be utilized by plants. Similarly, biofilms play a role in the solubilization of phosphorus, making this critical nutrient more available to plants. These processes are essential for maintaining soil fertility and supporting sustainable agricultural practices (Richardson et al., 2009).

14.2.3.3 Role in Nutrient Cycling and Bioavailability

Biofilms are actively involved in various nutrient cycling processes, which are critical for maintaining soil fertility and promoting plant growth. One of the key roles of biofilms is in the nitrogen cycle, where they facilitate processes such as nitrogen fixation, nitrification, and denitrification.

Nitrogen-fixing bacteria within biofilms, such as members of the genera *Rhizobium* and *Azotobacter*, convert atmospheric nitrogen into ammonia through the process of biological nitrogen fixation. This ammonia is then transformed into nitrites and nitrates by nitrifying bacteria, making it available for plant uptake. Denitrifying bacteria within biofilms convert nitrates back into nitrogen gas, completing the nitrogen cycle and maintaining a balance of nitrogen in the soil (Galloway et al., 2003).

Biofilms also play a significant role in the phosphorus cycle. Phosphorus-solubilizing microorganisms within biofilms release organic acids that solubilize insoluble phosphorus compounds, converting them into forms that are readily available for plant uptake. This solubilization process is crucial for ensuring an adequate supply of phosphorus, which is a limiting nutrient in many soils (Rodŕiguez & Fraga, 1999).

In addition to nitrogen and phosphorus, biofilms contribute to the cycling of other essential nutrients, such as sulfur, potassium, and micronutrients. The metabolic diversity and cooperative interactions within biofilms enable the transformation and mobilization of these nutrients, enhancing their bioavailability and supporting plant nutrition.

14.2.3.4 Promotion of Plant Root Health and Growth

Biofilms play a pivotal role in promoting plant root health and growth. The microbial communities within biofilms interact with plant roots, forming beneficial associations that enhance nutrient uptake and protect against soil-borne pathogens.

One of the primary mechanisms by which biofilms promote root health is through the production of plant growth-promoting substances, such as indole-3-acetic acid (IAA), gibberellins, and cytokinins. These phytohormones stimulate root elongation, lateral root formation, and root hair development, thereby increasing the root surface area and improving nutrient absorption (Bhattacharyya & Jha, 2012).

Biofilms also enhance plant root health by protecting against soil pathogens. The microorganisms within biofilms produce various antimicrobial compounds, such as antibiotics, siderophores, and lytic enzymes, which inhibit the growth of pathogenic bacteria and fungi. Additionally, biofilms can outcompete pathogens for space and resources, reducing the incidence of soil-borne diseases (Raaijmakers et al., 2009).

Moreover, biofilms improve the soil environment around plant roots by enhancing soil structure and increasing water retention. The formation of stable soil aggregates and the retention of moisture create a favorable environment for root growth and development. These improvements in soil physical properties are critical for supporting healthy root systems and ensuring optimal plant growth.

14.3 The Heroes Beneath: How Biofilms Enhance Soil Health

3.1 Nutrient Cycling and Bioavailability

Biofilms are pivotal in enhancing soil health through their role in nutrient cycling and bioavailability. They facilitate critical processes such as nitrogen fixation, phosphorus solubilization, and organic matter mineralization, which are essential for maintaining soil fertility and supporting plant growth.

14.3.1.1 Nitrogen Fixation

Nitrogen is a vital nutrient for plant growth, but its availability in soil is often limited. Biofilms harbor nitrogen-fixing bacteria, such as those from the genera *Rhizobium* and *Azotobacter*, which convert atmospheric nitrogen (N_2) into ammonia (NH_3) through the process of biological nitrogen fixation. This ammonia is then further transformed into nitrate (NO_3^-) and ammonium (NH_4^+), forms that plants can readily uptake and utilize for their growth and development (Galloway et al., 2003). The symbiotic relationships between these nitrogen-fixing bacteria and plant roots, particularly in leguminous plants, enhance the nitrogen content in soil, promoting healthier and more robust plant growth.

In symbiotic nitrogen fixation, bacteria such as *Rhizobium* form mutualistic associations with leguminous plants. These bacteria colonize root nodules, specialized structures where they fix atmospheric nitrogen in exchange for carbohydrates supplied by the plant. The fixed nitrogen is converted into ammonia and then to ammonium (NH_4^+), which the plant can utilize for growth. This process significantly enhances soil nitrogen levels, benefiting both the host plant and subsequent crops in crop rotation systems (Galloway et al., 2003).

Biofilms also support free-living nitrogen-fixing bacteria, such as *Azotobacter* and *Clostridium*, which reside in the rhizosphere and within soil aggregates. These bacteria fix nitrogen independently of plant hosts, contributing to the overall nitrogen pool in the soil. The ammonia produced by these bacteria can be further nitrified by other soil microorganisms, converting it into nitrate (NO_3^-), another form readily taken up by plants (Philippot et al., 2013).

14.3.1.2 Phosphorus Solubilization

Biofilms enhance phosphorus availability in soil through phosphorus-solubilizing microorganisms. These microbes release organic acids and phosphatases that convert insoluble phosphorus into soluble forms, like orthophosphate, which plants can absorb. This process is vital for plant nutrition, supporting energy transfer, photosynthesis, and cell division, especially in the rhizosphere where biofilms aid phosphorus uptake by plant roots.

14.3.1.3 Organic Matter Mineralization

Biofilms play a key role in nutrient cycling by decomposing organic matter. Microorganisms in biofilms release enzymes that break down complex compounds, releasing essential nutrients like carbon, nitrogen, and sulfur into the soil for plant uptake (Nannipieri et al., 2017). Decomposed organic matter forms humus, a stable organic carbon that boosts soil fertility. Humus improves soil structure, water retention, and nutrient storage while enhancing the soil's cation exchange capacity (CEC) to better supply nutrients to plants (Lehmann & Kleber, 2015).

14.3.2 Water Retention and Soil Moisture Management

Biofilms play a crucial role in water retention and soil moisture management, which are essential aspects of soil health and plant productivity. Through their unique physical and chemical properties, biofilms enhance the soil's ability to retain water, improve soil structure, and maintain moisture levels, thereby supporting plant growth and mitigating the impacts of drought.

According to one of the studies done by Sharma et al. (2023) using EPS-producing bacterial strains treated with *Klebsiella* SSN1 PGPR can reduce drought. Plants treated with the PGPR strain were then subjected to neutron radiography to visualize the water retention activity in order to demonstrate the ability of bacteria in the soil to retain water. The comparison was done between three strains of *Klebsiella* from M5al which is a non-EPS-producing strain, SGM81, and SSN1 which is an EPS-producing strain. M5a1 was used for negative control whereas SGM81 was used for positive control. The results were calculated on the basis of relative water content obtained from neutron radiography (Fig. 14.2).

14.3.2.1 Hydrophilic Nature of the EPS Matrix

Biofilms significantly enhance soil health by improving water retention and soil moisture management. The EPS produced by biofilms are highly hydrophilic, meaning they have a strong affinity for water. This property allows biofilms to retain water within the soil matrix, enhancing the soil's water-holding capacity.

The EPS matrix acts like a sponge, absorbing and holding water, which is then slowly released to plant roots during periods of drought or low soil moisture. This water-retention capability is particularly important in arid and semi-arid regions where water availability is a limiting factor for plant growth (Chenu & Cosentino, 2011). This hydrophilic nature allows biofilms to trap moisture within the soil matrix, effectively increasing the soil's water-holding capacity (Flemming et al., 2016). By maintaining soil moisture levels, biofilms help mitigate the effects of water stress on plants, promoting healthier and more resilient plant communities.

14.3.2.2 Soil Structure and Aggregate Stability

Biofilms contribute to the formation and stabilization of soil aggregates, which are clusters of soil particles bound together by organic matter, clay, and microbial secretions. The EPS produced by biofilms acts as a binding agent that glues soil particles together, forming stable aggregates. These aggregates improve soil structure by creating a porous and friable soil matrix that enhances water infiltration and retention (Six et al., 2004).

Stable soil aggregates prevent soil compaction and crusting, which can impede water infiltration and root penetration. By maintaining a loose and porous soil structure, biofilms facilitate the movement of water through the soil profile, ensuring that water reaches deeper soil layers where it can be accessed by plant roots. Improved soil structure also reduces surface runoff and erosion, further enhancing the soil's ability to retain water (Bronick & Lal, 2005).

14.3.2.3 Enhanced Water Availability

The presence of biofilms in the soil can significantly enhance the availability of water to plants. The EPS matrix not only retains water but also slows down its evaporation from the soil surface. This retained water is gradually released to plant

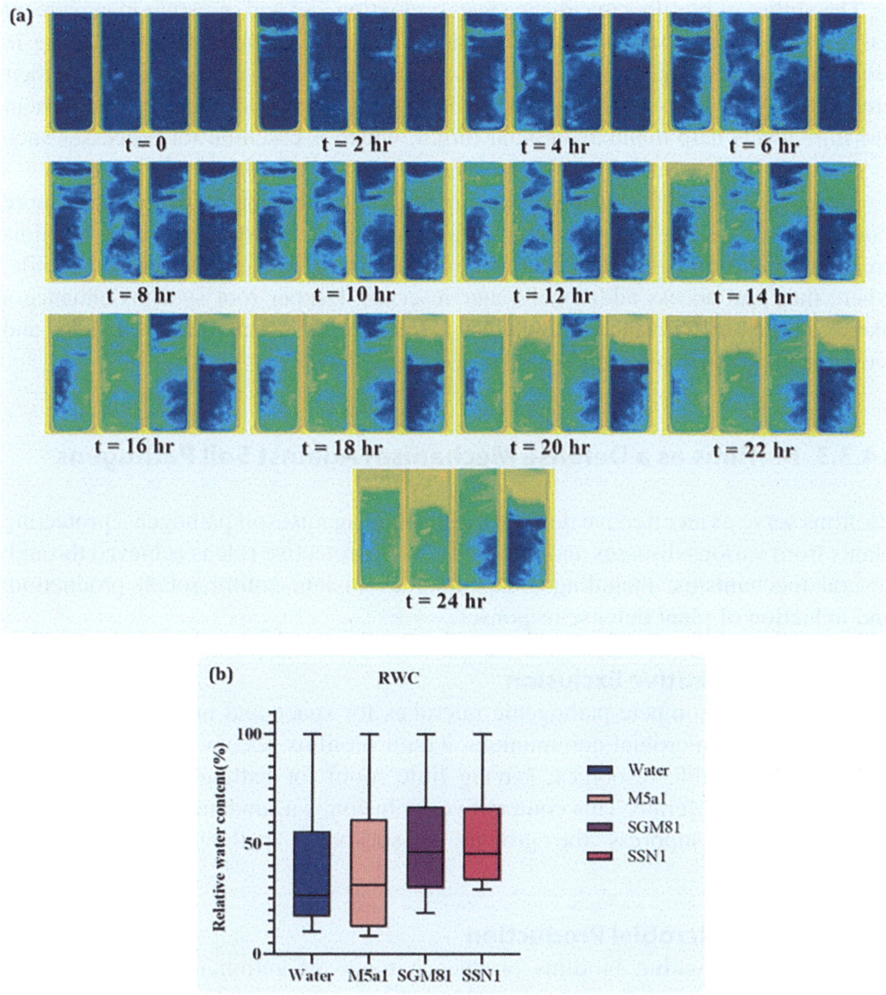

Fig. 14.2 (**a**) Neutron radiography showing the drying-out process at various time points. The samples are grouped as follows: Control water, M5a1, SGM81, and SSN1. (**b**) The relative water content (RWC) of each sample that was taken during neutron radiography, which was done every 4 hours. Whereas SGM81 and SSN1 displayed the largest RWC ratio, M5a1's RWC was nearly identical to that of water

roots, providing a steady supply of moisture and reducing the need for frequent irrigation (Or et al., 2006).

Biofilms also improve the soil's capacity to capture and store water from precipitation or irrigation. During rainfall events, the porous structure of biofilm-stabilized soil allows for rapid infiltration of water, reducing surface runoff and maximizing water capture. This stored water can then be gradually utilized by plants, ensuring that they have access to moisture even during dry periods (Morales et al., 2010).

The ability of biofilms to enhance water retention and soil moisture management has a direct impact on plant growth and resilience to drought. Plants growing in biofilm-rich soils benefit from improved water availability, which supports their physiological processes and enhances their growth and productivity. Consistent moisture levels help maintain cellular turgor, which is essential for processes such as photosynthesis, nutrient uptake, and root expansion (Bakker et al., 2012).

Moreover, biofilms can contribute to the establishment of deeper and more extensive root systems. By improving soil structure and moisture retention, biofilms create an environment that encourages roots to grow deeper into the soil profile, where they can access additional water reserves. Deeper root systems enhance a plant's ability to withstand drought by increasing its water uptake capacity and reducing its reliance on surface moisture (Lynch, 2007).

14.3.3 Biofilms as a Defense Mechanism Against Soil Pathogens

Biofilms serve as an effective defense mechanism against soil pathogens, protecting plants from various diseases and infections. This protective role is achieved through several mechanisms, including competitive exclusion, antimicrobial production, and induction of plant defense responses.

14.3.3.1 Competitive Exclusion

Biofilms can outcompete pathogenic microbes for space and nutrients within the soil. The dense microbial communities within biofilms occupy ecological niches and utilize available resources, leaving little room for pathogenic organisms to establish and proliferate. This competitive exclusion is a fundamental way through which biofilms suppress the growth of soil-borne pathogens (Raaijmakers et al., 2009).

14.3.3.2 Antimicrobial Production

Microorganisms within biofilms produce a range of antimicrobial compounds, including antibiotics, siderophores, and lytic enzymes, which inhibit the growth of pathogenic bacteria and fungi. For example, certain strains of *Bacillus* and *Pseudomonas* within biofilms are known to produce antibiotics that are effective against common soil pathogens. Additionally, siderophores sequester iron, a critical nutrient, making it unavailable to pathogens and thereby limiting their growth (Raaijmakers et al., 2009).

14.3.3.3 Induction of Plant Defense Responses

Biofilms can also enhance plant resistance to pathogens by inducing systemic resistance in plants. Certain beneficial microorganisms within biofilms trigger the plant's immune system, leading to the production of defensive compounds such as phytoalexins and pathogenesis-related proteins. This induced systemic resistance (ISR) helps plants to better withstand attacks from a variety of pathogens (Van Wees et al., 2008).

14.3.4 Promotion of Plant Root Health and Growth

The interaction between biofilms and plant roots is a cornerstone of soil health and plant productivity. Biofilms form intimate associations with plant roots, colonizing the rhizosphere, which is the soil region influenced by root exudates. This symbiotic relationship enhances nutrient availability, promotes root growth, and improves overall plant health.

14.3.4.1 Rhizosphere Colonization

Biofilms colonize the rhizosphere by adhering to root surfaces and utilizing root exudates as a carbon and energy source. Root exudates, which include sugars, amino acids, and organic acids, provide a rich nutrient source that supports the growth and activity of biofilm-forming microorganisms (Bais et al., 2006). The presence of biofilms in the rhizosphere enhances nutrient cycling and availability, leading to improved plant nutrition.

14.3.4.2 Root Exudate Utilization

Microorganisms within biofilms metabolize root exudates and, in turn, produce compounds that promote plant growth. For instance, biofilms can produce plant growth-promoting hormones such as indole-3-acetic acid (IAA) and gibberellins, which stimulate root elongation and lateral root formation. These phytohormones increase the root surface area, allowing plants to absorb more water and nutrients from the soil (Vacheron et al., 2013).

Additionally, biofilms can improve root architecture by producing extracellular enzymes that modify the soil environment around roots. These enzymes can degrade organic matter and release nutrients that are otherwise unavailable to plants. The enhanced nutrient availability and improved root growth lead to healthier and more robust plants.

14.4 Biofilms and Crop Productivity: A Symbiotic Relationship

Biofilms play a critical role in enhancing crop productivity through various mechanisms. By facilitating nutrient uptake, improving plant resistance to abiotic and biotic stresses, and providing practical applications in agriculture, biofilms contribute significantly to sustainable crop production. The successful application of biofilm-based inoculants and biostimulants in different crops underscores their potential to revolutionize agricultural practices and support global food security.

14.4.1 Enhancing Nutrient Uptake in Crops

Biofilms enhance nutrient uptake in crops through multiple mechanisms, including root colonization, enzyme production, and the creation of beneficial microenvironments. These processes collectively improve the availability and absorption of essential nutrients, leading to enhanced plant growth and productivity.

Biofilms enhance the efficiency of nutrient uptake by increasing the root surface area through the production of plant growth-promoting substances such as indole-3-acetic acid (IAA). IAA stimulates root elongation and lateral root formation, thereby increasing the root system's capacity to absorb water and nutrients (Vacheron et al., 2013). Additionally, biofilm-forming bacteria can modify root architecture, promoting the development of fine root hairs that further increase the root surface area and enhance nutrient absorption.

The production of siderophores by biofilm-forming bacteria enhances iron uptake by chelating iron from the soil and making it available to plants. Siderophores are small, high-affinity iron-chelating compounds that bind to iron, solubilizing it from insoluble mineral phases and facilitating its transport to the plant roots (Ahmed & Holmström, 2014). These enzymes and siderophores increase the availability of essential nutrients in the soil, promoting healthier and more robust plant growth.

Biofilms create microenvironments with optimized pH, redox potential, and nutrient gradients that boost nutrient availability. For example, organic acids produced by biofilms lower pH, solubilizing phosphate and micronutrients. Their spatial organization enables efficient nutrient cycling and exchange between microbes and plant roots. Biofilms create zones of distinct metabolic activity, with outer layers housing aerobic bacteria for nitrification and inner layers containing anaerobic bacteria for denitrification. This spatial organization optimizes nutrient cycling and improves nutrient uptake efficiency for plants (Stewart & Franklin, 2008).

14.4.2 Biofilms and Their Role in Plant Stress Resistance

Biofilms contribute to plant resistance against both abiotic and biotic stresses, thereby improving crop resilience and productivity. Through various mechanisms, biofilms help plants withstand drought, salinity, and pathogen attacks.

Biofilms enhance plant resistance to abiotic stresses such as drought and salinity by improving soil moisture retention and nutrient availability. The hydrophilic nature of the EPS matrix produced by biofilms allows the soil to retain more water, reducing drought stress on plants. Additionally, biofilms can produce osmoprotectants and compatible solutes that help plants cope with osmotic stress caused by high salinity (Premalatha et al., 2022).

In saline soils, biofilms can sequester and immobilize salts, reducing their toxic effects on plants. The production of exopolysaccharides by biofilms helps to buffer soil salinity and maintain a favorable microenvironment for root growth and nutrient uptake (Flemming et al., 2016). Furthermore, biofilms can enhance the soil's cation exchange capacity, which improves the soil's ability to retain essential nutrients and reduce the negative impact of salinity on plant growth.

Biofilms also play a role in enhancing plant resilience to temperature extremes. For instance, certain biofilm-forming bacteria produce antifreeze proteins that help plants withstand freezing temperatures. These proteins inhibit the formation of ice

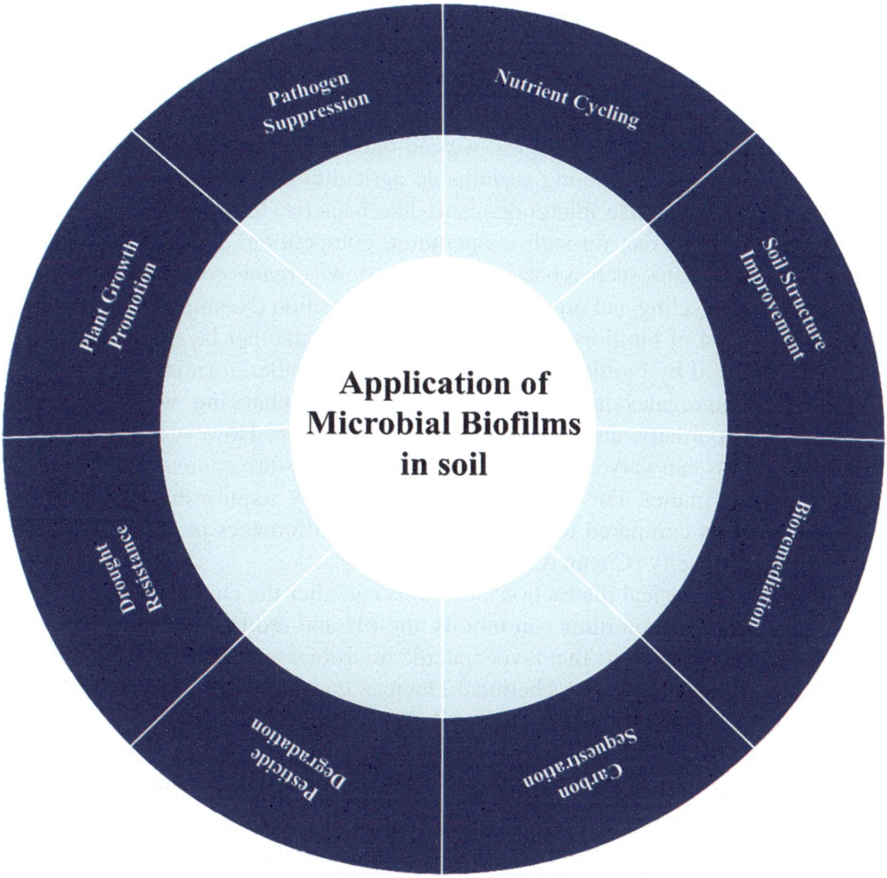

Fig. 14.3 Diverse applications of microbial biofilm in soil

crystals within plant tissues, reducing frost damage and improving plant survival in cold climates (Davies, 2014).

Microorganisms in biofilms produce a variety of antimicrobial compounds, such as antibiotics, lytic enzymes, and siderophores, which inhibit the growth of pathogens. These compounds create a protective barrier around plant roots, enhancing their resistance to infections. For example, certain strains of *Pseudomonas* within biofilms produce phenazines, a class of antibiotics that are effective against a wide range of soil-borne pathogens (Raaijmakers et al., 2006).

Biofilms can also induce systemic resistance in plants by triggering the production of defense-related hormones and enzymes. This induced systemic resistance (ISR) primes the plant's immune system, enabling it to respond more effectively to pathogen attacks. For instance, biofilm-forming *Bacillus* strains can induce the production of salicylic acid and jasmonic acid, key signaling molecules in plant defense, thereby enhancing the plant's ability to resist pathogen infections (Van Wees et al., 2008). Figure 14.3 shows diverse applications in the soil environment.

14.5 Challenges and Considerations

While biofilms benefit soil health and crop productivity, their complex interactions and potential risks require careful management. By understanding biofilm dynamics and implementing balanced strategies, we can maximize their positive effects while minimizing risks, and promoting sustainable agriculture and long-term soil health. Biofilms consist of diverse microorganisms like bacteria, fungi, archaea, protozoa, and algae, which interact through cooperation, competition, and quorum sensing. Synergistic interactions, such as bacteria breaking down organic compounds for fungi, enhance nutrient cycling and organic matter decomposition (Nannipieri et al., 2017).

The interaction of biofilms with soil particles adds another layer of complexity. The EPS produced by biofilms binds soil particles together, forming stable aggregates. These aggregates improve soil structure by enhancing water retention, increasing soil porosity, and promoting root penetration. However, the nature of these interactions can vary depending on soil type, moisture content, and the presence of organic matter. For example, sandy soils may require different biofilm-forming strategies compared to clayey soils due to differences in particle size and water-holding capacity (Chenu & Cosentino, 2011).

In addition to physical interactions, biofilms can alter the chemical properties of the soil. For example, biofilms can modify the pH and redox potential of the soil, creating microenvironments that favor specific microbial processes such as nitrification and denitrification. These chemical changes can influence the availability of nutrients and the overall fertility of the soil (Flemming et al., 2016).

In some cases, biofilms can harbor pathogenic microorganisms that pose a risk to plant health. For example, biofilms formed by plant-pathogenic bacteria such as *Xanthomonas* or *Pseudomonas syringae* can persist in the soil and on plant surfaces, leading to chronic infections and reduced crop yields. The protective environment of the biofilm can shield these pathogens from environmental stresses and antimicrobial agents, making them difficult to control (Mansfield et al., 2012).

Soil clogging can also affect the efficiency of irrigation systems. Biofilms that develop within irrigation pipes and emitters can obstruct water flow, reducing the distribution uniformity and overall effectiveness of irrigation. This can lead to uneven water application, water wastage, and increased maintenance costs (Satyanarayana et al., 2012).

Moreover, biofilms can influence the spatial distribution of nutrients within the soil. While biofilms may enhance nutrient availability in the immediate vicinity of roots, they can also create localized zones of nutrient depletion or accumulation, leading to heterogeneity in nutrient distribution. This uneven distribution can affect root growth patterns and nutrient uptake efficiency (Kuzyakov & Blagodatskaya, 2015).

Regular monitoring of biofilm activity in soil is essential for understanding their impact on soil health and crop productivity. Techniques such as microbial profiling, enzyme assays, and soil respiration measurements can provide insights into the composition and metabolic activity of biofilms. Advances in molecular biology, such as metagenomics and transcriptomics, allow for detailed analysis of biofilm communities and their functional potential (Zarraonaindia et al., 2015).

Microbial profiling using next-generation sequencing (NGS) can identify the diversity and abundance of microbial taxa within biofilms, providing a comprehensive view of the microbial community structure. Enzyme assays can measure the activity of key enzymes involved in nutrient cycling, such as phosphatases and nitrogenases, offering insights into the functional dynamics of biofilms. Soil respiration measurements can assess the overall metabolic activity of biofilms, indicating their impact on soil carbon cycling and microbial biomass (Nannipieri et al., 2017).

Balancing biofilm formation involves promoting beneficial biofilms while preventing excessive growth that can lead to negative impacts. This can be achieved through the use of biofilm-forming inoculants that enhance nutrient cycling and plant growth. For instance, the application of biofilm-forming PGPR can improve nutrient uptake and plant resilience to stresses. However, care must be taken to avoid over-inoculation, which can lead to soil clogging and other issues (Bashan et al., 2014).

Inoculant formulations should be optimized to ensure effective colonization of plant roots and soil particles without causing excessive biofilm growth. This can involve selecting strains with desired traits, such as efficient nutrient solubilization or pathogen suppression, and developing carrier materials that support the establishment and activity of biofilms. Additionally, the timing and method of inoculant application should be tailored to specific crop and soil conditions to maximize benefits and minimize risks (Bashan et al., 2014).

Soil management practices that support biofilm activity can improve soil fertility and crop productivity. Reduced tillage preserves biofilm-stabilized soil aggregates, while cover cropping provides organic inputs and root exudates that promote biofilm growth. Organic amendments like compost and manure enrich the soil, fostering beneficial biofilms and enhancing fertility, leading to healthier and more sustainable agricultural systems (Lal, 2015).

14.6 Leveraging Biofilms for Sustainable Agriculture

Leveraging biofilms for sustainable agriculture offers significant potential for enhancing soil health, crop productivity, and environmental resilience. Through biofilm engineering, innovative applications in precision agriculture, and integration into organic farming practices, biofilms can contribute to sustainable and productive agricultural systems. By harnessing the power of biofilms, farmers can improve nutrient cycling, disease resistance, and stress resilience, supporting the long-term sustainability of agriculture.

14.6.1 Biofilm Engineering for Soil and Crop Health

Biofilm engineering involves the use of genetic and metabolic modifications to enhance the beneficial properties of biofilms, thereby improving soil health and crop productivity. Genetically engineered biofilms hold significant potential for sustainable agriculture by enhancing nutrient cycling, disease resistance, and environmental resilience.

One of the primary goals of biofilm engineering is to enhance nutrient availability in the soil. By introducing genes that encode nutrient-solubilizing enzymes, such as phosphatases and nitrogenases, biofilms can be engineered to increase the bioavailability of essential nutrients like phosphorus and nitrogen. For instance, genetically modified strains of Bacillus and Pseudomonas have been developed to express higher levels of phosphatases, leading to improved phosphorus solubilization and increased plant growth (Bashan et al., 2014).

In addition to phosphatases, nitrogenase enzymes in nitrogen-fixing bacteria are crucial for converting atmospheric nitrogen into ammonia, which plants can readily uptake. By enhancing these nitrogen-fixing capabilities through genetic engineering, biofilms can provide a more consistent and abundant source of nitrogen for crops, reducing the reliance on synthetic fertilizers. This not only boosts plant growth but also promotes sustainable farming practices by minimizing the environmental impact of chemical fertilizers.

Biofilm engineering can also enhance the ability of biofilms to protect plants from pathogens. By incorporating genes that encode antimicrobial compounds, such as antibiotics and lytic enzymes, biofilms can be designed to suppress soil-borne diseases. For example, genetically engineered *Pseudomonas fluorescens* strains that produce the antibiotic phenazine-1-carboxylic acid have been shown to effectively control root diseases in various crops (Haas & Défago, 2005). Additionally, biofilms can be engineered to express siderophores that sequester iron, limiting its availability to pathogenic microbes and thereby reducing their virulence (Ahmed & Holmström, 2014).

Biofilm engineering can enhance the resilience of agricultural systems to environmental stresses such as drought and salinity. By introducing genes that enable the production of osmoprotectants and stress-responsive proteins, biofilms can help plants withstand abiotic stresses. For instance, biofilms engineered to produce trehalose, a sugar that protects cells from dehydration, can improve plant tolerance to drought conditions (Jain & Roy, 2009). Similarly, biofilms that produce antifreeze proteins can help plants survive freezing temperatures, extending the growing season and reducing crop losses due to frost (Davies, 2014).

Additionally, biofilms can be engineered to produce exopolysaccharides (EPS) with enhanced water-retention capabilities. These EPS can improve soil structure and increase the soil's water-holding capacity, making moisture more available to plants during dry periods. This attribute is particularly beneficial in arid and semi-arid regions where water scarcity is a major constraint to agriculture.

14.6.2 Innovative Biofilm Applications in Precision Agriculture

The integration of biofilm technologies with precision agriculture practices offers new opportunities for targeted delivery and real-time monitoring of soil and crop health. Precision agriculture aims to optimize field-level management through the use of technology and data analysis, and biofilms can play a crucial role in this approach.

Biofilm-based formulations can be precisely applied to specific areas of the field where they are needed the most. This targeted delivery can be achieved through advanced application techniques such as seed coatings, soil injections, and foliar sprays. For instance, seed coatings with biofilm-forming bacteria ensure that beneficial microbes are in close contact with the emerging roots, promoting early root colonization and nutrient uptake (González-Rodríguez et al., 2023). Soil injections of biofilm formulations can directly deliver beneficial microbes to the rhizosphere, enhancing their effectiveness in nutrient solubilization and pathogen suppression.

Biofilms can be integrated with sensor technologies to monitor soil and crop health in real time. Sensors embedded in the soil can detect changes in biofilm activity, such as shifts in microbial metabolism, nutrient availability, and moisture levels. This information can be transmitted to farmers, enabling them to make informed decisions about irrigation, fertilization, and pest management. For example, biosensors that detect the production of specific microbial metabolites can indicate the presence of beneficial or pathogenic biofilms, allowing for timely interventions (Simoes et al., 2020).

Real-time monitoring systems can also track environmental conditions that affect biofilm activity, such as temperature, humidity, and soil pH. By integrating this data with biofilm activity measurements, farmers can gain a comprehensive understanding of the factors influencing soil health and crop performance. This holistic approach enables proactive management strategies that enhance agricultural sustainability.

The data collected from biofilm sensors can be analyzed using machine learning algorithms to predict crop performance and optimize management practices. By correlating biofilm activity with crop yield and soil health metrics, farmers can identify the most effective strategies for enhancing productivity and sustainability. Precision agriculture platforms that integrate biofilm data with other agronomic information, such as weather forecasts and soil nutrient levels, can provide comprehensive recommendations for field management (Kumar et al., 2022).

14.6.3 Integration of Biofilms into Organic Farming Practices

Biofilms align well with the principles of organic farming, which emphasize the use of natural inputs and ecological processes to enhance soil fertility and crop health. The integration of biofilms into organic farming practices can enhance sustainability and productivity in several ways.

Biofilms contribute to soil fertility by promoting nutrient cycling and organic matter decomposition. Organic farming relies on the use of compost, manure, and cover crops to maintain soil health, and biofilms can enhance the effectiveness of these inputs. For example, biofilm-forming bacteria in compost can accelerate the breakdown of organic matter, releasing nutrients that are readily available to plants (Hamilton et al., 2019). Similarly, biofilms in the rhizosphere of cover crops can improve nitrogen fixation and phosphorus solubilization, enhancing soil fertility for subsequent crops.

The use of biofilm inoculants can also enhance the efficiency of green manures and other organic amendments. By promoting microbial activity and nutrient release, biofilms ensure that plants have access to essential nutrients throughout the growing season. This leads to healthier plants and higher yields, aligning with the goals of organic farming.

Organic farming emphasizes the importance of plant-microbe symbioses, such as mycorrhizal associations and nitrogen-fixing bacteria. Biofilms can enhance these symbiotic relationships by providing a stable and conducive environment for microbial colonization. For example, biofilms can facilitate the establishment of mycorrhizal fungi on plant roots, improving nutrient uptake and plant resilience to stress (Rillig et al., 2015). Similarly, biofilms that harbor nitrogen-fixing bacteria can enhance the availability of nitrogen to plants, reducing the need for synthetic fertilizers.

Biofilms support sustainable soil management practices by improving soil structure, water retention, and organic matter content. In organic farming, maintaining healthy soil is crucial for long-term productivity, and biofilms contribute to this goal by forming stable soil aggregates and enhancing soil microbial activity. Practices such as reduced tillage and the use of organic amendments can further support the formation and activity of beneficial biofilms, leading to improved soil health and crop yields (Lal, 2015).

14.7 Future Directions and Research Needs

Future biofilm research offers great potential for advancing sustainable agriculture. Emerging tools like high-resolution imaging, sequencing, microfluidics, biosensors, and synthetic biology provide new ways to study and manipulate biofilms. Biofilms can also enhance soil carbon sequestration and resilience to environmental stress. Collaborative efforts across disciplines, including farmer and industry partnerships, are crucial for unlocking biofilms' full potential in agriculture, leading to more sustainable and resilient systems.

14.7.1 Emerging Technologies in Biofilm Research and Monitoring

Emerging technologies have revolutionized the study and manipulation of soil biofilms, enabling new methods to improve soil health and crop productivity. Advances in imaging techniques like confocal laser scanning microscopy (CLSM) and atomic force microscopy (AFM) allow researchers to visualize biofilms in high detail, revealing their 3D structure, cell arrangement, and EPS distribution. These tools are essential for understanding biofilm interactions with soil particles (Neu & Lawrence, 2015).

Microfluidic devices have emerged as powerful tools for studying biofilm dynamics under controlled conditions. These devices create microenvironments that mimic soil conditions and allow researchers to observe biofilm formation, growth, and response to environmental changes in real time. Microfluidic platforms can be used to test the effects of different nutrients, antimicrobial agents, and environmental stressors on biofilm development, providing valuable data for optimizing biofilm-based agricultural applications (Simoes et al., 2020).

Biosensors integrated with microfluidic devices allow real-time monitoring of biofilm activity, detecting metabolites, signaling molecules, and environmental changes. For instance, biosensors tracking quorum-sensing molecules reveal bacterial communication, aiding in managing biofilms. This real-time data helps optimize practices like irrigation and fertilization to improve biofilm performance and soil health (Simoes et al., 2020).

Synthetic biology approaches allow for the design and construction of synthetic biofilms with customized properties. By combining genetic elements from different organisms, researchers can create biofilms that perform specific functions, such as enhanced nitrogen fixation or phosphate solubilization. Synthetic biofilms can be programmed to respond to environmental cues, such as nutrient availability or pathogen presence, by activating or deactivating certain metabolic pathways. This level of control enables the development of biofilm-based solutions that are highly adaptable and responsive to changing agricultural conditions (Bargaz et al., 2021).

14.7.2 Potential for Biofilms in Climate Change Mitigation and Adaptation

Biofilms have significant potential in climate change mitigation and adaptation by enhancing soil carbon sequestration and improving soil resilience to environmental stressors associated with climate change.

Biofilms play a crucial role in soil carbon sequestration by stabilizing organic matter and promoting the formation of soil aggregates. The EPS produced by biofilms binds soil particles and organic matter together, forming stable aggregates that protect organic carbon from decomposition. This process increases the storage of carbon in the soil, reducing the amount of carbon dioxide released into the atmosphere. The stable aggregates formed by biofilms also create microhabitats that protect organic matter from microbial degradation, further enhancing soil carbon sequestration (Chenu & Cosentino, 2011).

Biofilms also enhance the soil's ability to buffer against extreme temperature fluctuations and soil erosion. The formation of stable soil aggregates by biofilms reduces soil erosion and loss of topsoil, which are critical for maintaining soil fertility and productivity under changing climatic conditions. Stable soil aggregates also enhance root penetration and gas exchange, promoting healthier plant growth. By improving soil structure and stability, biofilms contribute to the overall resilience of agricultural systems to climate change (Rillig et al., 2015).

14.7.3 Collaborative Approaches for Advancing Biofilm Applications in Agriculture

Advancing the application of biofilms in agriculture requires interdisciplinary research and collaboration among scientists, farmers, and industry stakeholders. Collaborative efforts can drive innovation, facilitate the adoption of biofilm technologies, and ensure that these solutions are effective, scalable, and sustainable.

Collaboration among researchers from different disciplines, such as microbiology, soil science, agronomy, and engineering, is essential for advancing biofilm research. Interdisciplinary teams can integrate knowledge and methodologies to address complex challenges related to biofilm formation, function, and application. For example, collaborations between microbiologists and engineers can lead to the development of advanced biofilm monitoring tools and precision delivery systems. These collaborations can also facilitate the integration of biofilm research with other areas of agricultural science, such as plant breeding and soil management (Baveye et al., 2018).

Field trials and demonstration projects can showcase the practical benefits of biofilm technologies, such as improved nutrient uptake, soil health, and crop yields, providing farmers with hands-on experience. Extension services and training programs can support adoption by offering technical assistance and education on biofilm management. Public-private partnerships can combine scientific and commercial expertise to accelerate biofilm solutions, with government support through funding, incentives, and regulations ensuring their safe and effective use in agriculture.

14.8 Conclusion

Biofilms, complex communities of microorganisms in an EPS matrix, play a key role in enhancing soil health and crop productivity. This chapter has explored their contributions to sustainable agriculture, emphasizing nutrient cycling, water retention, pathogen suppression, and soil structure improvement.

Biofilms promote nutrient solubilization and mineralization, enhancing the availability of nitrogen, phosphorus, and potassium for plant uptake, and leading to improved crop yields. They also boost soil water retention, making them crucial for drought resilience in arid regions (Liu et al., 2017; Anavadiya et al., 2024).

Biofilms protect plants from pathogens through competitive exclusion and antimicrobial production, reducing disease incidence and promoting higher productivity. They enhance soil structure by binding particles together, improving aeration, and water infiltration, and reducing erosion (Gowthaman et al., 2018).

Biofilms contribute to carbon sequestration by stabilizing organic matter and forming soil aggregates, helping mitigate climate change. Additionally, they reduce the need for chemical fertilizers by enhancing nutrient bioavailability, lowering production costs and environmental impacts. Biofilms improve water use efficiency in agriculture, optimizing irrigation and enhancing drought resilience. They offer a natural alternative to chemical pesticides, supporting sustainable pest management and aligning with organic farming principles (Zhang et al., 2017).

Biofilms promote sustainable soil management through practices like reduced tillage and organic amendments. Research and development investments are essential to optimize biofilm applications. Interdisciplinary collaboration and funding can drive innovations in biofilm-based agriculture.

Farmer education and training are crucial for adopting biofilm technologies. Extension services and workshops can equip farmers with practical knowledge. Policymakers can support biofilm adoption through incentives and supportive regulations that promote sustainable farming practices (Gouda et al., 2018).

Policymakers can play a significant role in promoting the adoption of biofilm technologies by providing supportive policies and incentives. Financial incentives, such as subsidies and grants, can encourage farmers to invest in biofilm-based solutions. Additionally, regulatory frameworks that recognize and promote the use of biofilms in agriculture can facilitate their integration into mainstream farming practices. Government policies that support sustainable agriculture and environmental stewardship will further enhance the adoption of biofilm technologies (Lal, 2015).

References

Ahmed, E., & Holmström, S. J. M. (2014). Siderophores in environmental research: Roles and applications. *Microbial Biotechnology, 7*, 196–208. https://doi.org/10.1111/1751-7915.12117

Ajijah, N., Fiodor, A., Pandey, A. K., et al. (2023). Plant Growth-Promoting Bacteria (PGPB) with biofilm-forming ability: A multifaceted agent for sustainable agriculture. *Diversity, 15*. https://doi.org/10.3390/D15010112

Anavadiya, B., Acharya, D., Saraf, M., & Goswami, D. (2023). Multispecies metabolomics interactions resulting in the development of resistance. In *Microbial biofilms: Challenges and advances in metabolomic study* (pp. 133–150). https://doi.org/10.1016/B978-0-323-95715-1.00016-9

Anavadiya, B., Chouhan, S., Saraf, M., & Goswami, D. (2024). Exploring endophytic actinomycetes: A rich reservoir of diverse antimicrobial compounds for combatting global antimicrobial resistance. *The Microbe, 4*, 100110. https://doi.org/10.1016/J.MICROB.2024.100110

Bais, H., Weir, T., Perry, L., et al. (2006). The role of root exudates in rhizosphere interactions with plants and other organisms. *Annual Review of Plant Biology, 57*, 233–266. https://doi.org/10.1146/annurev.arplant.57.032905.105159

Bakker, M. G., Manter, D. K., Sheflin, A. M., et al. (2012). Harnessing the rhizosphere microbiome through plant breeding and agricultural management. *Plant and Soil, 360*, 1–13. https://doi.org/10.1007/S11104-012-1361-X

Bargaz, A., Elhaissoufi, W., Khourchi, S., et al. (2021). Benefits of phosphate solubilizing bacteria on belowground crop performance for improved crop acquisition of phosphorus. *Microbiological Research, 252*, 126842.

Bashan, Y., de-Bashan, L., Prabhu, S., et al. (2014). Advances in plant growth-promoting bacterial inoculant technology: Formulations and practical perspectives (1998–2013). *Plant and Soil, 378*, 1–33. https://doi.org/10.1007/s11104-013-1956-x

Bates, S. T., Berg-Lyons, D., Caporaso, J. G., et al. (2011). Examining the global distribution of dominant archaeal populations in soil. *The ISME Journal, 5*, 908–917. https://doi.org/10.1038/ismej.2010.171

Baveye, P., Berthelin, J., Tessier, D., et al. (2018). The "4 per 1000" initiative: A credibility issue for the soil science community? *Geoderma, 309*(Supplement C), 118–123.
Berne, C., Ellison, C., Ducret, A., et al. (2018). Bacterial adhesion at the single-cell level. *Nature Reviews Microbiology, 16*(10), 616–627.
Bhattacharyya, P. N., & Jha, D. K. (2012). Plant growth-promoting rhizobacteria (PGPR): Emergence in agriculture. *World Journal of Microbiology and Biotechnology, 28*, 1327–1350. https://doi.org/10.1007/s11274-011-0979-9
Boles, B. R., Thoendel, M., & Singh, P. K. (2005). Rhamnolipids mediate detachment of Pseudomonas aeruginosa from biofilms. *Molecular Microbiology, 57*, 1210–1223. https://doi.org/10.1111/j.1365-2958.2005.04743.x
Bronick, C. J., & Lal, R. (2005). Soil structure and management: a review. *Geoderma, 124*(1–2), 3–22. https://doi.org/10.1016/j.geoderma.2004.03.005
Chenu, C., & Cosentino, D. (2011). Microbial regulation of soil structural dynamics. In *The architecture and biology of soils: Life in inner space* (pp. 37–70). CABI. https://doi.org/10.1079/9781845935320.0037
Davies, P. L. (2014). Ice-binding proteins: a remarkable diversity of structures for stopping and starting ice growth. *Trends in Biochemical Sciences, 39*(11), 548–555.
Dewi, W. S., Amalina, D. D., & Romadhon, M. R. (2023). Microbial biofilm for soil health, plant growth, and productivity under multi stress. A review. *IOP Conference Series: Earth and Environmental Science, 1162*. https://doi.org/10.1088/1755-1315/1162/1/012008
Flemming, H. C., & Wingender, J. (2010). The biofilm matrix. *Nature Reviews Microbiology, 8*(9), 623–633. https://doi.org/10.1038/nrmicro2415
Flemming, H. C., Wingender, J., Szewzyk, U., Steinberg, P., Rice, S. A., & Kjelleberg, S. (2016). Biofilms: an emergent form of bacterial life. *Nature Reviews Microbiology, 14*(9), 563–575. https://doi.org/10.1038/nrmicro.2016.94
Galloway, J., Aber, J., Erisman, J., et al. (2003). The nitrogen cascade. *Bioscience, 53*(4), 341–356.
González-Rodríguez, V. E., José, H., Anillo, B., et al. (2023). Biofilm-forming ability of phytopathogenic bacteria: A review of its involvement in plant stress. *Plants, 12*, 2207. https://doi.org/10.3390/PLANTS12112207
Gouda, S., Kerry, R., Das, G., et al. (2018). Revitalization of plant growth promoting rhizobacteria for sustainable development in agriculture. *Microbiological Research, 206*, 131.
Gowthaman, S., Nakashima, K., & Kawasaki, S. (2018). A state-of-the-art review on soil reinforcement technology using natural plant fiber materials: Past findings, present trends and future directions. *Materials, 11*(4), 553. https://doi.org/10.3390/ma11040553
Haas, D., & Défago, G. (2005). Biological control of soil-borne pathogens by fluorescent pseudomonads. *Nature Reviews Microbiology, 3*(4), 307–319.
Hall-Stoodley, L., Costerton, J. W., & Stoodley, P. (2004). Bacterial biofilms: from the natural environment to infectious diseases. *Nature Reviews Microbiology, 2*(2), 95–108.
Hamilton, K., Reyneke, B., Waso, M., et al. (2019). A global review of the microbiological quality and potential health risks associated with roof-harvested rainwater tanks. *NPJ Clean Water*. https://doi.org/10.1038/s41545-019-0030-5
Jain, N. K., & Roy, I. (2009). Effect of trehalose on protein structure. *Protein Science, 18*, 24–36. https://doi.org/10.1002/PRO.3
Kaplan, J. B. (2010). Biofilm dispersal: Mechanisms, clinical implications, and potential therapeutic uses. *Journal of Dental Research, 89*, 205–218. https://doi.org/10.1177/0022034509359403
Kumar, A., & Singh, J. (2020). Biofilms forming microbes: Diversity and potential application in plant–microbe interaction and plant growth. In *Plant microbiomes for sustainable agriculture* (pp. 173–197). https://doi.org/10.1007/978-3-030-38453-1_6
Kumar, S., Meena, R., Sheoran, S., et al. (2022). *Remote sensing for agriculture and resource management*. Elsevier.
Kuzyakov, Y., & Blagodatskaya, E. (2015). Microbial hotspots and hot moments in soil: Concept & review. *Soil Biology and Biochemistry, 83*, S29–S36. https://doi.org/10.1016/j.soilbio.2015.01.025

Lal, R. (2015). Restoring soil quality to mitigate soil degradation. *Sustainability, 7*, 5875–5895. https://doi.org/10.3390/SU7055875

Lehmann, J., & Kleber, M. (2015). The contentious nature of soil organic matter. *Nature*. https://doi.org/10.1038/nature16069

Liu, H., Carvalhais, L. C., Crawford, M., et al. (2017). Inner plant values: Diversity, colonization and benefits from endophytic bacteria. *Frontiers in Microbiology, 8*. https://doi.org/10.3389/FMICB.2017.02552/FULL

Lynch, J. P. (2007). Roots of the second green revolution. *Australian Journal of Botany, 55*, 493–512. https://doi.org/10.1071/BT06118

Mager, D. M., & Thomas, A. D. (2011). Extracellular polysaccharides from cyanobacterial soil crusts: A review of their role in dryland soil processes. *Journal of Arid Environments, 75*, 91–97. https://doi.org/10.1016/J.JARIDENV.2010.10.001

Mansfield, J., Genin, S., Magori, S., et al. (2012). Top 10 plant pathogenic bacteria in molecular plant pathology. *Molecular Plant Pathology, 13*, 614–629. https://doi.org/10.1111/J.1364-3703.2012.00804.X

Morales, V. L., Parlange, J. Y., & Steenhuis, T. S. (2010). Are preferential flow paths perpetuated by microbial activity in the soil matrix? A review. *Journal of Hydrology, 393*(1–2), 29–36.

Nannipieri, P., Ascher, J., Ceccherini, M., Landi, L., Pietramellara, G., & Renella, G. (2017). Microbial diversity and soil functions. *European Journal of Soil, 54*, 655.

Neu, T. R., & Lawrence, J. R. (2015). Innovative techniques, sensors, and approaches for imaging biofilms at different scales. *Trends in Microbiology, 23*(4), 233–242.

Or, D., Smets, B., Wraith, J., et al. (2006). Physical constraints affecting bacterial habitats and activity in unsaturated porous media–a review. *Advances in Water Resources*. https://doi.org/10.1016/j.advwatres.2006.05.025

Philippot, L., Raaijmakers, J. M., Lemanceau, P., & Van Der Putten, W. H. (2013). Going back to the roots: the microbial ecology of the rhizosphere. *Nature Reviews Microbiology, 11*(11), 789–799.

Premalatha, R. P., Malarvizhi, P., & Parameswari, E. (2022). Effect of biochar doses under various levels of salt stress on soil nutrient availability, soil enzyme activities and plant growth in a marigold crop. *Crop & Pasture Science, 74*, 66. https://doi.org/10.1071/CP21542

Raaijmakers, J. M., De Bruijn, I., & De Kock, M. J. D. (2006). Cyclic lipopeptide production by plant-associated Pseudomonas spp.: Diversity, activity, biosynthesis, and regulation. *Molecular Plant-Microbe Interactions, 19*, 699–710. https://doi.org/10.1094/MPMI-19-0699

Raaijmakers, J. M., Paulitz, T. C., Steinberg, C., et al. (2009). The rhizosphere: A playground and battlefield for soilborne pathogens and beneficial microorganisms. *Plant and Soil, 321*, 341–361. https://doi.org/10.1007/s11104-008-9568-6

Rather, M. A., Gupta, K., & Mandal, M. (2021). Impact of microbial biofilm on crop productivity and agricultural sustainability. In *Microbes in land use change management*. Elsevier.

Richardson, A. E., Barea, J.-M., & Mcneill, A. M. (2009). Acquisition of phosphorus and nitrogen in the rhizosphere and plant growth promotion by microorganisms. *Plant and Soil, 321*, 305–339. https://doi.org/10.1007/s11104-009-9895-2

Rillig, M., Aguilar-Trigueros, C., et al. (2015). Plant root and mycorrhizal fungal traits for understanding soil aggregation. *New Phytologist, 205*, 1385–1388. https://doi.org/10.1111/nph.13045

Rodríguez, H., & Fraga, R. (1999). Phosphate solubilizing bacteria and their role in plant growth promotion. *Biotechnology Advances, 17*(4–5), 319–339.

Satyanarayana, T., Johri, B., & Prakash, A. (2012). *Microorganisms in sustainable agriculture and biotechnology*. Springer Science & Business Media.

Sharma, S., Roy, T., Kashyap, Y., et al. (2023). Characterizing and demonstrating the role of Klebsiella SSN1 exopolysaccharide in osmotic stress tolerance using neutron radiography. *Scientific Reports*. https://doi.org/10.1038/s41598-023-37133-w

Simoes, M., Borges, A., & Simoes, L. (2020). *Recent trends in biofilm science and technology*. Academic Press.

Six, J., Bossuyt, H., Degryze, S., et al. (2004). A history of research on the link between (micro) aggregates, soil biota, and soil organic matter dynamics. *Soil and Tillage Research, 79*, 7.

Smith, S., & Read, D. (2010). *Mycorrhizal symbiosis*. Academic Press.

Stewart, P., & Franklin, M. J. (2008). Physiological heterogeneity in biofilms. *Nature Reviews Microbiology*. https://doi.org/10.1038/nrmicro1838

Stoodley, P., Sauer, K., Davies, D. G., & Costerton, J. W. (2002). Biofilms as complex differentiated communities. *Annual Reviews in Microbiology, 56*, 187–209. https://doi.org/10.1146/annurev.micro.56.012302.160705

Vacheron, J., Desbrosses, G., Bouffaud, M. L., et al. (2013). Plant growth-promoting rhizobacteria and root system functioning. *Frontiers in Plant Science, 4*. https://doi.org/10.3389/FPLS.2013.00356/FULL

Van Wees, S. C., Van der Ent, S., & Pieterse, C. M. (2008). Plant immune responses triggered by beneficial microbes. *Current Opinion in Plant Biology*. https://doi.org/10.1016/j.pbi.2008.05.005

Vos, M., Wolf, A. B., Jennings, S. J., & Kowalchuk, G. A. (2013). Micro-scale determinants of bacterial diversity in soil. *FEMS Microbiology Reviews*. https://doi.org/10.1111/1574-6976.12023

Whitchurch, C., Tolker-Nielsen, T., Ragas, P., et al. (2014). Extracellular DNA required for bacterial biofilm formation. *Science, 295*, 1487. https://doi.org/10.1126/science.295.5559.1487

Zarraonaindia, C., Owens, I. M., Weisenhorn, S. M., et al. (2015). The soil microbiome influences grapevine-associated microbiota. *MBio, 6*, 10.1128/mBio.02527-14.

Zhang, B., Penton, C., Xue, C., et al. (2017). Soil depth and crop determinants of bacterial communities under ten biofuel cropping systems. *Soil Biology and Biochemistry, 112*, 140–152.

Metabolic Mysteries: New Perspectives on Soil Bacterial Metabolism

15

Rohit Patel and Shraddha Gang

Abstract

The intricate interplay between soil, plants, and bacteria forms a fundamental component of terrestrial ecosystems, with bacterial metabolism playing a pivotal role in maintaining this delicate balance. Soil bacterial metabolism is heavily influenced by plant-root interactions and environmental factors. Metabolomics, a powerful tool for analyzing bacterial metabolites, is utilized to unravel the diverse metabolic pathways and responses of soil bacteria to various environmental stimuli. Bacterial metabolism is also susceptible to the impact of contemporary environmental challenges such as plastic pollution and heavy metal contamination, highlighting how these stressors affect bacterial metabolic processes and, consequently, soil and plant health. The role of root exudates in shaping the rhizosphere is crucial, emphasizing their influence on the microbial community structure and soil nutrient dynamics is essential for a complete understanding of soil-microbe-plant interplay. Additionally, the impact of different cropping systems on the rhizosphere and soil fertility is also significant; such agricultural practices can alter soil microbial communities and affect plant growth and health. Furthermore, the rhizosphere is a dynamic environment housing both pathogens and symbionts, exploring how these microbial interactions influence plant health and soil quality is of great importance.

R. Patel
Department of Microbiology & Biotechnology, University School of Sciences, Gujarat University, Ahmedabad, Gujarat, India

S. Gang (✉)
School of Life Sciences, College of Engineering, Environment and Science, Coventry University, Coventry, UK

M. Saraf et al. (eds.), *Climate Change and Soil Microorganisms for Environmental Sustainability*, Microorganisms for Sustainability 28,
https://doi.org/10.1007/978-981-96-3425-5_15

Keywords

Metabolomics · Cropping systems · Root exudates · Soil

15.1 Introduction

The exploration of metabolic mysteries in soil bacteria opens a fascinating window into the intricate and diverse processes that drive the health and sustainability of our planet's ecosystems. Metabolism, fundamentally, is the sum of all chemical reactions within a cell or organism, orchestrating the conversion of nutrients into energy and building blocks for growth. In the context of soil bacteria, metabolism takes on an added layer of complexity and significance, driven by their microbial diversity and the myriad interactions within their habitats. Metabolomics, an extension of this concept, represents a comprehensive analysis of metabolites within a biological system (Wu et al., 2022). It provides a dynamic snapshot of the metabolic state and physiological processes occurring in organisms at any given time. In soil bacterial communities, metabolomics has emerged as a pivotal tool, unveiling the metabolic diversity and adaptations that bacteria undergo in response to their ever-changing environment. This approach offers a more nuanced understanding of microbial metabolism than traditional methods, revealing the intricate biochemical pathways and their regulation in the soil microbiome (Dongdong et al., 2023).

Furthermore, the ecological significance of bacteria in soil is profound. These microorganisms are not only crucial for maintaining soil health and fertility but also act as key players in global biogeochemical cycles. They influence the carbon cycle through their roles in decomposition and carbon sequestration, affecting climate change dynamics (Dongdong et al., 2023). Soil bacteria also engage in symbiotic relationships with plants, enhancing nutrient uptake and providing resistance against pathogens (Molefe et al., 2023). The quest to unravel the metabolic pathways in soil microbiomes is fraught with challenges, owing to the complexity and diversity of these microbial communities (Fig. 15.1). Apart from bacteria, the soil microbiome also includes fungi, archaea, and viruses, that play a critical role in maintaining soil health and ecosystem functioning. Functional redundancy refers to the presence of multiple microorganisms within a community that can perform the same or similar ecological functions. This redundancy is important because it ensures that if one microbial species is lost or its function is diminished due to environmental changes or disturbances, other species can step in to perform the necessary functions, thereby maintaining ecosystem stability (Bardgett & Van Der Putten, 2014). Understanding how we study current bacteria and metabolites in soil, and the relationship between metagenomics and metabolomics, is crucial to overcoming these challenges. To study current bacteria in soil, researchers employ a variety of techniques ranging from traditional culturing methods to advanced genomic approaches.

Traditional methods involve isolating and culturing bacteria in the laboratory, which allows for the study of their metabolic capabilities under controlled conditions. However, a significant limitation of this approach is that many soil bacteria

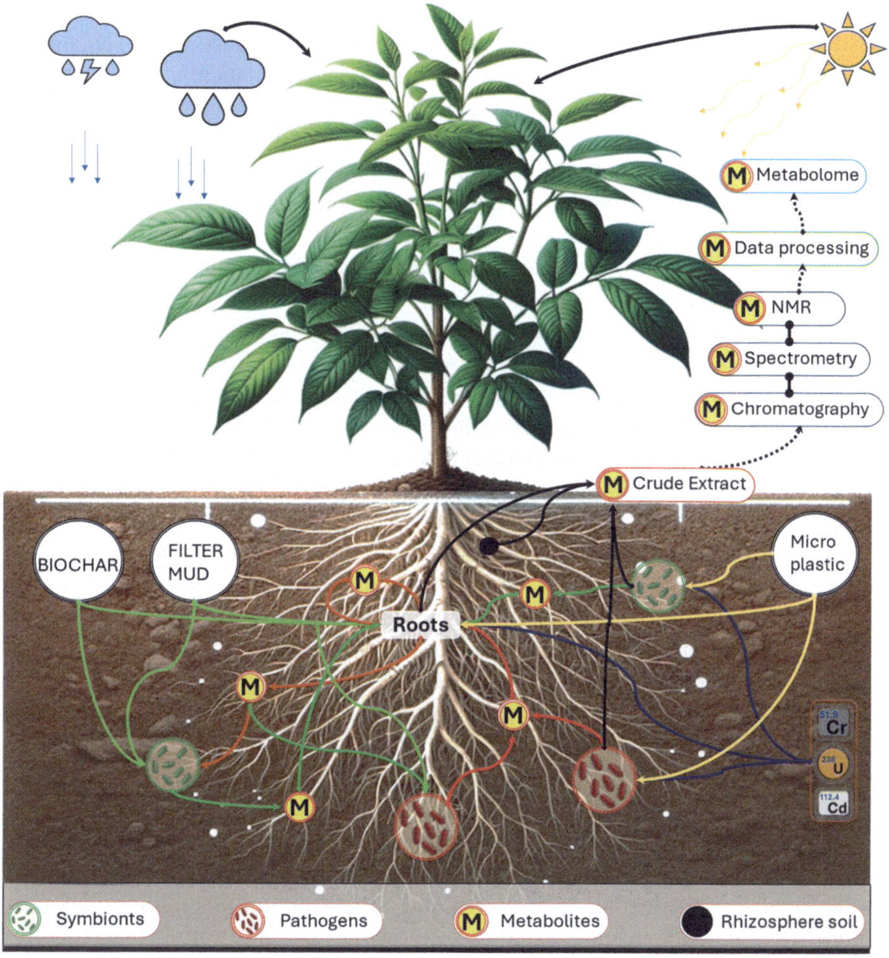

Fig. 15.1 Representation of the sequence of steps in a metabolomic study, starting from environmental and soil factors affecting plant metabolism to the extraction and analysis of metabolites using chromatography, mass spectrometry, and NMR. Here (M) represents the various interactions between plants, soil, microbes, and pollutants that can lead to the production of different metabolites

are not culturable in laboratory settings (Bodor et al., 2020). This limitation has been largely overcome by the advent of molecular techniques, particularly DNA sequencing. Metagenomics, a powerful tool in this regard, involves sequencing the collective genome of microbial communities directly from environmental samples (Taş et al., 2021). This approach provides insights into the genetic potential and diversity of soil bacteria, enabling researchers to identify not only the microbial inhabitants but also predict their functional capabilities (Caicedo et al., 2020).

The landscape of soil microbiome research is continually reshaped by advancements in metagenomics and metabolomics technologies, accompanied by the growing role of bioinformatics. In metagenomics, technological changes have been transformative. The shift from Sanger sequencing to next-generation sequencing (NGS) technologies marked a significant milestone, dramatically increasing the speed and reducing the cost of DNA sequencing (Hu et al., 2021). This advancement has enabled the sequencing of entire microbial communities in soil samples with unprecedented depth and resolution. Techniques like shotgun metagenomics, where all the DNA in a sample is sequenced, provide a comprehensive view of the microbial genomic landscape (Joseph & Pe'er, 2021). Additionally, long-read sequencing technologies, such as those developed by Pacific Biosciences and Oxford Nanopore, are gaining traction. These technologies allow for the sequencing of longer DNA fragments, aiding in the assembly of complex microbial genomes and the identification of novel microbial species and genes (Brown & Clarke, 2016; Mccarthy, 2010). High-resolution mass spectrometry (HRMS) and nuclear magnetic resonance (NMR) spectroscopy have become more sensitive and accurate, allowing for the detection and quantification of a broader range of metabolites with greater precision (Rathahao-Paris et al., 2016). The development of liquid chromatography-mass spectrometry (LC-MS) and gas chromatography-mass spectrometry (GC-MS) techniques has also revolutionized the field. These methods enable the separation and identification of complex mixtures of metabolites, providing a detailed metabolic fingerprint of soil microbial communities (Putri et al., 2022). Unlike traditional transcriptomics, which focuses on gene expression within a single organism, metatranscriptomics captures the functional gene expression profile of entire microbial communities. This approach allows for the identification and quantification of actively expressed genes, giving insights into the functional potential and metabolic activities of microbial communities in their natural environments (Rippin et al., 2019). Metatranscriptomics can complement metabolomics by providing information on the active metabolic pathways and the genes involved in metabolite synthesis and degradation in microbial communities (Li et al., 2023a). Bioinformatics provides the frameworks and algorithms for assembling genomes, annotating genes, identifying metabolic pathways, and integrating metagenomic and metabolomic data (Aguiar-Pulido et al., 2016). Machine learning and artificial intelligence (AI) are increasingly being applied to decipher complex patterns and interactions within these datasets, offering new insights into the metabolic functioning of soil microbiomes (Liebal et al., 2020). As these technologies continue to evolve, they promise to unveil deeper insights into the metabolic mysteries of soil bacteria, with far-reaching implications for ecology, agriculture, and environmental management.

This chapter explores recent findings in soil bacterial metabolism using metagenomics, metabolomics, and bioinformatics. We examine how these discoveries contribute to understanding global issues like pollution, climate change, and agricultural productivity. The chapter also discusses how these findings contribute to a broader ecological understanding. Soil bacteria play a crucial role in maintaining ecosystem balance through their metabolic activities. Deciphering these roles provides insights into ecosystem functioning and resilience amid environmental changes.

15.2 Environmental Stressors and Bacterial Metabolism

Environmental stressors are external factors that disrupt the natural state of an ecosystem, exerting pressure on living organisms and altering their normal functioning, growth, and survival. Environmental stressors can significantly impact various components of ecosystems, including microbial communities, plants, animals, and the abiotic environment (Killen et al., 2013). Environmental stressors significantly impact bacterial metabolism, influencing both the structure and functional dynamics of microbial communities. These stressors include a range of physical, chemical, and biological factors that alter microbial activity and survival. Among the prominent stressors are microplastics, heavy metals, and radioactive substances, each posing unique challenges to bacterial ecosystems (Bonilla, 2020). Understanding how these environmental stressors affect bacterial metabolism is crucial for assessing their impact on ecosystem health and stability, as well as for developing strategies to mitigate these effects. This knowledge is essential for addressing environmental challenges and promoting sustainable management of microbial resources.

15.2.1 Microplastics

Microplastics, small plastic fragments typically less than 5 millimeters in size, have emerged as a pervasive and insidious form of pollution, significantly impacting ecosystems worldwide. These tiny particles originate from a variety of sources, including the breakdown of larger plastic debris, cosmetic products, and synthetic textiles. Their small size allows them to evade filtration systems and accumulate in both terrestrial and aquatic environments, posing a substantial threat to ecological health (Ivar do Sul & Costa, 2014). Once ingested, they can cause physical harm, such as blockages and abrasions, or act as vectors for other pollutants, including heavy metals and organic contaminants, which adhere to their surfaces. In soil ecosystems, microplastics disrupt the delicate balance by affecting soil structure, water retention, and nutrient cycles. Their presence has been shown to impact soil microbial communities, altering metabolic processes and potentially affecting plant growth and health. This disruption can have cascading effects on soil fertility and agricultural productivity (Xu et al., 2020). The persistence of microplastics in the environment is another major concern, as they are resistant to degradation. Their longevity means that they accumulate over time, making remediation efforts challenging. Addressing the issue of microplastic pollution requires a multifaceted approach, including reducing plastic use, improving waste management, and developing biodegradable alternatives.

Polylactic acid (PLA) based plastics produce a more significant impact on soil enzyme activities and nutrition substances availability compared to polyethylene (PE). Plastic pollutants also reduce the activities of urease, phosphatase, and catalase. The availability of certain metabolites, such as sugars, amino acids, and fatty acids also suffers significantly due to plastic stress (Dong et al., 2024). Another study aimed to understand the impact of microplastics (MPs) on maize growth revealed that MPs with small sizes have a more detrimental impact on maize growth compared to larger ones also the side effects are influenced by the type of soil. In the study, black soil with larger MPs enhanced maize growth, whereas in silty loam soil, growth was inhibited. Among observed effects, the most significant differences were observed in the form of increased oxidative stress in maize roots which reduced metabolic activity. Such reduction in metabolic activities resulted in disruption of multiple amino acid metabolic pathways, including those for phenylalanine and tyrosine, and inhibition of lignin biosynthesis in maize roots. Root and rhizosphere form the communication hub for microbes and plants such nutritional disruption is capable of altering plant-microbe communication (Zhang et al., 2024). Polyhydroxybutyrate-co-hydroxyvalerate (PHBV), a biodegradable polymer from the polyhydroxyalkanoate family, is made by bacteria and is considered eco-friendly. However, studies reveal its ecological impacts, especially on soil and plant systems, cautioning the need to fully assess the environmental effects of such biodegradable materials before extensive use. The impact of biobased microplastic PHBV on soil and maize health has been studied which revealed that PHBV concentrations from 0.01% to 10% lead to reduced maize growth and lower foliar nitrogen. It highlights disruptions in maize metabolic processes, including the TCA cycle and carbohydrate metabolism (Brown et al., 2023).

15.2.2 Heavy Metal and Radioactive Elements

Heavy metals and radioactive elements present a unique and critical challenge in environmental management due to their persistence and toxicity (Ugwu et al., 2022). The study on uranium's impact on the rhizosphere and non-rhizosphere soil of ryegrass found that uranium inhibits urease and sucrase activities in both soils, with less severe effects in the rhizosphere. This suggests ryegrass roots may offer some protection. Uranium stress in the rhizosphere increased gibberellin A1 production and stimulated nitrogen metabolism and the Phosphotransferase System. However, uranium reduced the abundance and diversity of key microbes like *Aspergillus* and *Acidobacter* in both soils. A positive correlation was found between *Chujaibacter* and certain lipids and organic compounds under uranium stress in the rhizosphere. These insights offer valuable perspectives for managing uranium-contaminated soils (Wang et al., 2023b).

Chromium-VI [Cr(VI)] is among the prominent contaminants used for industrial purposes which often results in environmental contamination. Being a heavily oxidizing agent Cr(VI) promotes mutagenesis and carcinogenesis in living organisms. Microbial remediation of such metals is sought after. Metabolomics examination of differential metabolites and metabolic pathways during the Cr(VI) removal by bacteria *Bacillus mobilis CR3* reveals that the metabolic pathways involved in energy supply, amino acid synthesis and metabolism, and nucleotide metabolism are the most altered pathways (Ye et al., 2024). In examining the interactions between *Canna indica* and its rhizosphere bacteria under chromium (Cr) stress, the research uncovers key responses and adaptations. It reveals that Cr stress triggers an increase in malondialdehyde and reactive oxygen species in *C. indica*, indicative of oxidative stress. The plant combats this by boosting antioxidant enzyme activities and levels of glutathione and soluble sugar. The metabolomic analysis identifies 79 differentially expressed metabolites, playing critical roles in pathways like glycerophospholipid and N-glycan biosynthesis, essential for the plant's adaptation to Cr stress (Mao et al., 2023). Hence, the environmental presence of Cr contamination is also capable of impacting plant-microbe interactions via modifications of bacterial metabolic pathways.

Cadmium (Cd) contamination's impact on the bacterial community in blueberry (*Vaccinium* spp.) roots and soil metabolite composition is profound. The presence of Cd drastically reduces bacterial diversity across the rhizosphere, rhizoplane, and endophytic regions. This reduction is coupled with a notable shift in dominant bacterial species, with an increase in acid-tolerant taxa like *Acidiphilium* and *Granulicella*. Concurrently, the soil metabolite profile undergoes significant changes, especially an increase in lipids and lipid-like molecules (Chen et al., 2023). *Enterobacter* sp. FM-1 inoculation in *Bidens pilosa* L. rhizosphere in heavy metal-contaminated soil significantly enhances soil enzyme activities and alters microbial dynamics. FM-1 boosts carbon and nitrogen cycling pathways while reducing heavy metal resistance genes in low-contamination soil. The inoculation also varies the rhizosphere metabolite secretion, with an emphasis on lipids and amino acids, key in phytoremediation. This interaction between FM-1 and *B. pilosa* L. shows the potential to improve soil health and aid in heavy metal phytoremediation (Li et al., 2023b). It has been established that contaminants such as uranium, chromium, and cadmium significantly impair soil enzyme functionality, disrupt microbial diversity, and alter metabolic pathways.

15.3 Rhizosphere Metabolome Determining Factors

15.3.1 Root Exudates

Root exudates are compounds excreted by plant roots and can be classified as primary or secondary metabolites based on their role in the plant (More et al., 2019). The root exudation profile study of maize shows significant changes across different plant developmental stages, influenced by soil type and plant genotype. The main driver of variation in root exudation was found to be the plant's developmental stage, as the trend of a general decrease in carbon exudation rate per unit root surface area was observed as the plant matures (Santangeli et al., 2024). Additionally, the exudation rates are reported higher in loam soil compared to sand which goes on to represent the impact of soil type. On the other hand, it is also reported that root hairless rth3 maize mutant exhibits higher exudation rates than the root hair-forming wild type, despite having a lower root biomass. Such changes in root exudate rates also have a significant impact on root and rhizosphere colonizers as many of these exudates act as nutrients or poison for microbes and ultimately contribute to forming the rhizobiome (Santangeli et al., 2024). In another study, Benzoxazinoids

(BXs) have been examined for their effect on root exudates. BXs act as signaling molecules, attractants, stimulants, inhibitors, or repellents for rhizosphere microbes, affecting their growth and interactions with plants. The presence or absence of BXs in root exudates can significantly alter the root-associated microbiota. For example, a BX-deficient maize mutant exuded significantly fewer BXs compared to wild-type plants, resulting in a different root-associated microbiota composition that has implications for plant growth and defense (Hu et al., 2018).

A study of pecan rhizosphere has identified the correlation between the age of plants and metabolite profile shift in the rhizosphere, highlighting an elevated presence of specific metabolites like orientaloside and marmesin rutinoside, which are indicative of a robust plant-microbial interface. Furthermore, the initial microbial activity of the rhizosphere receives significantly higher contributions from bacteria rather than fungi as plants grow older fungal interactions become more prominent which also results in an altered metabolite profile of soil (Liu et al., 2023). Another study analyzed soil samples from 36 plots using genome sequencing and metabolomics assays. It found higher soil nutrients in temperate marshes, while root exudates were more pronounced in subtropical marshes, indicating diverse nutrient availability and root-soil interactions across climates. The study observed greater bacterial diversity and complex network structures in temperate marshes, suggesting intense bacterial competition. Climatic, edaphic factors and root exudates were identified as major influencers on bacterial communities, with plant species having a limited effect (Wang et al., 2023a).

The complexity of the exudation process makes it challenging to study root exudates metabolically (Vives-Peris et al., 2020). Different sampling strategies are used to collect root exudates, such as hydroponics or in vitro growth conditions and soil or solid-matrix growth. However, hydroponic and in vitro conditions may not reflect real exudation conditions for terrestrial plants, leading to less reliable results. While hydroponic growth enables reproducible experiments, it is often offset by low reliability (Escolà Casas & Matamoros, 2021). Among the suggested solutions for metabolomics analysis of root exudates are the use of ^{13}C-NMR and biospectroscopy techniques, which can provide quick and inexpensive analysis with minimal sample preparation, allowing for an immediate overview of root exudate composition. Future research can benefit from ^{13}C-NMR in analyzing plants grown in ^{13}C-enriched CO_2, and various mass spectrometry techniques for root exudate analysis. GC-MS is ideal for volatile compounds, LC-MS for a broad range, CE-MS for nonvolatile polar compounds, and MALDI imaging for rapid analysis and localization of exudates. A combination of LC-MS with GC and biospectroscopy offers a more complete understanding of root exudates (Escolà Casas & Matamoros, 2021).

15.3.2 Cropping Systems

Continuous cropping refers to the agricultural practice where the same crop species is grown repeatedly on the same land over multiple growing seasons (Ma et al., 2023b). Intercropping, on the other hand, involves growing two or more crop species simultaneously on the same field during a growing season. Intercropping can enhance biodiversity, improve soil health, and increase overall crop yield. It is a sustainable agricultural practice that helps in maintaining soil fertility and reducing the need for chemical fertilizers and pesticides (Khanal et al., 2021). Cropping techniques often alter the nutritional profiles of soil which ultimately results in the modification of the rhizosphere microbiome mleading to metabolome modification.

Continuous cropping significantly affects plant physiology and agronomic traits across various crops. Specifically, in the case of ramie, a plant often subjected to this practice, continuous cropping has been linked to a decline in both yield and quality. This decrease is attributed to changes in the soil microbiome's metabolic processes (Wang et al., 2020). Key metabolites identified as influential in ramie growth under continuous cropping conditions include sinapic acid, along with others like jasmonic acid, glutamine, and inositol, which have shown strong correlations with vital plant growth parameters such as height, stem diameter, and overall yield. Major altered pathways include phenylpropanoid biosynthesis, fatty acid, amino acid metabolism, and ascorbate-aldarate metabolism. Here, the study highlights the role of lipid metabolism in plant response to cropping stress and the importance of glutamine in nitrogen assimilation and stress response. Myo-inositol's role in the ascorbate and aldarate metabolism pathway was noted for its antioxidant effects (Fu et al., 2023). In another study, continuous cropping was found to increase potential autotoxins in pea roots, notably cinnamic acid, impacting plant growth. Pea plants responded to continuous cropping and the altered soil microbial environment by modifying their metabolite composition (Nayyar et al., 2009). Key pathways, like flavonoid biosynthesis and linoleic acid metabolism, were enriched, highlighting their role in the plant's immune response. Continuous cropping also activated lignin and flavonoid biosynthesis genes and ethylene and jasmonic acid signaling pathways, essential for the plant's defense. While continuous cropping does not significantly alter soil bacterial diversity, it leads to changes in bacterial abundance and community structure, affecting pea plant metabolism and health (Ma et al., 2023a).

Continuous cropping, often necessary for certain crops like peanuts due to market demands and farming practices, can lead to challenges such as soil nutrient depletion and reduced yields. Green manure (GM), which involves growing and then plowing under certain crops like winter wheat and oilseed rape during fallow periods, emerges as a solution to these issues. Research has demonstrated that GM in continuous cropping systems significantly enhances soil health. Green manure crops, when plowed back into the soil, add organic matter, improve soil structure, and increase nutrient availability. They play a crucial role in enhancing soil bacterial communities, notably increasing beneficial bacteria essential for nitrogen cycling (Northup & Rao, 2015). Metabolomics analysis shows that GM leads to improved nitrogen and carbon metabolisms, indicated by increased levels of beneficial metabolites such as sucrose and malic acid. Consequently, enhanced soil quality and a notable increase in peanut pod yield are observed. By incorporating GM into continuous cropping systems, it is possible to mitigate the adverse effects of these practices, improving both soil health and crop productivity. This approach not only addresses the immediate issues of soil degradation and reduced yield but also contributes to the long-term sustainability of agricultural practices (Xu et al., 2023).

Intercropping and rotational cropping offer alternative solutions to mitigate the adverse effects of continuous cropping on soil health. Intercropping involves growing two or more different crops in proximity, which can lead to a more efficient use of resources, reduced pest and disease incidence, and improved biodiversity. This diversity in planting helps maintain soil fertility and structure, reducing the risk of soil degradation (Khanal et al., 2021). Rotational cropping, on the other hand, entails alternating the type of crop grown in a particular field across different seasons or years. This practice helps in breaking pest and disease cycles, balancing nutrient demands, and reducing soil erosion. Both intercropping and rotational cropping contribute to maintaining soil health, enhancing biodiversity, and improving overall crop yield and sustainability in agricultural systems (Dury et al., 2012). The experiment investigating rotational strip intercropping of peanut and cotton offers valuable insights into sustainable agricultural practices. While intercropping did not enhance peanut yield, it significantly increased seed cotton yield by up to 52.63% compared to monoculture. This system also enriched differential metabolites from tryptophan and carbon pathways in both crops, impacting nutrient assimilation. Furthermore, it led to changes in the microbial community structure and composition in the soil (Lu et al., 2024). Similarly, the study on apple-marigold intercropping demonstrates significant improvements in soil health. Intercropping resulted in increased soil moisture, nitrogen content, and enzyme activity compared to clean tillage. Soil metabolomics showed a higher concentration of carbohydrates, especially in starch and sucrose metabolism, indicating improved soil quality and fertility. The bacterial community in the apple rhizosphere was found to be positively impacted, with an increase in beneficial bacterial groups like *Rhizobiales* and *Pseudomonadales*, enhancing plant growth. This intercropping system also improved soil organic matter, available phosphorus, and nitrogen, crucial for sustainable fruit production. Overall, marigold intercropping with apple trees enriches soil metabolites and supports a healthier bacterial ecosystem, offering an effective method for sustainable orchard management (Xue et al., 2023). Sorghum is a crucial crop globally, the study investigating the response of peanuts to salt stress when intercropped with sorghum, using soil metabolomics and metagenomic sequencing identifies 124 metabolites, predominantly carbohydrates, that are significantly increased in peanuts-sorghum intercropping under salt stress. Key carbohydrates like D-allose, sucrose, sorbitol, and fructose show higher levels in intercropped peanuts compared to solo-cropped ones. Additionally, the intercropping leads to a richer soil microbial community with beneficial microbes such as *Actinobacteria* and *Nitrospira*, which are associated with key metabolites like sucrose and sorbitol. This suggests their role in carbohydrate metabolism and the link between metabolite composition and microbial community in managing salt stress (Shi et al., 2023). In conclusion, the studies highlight the intricate relationship between soil metabolites and bacterial communities in response to cropping systems. Continuous cropping impacts soil metabolomics and microbes, affecting plant health. Green manure enriches soil metabolites and beneficial bacteria, enhancing soil and yield.

Intercropping and rotational cropping improve soil health through metabolite changes and diverse microbes, promoting sustainability. Understanding this interplay is crucial for sustainable agriculture.

15.3.3 Microbial Community Composition

The interaction between pathogens and symbionts in the rhizosphere is a pivotal aspect of plant biology, with profound implications for both natural ecosystems and agricultural productivity. This dynamic interplay significantly influences soil metabolome regulation, thereby impacting plant health and growth. Pathogens can disrupt plant metabolic processes, leading to diseases and reduced yields (Chaloner et al., 2021). Conversely, symbiotic relationships with certain soil microorganisms can enhance plant health, often through the production of beneficial compounds that help in disease resistance or nutrient acquisition (Harman et al., 2021).

The exploration of microbial communities for pathogen management is a pivotal component in the dynamics of plant-microbe interactions, especially in the realm of sustainable agriculture. This concept is exemplified by recent studies on *Streptomyces corchorusii* CG-G2, a strain extracted from the rhizosphere soil of *Calotropis gigantea*. This research has brought to light the ability of *S. corchorusii* to produce certain volatile organic compounds (VOCs) such as methyl 2-methyl butyrate, hexanenitrile, and methyl 2-Ethyl hexanoate. These compounds have been observed to exert antagonistic effects on *Colletotrichum gloeosporioides*, a pathogen responsible for strawberry anthracnose (Sharma & Kulshrestha, 2015). Detailed metabolomic investigations have uncovered a complex interaction, wherein these VOCs not only directly target the pathogen but also intricately influence the metabolic processes of the host plant, particularly in flavonoid metabolism. This interaction demonstrates a multifaceted mechanism where the microbial agent serves to directly combat the pathogen while simultaneously modifying the host plant's metabolic pathways to bolster its resistance to disease. Furthermore, the VOC methyl 2-methyl butyrate has been identified as having an impact on the fruit's quality, offering the dual advantage of pathogen suppression and enhancement of post-harvest fruit preservation (Li et al., 2024).

The comparative analysis of rhizosphere interaction modification by root exudates from two peanut varieties—*Arachis correntina*, a wild species with robust adaptability, and GH85, a cultivar that yields high but is prone to pathogen attack, highlights the critical role of root exudates in shaping the rhizosphere environment. Soil pathogens such as *R. solanacearum* (Xue et al., 2020) and *F. moniliforme* (Hossain et al., 2021) colonization are heavily impacted by root exudates. It is observed that *A. correntina*'s root exudates exhibit a more potent inhibitory effect on these pathogens compared to GH85. This difference is primarily linked to the unique composition of the root exudates, especially concerning phenolic acids and amino acids. Delving into the metabolic underpinnings, a notable trend in the downregulation of differentially expressed genes and metabolites related to pathways such as carbon metabolism, phenylpropanoid biosynthesis, and amino acid biosynthesis is observed within the host. These pathways play a crucial role in the plant's defensive mechanisms against pathogens. The findings suggest that the phenolic acids, abundant in GH85's exudates, could inadvertently support pathogen growth, whereas the lower amino acid content in *A. correntina*'s exudates is associated with stronger pathogen inhibition (Li et al., 2023c). Such results create a conflict with known research which reports that phenolic acid contributes to symbiosis promotion (Mandal et al., 2010) and pathogen death (Shalaby et al., 2016). For peanuts, the composition of phenolic acids and amino acids in peanut root exudates significantly influences the growth dynamics of soil pathogens. Thus the importance of rhizosphere metabolome composition in determining plant-pathogen interactions is an essential consideration in sustainable agricultural practices (Li et al., 2023c). Coumarins, a group of secondary metabolites produced via the phenylpropanoid pathway in plants, play a crucial role in iron uptake from iron-deficient soils and exhibit diverse biological activities, including antimicrobial, antiviral, anticoagulant, and anticancer properties (Stringlis et al., 2019). Organic farming has also been identified as a regulatory factor in plant-microbe interaction. Investigation into the role of *Pseudomonas* species in conferring disease suppressiveness along with organic farming practices reveals a higher diversity of antagonistic *Pseudomonas* species in soils from organic fields compared to conventional fields. Specific Pseudomonas species such as *Pseudomonas lini*, *Pseudomonas indica*, *Pseudomonas chlororaphis*, and *Pseudomonas alcaliphila*, known for their biocontrol capabilities against plant pathogens, were identified in organic soils (Khatri et al., 2023).

Environmental factors significantly influence microbial diversity, shaping the composition and functionality of microbial communities in various ecosystems. Temperature is a key environmental factor affecting microbial diversity. For instance, extreme temperatures can favor thermophilic or psychrophilic organisms while suppressing others less adapted to such conditions (Shu & Huang, 2022). Moisture availability is another crucial factor, especially in soil environments, where it affects microbial mobility, nutrient diffusion, and metabolic activities. Drought conditions can reduce microbial diversity by creating a more selective environment that favors drought-resistant species while eliminating more sensitive ones (Cui et al., 2020). Microorganisms have specific pH ranges in which they thrive; deviations from these ranges can inhibit their growth and metabolism. Acidic or alkaline conditions can lead to reduced microbial diversity by selecting pH-tolerant species while excluding others (Lund et al., 2020). Additionally, the presence of chemical pollutants like heavy metals, pesticides, and emerging contaminants such as microplastics can impose selective pressures on microbial communities, often reducing overall diversity and favoring species with resistance mechanisms (Ahmad et al., 2022).

Along with environmental factors, time also shows significant impact on microflora. The dynamics of the soil microbiome are heavily influenced by temporal changes such as seasonal variations, daily cycles, and long-term environmental shifts. For instance, temperature and moisture fluctuations across seasons lead to shifts in microbial populations, with certain taxa thriving during specific conditions. In temperate climates, bacterial and fungal communities often show distinct seasonal patterns; microbial diversity tends to be higher during warmer and wetter months when conditions are favorable for growth and metabolic activity. Conversely, colder months or periods of drought may reduce microbial activity, leading to lower diversity and shifts in community composition toward more stress-tolerant organisms (Núñez et al., 2021). Daily cycles, such as diurnal temperature changes and fluctuations in moisture due to dew formation and evaporation, also influence soil microbial dynamics. These short-term variations can affect microbial metabolism and community interactions (Hu et al., 2019). Overall, the soil microbiome is dynamic, responding continuously to temporal changes in its environment. These changes shape the functional potential of microbial communities, affecting soil health and ecosystem resilience.

15.4 Techniques in Metabolomics

An array of analytical methodologies has been deployed in the realm of plant-microbe interactions, focusing on their application in metabolomics studies. Table 15.1 below succinctly presents a collection of research endeavors that utilize a spectrum of advanced techniques, such as high-performance liquid chromatography (HPLC), gas chromatography/mass spectrometry (GC-MS), and liquid chromatography-mass spectrometry (LC-MS), various modifications of these methods (Gupta et al., 2022). These studies encompass a diverse range of plant species and microbial interactions, offering insights into key metabolic processes and their implications. Examination of the research focus, techniques employed, plant species involved, microbial interactions, and the pivotal findings of each study reveals the capabilities of these techniques across diverse species and conditions. The synthesis of these studies not only contributes to our understanding of plant-microbe symbiosis but also highlights the versatility and efficacy of metabolomics as a tool for elucidating complex biological interactions.

15.4.1 Microbial Metabolic Modeling

Microbial metabolic modeling is a computational approach used to understand and predict the metabolic behavior of microorganisms. It involves the use of mathematical models to represent the complex networks of biochemical reactions within a microbial cell, allowing researchers to simulate and analyze the metabolic activities of these organisms under various conditions. These models are built on genomic, transcriptomic, proteomic, and metabolomic data, integrating various biological levels to provide a comprehensive view of microbial metabolism (Gu et al., 2019). Microbial metabolic modeling has become an essential tool in systems biology, synthetic biology, and biotechnology, offering insights into microbial physiology, optimizing metabolic pathways, and designing microorganisms for industrial applications.

Table 15.1 Various techniques used for metabolomics analysis

Research focus	Techniques used[a]	Host species	Microbial interactions	Key findings	Reference
Attraction of *B. amyloliquefaciens* by cucumber roots	HPLC, LC-MS	*Cucumis sativus*	*Bacillus amyloliquefaciens* SQR9	Citric acid and Fumaric acid in exudates cause biofilm formation	Zhang et al. (2014)
Metabolism study of maize cultivars with *Azospirillum* strains	LC–MS, NMR	*Zea mays*	*Azospirillum* strains	Bacteria strain-dependent effects on secondary metabolism of maize	Walker et al. (2011)
Plant compounds affected by bacterial nitrogen fixation	GC-MS	*Zea mays*	*Herbaspirillum seropedicae* SmR1 and *Azospirillum brasilense* FP2	Metabolome is affected by type of strain of bacteria	Brusamarello-Santos et al. (2017)
Metabolite analysis in rice roots with PGPB	UHPLC-DAD/ESI-QTOF, gene expression data	*Oryza sativa*	*Azospirillum* sp.	Reduced accumulation of three alkylresorcinols and increased accumulation of two hydroxycinnamic acid amides (HCAA)	Valette et al. (2020)
Plant defense mechanisms in tomato and PGPB interactions	UHPLC-QqQ-MS	*Solanum lycopersicum*	*Pseudomonas fluorescens* (N04), *Ps. koreensis* (N19), *Paenibacillus alvei* (T22), and *Lysinibacillus sphaericus* (T19)	Induction of defense related metabolic reprogramming	Mhlongo et al. (2020)
Spatial metabolite information in *Medicago truncatula* roots and nodules	MALDI-MSI	*Medicago truncatula*	*Sinorhizobium meliloti*	Detection of various metabolites produced during nitrogen fixation in roots and nodules	Ye et al. (2013)
Spatiotemporal production of antimicrobial compounds by *Bacillus*	MALDI-MSI, IC-MALDI-TOF-MS	–	*B. subtilis* and *B. amyloliquefaciens*	Identification of antimicrobial compounds in root-colonizing *Bacillus*	Moree et al. (2013) and Pathak and Keharia (2013)
Identification of *antibiotic* compounds in *Solanum lycopersicum*	MALDI-MSI, UHPLC-MS	*Solanum lycopersicum*	*B. amyloliquefaciens*	Discovery of a new surfactin variant released by roots post-inoculation	Debois et al. (2014)

Metabolic effect of endophytic *Neotyphodium lolii* on *Lolium perenne*	DI-ESI-MS	*Lolium perenne*	*Neotyphodium lolii*	Identification of metabolites involved in symbiosis induction	Cao et al. (2008)
Metabolic response in maize to PGPB and beneficial fungi	UHPLC/Q-TOF-MS	*Zea mays*	*Bacillus megaterium* MHBM77, along with beneficial fungi	Distinct metabolic responses observed in treated maize plants	Rouphael et al. (2020)
Effect of nutritional status on plant-*Azospirillum* interaction	GC–MS, LC–MS	*Brachypodium distachyon*	*Azospirillum brasilense*	Nutritional status influences bacterial interaction with the plant	Schillaci et al. (2021)

[a]*HPLC* High-performance liquid chromatography, *GC-MS* Gas chromatography/mass spectrometry, *LC-MS* Liquid chromatography-mass spectrometry, *NMR* Nuclear magnetic resonance spectroscopy, *HPLC-DAD-ESI-QTOF-MS* Ultrahigh performance liquid chromatography-diode array detection-electrospray ionization-quadrupole time-of-flight-mass spectrometry, *UHPLC-QqQ-MS* Ultra high-performance liquid chromatography coupled with triple quadrupole mass spectrometry, *MALDI MSI* Matrix-assisted laser desorption/ionization mass spectrometry imaging, *IC-MALDI-TOF-MS* Intact-cell matrix-assisted laser desorption/ionization time-of-flight mass spectrometry, *UPLC–MS* Ultrahigh-performance liquid chromatography–mass spectrometry, *DI-ESI-MS* Direct infusion electrospray ionization mass spectrometry, *UHPLC/Q-TOF-MS* Ultra-high performance liquid chromatography-quadrupole time-of-flight mass spectrometry

Use cases for such models exist in various fields including environment and ecology, engineering, and synthetic biology, along with human health. One of the primary applications of microbial metabolic modeling in metabolomics is to explore the metabolic capabilities of microbial communities in different environments. By modeling the metabolism of diverse microbial species, researchers can predict how these organisms interact with each other and their environment. For instance, models can simulate nutrient fluxes within a microbial community, revealing how different species contribute to the cycling of carbon, nitrogen, sulfur, and other elements in soil, oceans, or the human gut (Zomorrodi & Segrè, 2016). Microbial metabolic models are extensively used in metabolic engineering to design microorganisms for the production of biofuels, pharmaceuticals, and other valuable chemicals. By simulating various genetic modifications and pathway optimizations, researchers can predict the impact of these changes on microbial metabolism and identify strategies to enhance the production of desired metabolites. For example, metabolic models have been used to engineer *Escherichia coli* strains capable of efficiently producing high-value chemicals such as succinic acid, ethanol, and isoprenoids (Chubukov et al., 2018). In the context of human health, microbial metabolic modeling helps understand the metabolic interactions between the human host and its microbiome. Models can predict how changes in the gut microbiota, influenced by diet, antibiotics, or diseases, affect host metabolism and contribute to health or disease states. This approach has been used to study metabolic disorders, such as obesity and diabetes, and to design personalized therapeutic strategies targeting the gut microbiome (Thiele et al., 2020).

Several computational approaches are employed in microbial metabolic modeling to achieve these insights. Constraint-based modeling (CBM), particularly flux balance analysis (FBA), is the most commonly used method. FBA involves constructing a stoichiometric matrix that represents all metabolic reactions within a cell and applying constraints like mass balance, energy balance, and nutrient availability. This method allows researchers to predict the distribution of metabolic fluxes through various pathways, optimizing a specific objective such as growth rate or metabolite production (Orth et al., 2010). Another approach, kinetic modeling, incorporates enzyme kinetics and regulatory mechanisms to simulate the dynamic behavior of metabolic networks over time. Although more computationally demanding and requiring detailed kinetic parameters, kinetic modeling provides a more realistic representation of metabolic processes and can simulate transient dynamics and regulatory responses to environmental changes (Chassagnole et al., 2002). Genome-scale metabolic models (GEMs) are comprehensive models that encompass all known metabolic reactions within an organism, reconstructed from its genomic data. These models allow researchers to explore system-level properties, identify essential genes and metabolic bottlenecks, and predict the effects of genetic modifications on the overall metabolism (Gu et al., 2019). Recently, machine learning and data-driven approaches have been integrated into microbial metabolic modeling, utilizing large-scale omics data to predict metabolic functions and interactions. These advanced methods can identify novel metabolic pathways, forecast the impacts of environmental changes on microbial metabolism, and optimize models based on empirical data (Zampieri et al., 2019). Together, these computational techniques enhance our ability to understand and manipulate microbial metabolic networks, providing valuable insights for various applications in science and industry.

15.5 Soil Amendments and Bacterial Metabolism

In the intricate web of terrestrial ecosystems, the composition of soil plays a pivotal role in shaping bacterial diversity and metabolism, thereby profoundly influencing ecological balances and plant health. Recent studies have illuminated the significant impact of soil amendments, particularly biochar, on the improvement of soil structure and microbial communities. This influence is exemplified in agricultural and forest soils, where biochar application has been shown to modify soil properties, enhance microbial diversity, and induce metabolic shifts, leading to healthier, disease-resistant plants (Abujabhah et al., 2016).

Biochar is a carbon-rich material produced through the pyrolysis of organic biomass under limited oxygen conditions. Biochar has emerged as a key element in the enhancement of soil structure and the bolstering of microbial communities within agricultural and forestry ecosystems. Plants of economic importance, such as *Myrica rubra*, are frequently plagued by disease outbreaks, causing substantial financial losses. The application of biochar has been documented to alter soil characteristics and enrich microbial diversity which contributes to the cultivation of plants that are resistant to diseases (Ren et al., 2023). Apart from promoting disease resistance, biochar has also been applied to saline-alkali soil to promote the growth of winter wheat using modified biochar. Modified forms of biochar such as acidified biochar (HBC) (Bolan et al., 2023) and nanoparticle-size biochar, play a significant role in metabolome modification in plants and microbes (Chen et al., 2022). Significant increases in metabolites such as amino acids, lipids, flavonoids, and organic acids in winter wheat leaves after treatment with modified biochar are observed. Modifications in the regulation of many pathways are also observed most significant ones are, the pyrimidine metabolism pathway, vital for nucleotide synthesis and DNA repair under stress, and the enhancement of β-alanine metabolism and branched-chain amino acids (valine, leucine, isoleucine), important for stress resistance and protein synthesis (Duan et al., 2023). In another study, biochar and nitrogen fertilizer interactions have been observed to substantially enhance soil inorganic nitrogen levels, accelerate ammonium ion production, and later boost nitrate ion formation. Furthermore, the presence of biochar promotes changes in 756 metabolites, highlighting an increase in lipids and organic acids (Xia et al., 2023).

Apart from biochar, filter mud (FM) has been another famous organic soil amendment for improvement of crop growth and soil quality. FM is actively applied in ratoon cropping. Ratoon cropping, a common practice in sugarcane and other crops, involves harvesting the crop while leaving the roots and growing shoot apices intact for regrowth in the subsequent season. However, this technique faces challenges like nutrient depletion and disease susceptibility (Xu et al., 2021). The use of FM and biochar amendment provides significant improvements in ratoon crop weight, stem diameter, sugar content, and soil physico-chemical properties such as soil nitrate, organic matter, total nitrogen, and carbon. Other notable changes include soil-available potassium and nitrate, cellulase activity, and phosphatase enhancement. BC and FM amendments enrich specific stress-regulating metabolites, including jasmonic acid, melatonin, abscisic aldehyde, and others (Fallah et al., 2023). In summary, soil amendments like biochar and filter mud can revolutionize agricultural practices. They are not just soil enhancers and catalysts that trigger a chain reaction of positive changes in soil ecosystems.

15.6 Future Directions

The future of soil bacterial metabolism research is rich with opportunities and challenges crucial for advancing this vital field. Identifying and exploring knowledge gaps is paramount, especially in understanding the complex interactions among different soil bacterial species and their relationships with other soil organisms. Delving deeper into these dynamics is essential for a comprehensive grasp of nutrient cycling and soil health. Significant focus is also needed on the impact of climate change on soil bacterial metabolism. The effects of extreme temperatures altered precipitation patterns, and increased CO_2 levels on microbial metabolic processes—and their implications for soil ecosystem functions—remain largely unexplored. The potential for metabolic engineering of soil bacteria to enhance ecosystem services presents an exciting frontier. Engineering bacteria for more efficient nitrogen fixation could reduce dependency on synthetic fertilizers, leading to more sustainable agricultural practices. Similarly, genetically modifying bacteria to degrade or immobilize pollutants can be pivotal in phytoremediation efforts, aiding in the cleanup of contaminated soils. Enhancing the natural ability of soil bacteria to suppress plant pathogens opens the door to developing biological methods for disease control in crops, which could revolutionize agriculture. Lastly, the prospects of discovering novel metabolites from soil bacteria hold immense potential. The soil microbiome is a treasure trove of bioactive compounds, many yet to be discovered or fully characterized. These novel metabolites could play a transformative role in medicine and biotechnology, particularly in developing new antibiotics and enzymes. In agriculture, identifying metabolites that promote plant growth, improve stress tolerance, or enhance nutrient uptake could revolutionize farming methods. Furthermore, certain bacterial metabolites could serve as biomarkers for soil health or environmental stress, providing invaluable tools for monitoring ecosystem changes or assessing the impact of human activities.

References

Abujabhah, I. S., Doyle, R., Bound, S. A., & Bowman, J. P. (2016). The effect of biochar loading rates on soil fertility, soil biomass, potential nitrification, and soil community metabolic profiles in three different soils. *Journal of Soils and Sediments, 16*(9), 2211–2222.

Aguiar-Pulido, V., et al. (2016). Metagenomics, metatranscriptomics, and metabolomics approaches for microbiome analysis. *Evolutionary Bioinformatics, 12*, 5–16. https://journals.sagepub.com/doi/full/10.4137/EBO.S36436

Ahmad, H. A., et al. (2022). The environmental distribution and removal of emerging pollutants, highlighting the importance of using microbes as a potential degrader: A review. *Science of the Total Environment, 809*, 151926.

Bardgett, R. D., & Van Der Putten, W. H. (2014). Belowground biodiversity and ecosystem functioning. *Nature, 515*(7528), 505–511.

Bodor, A., et al. (2020). Challenges of unculturable bacteria: Environmental perspectives. *Reviews in Environmental Science and Biotechnology, 19*(1), 1.

Bolan, N., et al. (2023). Soil acidification and the liming potential of biochar. *Environmental Pollution, 317*, 120632.

Bonilla, C. Y. (2020). Generally stressed out bacteria: Environmental stress response mechanisms in gram-positive bacteria. *Integrative and Comparative Biology, 60*(1), 126–133.

Brown, C. G., & Clarke, J. (2016). Nanopore development at Oxford nanopore. *Nature Biotechnology, 34*(8), 810–811.

Brown, R. W., et al. (2023). Bioplastic (PHBV) addition to soil alters microbial community structure and negatively affects plant-microbial metabolic functioning in maize. *Journal of Hazardous Materials, 441*, 129959.

Brusamarello-Santos, L. C., et al. (2017). Metabolic profiling of two maize (Zea Mays L.) inbred lines inoculated with the nitrogen fixing plant-interacting bacteria Herbaspirillum seropedicae and Azospirillum brasilense. *PLoS One, 12*(3).

Caicedo, H. H., et al. (2020). PICRUSt2 for prediction of metagenome functions. *Nature Biotechnology, 38*(6), 669–673.

Cao, M., et al. (2008). Advanced data-mining strategies for the analysis of direct-infusion ion trap mass spectrometry data from the association of perennial ryegrass with its endophytic fungus, Neotyphodium lolii. *Plant Physiology, 146*(4), 1501–1514.

Chaloner, T. M., Gurr, S. J., & Bebber, D. P. (2021). Plant pathogen infection risk tracks global crop yields under climate change. *Nature Climate Change, 11*(8), 710–715.

Chassagnole, C., et al. (2002). Dynamic modeling of the central carbon metabolism of Escherichia coli. *Biotechnology and Bioengineering, 79*(1), 53–73.

Chen, X., Duan, M., Zhou, B., & Cui, L. (2022). Effects of biochar nanoparticles as a soil amendment on the structure and hydraulic characteristics of a sandy loam soil. *Soil Use and Management, 38*(1), 836–849.

Chen, S. P., et al. (2023). Cadmium alters the rhizosphere bacterial community structure in blueberry roots and increases the content of lipids in the soil. *Rhizosphere, 27*, 100755.

Chubukov, V., et al. (2018). Synthetic and systems biology for microbial production of commodity chemicals. *npj Systems Biology and Applications, 2*.

Cui, Y., et al. (2020). Soil moisture mediates microbial carbon and phosphorus metabolism during vegetation succession in a semiarid region. *Soil Biology and Biochemistry, 147*, 107814.

Debois, D., et al. (2014). Spatiotemporal monitoring of the antibiome secreted by Bacillus biofilms on plant roots using MALDI mass spectrometry imaging. *Analytical Chemistry, 86*(9), 4431–4438.

Dong, D. Z., et al. (2024). Plastic residues alter soil microbial community compositions and metabolite profiles under realistic conditions. *Science of the Total Environment, 906*, 167352.

Dongdong, C., et al. (2023). Soil nutrients directly drive soil microbial biomass and carbon metabolism in the sanjiangyuan alpine grassland. *Journal of Soil Science and Plant Nutrition, 23*(3), 3548–3560.

Duan, M. L., et al. (2023). Integrated microbiological and metabolomics analyses to understand the mechanism that allows modified biochar to affect the alkalinity of saline soil and winter wheat growth. *Science of the Total Environment, 866*.

Dury, J., et al. (2012). Models to support cropping plan and crop rotation decisions. A review. *Agronomy for Sustainable Development, 32*(2), 567–580.

Escolà Casas, M., & Matamoros, V. (2021). Analytical challenges and solutions for performing metabolomic analysis of root exudates. *Trends in Environmental Analytical Chemistry, 31*, e00130.

Fallah, N., et al. (2023). Sustained organic amendments utilization enhances ratoon crop growth and soil quality by enriching beneficial metabolites and suppressing pathogenic bacteria. *Frontiers in Plant Science, 14*.

Fu, Y. F., et al. (2023). Untargeted metabolomics reveal rhizosphere metabolites mechanisms on continuous ramie cropping. *Frontiers in Plant Science, 14*.

Gu, C., et al. (2019). Current status and applications of genome-scale metabolic models. *Genome Biology, 20*(1), 121.

Gupta, S., Schillaci, M., & Roessner, U. (2022). Metabolomics as an emerging tool to study plant-microbe interactions. *Emerging Topics in Life Sciences, 6*(2), 175–183.

Harman, G., Khadka, R., Doni, F., & Uphoff, N. (2021). Benefits to plant health and productivity from enhancing plant microbial symbionts. *Frontiers in Plant Science, 11*.

Hossain, M. S., Ali, M. A., Khan, M. A. I., & Maniruzzaman, S. (2021). Prevalence and transmission of Fusarium moniliforme: A seed borne pathogen of rice. *Bangladesh Rice Journal, 24*(1), 11–19.

Hu, L., et al. (2018). Root exudate metabolites drive plant-soil feedbacks on growth and defense by shaping the rhizosphere microbiota. *Nature Communications, 9*(1).

Hu, A., et al. (2019). Diurnal temperature variation and plants drive latitudinal patterns in seasonal dynamics of soil microbial community. *Frontiers in Microbiology, 10*(APR).

Hu, T., Chitnis, N., Monos, D., & Dinh, A. (2021). Next generation sequencing technologies: An overview. *Human Immunology, 82*(11), 801–811.

Ivar do Sul, J. A., & Costa, M. F. (2014). The present and future of microplastic pollution in the marine environment. *Environmental Pollution, 185*, 352–364.

Joseph, T. A., & Pe'er, I. (2021). An introduction to whole-metagenome shotgun sequencing studies. *Methods in Molecular Biology, 2243*, 107–122.

Khanal, U., et al. (2021). Intercropping—Evaluating the advantages to broadacre systems. *Agriculture (Switzerland), 11*(5).

Khatri, S., et al. (2023). *Pseudomonas is a key player in conferring disease suppressiveness in organic farming* (Vol. 503, p. 85). Plant and Soil.

Killen, S. S., et al. (2013). Environmental stressors Alter relationships between physiology and behaviour. *Trends in Ecology and Evolution, 28*(11), 651–658.

Li, C., et al. (2023a). Integrated metabolomics, transcriptomics, and proteomics analyses reveal co-exposure effects of polycyclic aromatic hydrocarbons and cadmium on ryegrass (Lolium perenne L.). *Environment International, 178*, 108105.

Li, Y., et al. (2023b). Metagenomics combined with metabolomics reveals the effect of Enterobacter Sp. inoculation on the rhizosphere microenvironment of Bidens Pilosa L. in heavy metal contaminated soil. *Journal of Hazardous Materials, 458*, 132033.

Li, Z., et al. (2023c). Root metabolism and effects of root exudates on the growth of *Ralstonia solanacearum* and *Fusarium moniliforme* were significantly different between the two genotypes of peanuts. *Genes, 14*(2).

Li, X. J., et al. (2024). Biocontrol potential of volatile organic compounds produced by Streptomyces corchorusii CG-G2 to strawberry anthracnose caused by Colletotrichum gloeosporioides. *Food Chemistry, 437*.

Liebal, U. W., et al. (2020). Machine learning applications for mass spectrometry-based metabolomics. *Metabolites, 10*(6), 1–23.

Liu, J., et al. (2023). Integrated microbiome and metabolomics analysis reveal a closer relationship between soil metabolites and bacterial community than fungal community in pecan plantations. *Land Degradation and Development, 34*(10), 2812–2824.

Lu, J. H., et al. (2024). Rotational strip peanut/cotton intercropping improves agricultural production through modulating plant growth, root exudates, and soil microbial communities. *Agriculture Ecosystems & Environment, 359*, 108767.

Lund, P. A., et al. (2020). Understanding how microorganisms respond to acid PH is central to their control and successful exploitation. *Frontiers in Microbiology, 11*.

Ma, L., et al. (2023a). New insights into the occurrence of continuous cropping obstacles in pea (*Pisum Sativum* L.) from soil bacterial communities, root metabolism and gene transcription. *BMC Plant Biology, 23*(1), 226.

Ma, Z., et al. (2023b). Obstacles in continuous cropping: Mechanisms and control measures. *Advances in Agronomy, 179*, 205–256.

Mandal, S. M., Chakraborty, D., & Dey, S. (2010). Phenolic acids act as signaling molecules in plant-microbe symbioses. *Plant Signaling and Behavior, 5*(4), 359–368.

Mao, H., et al. (2023). Recruitment and metabolomics between *Canna indica* and rhizosphere bacteria under Cr stress. *Frontiers in Microbiology, 14*.

Mccarthy, A. (2010). Third generation DNA sequencing: Pacific biosciences' single molecule real time technology. *Chemistry and Biology, 17*(7), 675–676.

Mhlongo, M. I., et al. (2020). Metabolic profiling of PGPR-treated tomato plants reveal priming-related adaptations of secondary metabolites and aromatic amino acids. *Metabolites, 10*(5).

Molefe, R. R., Amoo, A. E., & Babalola, O. O. (2023). Communication between plant roots and the soil microbiome; involvement in plant growth and development. *Symbiosis, 90*(3), 231–239.

More, S. S., Shinde, S. E., & Kasture, M. C. (2019). Root exudates a key factor for soil and plant: An overview. *The Pharma Innovation Journal, 8*(6), 449–459. https://www.thepharmajournal.com/archives/?year=2019&vol=8&issue=6&ArticleId=3841

Moree, W. J., et al. (2013). Imaging mass spectrometry of a coral microbe interaction with fungi. *Journal of Chemical Ecology, 39*(7), 1045–1054.

Nayyar, A., et al. (2009). Soil microbial quality associated with yield reduction in continuous-pea. *Applied Soil Ecology, 43*(1), 115–121.

Northup, B. K., & Rao, S. C. (2015). Green manures in continuous wheat systems affect grain yield and nitrogen content. *Agronomy Journal, 107*(5), 1666–1672.

Núñez, A., García, A. M., Moreno, D. A., & Guantes, R. (2021). Seasonal changes dominate long-term variability of the urban air microbiome across space and time. *Environment International, 150*, 106423.

Orth, J. D., Thiele, I., & Palsson, B. O. (2010). What is flux balance analysis? *Nature Biotechnology, 28*(3), 245–248.

Pathak, K. V., & Keharia, H. (2013). Characterization of fungal antagonistic bacilli isolated from aerial roots of banyan (Ficus benghalensis) using intact-cell MALDI-TOF mass spectrometry (ICMS). *Journal of Applied Microbiology, 114*(5), 1300–1310.

Putri, S. P., et al. (2022). Application of gas chromatography-mass spectrometry-based metabolomics in food science and technology. *Journal of Bioscience and Bioengineering, 133*(5), 425–435.

Rathahao-Paris, E., Alves, S., Junot, C., & Tabet, J. C. (2016). High resolution mass spectrometry for structural identification of metabolites in metabolomics. *Metabolomics, 12*(1), 1–15.

Ren, H. Y., et al. (2023). Improvement effect of biochar on soil microbial community structure and metabolites of decline disease bayberry. *Frontiers in Microbiology, 14*.

Rippin, M., et al. (2019). Metatranscriptomic and metabolite profiling reveals vertical heterogeneity within a Zygnema green algal mat from Svalbard (High Arctic). *Environmental Microbiology, 21*(11), 4283–4299.

Rouphael, Y., et al. (2020). Metabolomic responses of maize shoots and roots elicited by combinatorial seed treatments with microbial and non-microbial biostimulants. *Frontiers in Microbiology, 11*.

Santangeli, M., et al. (2024). Maize (*Zea Mays* L.) root exudation profiles change in quality and quantity during plant development – A field study. *Plant Science, 338*, 111896.

Schillaci, M., et al. (2021). The metabolic response of Brachypodium roots to the interaction with beneficial bacteria is affected by the plant nutritional status. *Metabolites, 11*(6).

Shalaby, S., et al. (2016). Plant phenolic acids induce programmed cell death of a fungal pathogen: MAPK signaling and survival of Cochliobolus heterostrophus. *Environmental Microbiology, 18*(11), 4188–4199.

Sharma, M., & Kulshrestha, S. (2015). Colletotrichum gloeosporioides: An anthracnose causing pathogen of fruits and vegetables. *Biosciences Biotechnology Research Asia, 12*(2), 1233–1246.

Shi, X. L., et al. (2023). Soil metagenome and metabolome of peanut intercropped with sorghum reveal a prominent role of carbohydrate metabolism in salt-stress response. *Environmental and Experimental Botany, 209*.

Shu, W. S., & Huang, L. N. (2022). Microbial diversity in extreme environments. *Nature Reviews Microbiology, 20*(4), 219–235.

Stringlis, I. A., De Jonge, R., & Pieterse, C. M. J. (2019). The age of coumarins in plant-microbe interactions. *Plant and Cell Physiology, 60*(7), 1405–1419.

Taş, N., et al. (2021). Metagenomic tools in microbial ecology research. *Current Opinion in Biotechnology, 67*, 184–191.

Thiele, I., et al. (2020). Personalized whole-body models integrate metabolism, physiology, and the gut microbiome. *Molecular Systems Biology, 16*(5), e8982.

Ugwu, E. I., et al. (2022). Application of green nanocomposites in removal of toxic chemicals, heavy metals, radioactive materials, and pesticides from aquatic water bodies. *Sustainable Nanotechnology for Environmental Remediation*, 321–346.

Valette, M., et al. (2020). A common metabolomic signature is observed upon inoculation of rice roots with various rhizobacteria. *Journal of Integrative Plant Biology, 62*(2), 228–246.

Vives-Peris, V., de Ollas, C., Gómez-Cadenas, A., & Pérez-Clemente, R. M. (2020). Root exudates: From plant to rhizosphere and beyond. *Plant Cell Reports, 39*(1), 3–17.

Walker, V., et al. (2011). Host plant secondary metabolite profiling shows a complex, strain-dependent response of maize to plant growth-promoting rhizobacteria of the genus Azospirillum. *New Phytologist, 189*(2), 494–506.

Wang, Y., et al. (2020). Identification of the rhizospheric microbe and metabolites that led by the continuous cropping of ramie (Boehmeria Nivea L. Gaud). *Scientific Reports, 10*(1).

Wang, R. M., Cui, L. J., Li, J., & Li, W. (2023a). Factors driving the halophyte rhizosphere bacterial communities in coastal salt marshes. *Frontiers in Microbiology, 14*.

Wang, Y. L., et al. (2023b). Non-targeted metabolomics and 16s RDNA reveal the impact of uranium stress on rhizosphere and non-rhizosphere soil of ryegrass. *Journal of Environmental Radioactivity, 258*.

Wu, W., et al. (2022). In situ diversity of metabolism and carbon use efficiency among soil bacteria. *Science Advances, 8*(44).

Xia, H., et al. (2023). Insight into mechanisms of biochar-fertilizer induced of microbial community and microbiology of nitrogen cycle in acidic soil. *Journal of Environmental Management, 336*, 117602.

Xu, C., et al. (2020). Are we underestimating the sources of microplastic pollution in terrestrial environment? *Journal of Hazardous Materials, 400*, 123228.

Xu, F., et al. (2021). Sugarcane ratooning ability: Research status, shortcomings, and prospects. *Biology, 10*(10), 1052.

Xu, Y., et al. (2023). Green manure increases peanut production by shaping the rhizosphere bacterial community and regulating soil metabolites under continuous peanut production systems. *BMC Plant Biology, 23*(1), 69.

Xue, H., Lozano-Durán, R., & Macho, A. P. (2020). Insights into the root invasion by the plant pathogenic bacterium Ralstonia solanacearum. *Plants, 9*(516), 1–9.

Xue, X. M., et al. (2023). Apple-marigold intercropping improves soil properties by changing soil metabolomics and bacterial community structures. *Frontiers in Microbiology, 14*, 1195985.

Ye, H., et al. (2013). MALDI mass spectrometry-assisted molecular imaging of metabolites during nitrogen fixation in the Medicago truncatula-Sinorhizobium meliloti symbiosis. *Plant Journal, 75*(1), 130–145.

Ye, Y. B., et al. (2024). Mechanism of Cr(VI) removal by efficient Cr(VI)-resistant *Bacillus mobilis* CR3. *World Journal of Microbiology & Biotechnology, 40*(1), 21.

Zampieri, G., Vijayakumar, S., Yaneske, E., & Angione, C. (2019). Machine and deep learning meet genome-scale metabolic modeling. *PLoS Computational Biology, 15*(7), e1007084.

Zhang, N., et al. (2014). Effects of different plant root exudates and their organic acid components on chemotaxis, biofilm formation and colonization by beneficial rhizosphere-associated bacterial strains. *Plant and Soil, 374*(1–2), 689–700.

Zhang, Y. H., et al. (2024). Impact of microplastic particle size on physiological and biochemical properties and rhizosphere metabolism of *Zea Mays* L.: Comparison in different soil types. *Science of the Total Environment, 908*, 168219.

Zomorrodi, A. R., & Segrè, D. (2016). Synthetic ecology of microbes: Mathematical models and applications. *Journal of Molecular Biology, 428*(5), 837–861.

Unseen Allies: Exploring Cooperative Behavior Among Soil Microbes

16

Karan Prajapati, Bhavdip Anavadiya, and Milan Dabhi

Abstract

Soil microbes, through their dynamic and intricate interactions, form the backbone of terrestrial ecosystems and are pivotal in enhancing agricultural productivity. This study delves deep into the complex web of relationships among bacteria, fungi, archaea, and protozoa. It emphasizes their cooperative behaviors such as mutualistic, syntrophic, and biofilm interactions that are crucial for nutrient cycling, improving soil fertility, supporting plant growth, and maintaining ecosystem resilience. Chemical signaling and quorum sensing mechanisms enable microbes to coordinate activities like biofilm formation, nutrient exchange, and pathogen suppression. Environmental factors, such as soil type, moisture, pH, and human activities, also influence these microbial interactions. Additionally, this chapter discusses the practical implications of understanding microbial interactions, advocating for their integration into sustainable agricultural practices to enhance crop management and soil conservation. This comprehensive investigation not only advances scientific knowledge of soil microbiology but also highlights its applications in improving soil health and resilience amidst environmental challenges.

Keywords

Soil microbe · Microbial interaction · Cooperative behavior · Mutualism · Syntrophism

K. Prajapati (✉) · B. Anavadiya · M. Dabhi
Department of Microbiology and Biotechnology, University School of Sciences, Gujarat University, Ahmedabad, Gujarat, India
e-mail: karanprajapati@gujaratuniversity.ac.in

M. Saraf et al. (eds.), *Climate Change and Soil Microorganisms for Environmental Sustainability*, Microorganisms for Sustainability 28,
https://doi.org/10.1007/978-981-96-3425-5_16

16.1 Introduction

Soil microbiology is the study of microorganisms in the soil, including bacteria, fungi, archaea, and protozoa, and their interactions within the soil environment (Jassim et al., 2023). This field is integral to understanding the complex web of life that exists beneath the Earth's surface, where microorganisms play a crucial role in nutrient cycling, organic matter decomposition, soil structure formation, and the overall functioning of terrestrial ecosystems (Prasad et al., 2021; Hartmann & Six, 2023). The relevance of soil microbiology extends to both ecological and agricultural systems; in ecological contexts, soil microbes are fundamental to processes that sustain plant growth and regulate global biogeochemical cycles (Basu et al., 2020; Paul & Frey, 2023). In agricultural systems, these microorganisms are vital for soil fertility, crop productivity, and sustainable farming practices (Dilnashin et al., 2020; Shah et al., 2021).

Cooperative behavior among soil microbes refers to interactions that lead to mutual benefits or enhance the survival and functioning of the microbial community in the soil environment (Besset-Manzoni et al., 2018). This includes mutualistic relationships, such as those between mycorrhizal fungi and plant roots, where the fungi enhance nutrient uptake for the plant in exchange for carbohydrates, as well as syntrophic interactions where different microbial species work together to decompose complex organic compounds that neither could break down alone (Lanfranco et al., 2016; Sarsan et al., 2022). Cooperation can also manifest in biofilm formation and the establishment of microbial consortia, where microbes coordinate their activities to better withstand environmental stresses and outcompete pathogens (Oliveira et al., 2015; Maciag et al., 2023).

The field of soil microbiology has evolved significantly since its inception. Early studies focused primarily on identifying individual soil microorganisms and understanding their roles in isolation. However, with the advent of molecular biology techniques and the development of metagenomics and bioinformatics tools, the field has shifted toward a more holistic understanding of soil as a dynamic ecosystem where microorganisms interact in complex networks (Nkongolo & Narendrula-Kotha, 2020; Zhang et al., 2021a). This evolution has allowed researchers to uncover the intricate cooperative behaviors that underpin soil health and productivity, leading to new insights into how these behaviors contribute to ecosystem stability and agricultural sustainability (Wu et al., 2023).

The cooperative behaviors among soil microbes are ecologically significant because they contribute to several essential soil functions, such as nutrient cycling, organic matter decomposition, and soil (Van Elsas et al., 2019; Paul & Frey, 2023). For example, the cooperative interaction between nitrogen-fixing bacteria and leguminous plants is vital for converting atmospheric nitrogen into ammonia, a form that plants can readily use, thereby enhancing soil fertility without the need for synthetic fertilizers (Mahmud et al., 2020). Similarly, microbial consortia in the soil, which involve cooperation among various bacteria and fungi, play a critical role in breaking down complex organic matter into simpler compounds, releasing essential nutrients back into the soil, and maintaining soil structure (El Hamss et al., 2023).

This chapter aims to explore the various dimensions of cooperative behavior among soil microbes, focusing on their mechanisms, ecological significance, and impact on soil health and plant productivity. The primary objectives are to elucidate the genetic, biochemical, and ecological bases of microbial cooperation, to discuss the theoretical frameworks that explain these behaviors, and to highlight the implications for sustainable agriculture and environmental management. The discussion will cover key examples of microbial cooperation, such as mutualistic relationships, syntrophic interactions, and biofilm formation, and will examine the factors influencing these interactions, including environmental conditions and human activities.

By focusing on these areas, this chapter will provide a comprehensive understanding of the unseen alliances among soil microbes that shape soil ecosystems, offering insights into their roles in promoting soil health, supporting plant growth, and maintaining ecosystem resilience. This exploration will contribute to a broader understanding of soil microbiology and its applications in agriculture and environmental sustainability.

16.2 The Soil Microbial Community

The study of soil microbial communities is a fascinating and complex field that intertwines various aspects of biology, ecology, chemistry, and environmental science. These communities represent a dynamic and intricate world beneath our feet, playing a pivotal role in numerous ecological processes (Gentry et al., 2015; Zhalnina et al., 2018).

The soil is a rich repository of microbial diversity. The diversity and density of microbial life in soil are staggering, a single gram of soil can contain billions of microorganisms, belonging to thousands of different species (Chernov & Semenov, 2021). This immense diversity is not just a numerical marvel but also reflects a vast array of metabolic and functional capabilities. The soil microbial community, a dynamic and intricate ecosystem, plays a pivotal role in the Earth's biosphere (Crowther et al., 2019). This complex community, composed of a myriad of microorganisms including bacteria, fungi, protozoa, and archaea, functions as the backbone of terrestrial ecosystems, influencing soil health, plant growth, and global biogeochemical cycles (Mohanty et al., 2021). Soil microbes are essential for the cycling of nutrients, decomposition of organic matter, and formation of soil structure. They are the driving force behind the conversion of complex organic compounds into simpler molecules, making nutrients available to plants. This decomposition process is crucial for the recycling of carbon, nitrogen, phosphorus, and other essential elements in the ecosystem (Haruna & Yahaya, 2021; Prasad et al., 2021). Additionally, soil microbes play a significant role in regulating the Earth's climate. Through processes such as respiration, methanogenesis, and nitrogen fixation, they directly and indirectly contribute to the production and consumption of greenhouse gases like carbon dioxide, methane, and nitrous oxide (Robertson, 2014; Alengebawy et al., 2024).

The composition of soil microbial communities is influenced by various factors, including soil type, climate, vegetation, and land use practices (Chernov & Zhelezova, 2020; Ramírez et al., 2020). Typically, these communities are dominated by bacteria and fungi, with each group contributing uniquely to soil processes. Bacteria, for example, are crucial in nutrient cycling, while fungi are vital in decomposing complex organic matter like cellulose and lignin (Horwath, 2017). Soil microbes do not exist in isolation but interact with each other and their environment in complex ways. These interactions can be competitive, as microbes vie for resources, or cooperative, such as in the formation of microbial consortia that collectively perform tasks no single organism could achieve alone (Little et al., 2008; El Hamss et al., 2023). An example of such cooperative behavior is in the nitrogen cycle, where different microbes work in concert to convert atmospheric nitrogen into forms usable by plants (Chen et al., 2023).

Recent advances in molecular biology, genomics, and bioinformatics have revolutionized our understanding of soil microbial communities. Techniques such as DNA sequencing and metagenomics have enabled scientists to unravel the genetic and functional diversity of soil microbes, providing insights into their roles in soil ecosystems (Nkongolo & Narendrula-Kotha, 2020).

16.2.1 Composition of Soil Microbial Communities

16.2.1.1 Bacteria

Soil bacteria constitute an incredibly diverse group both in terms of genetic composition and functional roles within soil ecosystems. In a single gram of soil, there can be upward of a billion bacteria, with estimates ranging from 10^8 to 10^{10} cells per gram of soil, encompassing thousands of different species (Pandey et al., n.d.; Gentry et al., 2021). This remarkable diversity includes a range of morphologies from simple spheres (cocci) and rods (bacilli) to spirals (spirilla) and complex filamentous forms (Boymurod, 2024). Their metabolic activities are equally diverse, with some bacteria functioning as heterotrophs that decompose organic matter, autotrophs that fix carbon, chemotrophs that derive energy from chemical reactions, and nitrogen-fixers that convert atmospheric nitrogen into forms usable by plants (Fenchel et al., 2012; Subba, 2017). For example, nitrogen-fixing *Rhizobium* spp. form symbiotic relationships with legume roots, enhancing plant growth and soil fertility (Basu & Kumar, 2020). Nitrifying bacteria like *Nitrosomonas* and *Nitrobacter* are pivotal in transforming ammonia into nitrate, a key step in the nitrogen cycle (Robertson & Groffman, 2023). Soil bacteria also play a critical role in decomposing organic matter; genera such as *Streptomyces* not only break down complex compounds but are also known for producing the distinct earthy smell of fertile soil and a range of antibiotics (Beroigui & Errachidi, 2023). Bacteria like *Pseudomonas fluorescens* and *Bacillus* spp. contribute to plant health by suppressing pathogens (Dimkić et al., 2022). Furthermore, some species, including

Azotobacter and certain *Mycobacterium* strains, are renowned for their roles in free-living nitrogen fixation and pollutant degradation, respectively (Gurikar et al., 2016; Foght, 2018). Across various soil samples, Proteobacteria, Acidobacteria, and Actinobacteria are the most abundant phyla (Mhete et al., 2020; Chang et al., 2021). These phyla play a crucial role in the soil ecosystem and significantly influence the physico-chemical properties of the soil. The vast array of bacteria in soil, with their varied roles, underscores their indispensable contribution to maintaining soil health, supporting plant growth, and driving crucial biogeochemical cycles in terrestrial ecosystems (Prasad et al., 2021).

16.2.1.2 Archaea

Archaea, once thought to be confined to extreme environments, have now been recognized as ubiquitous and influential members of soil microbial communities (Wang et al., 2020). These microorganisms, distinct from bacteria and eukaryotes, in terms of abundance, archaea are highly variable, ranging from a minor fraction to a significant portion of the microbial biomass in soil, depending on environmental conditions. Their roles in soil are multifaceted and crucial for ecosystem functioning. Archaea are particularly important in the global nitrogen cycle; for example, ammonia-oxidizing archaea (AOA) are key players in nitrification, converting ammonia into nitrite, a process vital for soil fertility and plant growth. Additionally, certain archaeal species are involved in methanogenesis and methane oxidation, processes critical in global carbon cycling and climate regulation. Recent studies have shed light on the diversity of archaea in soil, revealing a vast array of species adapted to various soil environments, from agricultural lands to arid deserts. This diversity is not just numerical but also functional, as soil archaea exhibit a wide range of metabolic pathways adapted to different environmental conditions. Their resilience and adaptability make them crucial for maintaining soil health, particularly in the face of changing environmental conditions. The study of soil archaea, propelled by advances in molecular techniques and metagenomics, continues to reveal their integral role in sustaining the balance and productivity of soil ecosystems, marking them as key players in the unseen world beneath our feet (Offre et al., 2013; Baker et al., 2020).

16.2.1.3 Fungi

Soil fungi, an integral component of terrestrial ecosystems, exhibit a vast diversity in both form and function, significantly contributing to soil health and ecosystem dynamics. Estimates suggest that one gram of soil can contain anywhere from 100,000 to several million fungal spores or cells, reflecting a broad spectrum of species. These organisms vary from microscopic yeasts to complex mycelial networks, and their roles in soil are multifaceted. Fungi are primarily recognized for their ability to decompose organic materials, especially more complex compounds like cellulose and lignin, a process crucial for carbon cycling and soil structure formation (Treseder & Lennon, 2015). Key decomposers like the genus *Aspergillus* and *Penicillium* break down dead plant material like decomposition of organic matter, releasing nutrients back into the soil (Srinivasan et al., 2020). Mycorrhizal fungi, such as those in the

Glomus genus, form symbiotic associations with plant roots, facilitating the absorption of water and nutrients like phosphorus, and in return, receiving carbohydrates (Nadeem et al., 2017). This symbiosis is vital for the health and productivity of many plants, including key agricultural crops. Fungi also play a role in disease suppression, competing with plant pathogens and sometimes producing antimicrobial compounds. Furthermore, certain fungi are involved in bioremediation processes, with species like *Trichoderma* being effective in degrading environmental pollutants (Kidwai et al., 2022). The functional diversity of soil fungi is immense, ranging from nutrient cycling and supporting plant growth to contributing to global carbon budgets and environmental clean-up. This diversity not only highlights the ecological importance of soil fungi but also underscores their potential applications in sustainable agriculture and environmental management (Kües et al., 2018).

16.2.1.4 Protozoa

Soil protozoa, encompassing a diverse array of species such as flagellates, amoebae, and ciliates, are increasingly recognized as pivotal actors in the soil's microbial tapestry (Geisen et al., 2018). These microorganisms are integral to the soil's nutrient dynamics, particularly in the cycling of nitrogen and carbon. They exert control over bacterial populations through predation, a process that not only regulates microbial numbers but also facilitates the liberation of nutrients in forms more readily assimilable by plants, thus bolstering soil fertility (Bonkowski, 2004). Additionally, their physical movement through the soil matrix has been identified as a key factor in influencing soil structure, impacting vital aspects like aeration, water infiltration, and nutrient distribution (Geisen et al., 2018).

16.2.2 The Factors Affecting Complexity and Dynamic Nature of Soil Microbial Communities

16.2.2.1 Inter-Species Interactions

Soil ecosystems are extraordinarily diverse, hosting an array of microorganisms including bacteria, fungi, protozoa, and archaea. The interactions among these species form a complex web that significantly influences the structure and functions of microbial communities. These interactions, encompassing competition, mutualism, parasitism, predation, and other nuanced relationships, are crucial in determining soil health and plant growth (Sarsan et al., 2022). Effective nutrient cycling, disease suppression, and plant stress tolerance are heavily influenced by the diversity and nature of microbial interactions (Trivedi et al., 2020). Maintaining a diverse and balanced microbial community is therefore essential for optimal plant health and soil functionality.

Competition is a fundamental interaction where microbes compete for limited resources like nutrients and space. This competition can drive significant changes in the community composition. For instance, in nitrogen-limited soil, bacteria that utilize nitrate often outcompete those dependent on ammonia, leading to shifts in community dynamics. Such competitive interactions can have profound implications for nutrient cycling and soil fertility (Norton & Ouyang, 2019).

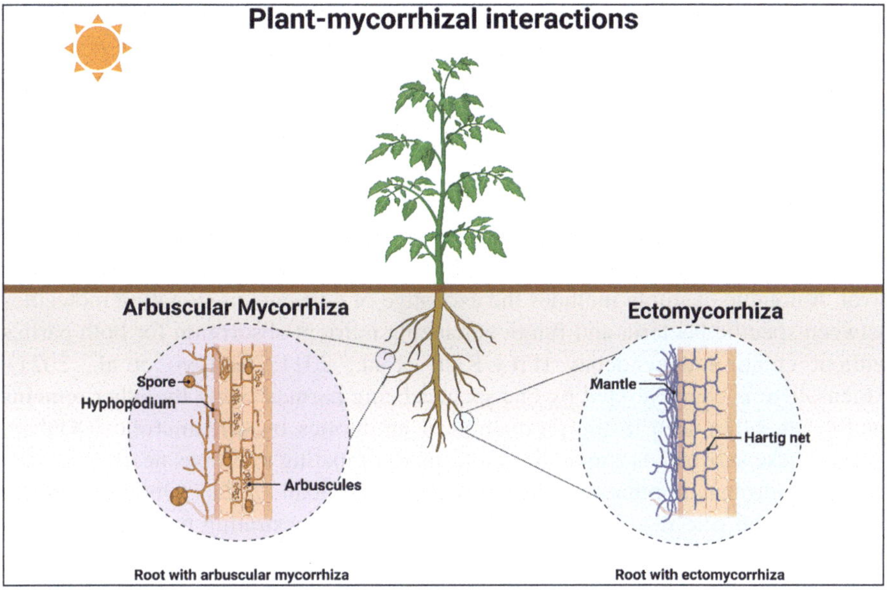

Fig. 16.1 Diagram of plant-mycorrhizal interactions. (Created in BioRender.com)

Mutualism plays a key role in soil ecosystems. The symbiosis between mycorrhizal fungi and plant roots is a prime example. In Fig. 16.1, two types of mycorrhizal associations are illustrated: arbuscular mycorrhiza, where fungal hyphae penetrate root cortical cells to form arbuscules, and ectomycorrhiza, where fungal hyphae create a mantle around the root and a Hartig net within the intercellular spaces. These symbiotic interactions facilitate nutrient exchange between plants and fungi, leading to enhanced plant growth and improved soil health. These fungi enhance plant nutrient uptake, particularly phosphorus, and in return, receive carbohydrates from the plant (Peterson & Massicotte, 2004). This mutual exchange not only stimulates plant growth but also improves soil structure and integrity. Beyond mycorrhizae, other mutualistic interactions include bacteria that fix atmospheric nitrogen in symbiosis with legumes, providing essential nutrients to the plants and receiving shelter and food sources in return (Tedersoo et al., 2020).

Parasitism and predation add complexity to soil microbial dynamics. Bacteriophages, viruses that infect bacteria, play a dual role. They control bacterial populations, preventing any single species from dominating, and contribute to genetic diversity through mechanisms like horizontal gene transfer (Blakely, 2023; Puxty & Millard, 2023). This predatory relationship is crucial for maintaining a balanced microbial ecosystem.

Soil microbial communities are not only complex but also engage in a myriad of interactions that, while less conspicuous, are crucial for their resilience and adaptability. Among these, commensalism represents a relationship where one organism reaps benefits without impacting the other (Morris & Schniter, 2018). For instance, certain fungi produce waste products that are utilized as nutrients by specific bacteria, enhancing bacterial growth without affecting the fungi. Allelopathy is another intricate interaction where organisms release biochemicals that significantly affect the growth and survival of surrounding life, shaping both soil and plant community structures (Polyak & Sukcharevich, 2019). This is exemplified by plants like black walnut trees, which secrete juglone, a compound toxic to many competing plant species, thereby influencing the microbial and plant landscape around them (Lal & Biswas, 2023). In protocooperation, mutual benefits arise from the interaction between species, though these species do not depend on this relationship for survival. A notable example includes the exchange of nutrients or signaling molecules between specific bacteria and fungi, enhancing nutrient absorption for both parties without creating dependency (Frey-Klett et al., 2011; Santoyo et al., 2021). Amensalism is characterized by one species being harmed while the other remains unaffected, often seen in the production of antibiotics by soil microbes (Klaban, 2021). These substances can inhibit or destroy competing microbes nearby, regulating the microbial population and composition (Klaban, 2021). Quorum sensing plays a pivotal role in soil bacterial communities, orchestrating behaviors such as biofilm formation. This coordination mechanism ensures that bacterial actions, like the secretion of enzymes or toxins, are optimized for the collective benefit, influencing both the structure of microbial communities and their interactions with plant roots. Facilitation involves one species making the environment more suitable for another (Hart, 2023). In soil, certain microbes can alter the pH or secrete enzymes that make nutrients more available, indirectly benefiting other microbes or even plant roots.

16.2.2.2 Environmental Influences

The composition and activity of soil microbial communities are profoundly influenced by environmental factors, including soil type, moisture, pH, temperature, and broader changes such as climate variation and human-induced alterations (Bissett et al., 2013; Singh & Yadav, 2021). The dynamic nature of these communities means they are constantly adapting to these environmental cues. A comprehensive overview of these environmental factors is presented in Table 16.1. It details their specific influence on soil microbial populations, and the consequent impact on ecological systems, and provides illustrative examples of each factor's effect.

16.2.2.3 Human Activities

Human activities, particularly in agriculture, land use, and pollution, have a significant impact on soil microbial communities (Zhang et al., 2016; McKinley, 2019). These impacts have far-reaching consequences for ecosystem services such as nutrient cycling, soil fertility, and plant health. In the realm of agriculture, the use of agrochemicals like pesticides, herbicides, and fertilizers has been shown to directly affect soil microbial diversity (Ataikiru et al., 2019). Studies in the field of soil biology and environmental microbiology have demonstrated that there can be notable declines, in microbial diversity in soils treated with nitrogen fertilizers (Tian et al., 2015). Additionally, certain herbicides have been found to substantially reduce

Table 16.1 Environmental factors influencing soil microbial communities and their ecosystem impacts

Environmental factor	Influence on soil microbes	Impact on ecosystem	Examples
Soil type	Determines the physical and chemical properties of the soil, affecting microbial distribution and function.	Leads to diverse microbial ecosystems, influencing nutrient cycling and soil fertility.	Sandy soils typically have lower microbial diversity due to poor nutrient retention, while clay-rich soils support a more diverse microbial community.
Moisture	Affects microbial metabolic rates and nutrient diffusion.	Changes in moisture can shift microbial community composition and function, impacting decomposition and nutrient availability.	Prolonged drought conditions can reduce microbial activity, impacting soil nutrient cycles, whereas excessive moisture can lead to anaerobic conditions, favoring microbes like methanogens.
pH	Influences nutrient solubility and the availability of toxic substances, affecting microbial growth.	Alterations in pH can lead to changes in microbial community structure, affecting soil health and nutrient cycles.	Acidic soils (low pH) may harbor acidophilic microbes, impacting plant growth and metal solubility, while alkaline soils (high pH) support different microbial populations affecting nutrient availability.
Temperature	Directly impacts microbial metabolic rates and enzyme activities.	Temperature extremes stress microbial communities, potentially affecting carbon cycling and storage, and overall ecosystem resilience.	Rising global temperatures can shift microbial communities toward those that decompose organic matter faster, impacting carbon storage in soils.
Climate change and extreme weather events	Causes drastic changes in environmental conditions, impacting microbial diversity and function.	Can lead to a loss of microbial diversity and shifts in community composition, affecting soil health, plant growth, and carbon sequestration capabilities.	Extreme events like wildfires can drastically alter soil microbial communities, reducing diversity and impacting nutrient cycling and regeneration post-disturbance.

beneficial mycorrhizal fungi. These changes can disrupt crucial ecological processes including nutrient cycling.

Changes in land use, especially the transformation of forests into agricultural or urban areas, also significantly impact soil structure and microbial habitats. Research in this area has revealed a decrease in microbial biomass in urban soils compared to forested areas. Moreover, a reduction in microbial functional diversity has been reported due to land use changes (Le Provost et al., 2020). These reductions in microbial diversity and biomass can impair soil health, affecting its ability to support plant growth and carbon sequestration (Zhang et al., 2016; Ramesh et al., 2019).

Furthermore, pollution from industrial activities, heavy metals, and plastics poses a severe threat to soil microbes. Studies have shown that heavy metal contamination can decrease microbial diversity. These changes hinder the natural processes of soil, such as decomposition and nutrient cycling, and reduce its capacity for pollution remediation (Abdu et al., 2017; Zhou et al., 2021).

The cumulative evidence from various scientific studies highlights the extensive impact of human activities on soil microbial communities. Adopting sustainable agricultural practices, urban planning that considers soil health, and effective pollution mitigation strategies are crucial for maintaining microbial diversity and functionality (Nadarajah & Abdul Rahman, 2023). These strategies are essential not only for ecosystem health but also for addressing broader challenges like food security and climate change.

16.2.2.4 Evolutionary Adaptations

Soil ecosystems are constantly evolving landscapes, where microbial communities dynamically adapt to a range of environmental pressures, ensuring ecosystem resilience and functionality (Philippot et al., 2021). These adaptations occur at various levels, including genetic changes like mutations, horizontal gene transfer, and recombination. These genetic shifts allow soil microbes to rapidly adjust to diverse and challenging conditions, such as developing resistance to antibiotics or adapting to heavy metal-contaminated soils (Wiedenbeck & Cohan, 2011).

Beyond genetic changes, soil microbes also undergo physiological adaptations to environmental stresses. For example, in response to high salinity, certain bacteria modify their metabolic pathways, a crucial survival strategy in saline environments (Gamalero et al., 2020). These physiological changes are pivotal in maintaining soil health and fertility, particularly in environments subject to rapid changes.

Community-level adaptations are also evident in soil microbial populations. These adaptations manifest as shifts in the composition and interactions within microbial communities, allowing them to respond to environmental challenges like drought. Adaptations at this level ensure the continued performance of essential ecosystem functions, such as decomposition and nutrient cycling, even under stressful conditions (Kumar et al., 2024).

The ability of soil microbial communities to evolve and adapt is a testament to their dynamic nature and critical role in ecosystem health. Understanding these evolutionary adaptations is vital for effective soil ecosystem management, especially in the face of rapid environmental changes and human impact (Angulo et al., 2022). Insights into these processes are instrumental in shaping sustainable land management practices that nurture and protect microbial diversity and functionality.

16.3 Cooperative Interactions in Soil Microbial Communities

Cooperation among microorganisms in soil ecosystems is fundamental to ecological functioning, greatly contributing to nutrient cycling, soil fertility, and plant health (Yadav et al., 2021). Unlike competitive or antagonistic interactions, cooperative relationships involve mutual benefit or support among different microbial species.

16.3.1 Mutualism

Mutualism is a type of symbiotic relationship where both species involved benefit from the interaction. Mutualistic interactions are a cornerstone of cooperation in soil ecosystems. A prime example is the symbiosis between rhizobia bacteria and leguminous plants, such as peas, beans, and clovers. In this relationship, rhizobia fix atmospheric nitrogen into a usable form for plants, while in return, they receive carbohydrates and a protective niche (Chellem et al., 2024). This interaction is pivotal for enhancing soil fertility without the need for artificial fertilizers.

16.3.2 Syntrophy

Syntrophy is the phenomenon where two or more microorganisms cooperate to degrade a substance that neither can degrade alone, often involving the exchange of intermediate metabolic products (Morris et al., 2013). Syntrophy, involving the sequential metabolism of substrates by different organisms, is vital in anaerobic environments. A notable instance occurs in wetlands, where fermentative bacteria degrade complex organic compounds, producing simpler molecules that methanogens then use to produce methane (Yarwood, 2018). This process is crucial for organic matter decomposition and methane production in anaerobic soil conditions.

16.3.3 Cross-Feeding and Metabolic Cooperation

Cross-feeding and metabolic cooperation occur when one species of microorganism produces a by-product that is used by another species, creating a network of nutrient exchange (D'Souza et al., 2018). Cross-feeding and metabolic cooperation are widespread, as seen in the interaction between mycorrhizal fungi and phosphate-solubilizing bacteria. The fungi extend the root system of plants, aiding in nutrient absorption, while the bacteria convert insoluble phosphates into absorbable forms (Jin et al., 2024). This cooperation enhances the availability of nutrients for both plants and fungi, illustrating a perfect example of mutual benefit.

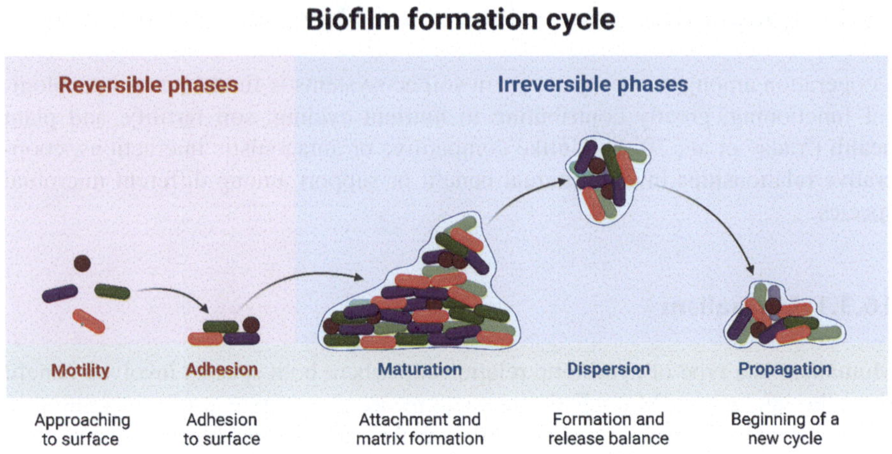

Fig. 16.2 Mechanism of biofilm formation. (Created in BioRender.com)

16.3.4 Biofilms and Microbial Consortia

Biofilms and microbial consortia refer to communities of microorganisms living together in a self-produced matrix, often involving complex interactions and cooperation for survival and functionality (Qian et al., 2020). Biofilms and microbial consortia in the plant rhizosphere showcase structured communities where cooperation is essential. These consortia consist of bacteria, fungi, and other microbes that collectively promote plant growth (Wu et al., 2023). Figure 16.2 illustrates the biofilm formation process, which progresses through distinct reversible and irreversible stages. In the reversible phase, microorganisms first exhibit motility, moving toward a surface, followed by adhesion, where they temporarily attach. The process then transitions into the irreversible phase, beginning with maturation, during which microbial cells become encapsulated within a protective extracellular matrix. Next is the dispersion phase, where some cells detach and are released from the biofilm, striking a balance between attachment and release. Finally, propagation occurs, initiating a new cycle of biofilm development. This cyclical mechanism plays a vital role in microbial survival, influencing ecological balance and pathogenic behavior in both natural environments and clinical contexts (Ghannoum et al., 2020). Some bacteria produce growth hormones, while others aid in disease resistance, exemplifying how these microbial assemblages facilitate nutrient exchange and improve plant health (Rosier et al., 2016).

16.4 Mechanisms of Cooperative Behavior

16.4.1 Chemical Signaling and Quorum Sensing

Chemical signaling is a foundational mechanism that allows soil microbes to perceive their environment and communicate with each other. Among these mechanisms, quorum sensing is one of the most pivotal processes that regulate cooperative behavior among bacterial populations (Sharma et al., 2020). This phenomenon was first identified in the marine bacteria *Vibrio fischeri* and has since been observed in various soil microbial communities (Essock-Burns et al., 2021).

Quorum sensing is crucial for coordinating complex behaviors that are beneficial at the community level but may be costly to individual cells. For example, the production of extracellular enzymes required for breaking down organic matter is often regulated by quorum sensing (Israel et al., 2023). This allows microbes to efficiently utilize resources and prevent premature depletion of energy reserves. In the rhizosphere, quorum sensing influences interactions between bacteria and plants, as well as among bacterial species (Chepsergon & Moleleki, 2023). For example, *Pseudomonas* spp. can produce siderophores, iron-chelating compounds under quorum sensing control, which suppress plant pathogens by depriving them of iron (Deb & Tatung, 2024). Recent research has uncovered new insights into the complexity and diversity of quorum sensing systems in soil microbes. Advances in molecular biology and genomics have revealed novel quorum sensing pathways, regulatory networks, and signaling molecules that contribute to microbial cooperation and community dynamics (Davis et al., 2015; Chiquito-Contreras et al., 2024).

16.4.1.1 Mechanisms of Quorum Sensing

Quorum sensing involves the synthesis, release, and detection of small diffusible signaling molecules known as autoinducers (Nag et al., 2021). Bacteria produce these molecules at a basal level, and as the population density increases, the concentration of autoinducers rises. Once a threshold concentration is reached, these molecules bind to specific receptor proteins, triggering a signal transduction cascade that alters gene expression (Bridges et al., 2022). This process enables the population to synchronize activities such as biofilm formation, bioluminescence, virulence, and secondary metabolite production.

The quorum sensing process begins with the synthesis of signaling molecules called autoinducers. These small molecules are produced constitutively by bacteria and accumulate in the environment as the population density increases. N-acyl homoserine lactones (AHLs) are the most common autoinducers in gram-negative bacteria. They are synthesized by enzymes called LuxI-type synthases, which use S-adenosylmethionine (SAM) and acyl-acyl carrier proteins (acyl-ACPs) as substrates to produce AHLs. The acyl chain length and structure of the AHLs can vary, affecting their specificity and activity (Miller & Bassler, 2001; Kai, 2018). In gram-positive bacteria, autoinducers are often short peptides (Oligopeptides) that are processed and secreted by specialized transporters. These peptides are recognized by two-component regulatory systems consisting of a membrane-bound sensor kinase and a cytoplasmic response regulator (Bhatt, 2019). Autoinducer-2 (AI-2) is a universal signaling molecule produced by both gram-negative and gram-positive bacteria (Zhang et al., 2020). It is synthesized by the LuxS enzyme from the precursor molecule 4,5-dihydroxy-2,3-pentanedione (DPD) and is involved in inter-species communication (Pereira et al., 2013). Studies have identified new classes of signaling molecules, such as γ-butyrolactones in *Streptomyces* species, which regulate antibiotic production and sporulation (Bai & Rai, 2022). These molecules highlight the diversity of quorum sensing systems beyond the traditional AHLs and oligopeptides.

As the bacterial population grows, the concentration of autoinducers increases in the environment. The accumulation of these molecules is a key aspect of quorum sensing, as it reflects the population density of the microbial community. In many bacteria, autoinducers diffuse freely across the cell membrane and accumulate in the extracellular environment. The rate of diffusion and accumulation depends on factors such as cell density, environmental conditions, and the properties of the autoinducer (Hense & Schuster, 2015; Moreno-Gámez et al., 2017). A critical concentration threshold of autoinducers must be reached for quorum sensing to be activated. This threshold ensures that the response is initiated only when a sufficient number of cells are present, allowing the population to coordinate their behavior effectively (Pérez-Velázquez et al., 2016).

Once the concentration of autoinducers reaches a threshold, these molecules are detected by specific receptor proteins, initiating the quorum sensing signaling cascade. In gram-negative bacteria, AHLs typically bind to cytoplasmic LuxR-type receptor proteins. The binding of AHLs to LuxR induces a conformational change, enabling the LuxR-AHL complex to bind to promoter regions of target genes and activate their transcription (Della Sala et al., 2019). In gram-positive bacteria, oligopeptide autoinducers bind to membrane-bound sensor kinases, activating a two-component signal transduction system. This activation results in the phosphorylation of a response regulator, which then binds to DNA and modulates gene expression (Rocha-Estrada et al., 2010; Haque et al., 2019). AI-2 detection involves the Lsr (luxS-regulated) transport system, which imports AI-2 into the cell. Once inside, AI-2 is phosphorylated and interacts with regulatory proteins to modulate gene expression (Pereira et al., 2013).

The binding of autoinducers to their receptors triggers a cascade of molecular events that lead to changes in gene expression and the regulation of cooperative behaviors. The LuxR-AHL complex in gram-negative bacteria binds to specific DNA sequences known as lux boxes, activating the transcription of quorum sensing-regulated genes. These genes often encode for virulence factors, biofilm components, and secondary metabolites (Garg et al., 2014; Tommonaro, 2019). Quorum sensing systems can be highly complex, involving multiple interconnected regulatory networks. In some bacteria, secondary quorum sensing systems exist, allowing for fine-tuned control of gene expression based on environmental cues and community composition (Mukherjee & Bassler, 2019). Positive feedback loops are common in quorum sensing systems, where the activation of quorum sensing genes leads to increased production of autoinducers, further amplifying the response. This ensures robust and coordinated community-level behavior (Boo et al., 2021).

The culmination of quorum sensing is the coordinated expression of genes that lead to community-level behaviors beneficial for the microbial population. Quorum sensing often regulates biofilm formation, where bacterial cells adhere to surfaces and produce extracellular polymeric substances (EPS) that encase the community (Erkihun et al., 2024). Biofilms provide protection from environmental stressors and enhance resource acquisition. Pathogenic bacteria use quorum sensing to regulate the production of virulence factors, such as toxins and enzymes, allowing them to establish infections and evade host defenses (Castillo-Juárez et al., 2015). Quorum sensing controls the production of secondary metabolites, such as antibiotics and siderophores, which can inhibit competitors and enhance nutrient acquisition (Meryem et al., 2022). These metabolites play critical roles in microbial interactions and community dynamics.

16.4.1.2 Example: Quorum Sensing in *Pseudomonas aeruginosa*

P. aeruginosa is a well-studied model organism for quorum sensing research due to its complex quorum sensing networks that regulate various behaviors, including virulence, biofilm formation, and antibiotic resistance (Sikdar & Elias, 2022).

The las system is the primary quorum sensing circuit in *P. aeruginosa* and uses the autoinducer N-(3-oxododecanoyl)-homoserine lactone (3OC12-HSL), synthesized by the enzyme LasI (Alayande et al., 2018). 3OC12-HSL binds to the LasR receptor, forming a complex that activates the expression of genes involved in virulence, such as lasB (elastase) and aprA (alkaline protease). These enzymes degrade host tissues and facilitate infection (Ghanem et al., 2023; Vetrivel et al., 2023). The Rhl system operates in conjunction with the las system and uses the autoinducer N-butyryl homoserine lactone (C4-HSL), synthesized by RhlI. C4-HSL binds to the RhlR receptor, leading to the activation of genes involved in rhamnolipid production and other cooperative behaviors. Rhamnolipids play roles in biofilm formation, motility, and antimicrobial activity (Dusane et al., 2010; Dolan et al., 2020). In addition to the las and Rhl systems, *P. aeruginosa* utilizes the Pseudomonas quinolone signal (PQS) system, which interacts with the other quorum sensing networks. PQS modulates the production of secondary metabolites and is involved in the regulation of iron acquisition and oxidative stress response (Sams et al., 2016; Vadakkan et al., 2024).

P. aeruginosa utilizes quorum sensing to regulate biofilm development, a key factor in its pathogenicity and resistance to antibiotics (Furiga et al., 2016). Biofilm formation involves several stages, including initial attachment, microcolony formation, maturation, and dispersion (Sauer et al., 2022). Quorum sensing controls the expression of genes associated with each stage, ensuring that biofilms are formed in response to appropriate environmental cues and population densities (Moreno-Gámez et al., 2017).

Quorum sensing regulates the expression of pili and flagella, which are essential for initial attachment to surfaces (Chang, 2018). It also controls the production of EPS, which facilitates microcolony formation and stability (Wang et al., 2022). During biofilm maturation, quorum sensing activates genes involved in the synthesis of matrix components and enzymes that remodel the biofilm structure (Mann & Wozniak, 2012). This ensures the biofilm remains robust and adaptive to environmental changes. Quorum sensing modulates the expression of dispersal factors, allowing cells to detach from the biofilm and colonize new environments when conditions become unfavorable (Solano et al., 2014; Emerenini et al., 2015).

16.4.2 Interactions Between Different Microbial Species and Kingdoms

Soil ecosystems are characterized by intricate networks of interactions that involve multiple microbial species and extend across different microbial kingdoms (Hassani et al., 2018; Guseva et al., 2022). These interactions can be mutualistic, commensal, or competitive, and they play critical roles in maintaining soil health and ecosystem function.

Mycorrhizal fungi form mutualistic associations with approximately 90% of terrestrial plants (Lanfranco et al., 2016). There are two main types of mycorrhizal associations, arbuscular mycorrhizal fungi, which penetrate plant root cells, and ectomycorrhizal fungi, which form a sheath around plant roots (Bahadur, 2021). Both types facilitate nutrient exchange, with fungi providing plants with enhanced access to soil nutrients, particularly phosphorus, in exchange for carbon compounds from the plant (Tedersoo et al., 2020). Recent research has revealed that mycorrhizal networks can connect different plant species, allowing for the transfer of nutrients and signaling molecules (Boyno & Demir, 2022). This connectivity can enhance plant resilience to stress and facilitate communication between plants, influencing plant community dynamics and biodiversity (Wahab et al., 2023).

The symbiosis between rhizobia bacteria and legumes is a classic example of interkingdom cooperation. Rhizobia infects the root nodules of legume plants and fixes atmospheric nitrogen into ammonia, which the plant can utilize for growth. In return, the plant provides the bacteria with carbohydrates and a protective niche. This interaction is vital for improving soil fertility and reducing the need for synthetic nitrogen fertilizers (Gopalakrishnan et al., 2015). The establishment of this symbiosis involves a complex exchange of signaling molecules. Legumes secrete flavonoids that induce the expression of nodulation (nod) genes in rhizobia, leading to the production of nod factors. These factors trigger nodule formation in plant roots, facilitating the establishment of the symbiotic relationship (Skorupska et al., 2017).

Microbial consortia, composed of diverse microbial communities working synergistically, are essential for nutrient cycling in soil ecosystems. These consortia often exhibit a division of labor, where different species perform specific roles in the degradation of complex organic matter. For instance, during the decomposition of plant litter, fungi initially break down complex polymers like cellulose and lignin into simpler compounds (Findlay, 2021). Bacteria further degrade these compounds into inorganic nutrients, such as nitrogen and phosphorus, which are then available for plant uptake (Rashid et al., 2016). This cooperative decomposition is crucial for maintaining soil fertility and supporting plant growth.

Microbial consortia also play a role in bioremediation, where they work together to degrade pollutants in the soil. For example, certain bacteria and fungi can degrade hydrocarbons in contaminated soils, highlighting the potential of these consortia for environmental clean-up efforts (Daccò et al., 2020).

16.4.3 Influence of Environmental Factors on Microbial Communication

Environmental factors, including pH, temperature, moisture, and nutrient availability, significantly influence microbial signaling and behavior (Islam et al., 2020). These factors can modulate the production and stability of signaling molecules, affecting the efficiency of microbial communication and cooperation (Singh et al., 2022).

16.4.3.1 pH and Temperature

Soil pH affects microbial community composition and function by influencing enzyme activity, nutrient availability, and the solubility of signaling molecules. Extreme pH levels can inhibit microbial growth and alter signaling pathways, leading to shifts in microbial interactions (Naz et al., 2022). For example, acidic soils may reduce the efficacy of quorum sensing in certain bacterial species, affecting their cooperative behaviors (Ayaz & Gothalwal, 2014).

Temperature is another critical factor that influences microbial metabolism and signaling. Temperature fluctuations can affect the expression of quorum sensing genes and the stability of signaling molecules. Warmer temperatures may enhance microbial activity but can also disrupt delicate balances within microbial communities, potentially leading to shifts in cooperation and competition (Wahid et al., 2020).

16.4.3.2 Moisture and Nutrient Availability

Soil moisture plays a crucial role in microbial communication by affecting the diffusion of signaling molecules. In dry soils, the limited diffusion of these molecules may necessitate closer physical proximity for effective communication. Conversely, in well-hydrated soils, signaling molecules can diffuse more freely, facilitating interactions over larger distances (Singh et al., 2022).

Nutrient availability also influences microbial cooperation. In nutrient-rich environments, competition for resources may be reduced, allowing for more cooperative interactions. Conversely, in nutrient-poor soils, microbes may evolve strategies to optimize nutrient acquisition, such as forming consortia that specialize in different aspects of nutrient cycling (Cao et al., 2024).

Recent studies have highlighted the role of specific nutrients, such as iron, in modulating microbial interactions. Iron availability can influence the production of siderophores, which are iron-chelating compounds produced by bacteria to scavenge iron from the environment. Siderophore production is often regulated by quorum sensing and can impact competitive interactions among microbial species (Popat et al., 2017; Kramer et al., 2020).

In conclusion, the mechanisms of cooperative behavior among soil microbes are complex and influenced by a variety of biotic and abiotic factors (Jain et al., 2020). Understanding these interactions is crucial for managing soil health and agricultural productivity, as well as for developing sustainable practices that harness the benefits of microbial cooperation. Advances in molecular and ecological research continue to reveal the intricate networks of communication and cooperation that underpin soil microbial communities, offering new insights into their roles in ecosystem function and resilience.

16.5 Implications for Agriculture and Environmental Management

16.5.1 Application of Cooperative Soil Microbes in Sustainable Agriculture

Understanding microbial cooperation in soil ecosystems has profound implications for sustainable agricultural practices. Cooperative interactions among soil microbes, including mutualistic relationships and syntrophic partnerships, can significantly enhance plant growth, nutrient cycling, and soil health (De Mandal et al., 2021). Recent studies have highlighted the importance of microbial consortia in promoting plant resilience to biotic and abiotic stresses (Khan, 2022; Carneiro et al., 2023). For example, mycorrhizal fungi and rhizobia are well-known symbionts that improve plant nutrient uptake and enhance soil fertility through nitrogen fixation and phosphate solubilization, respectively (Raklami et al., 2019). Moreover, these cooperative interactions can suppress soil-borne pathogens through mechanisms like competitive exclusion and the production of antimicrobial compounds (Hussain et al., 2020; Zhang et al., 2021b).

The application of microbial consortia in sustainable agriculture is increasingly being recognized as a potential strategy to reduce the reliance on chemical fertilizers and pesticides (Seenivasagan & Babalola, 2021). For instance, co-inoculation of plants with beneficial bacteria and fungi has been shown to synergistically improve plant growth and health. A recent meta-analysis demonstrated that microbial consortia could enhance crop yield by up to 48% compared to single-species inoculants (Liu et al., 2023). This increase is attributed to the complementary functions of different microbial species, such as nitrogen fixation, phosphate solubilization, and phytohormone production, which collectively contribute to improved nutrient acquisition and plant growth.

Furthermore, understanding the principles of microbial cooperation can help in designing more effective biofertilizers and biopesticides. For example, biofertilizers formulated with diverse microbial communities can provide a broader range of nutrients and growth-promoting factors than those with single strains (Mahanty et al., 2017). These biofertilizers can also enhance soil structure and organic matter content, further contributing to soil health and fertility (Khan et al., 2023). Similarly, biopesticides containing multiple microbial strains can target a wider range of pathogens, reducing the likelihood of resistance development (Vero et al., 2023).

In addition to direct applications, the study of cooperative soil microbes has also informed practices like crop rotation and cover cropping. These practices are known to influence the composition and diversity of soil microbial communities, promoting beneficial interactions that can improve soil health and plant productivity (Singh & Kumar, 2021; Wu et al., 2023). For instance, legume-based cover crops can enhance the abundance of nitrogen-fixing bacteria, while diverse crop rotations can promote microbial diversity, resilience, and stability in agroecosystems (Kebede, 2021; Liu et al., 2022).

16.5.2 Role in Bioremediation and Environmental Restoration

Soil microbes, particularly those involved in cooperative interactions, play a crucial role in bioremediation and environmental restoration. Cooperative microbial consortia can degrade complex pollutants more efficiently than individual strains due to their ability to share metabolic pathways and resources (Cao et al., 2022; Lü et al., 2024). This cooperative metabolism is particularly useful in the degradation of recalcitrant organic pollutants, such as PAHs, pesticides, and heavy metals (Kumari & Das, 2023).

Microbial consortia have been used to remediate contaminated soils through bioaugmentation and biostimulation approaches (Tribedi et al., 2018). In bioaugmentation, specific microbial consortia with known degradative capabilities are introduced to contaminated sites to accelerate the breakdown of pollutants. For instance, a study demonstrated the successful use of a microbial consortium to degrade polychlorinated biphenyls (PCBs) in contaminated soils, achieving a degradation efficiency of over ~85% (Lü et al., 2024).

In biostimulation, the native microbial communities in contaminated soils are stimulated by adding nutrients or substrates that enhance their metabolic activity and cooperative interactions (Romantschuk et al., 2023). This approach has been shown to be effective in degrading a wide range of organic pollutants and is often more sustainable and cost-effective than chemical remediation methods (Kumar et al., 2018). For example, the addition of organic amendments, such as compost or biochar, has been shown to enhance the activity of microbial consortia involved in the degradation of hydrocarbons and heavy metals in contaminated soils (Hussain et al., 2018; Tran et al., 2023).

Microbial cooperation also plays a role in the restoration of degraded ecosystems. For instance, soil microbes can facilitate the establishment and growth of vegetation in degraded lands by improving soil structure, enhancing nutrient availability, and suppressing soil-borne pathogens (Hartmann & Six, 2023). A study demonstrated that inoculating degraded soils with cooperative microbial consortia improved soil health and promoted the growth of native plant species, thereby accelerating the process of ecological restoration (Coban et al., 2022; Singh Rawat et al., 2023).

Moreover, the use of microbial consortia in the phytoremediation of contaminated soils is gaining attention. In this approach, plants are used to extract, stabilize, or degrade pollutants, and their growth is supported by cooperative soil microbes. These microbes enhance the efficiency of phytoremediation by improving plant growth, facilitating pollutant uptake, and directly degrading pollutants in the rhizosphere (Yang et al., 2020; Saeed et al., 2021). For example, mycorrhizal fungi and endophytic bacteria have been shown to enhance the phytoremediation of heavy metals by hyperaccumulator plants, such as *Brassica* and *Salix* species (El-Mahrouk et al., 2019; Kumar et al., 2020; Shah & Daverey, 2020).

16.5.3 Future Directions for Research and Application

Despite significant advancements in understanding microbial cooperation and its applications, several areas require further research and technological development. One promising area for future research is the elucidation of the molecular mechanisms underlying microbial cooperation. Advances in omics technologies, such as metagenomics, metatranscriptomics, and metabolomics, have provided insights into the complex interactions among soil microbes (Biswas & Sarkar, 2018). However, there is still a need for a deeper understanding of the signaling pathways, gene regulation networks, and metabolic interactions that govern cooperative behaviors. Such knowledge could inform the development of more targeted and efficient microbial-based products for agriculture and environmental management.

Another area for future research is the development of synthetic microbial consortia with tailored functions. Recent advances in synthetic biology have enabled the design and construction of synthetic consortia with specific metabolic capabilities and cooperative interactions (Escalante et al., 2015; Duncker et al., 2021). These synthetic consortia could be used to enhance crop growth, degrade specific pollutants, or restore degraded ecosystems more effectively than natural communities. However, challenges remain in ensuring the stability and functionality of these consortia in complex and dynamic soil environments.

Furthermore, research is needed to explore the potential of cooperative soil microbes in climate change mitigation. Soil microbes play a crucial role in soil carbon sequestration and greenhouse gas emissions (Arora & Chaudhry, 2021). Understanding how microbial cooperation influences these processes could inform the development of strategies to enhance soil carbon storage and reduce emissions of greenhouse gases, such as methane and nitrous oxide, from soils.

In terms of application, there is a need for the development of innovative technologies and formulations for delivering microbial consortia to soils. Current delivery methods, such as seed coatings and soil inoculants, often have limited efficacy due to challenges related to microbial survival, colonization, and activity in soils (O'Callaghan et al., 2022). Future research could focus on developing more effective delivery systems, such as encapsulation technologies or soil conditioners that protect and sustain microbial communities in soils.

Finally, interdisciplinary research that integrates microbiology, soil science, plant science, and environmental engineering is essential to fully realize the potential of cooperative soil microbes in agriculture and environmental management. Such collaborative efforts could lead to the development of holistic approaches that optimize the benefits of microbial cooperation for sustainable agriculture, bioremediation, and ecosystem restoration.

16.6 Conclusion

This chapter elucidated the intricate web of cooperative behavior among soil microbes, emphasizing their pivotal role in ecosystem functioning and sustainability. It explored the diverse forms of microbial cooperation in the soil, such as mutualistic relationships, syntrophy, cross-feeding, and the formation of microbial consortia and biofilms. These interactions enhance nutrient cycling, improve soil fertility, and support plant growth, highlighting the critical ecological services provided by soil microbes. The genetic, biochemical, and ecological bases of these cooperative behaviors have been detailed, showing how microbes communicate, share resources, and synchronize activities to maintain ecosystem balance and productivity. Looking ahead, the future of research in microbial cooperation in soil appears promising and ripe with potential. The development of more refined omics technologies and bioinformatics tools is expected to further unravel the complexities of microbial interactions. This progress promises to lead to innovative strategies for enhancing agricultural productivity and environmental sustainability through the application of microbial technologies. Emerging fields such as synthetic microbiology offer exciting prospects for constructing tailored microbial consortia with specific functions, potentially revolutionizing the approach to ecological restoration, bioremediation, and sustainable agriculture. Additionally, understanding the impacts of climate change and anthropogenic activities on microbial cooperation is crucial for devising strategies to mitigate these effects and harness microbial capabilities for carbon sequestration and reducing greenhouse gas emissions. The study of microbial cooperation within soil ecosystems is not merely an academic pursuit but a necessity for the advancement of sustainable ecological and agricultural practices. The intricate relationships among soil microbes are a testament to the sophistication of natural systems, offering blueprints for innovation in various biotechnological applications. By fostering a deeper understanding and appreciation of microbial interactions, better harnessing of these unseen allies can be achieved to address some of the most pressing environmental challenges of our time, such as food security, biodiversity conservation, and climate change mitigation. The continued exploration of this dynamic field remains essential for the well-being of our planet and future generations.

References

Abdu, N., Abdullahi, A. A., & Abdulkadir, A. (2017). Heavy metals and soil microbes. *Environmental Chemistry Letters, 15*, 65–84. https://doi.org/10.1007/s10311-016-0587-x

Alayande, A. B., Aung, M. M., & Kim, I. S. (2018). Correlation between quorum sensing signal molecules and Pseudomonas aeruginosa's biofilm development and Virulency. *Current Microbiology, 75*, 787–793. https://doi.org/10.1007/s00284-018-1449-5

Alengebawy, A., Deka, T. J., Pradhan, S. P., et al. (2024). Soil as a natural source of greenhouse gas emissions. In *Advances and technology development in greenhouse gases: Emission, capture and conversion* (pp. 227–263). Elsevier. https://doi.org/10.1016/b978-0-443-19231-9.00001-6

Angulo, V., Beriot, N., Garcia-Hernandez, E., et al. (2022). Plant–microbe eco-evolutionary dynamics in a changing world. *The New Phytologist, 234*, 1919–1928. https://doi.org/10.1111/nph.18015

Arora, P., & Chaudhry, S. (2021). Microbial mediated soil carbon sequestration and mitigation of green house gases. *COJ Reviews and Research, 3*, 562.

Ataikiru, T. L., Okpokwasili, G. S. C., & Okerentugba, P. O. (2019). Impact of pesticides on microbial diversity and enzymes in soil. *South Asian Journal of Research in Microbiology, 4*, 1–16. https://doi.org/10.9734/sajrm/2019/v4i230104

Ayaz, E., & Gothalwal, R. (2014). Effect of environmental factors on bacterial quorum sensing. *Cellular and Molecular Biology, 60*, 46–50.

Bahadur, A. (2021). Role of vesicular arbuscular mycorrhizae (VAM) fungi in leguminous crops. In *Diseases of pulse crops and their management* (pp. 463–481).

Bai, J. A., & Rai, R. V. (2022). Quorum sensing and quorum quenching metabolites in actinomycetes. In *Natural products from actinomycetes: Diversity, ecology and drug discovery* (pp. 223–265). Springer Singapore. https://doi.org/10.1007/978-981-16-6132-7_10

Baker, B. J., De Anda, V., Seitz, K. W., et al. (2020). Diversity, ecology and evolution of archaea. *Nature Microbiology, 5*, 887–900. https://doi.org/10.1038/s41564-020-0715-z

Basu, S., & Kumar, G. (2020). Nitrogen fixation in a legume-rhizobium symbiosis: The roots of a success story. In *Plant microbe symbiosis* (pp. 35–53). Springer. https://doi.org/10.1007/978-3-030-36248-5_3

Basu, S., Kumar, G., Chhabra, S., & Prasad, R. (2020). Role of soil microbes in biogeochemical cycle for enhancing soil fertility. In *New and future developments in microbial biotechnology and bioengineering* (pp. 149–157). Elsevier. https://doi.org/10.1016/B978-0-444-64325-4.00013-4

Beroigui, O., & Errachidi, F. (2023). Streptomyces at the heart of several sectors to support practical and sustainable applications: A review. *Progress in Microbes and Molecular Biology, 6*. https://doi.org/10.36877/pmmb.a0000345

Besset-Manzoni, Y., Rieusset, L., Joly, P., et al. (2018). Exploiting rhizosphere microbial cooperation for developing sustainable agriculture strategies. *Environmental Science and Pollution Research, 25*, 29953–29970. https://doi.org/10.1007/s11356-017-1152-2

Bhatt, V. S. (2019). Quorum sensing mechanisms in gram positive bacteria. In *Implication of quorum sensing system in biofilm formation and virulence* (pp. 297–311). Springer. https://doi.org/10.1007/978-981-13-2429-1_20

Bissett, A., Brown, M. V., Siciliano, S. D., & Thrall, P. H. (2013). Microbial community responses to anthropogenically induced environmental change: Towards a systems approach. *Ecology Letters, 16*, 128–139. https://doi.org/10.1111/ele.12109

Biswas, R., & Sarkar, A. (2018). 'Omics' tools in soil microbiology: The state of the art. In *Advances in soil microbiology: Recent trends and future prospects. Microorganisms for sustainability* (Vol. 3, pp. 35–64). Springer. https://doi.org/10.1007/978-981-10-6178-3_3

Blakely, G. W. (2023). Mechanisms of horizontal gene transfer and DNA recombination. In *Molecular medical microbiology* (pp. 309–324). https://doi.org/10.1016/B978-0-12-818619-0.00043-5

Bonkowski, M. (2004). Protozoa and plant growth: The microbial loop in soil revisited. *The New Phytologist, 162*, 617–631. https://doi.org/10.1111/j.1469-8137.2004.01066.x

Boo, A., Ledesma Amaro, R., & Stan, G. B. (2021). Quorum sensing in synthetic biology: A review. *Current Opinion in Systems Biology, 28*. https://doi.org/10.1016/j.coisb.2021.100378

Boymurod, Y. (2024). Microorganisms in nature, their shapes and sizes. *International Journal of Advanced Science and Research, 4*, 63–67. https://doi.org/10.37547/ijasr-04-05-13

Boyno, G., & Demir, S. (2022). Plant-mycorrhiza communication and mycorrhizae in inter-plant communication. *Symbiosis, 86*, 155–168. https://doi.org/10.1007/s13199-022-00837-0

Bridges, A. A., Prentice, J. A., Wingreen, N. S., & Bassler, B. L. (2022). Signal transduction network principles underlying bacterial collective behaviors. *Annual Review of Microbiology, 76*, 235–257. https://doi.org/10.1146/annurev-micro-042922-122020

Cao, Z., Yan, W., Ding, M., & Yuan, Y. (2022). Construction of microbial consortia for microbial degradation of complex compounds. *Frontiers in Bioengineering and Biotechnology, 10*. https://doi.org/10.3389/fbioe.2022.1051233

Cao, T., Luo, Y., Shi, M., et al. (2024). Microbial interactions for nutrient acquisition in soil: Miners, scavengers, and carriers. *Soil Biology and Biochemistry, 188*. https://doi.org/10.1016/j.soilbio.2023.109215

Carneiro, B., Cardoso, P., Figueira, E., et al. (2023). Forward-looking on new microbial consortia: Combination of rot fungi and rhizobacteria on plant growth-promoting abilities. *Applied Soil Ecology, 182*. https://doi.org/10.1016/j.apsoil.2022.104689

Castillo-Juárez, I., Maeda, T., Mandujano-Tinoco, E. A., et al. (2015). Role of quorum sensing in bacterial infections. *World Journal of Clinical Cases, 3*, 575. https://doi.org/10.12998/wjcc.v3.i7.575

Chang, C. Y. (2018). Surface sensing for biofilm formation in Pseudomonas aeruginosa. *Frontiers in Microbiology, 8*. https://doi.org/10.3389/fmicb.2017.02671

Chang, F., Jia, F., Lv, R., et al. (2021). Soil bacterial communities reflect changes in soil properties during the tillage years of newly created farmland on the loess plateau. *Applied Soil Ecology, 161*. https://doi.org/10.1016/j.apsoil.2020.103853

Chellem, S. R., Kishore, C. V. L., Reddy, G. S. V., et al. (2024). Symbiotic relationships between nitrogen-fixing bacteria and leguminous plants ecological and evolutionary perspectives: A review. *Uttar Pradesh Journal of Zoology, 45*, 145–160. https://doi.org/10.56557/upjoz/2024/v45i134143

Chen, H., Liu, K., Yang, E., et al. (2023). A critical review on microbial ecology in the novel biological nitrogen removal process: Dynamic balance of complex functional microbes for nitrogen removal. *Science of the Total Environment, 857*. https://doi.org/10.1016/j.scitotenv.2022.159462

Chepsergon, J., & Moleleki, L. N. (2023). Rhizosphere bacterial interactions and impact on plant health. *Current Opinion in Microbiology, 73*. https://doi.org/10.1016/j.mib.2023.102297

Chernov, T. I., & Semenov, M. V. (2021). Management of soil microbial communities: Opportunities and prospects (a review). *Eurasian Soil Science, 54*, 1888–1902. https://doi.org/10.1134/S1064229321120024

Chernov, T. I., & Zhelezova, A. D. (2020). The dynamics of soil microbial communities on different timescales: A review. *Eurasian Soil Science, 53*, 643–652. https://doi.org/10.1134/S106422932005004X

Chiquito-Contreras, C. J., Meza-Menchaca, T., Guzmán-López, O., et al. (2024). Molecular insights into plant–microbe interactions: A comprehensive review of key mechanisms. *Frontiers in Bioscience-Elite, 16*. https://doi.org/10.31083/j.fbe1601009

Coban, O., de Deyn, G. B., & van der Ploeg, M. (2022). Soil microbiota as game-changers in restoration of degraded lands. *Science, 80*, 375. https://doi.org/10.1126/science.abe0725

Crowther, T. W., van den Hoogen, J., Wan, J., et al. (2019). The global soil community and its influence on biogeochemistry. *Science, 80*, 365. https://doi.org/10.1126/science.aav0550

D'Souza, G., Shitut, S., Preussger, D., et al. (2018). Ecology and evolution of metabolic cross-feeding interactions in bacteria. *Natural Product Reports, 35*, 455–488. https://doi.org/10.1039/c8np00009c

Daccò, C., Girometta, C., Asemoloye, M. D., et al. (2020). Key fungal degradation patterns, enzymes and their applications for the removal of aliphatic hydrocarbons in polluted soils: A review. *International Biodeterioration and Biodegradation, 147*. https://doi.org/10.1016/j.ibiod.2019.104866

Davis, R. M., Muller, R. Y., & Haynes, K. A. (2015). Can the natural diversity of quorum-sensing advance synthetic biology? *Frontiers in Bioengineering and Biotechnology, 3*. https://doi.org/10.3389/fbioe.2015.00030

De Mandal, S., Sonali, S. S., et al. (2021). Plant–microbe association for mutual benefits for plant growth and soil health. In *Current trends in microbial biotechnology for sustainable agriculture* (pp. 95–121). Springer.

Deb, C. R., & Tatung, M. (2024). Siderophore producing bacteria as biocontrol agent against phytopathogens for a better environment: A review. *South African Journal of Botany, 165*, 153–162. https://doi.org/10.1016/j.sajb.2023.12.031

Della Sala, G., Teta, R., Esposito, G., & Costantino, V. (2019). The chemical language of gram-negative bacteria. *Quorum sensing*, 3–28. https://doi.org/10.1016/B978-0-12-814905-8.00001-0

Dilnashin, H., Birla, H., Hoat, T. X., et al. (2020). Applications of agriculturally important microorganisms for sustainable crop production. *Molecular aspects of plant beneficial microbes in agriculture*, 403–415. https://doi.org/10.1016/B978-0-12-818469-1.00033-X

Dimkić, I., Janakiev, T., Petrović, M., et al. (2022). Plant-associated bacillus and pseudomonas antimicrobial activities in plant disease suppression via biological control mechanisms – A review. *Physiological and Molecular Plant Pathology, 117*. https://doi.org/10.1016/j.pmpp.2021.101754

Dolan, S. K., Westmann, C. A., Pereira, G. K. B., et al. (2020). Quorum sensing in Pseudomonas aeruginosa: From gene and metabolic networks to bacterial pathogenesis. In *Trends in quorum sensing and quorum quenching* (pp. 79–88). CRC Press.

Duncker, K. E., Holmes, Z. A., & You, L. (2021). Engineered microbial consortia: Strategies and applications. *Microbial Cell Factories, 20*. https://doi.org/10.1186/s12934-021-01699-9

Dusane, D. H., Zinjarde, S. S., Venugopalan, V. P., et al. (2010). Quorum sensing: Implications on rhamnolipid biosurfactant production. *Biotechnology & Genetic Engineering Reviews, 27*, 159–184. https://doi.org/10.1080/02648725.2010.10648149

El Hamss, H., Radouane, N., Belabess, Z., & Lahlali, R. (2023). Microbial consortia: An approach to enhance the effectiveness of beneficial soil microbes. In *Novel biological and biotechnological applications in plant nematode management* (pp. 133–166). Springer.

El-Mahrouk, E. S. M., Eisa, E. A. H., Hegazi, M. A., et al. (2019). Phytoremediation of cadmium-, copper-, and lead-contaminated soil by Salix mucronata (synonym Salix safsaf). *HortScience, 54*, 1249–1257. https://doi.org/10.21273/HORTSCI14018-19

Emerenini, B. O., Hense, B. A., Kuttler, C., & Eberl, H. J. (2015). A mathematical model of quorum sensing induced biofilm detachment. *PLoS One, 10*. https://doi.org/10.1371/journal.pone.0132385

Erkihun, M., Asmare, Z., Endalamew, K., et al. (2024). Medical scope of biofilm and quorum sensing during biofilm formation: Systematic review. *Bacteria, 3*, 118–135. https://doi.org/10.3390/bacteria3030008

Escalante, A. E., Rebolleda-Gómez, M., Benítez, M., & Travisano, M. (2015). Ecological perspectives on synthetic biology: Insights from microbial population biology. *Frontiers in Microbiology, 6*. https://doi.org/10.3389/fmicb.2015.00143

Essock-Burns, T., Bennett, B. D., Arencibia, D., et al. (2021). Bacterial quorum-sensing regulation induces morphological change in a key host tissue during the Euprymna scolopes-Vibrio fischeri Symbiosis. *MBio, 12*. https://doi.org/10.1128/mBio.02402-21

Fenchel, T., Blackburn, H., & King, G. M. (2012). *Bacterial biogeochemistry: The ecophysiology of mineral cycling*. Academic press.

Findlay, S. E. G. (2021). Organic matter decomposition. In *Fundamentals of ecosystem science* (2nd ed., pp. 81–102). https://doi.org/10.1016/B978-0-12-812762-9.00004-6

Foght, J. (2018). Nitrogen fixation and hydrocarbon-oxidizing bacteria. In *Cellular ecophysiology of microbe: Hydrocarbon and lipid interactions. Handbook of hydrocarbon and lipid microbiology* (pp. 431–448). Springer.

Frey-Klett, P., Burlinson, P., Deveau, A., et al. (2011). Bacterial-fungal interactions: Hyphens between agricultural, clinical, environmental, and food microbiologists. *Microbiology and Molecular Biology Reviews, 75*, 583–609. https://doi.org/10.1128/mmbr.00020-11

Furiga, A., Lajoie, B., El, H. S., et al. (2016). Impairment of pseudomonas aeruginosa biofilm resistance to antibiotics by combining the drugs with a new quorum-sensing inhibitor. *Antimicrobial Agents and Chemotherapy, 60*, 1676–1686. https://doi.org/10.1128/AAC.02533-15

Gamalero, E., Bona, E., Todeschini, V., & Lingua, G. (2020). Saline and arid soils: Impact on bacteria, plants, and their interaction. *Biology (Basel), 9*, 1–27. https://doi.org/10.3390/biology9060116

Garg, N., Manchanda, G., & Kumar, A. (2014). Bacterial quorum sensing: Circuits and applications. *Antonie van Leeuwenhoek, International Journal of General and Molecular Microbiology, 105*, 289–305. https://doi.org/10.1007/s10482-013-0082-3

Geisen, S., Mitchell, E. A. D., Adl, S., et al. (2018). Soil protists: A fertile frontier in soil biology research. *FEMS Microbiology Reviews, 42*, 293–323. https://doi.org/10.1093/femsre/fuy006

Gentry, T. J., Pepper, I. L., & Pierson, L. S. (2015). Microbial diversity and interactions in natural ecosystems. In *Environmental microbiology* (3rd ed., pp. 441–460). https://doi.org/10.1016/B978-0-12-394626-3.00019-3

Gentry, T., Fuhrmann, J. J., & Zuberer, D. A. (2021). *Principles and applications of soil microbiology*. Elsevier.

Ghanem, S. M., Abd El-Baky, R. M., Abourehab, M. A. S., et al. (2023). Prevalence of quorum sensing and virulence factor genes among Pseudomonas aeruginosa isolated from patients suffering from different infections and their association with antimicrobial resistance. *Infection and Drug Resistance*, 2371–2385.

Ghannoum, M., Parsek, M., Whiteley, M., & Mukherjee, P. K. (2020). *Microbial biofilms*. John Wiley & Sons.

Gopalakrishnan, S., Sathya, A., Vijayabharathi, R., et al. (2015). Plant growth promoting rhizobia: Challenges and opportunities. *3 Biotech, 5*, 355–377. https://doi.org/10.1007/s13205-014-0241-x

Gurikar, C., Naik, M. K., & Sreenivasa, M. Y. (2016). Azotobacter: PGPR activities with special reference to effect of pesticides and biodegradation. In *Microbial inoculants in sustainable agricultural productivity: Vol. 1: Research perspectives* (pp. 229–244).

Guseva, K., Darcy, S., Simon, E., et al. (2022). From diversity to complexity: Microbial networks in soils. *Soil Biology and Biochemistry, 169*. https://doi.org/10.1016/j.soilbio.2022.108604

Haque, S., Yadav, D. K., Bisht, S. C., et al. (2019). Quorum sensing pathways in gram-positive and -negative bacteria: Potential of their interruption in abating drug resistance. *Journal of Chemotherapy, 31*, 161–187. https://doi.org/10.1080/1120009X.2019.1599175

Hart, S. P. (2023). How does facilitation influence the outcome of species interactions? *Journal of Ecology, 111*, 2094–2104. https://doi.org/10.1111/1365-2745.14189

Hartmann, M., & Six, J. (2023). Soil structure and microbiome functions in agroecosystems. *Nature Reviews Earth and Environment, 4*, 4–18. https://doi.org/10.1038/s43017-022-00366-w

Haruna, A., & Yahaya, S. M. (2021). Recent advances in the chemistry of bioactive compounds from plants and soil microbes: A review. *Chemistry Africa, 4*, 231–248. https://doi.org/10.1007/s42250-020-00213-9

Hassani, M. A., Durán, P., & Hacquard, S. (2018). Microbial interactions within the plant holobiont. *Microbiome, 6*, 58. https://doi.org/10.1186/s40168-018-0445-0

Hense, B. A., & Schuster, M. (2015). Core principles of bacterial autoinducer systems. *Microbiology and Molecular Biology Reviews, 79*, 153–169. https://doi.org/10.1128/mmbr.00024-14

Horwath, W. R. (2017). The role of the soil microbial biomass in cycling nutrients. In *Microbial biomass: A paradigm shift in terrestrial biogeochemistry* (pp. 41–66). World Scientific Publishing Company.

Hussain, F., Hussain, I., Khan, A. H. A., et al. (2018). Combined application of biochar, compost, and bacterial consortia with Italian ryegrass enhanced phytoremediation of petroleum hydrocarbon contaminated soil. *Environmental and Experimental Botany, 153*, 80–88. https://doi.org/10.1016/j.envexpbot.2018.05.012

Hussain, T., Akthar, N., Aminedi, R., et al. (2020). Role of the potent microbial based bioagents and their emerging strategies for the ecofriendly management of agricultural phytopathogens. In *Natural bioactive products in sustainable agriculture* (pp. 45–66). https://doi.org/10.1007/978-981-15-3024-1_4

Islam, W., Noman, A., Naveed, H., et al. (2020). Role of environmental factors in shaping the soil microbiome. *Environmental Science and Pollution Research, 27*, 41225–41247. https://doi.org/10.1007/s11356-020-10471-2

Israel, E., Ramganesh, S., Abia, A. L. K., & Chikere, C. B. (2023). Quorum sensing: Unravelling the intricacies of microbial communication for biofilm formation, biogeochemical cycling, and biotechnological applications. *Journal of Marine Science and Engineering, 11*. https://doi.org/10.3390/jmse11081586

Jain, A., Chakraborty, J., & Das, S. (2020). Underlying mechanism of plant–microbe crosstalk in shaping microbial ecology of the rhizosphere. *Acta Physiologiae Plantarum, 42*. https://doi.org/10.1007/s11738-019-3000-0

Jassim, Y. A., Chabuk, H. A.-H., & Al-Yassiry, Z. A. N. (2023). Soil microorganisms: Characteristics, importance and their functional role. *Journal of Biology and Nature*, 14–19. https://doi.org/10.56557/joban/2023/v15i18054

Jin, Z., Jiang, F., Wang, L., et al. (2024). Arbuscular mycorrhizal fungi and Streptomyces: Brothers in arms to shape the structure and function of the hyphosphere microbiome in the early stage of interaction. *Microbiome, 12*. https://doi.org/10.1186/s40168-024-01811-2

Kai, K. (2018). Bacterial quorum sensing in symbiotic and pathogenic relationships with hosts. *Bioscience, Biotechnology, and Biochemistry, 82*, 363–371. https://doi.org/10.1080/09168451.2018.1433992

Kebede, E. (2021). Contribution, utilization, and improvement of legumes-driven biological nitrogen fixation in agricultural systems. *Frontiers in Sustainable Food Systems, 5*. https://doi.org/10.3389/fsufs.2021.767998

Khan, S. T. (2022). Consortia-based microbial inoculants for sustaining agricultural activities. *Applied Soil Ecology, 176*. https://doi.org/10.1016/j.apsoil.2022.104503

Khan, A., Panthari, D., Sharma, R. S., et al. (2023). Biofertilizers: A microbial-assisted strategy to improve plant growth and soil health. In *Advanced microbial techniques in agriculture, environment, and health management* (pp. 97–118). https://doi.org/10.1016/B978-0-323-91643-1.00007-7

Kidwai, M. K., Malik, A., Dhull, S. B., et al. (2022). Bioremediation potential of Trichoderma species for metal (loid)s. In *Bioremediation of toxic metal (loid)s* (pp. 137–152). CRC Press.

Klaban, V. (2021). Microscopic world and the phenomenon of symbiosis in the natural environment. In *Microbes: The foundation stone of the biosphere* (pp. 243–261). https://doi.org/10.1007/978-3-030-63512-1_15

Kramer, J., Özkaya, Ö., & Kümmerli, R. (2020). Bacterial siderophores in community and host interactions. *Nature Reviews. Microbiology, 18*, 152–163. https://doi.org/10.1038/s41579-019-0284-4

Kües, U., Khonsuntia, W., & Subba, S. (2018). Complex fungi. *Fungal Biological Reviews, 32*, 205–218. https://doi.org/10.1016/j.fbr.2018.08.001

Kumar, V., Shahi, S. K., & Singh, S. (2018). Bioremediation: An eco-sustainable approach for restoration of contaminated sites. In *Microbial bioprospecting for sustainable development* (pp. 115–136). https://doi.org/10.1007/978-981-13-0053-0_6

Kumar, M., Dhaulaniya, A. S., Shree, P., & Singh, D. K. (2020). Endophytic bacterial applications in phytoremediation of organic pollutants and toxic metals. In *Microbial enzymes and biotechniques: Interdisciplinary perspectives* (pp. 131–160). https://doi.org/10.1007/978-981-15-6895-4_8

Kumar, P., Sharma, K., & Singh, J. (2024). Response of microbial community to environment changes. In *Microbial metagenomics in effluent treatment plant* (pp. 43–61). https://doi.org/10.1016/B978-0-443-13531-6.00006-9

Kumari, S., & Das, S. (2023). Bacterial enzymatic degradation of recalcitrant organic pollutants: Catabolic pathways and genetic regulations. *Environmental Science and Pollution Research, 30*, 79676–79705. https://doi.org/10.1007/s11356-023-28130-7

Lal, N., & Biswas, A. K. (2023). Allelopathic interaction and eco-physiological mechanisms in Agri-horticultural systems: A review. *Erwerbs-obstbau, 65*, 1861–1872. https://doi.org/10.1007/s10341-023-00864-1

Lanfranco, L., Bonfante, P., & Genre, A. (2016). The mutualistic interaction between plants and arbuscular mycorrhizal fungi. *Microbiology Spectrum, 4*. https://doi.org/10.1128/microbiolspec.funk-0012-2016

Le Provost, G., Badenhausser, I., Le Bagousse-Pinguet, Y., et al. (2020). Land-use history impacts functional diversity across multiple trophic groups. *Proceedings of the National Academy of Sciences of the United States of America, 117*, 1573–1579. https://doi.org/10.1073/pnas.1910023117

Little, A. E. F., Robinson, C. J., Peterson, S. B., et al. (2008). Rules of engagement: Interspecies interactions that regulate microbial communities. *Annual Review of Microbiology, 62*, 375–401. https://doi.org/10.1146/annurev.micro.030608.101423

Liu, C., Plaza-Bonilla, D., Coulter, J. A., et al. (2022). Diversifying crop rotations enhances agroecosystem services and resilience. *Advances in Agronomy, 173*, 299–335. https://doi.org/10.1016/bs.agron.2022.02.007

Liu, X., Mei, S., & Salles, J. F. (2023). Inoculated microbial consortia perform better than single strains in living soil: A meta-analysis. *Applied Soil Ecology, 190*. https://doi.org/10.1016/j.apsoil.2023.105011

Lü, H., Wei, J. L., Tang, G. X., et al. (2024). Microbial consortium degrading of organic pollutants: Source, degradation efficiency, pathway, mechanism and application. *Journal of Cleaner Production, 451*. https://doi.org/10.1016/j.jclepro.2024.141913

Maciag, T., Kozieł, E., Rusin, P., et al. (2023). Microbial consortia for plant protection against diseases: More than the sum of its parts. *International Journal of Molecular Sciences, 24*. https://doi.org/10.3390/ijms241512227

Mahanty, T., Bhattacharjee, S., Goswami, M., et al. (2017). Biofertilizers: A potential approach for sustainable agriculture development. *Environmental Science and Pollution Research, 24*, 3315–3335. https://doi.org/10.1007/s11356-016-8104-0

Mahmud, K., Makaju, S., Ibrahim, R., & Missaoui, A. (2020). Current progress in nitrogen fixing plants and microbiome research. *Plants, 9*. https://doi.org/10.3390/plants9010097

Mann, E. E., & Wozniak, D. J. (2012). Pseudomonas biofilm matrix composition and niche biology. *FEMS Microbiology Reviews, 36*, 893–916. https://doi.org/10.1111/j.1574-6976.2011.00322.x

McKinley, V. L. (2019). Effects of land use and restoration on soil microbial communities. In *Understanding terrestrial microbial communities*.

Meryem, S. S., Pervez, A., & Ebadi, A. G. (2022). Role of quorum sensing in nutrient acquisition and synergistic Plant-Microbe Association. In *Sustainable plant nutrition under contaminated environments* (pp. 287–308). Springer International Publishing. https://doi.org/10.1007/978-3-030-91499-8_14

Mhete, M., Eze, P. N., Rahube TO, & Akinyemi, F. O. (2020). Soil properties influence bacterial abundance and diversity under different land-use regimes in semi-arid environments. *Scientific African, 7*. https://doi.org/10.1016/j.sciaf.2019.e00246

Miller, M. B., & Bassler, B. L. (2001). Quorum sensing in bacteria. *Annual Review of Microbiology, 55*, 165–199. https://doi.org/10.1146/annurev.micro.55.1.165

Mohanty, B., Pradhan, D., Das, R., & Das, M. T. (2021). *Biogeochemical cycles in soil microbiomes in response to climate change* (pp. 491–519). https://doi.org/10.1007/978-3-030-76863-8_26

Moreno-Gámez, S., Sorg, R. A., Domenech, A., et al. (2017). Quorum sensing integrates environmental cues, cell density and cell history to control bacterial competence. *Nature Communications, 8*. https://doi.org/10.1038/s41467-017-00903-y

Morris, J. J., & Schniter, E. (2018). Black Queen markets: Commensalism, dependency, and the evolution of cooperative specialization in human society. *Journal of Bioeconomics, 20*, 69–105. https://doi.org/10.1007/s10818-017-9263-x

Morris, B. E. L., Henneberger, R., Huber, H., & Moissl-Eichinger, C. (2013). Microbial syntrophy: Interaction for the common good. *FEMS Microbiology Reviews, 37*, 384–406. https://doi.org/10.1111/1574-6976.12019

Mukherjee, S., & Bassler, B. L. (2019). Bacterial quorum sensing in complex and dynamically changing environments. *Nature Reviews. Microbiology, 17*, 371–382. https://doi.org/10.1038/s41579-019-0186-5

Nadarajah, K., & Abdul Rahman, N. S. N. (2023). The microbial connection to sustainable agriculture. *Plants, 12*. https://doi.org/10.3390/plants12122307

Nadeem, S. M., Khan, M. Y., Waqas, M. R., et al. (2017). Arbuscular mycorrhizas: An overview. In *Arbuscular mycorrhizas and stress tolerance of plants* (pp. 1–24). https://doi.org/10.1007/978-981-10-4115-0_1

Nag, M., Lahiri, D., Ghosh, A., et al. (2021). Quorum sensing. In *Biofilm-mediated diseases: Causes and controls* (pp. 21–45). https://doi.org/10.1007/978-981-16-0745-5_2

Naz, M., Dai, Z., Hussain, S., et al. (2022). The soil pH and heavy metals revealed their impact on soil microbial community. *Journal of Environmental Management, 321*. https://doi.org/10.1016/j.jenvman.2022.115770

Nkongolo, K. K., & Narendrula-Kotha, R. (2020). Advances in monitoring soil microbial community dynamic and function. *Journal of Applied Genetics, 61*, 249–263. https://doi.org/10.1007/s13353-020-00549-5

Norton, J., & Ouyang, Y. (2019). Controls and adaptive management of nitrification in agricultural soils. *Frontiers in Microbiology, 10*. https://doi.org/10.3389/fmicb.2019.01931

O'Callaghan, M., Ballard, R. A., & Wright, D. (2022). Soil microbial inoculants for sustainable agriculture: Limitations and opportunities. *Soil Use and Management, 38*, 1340–1369. https://doi.org/10.1111/sum.12811

Offre, P., Spang, A., & Schleper, C. (2013). Archaea in biogeochemical cycles. *Annual Review of Microbiology, 67*, 437–457. https://doi.org/10.1146/annurev-micro-092412-155614

Oliveira, N. M., Martinez-Garcia, E., Xavier, J., et al. (2015). Correction: Biofilm formation as a response to ecological competition (PLoS biol, 2015). *PLoS Biology, 13*. https://doi.org/10.1371/journal.pbio.1002232

Pandey, S., Aryal, S., & Byanjankar, A. *Microorganisms: The building blocks of biodiversity*.

Paul, E., & Frey, S. (2023). *Soil microbiology, ecology and biochemistry*. Elsevier.

Pereira, C. S., Thompson, J. A., & Xavier, K. B. (2013). AI-2-mediated signalling in bacteria. *FEMS Microbiology Reviews, 37*, 156–181. https://doi.org/10.1111/j.1574-6976.2012.00345.x

Pérez-Velázquez, J., Gölgeli, M., & García-Contreras, R. (2016). Mathematical modelling of bacterial quorum sensing: A review. *Bulletin of Mathematical Biology, 78*, 1585–1639. https://doi.org/10.1007/s11538-016-0160-6

Peterson, R. L., & Massicotte, H. B. (2004). Exploring structural definitions of mycorrhizas, with emphasis on nutrient-exchange interfaces. *Canadian Journal of Botany, 82*, 1074–1088. https://doi.org/10.1139/B04-071

Philippot, L., Griffiths, B. S., & Langenheder, S. (2021). Microbial community resilience across ecosystems and multiple disturbances. *Microbiology and Molecular Biology Reviews, 85*. https://doi.org/10.1128/mmbr.00026-20

Polyak, Y. M., & Sukcharevich, V. I. (2019). Allelopathic interactions between plants and microorganisms in soil ecosystems. *Biology Bulletin Reviews, 9*, 562–574. https://doi.org/10.1134/s2079086419060033

Popat, R., Harrison, F., da Silva, A. C., et al. (2017). Environmental modification via a quorum sensing molecule influences the social landscape of siderophore production. *Proceedings of the Royal Society B: Biological Sciences, 284*. https://doi.org/10.1098/rspb.2017.0200

Prasad, S., Malav, L. C., Choudhary, J., et al. (2021). Soil microbiomes for healthy nutrient recycling. In *Current trends in microbial biotechnology for sustainable agriculture*.

Puxty, R. J., & Millard, A. D. (2023). Functional ecology of bacteriophages in the environment. *Current Opinion in Microbiology, 71*. https://doi.org/10.1016/j.mib.2022.102245

Qian, X., Chen, L., Sui, Y., et al. (2020). Biotechnological potential and applications of microbial consortia. *Biotechnology Advances, 40*. https://doi.org/10.1016/j.biotechadv.2019.107500

Raklami, A., Bechtaoui, N., Tahiri, A., et al. (2019). Use of rhizobacteria and mycorrhizae consortium in the open field as a strategy for improving crop nutrition, productivity and soil fertility. *Frontiers in Microbiology, 10*, 1106.

Ramesh, T., Bolan, N. S., Kirkham, M. B., et al. (2019). Soil organic carbon dynamics: Impact of land use changes and management practices: A review. *Advances in Agronomy, 156*, 1–107. https://doi.org/10.1016/bs.agron.2019.02.001

Ramírez, P. B., Fuentes-Alburquenque, S., Díez, B., et al. (2020). Soil microbial community responses to labile organic carbon fractions in relation to soil type and land use along a climate gradient. *Soil Biology and Biochemistry, 141*. https://doi.org/10.1016/j.soilbio.2019.107692

Rashid, M. I., Mujawar, L. H., Shahzad, T., et al. (2016). Bacteria and fungi can contribute to nutrients bioavailability and aggregate formation in degraded soils. *Microbiological Research, 183*, 26–41. https://doi.org/10.1016/j.micres.2015.11.007

Robertson, G. P. (2014). Soil greenhouse gas emissions and their mitigation. In *Encyclopedia of agriculture and food systems* (pp. 185–196). https://doi.org/10.1016/B978-0-444-52512-3.00097-8

Robertson, G. P., & Groffman, P. M. (2023). Nitrogen transformations. In *Soil microbiology, ecology and biochemistry* (5th ed., pp. 407–438). https://doi.org/10.1016/B978-0-12-822941-5.00014-4

Rocha-Estrada, J., Aceves-Diez, A. E., Guarneros, G., & De La Torre, M. (2010). The RNPP family of quorum-sensing proteins in gram-positive bacteria. *Applied Microbiology and Biotechnology, 87*, 913–923. https://doi.org/10.1007/s00253-010-2651-y

Romantschuk, M., Lahti-Leikas, K., Kontro, M., et al. (2023). Bioremediation of contaminated soil and groundwater by in situ biostimulation. *Frontiers in Microbiology, 14*. https://doi.org/10.3389/fmicb.2023.1258148

Rosier, A., Bishnoi, U., Lakshmanan, V., et al. (2016). A perspective on inter-kingdom signaling in plant–beneficial microbe interactions. *Plant Molecular Biology, 90*, 537–548. https://doi.org/10.1007/s11103-016-0433-3

Saeed, Q., Xiukang, W., Haider, F. U., et al. (2021). Rhizosphere bacteria in plant growth promotion, biocontrol, and bioremediation of contaminated sites: A comprehensive review of effects and mechanisms. *International Journal of Molecular Sciences, 22*. https://doi.org/10.3390/ijms221910529

Sams, T., Baker, Y., Hodgkinson, J., et al. (2016). The pseudomonas quinolone signal (PQS). *Israel Journal of Chemistry, 56*, 282–294. https://doi.org/10.1002/ijch.201400128

Santoyo, G., Guzmán-Guzmán, P., Parra-Cota, F. I., et al. (2021). Plant growth stimulation by microbial consortia. *Agronomy, 11*. https://doi.org/10.3390/agronomy11020219

Sarsan, S., Pandiyan, A., Rodhe, A. V., & Jagavati, S. (2022). Synergistic interactions among microbial communities. In *Microbes in microbial communities: Ecological and applied perspectives* (pp. 1–37). https://doi.org/10.1007/978-981-16-5617-0_1

Sauer, K., Stoodley, P., Goeres, D. M., et al. (2022). The biofilm life cycle: Expanding the conceptual model of biofilm formation. *Nature Reviews Microbiology, 20*, 608–620. https://doi.org/10.1038/s41579-022-00767-0

Seenivasagan, R., & Babalola, O. O. (2021). Utilization of microbial consortia as biofertilizers and biopesticides for the production of feasible agricultural product. *Biology (Basel), 10*. https://doi.org/10.3390/biology10111111

Shah, V., & Daverey, A. (2020). Phytoremediation: A multidisciplinary approach to clean up heavy metal contaminated soil. *Environmental Technology and Innovation, 18*. https://doi.org/10.1016/j.eti.2020.100774

Shah, K. K., Tripathi, S., Tiwari, I., et al. (2021). Role of soil microbes in sustainable crop production and soil health: A review. *Agricultural Science and Technology, 13*, 109–118. https://doi.org/10.15547/ast.2021.02.019

Sharma, A., Singh, P., Sarmah, B. K., & Nandi, S. P. (2020). Quorum sensing: Its role in microbial social networking. *Research in Microbiology, 171*, 159–164. https://doi.org/10.1016/j.resmic.2020.06.003

Sikdar, R., & Elias, M. H. (2022). Evidence for complex interplay between quorum sensing and antibiotic resistance in Pseudomonas aeruginosa. *Microbiology Spectrum, 10*. https://doi.org/10.1128/spectrum.01269-22

Singh, J., & Kumar, S. (2021). Responses of soil microbial community structure and greenhouse gas fluxes to crop rotations that include winter cover crops. *Geoderma, 385*. https://doi.org/10.1016/j.geoderma.2020.114843

Singh Rawat, V., Kaur, J., Bhagwat, S., et al. (2023). Deploying microbes as drivers and indicators in ecological restoration. *Restoration Ecology, 31*. https://doi.org/10.1111/rec.13688

Singh, S., & Yadav, R. H. (2021). Influence of land use change on native microbial community and their response to the variations in micro environment. In *Microbes in land use change management* (pp. 325–340). https://doi.org/10.1016/B978-0-12-824448-7.00018-8

Singh, K., Chandra, R., & Purchase, D. (2022). Unraveling the secrets of rhizobacteria signaling in rhizosphere. *Rhizosphere, 21*. https://doi.org/10.1016/j.rhisph.2022.100484

Skorupska, A., Kidaj, D., & Wielbo, J. (2017). Flavonoids and nod factors: Importance in legume-microbe interactions and legume improvement. In *Microbes for legume improvement* (2nd ed., pp. 75–94). https://doi.org/10.1007/978-3-319-59174-2_3

Solano, C., Echeverz, M., & Lasa, I. (2014). Biofilm dispersion and quorum sensing. *Current Opinion in Microbiology, 18*, 96–104. https://doi.org/10.1016/j.mib.2014.02.008

Srinivasan, R., Prabhu, G., Prasad, M., et al. (2020). Penicillium. In *Beneficial microbes in agro-ecology* (pp. 651–667). https://doi.org/10.1016/B978-0-12-823414-3.00032-0

Subba, R. (2017). *Soil microbiology*. Oxford and IBH Publishing.

Tedersoo, L., Bahram, M., & Zobel, M. (2020). How mycorrhizal associations drive plant population and community biology. *Science, 80*, 367. https://doi.org/10.1126/science.aba1223

Tian, W., Wang, L., Li, Y., et al. (2015). Responses of microbial activity, abundance, and community in wheat soil after three years of heavy fertilization with manure-based compost and inorganic nitrogen. *Agriculture, Ecosystems and Environment, 213*, 219–227. https://doi.org/10.1016/j.agee.2015.08.009

Tommonaro, G. (2019). *Quorum sensing: Molecular mechanism and biotechnological application*. Academic Press.

Tran, H. T., Bolan, N. S., Lin, C., et al. (2023). Succession of biochar addition for soil amendment and contaminants remediation during co-composting: A state of art review. *Journal of Environmental Management, 342*. https://doi.org/10.1016/j.jenvman.2023.118191

Treseder, K. K., & Lennon, J. T. (2015). Fungal traits that drive ecosystem dynamics on land. *Microbiology and Molecular Biology Reviews, 79*, 243–262. https://doi.org/10.1128/mmbr.00001-15

Tribedi, P., Goswami, M., Chakraborty, P., et al. (2018). Bioaugmentation and biostimulation: A potential strategy for environmental remediation. *Journal of Microbiology & Experimentation, 6*. https://doi.org/10.15406/jmen.2018.06.00219

Trivedi, P., Leach, J. E., Tringe, S. G., et al. (2020). Plant–microbiome interactions: From community assembly to plant health. *Nature Reviews Microbiology, 18*, 607–621. https://doi.org/10.1038/s41579-020-0412-1

Vadakkan, K., Ngangbam, A. K., Sathishkumar, K., et al. (2024). A review of chemical signaling pathways in the quorum sensing circuit of Pseudomonas aeruginosa. *International Journal of Biological Macromolecules, 254*. https://doi.org/10.1016/j.ijbiomac.2023.127861

Van Elsas, J. D., Pratama, A. A., de Araujo, W. L., & Trevors, J. T. (2019). Microbial interactions in soil. In *Modern soil microbiology* (3rd ed., pp. 141–161). CRC Press.

Vero, S., Garmendia, G., Allori, E., et al. (2023). Microbial biopesticides: Diversity, scope, and mechanisms involved in plant disease control. *Diversity, 15*. https://doi.org/10.3390/d15030457

Vetrivel, A., Vetrivel, P., Dhandapani, K., et al. (2023). Inhibition of biofilm formation, quorum sensing and virulence factor production in Pseudomonas aeruginosa PAO1 by selected LasR inhibitors. *International Microbiology, 26*, 851–868.

Wahab, A., Muhammad, M., Munir, A., et al. (2023). Role of arbuscular mycorrhizal fungi in regulating growth, enhancing productivity, and potentially influencing ecosystems under abiotic and biotic stresses. *Plants, 12*. https://doi.org/10.3390/plants12173102

Wahid, F., Sharif, M., Ali, A., et al. (2020). Plant-microbes interactions and functions in changing climate. In *Environment, climate, plant and vegetation growth* (pp. 397–419). https://doi.org/10.1007/978-3-030-49732-3_16

Wang, H., Bier, R., Zgleszewski, L., et al. (2020). Distinct distribution of archaea from soil to freshwater to estuary: Implications of archaeal composition and function in different environments. *Frontiers in Microbiology, 11*. https://doi.org/10.3389/fmicb.2020.576661

Wang, Y., Bian, Z., & Wang, Y. (2022). Biofilm formation and inhibition mediated by bacterial quorum sensing. *Applied Microbiology and Biotechnology, 106*, 6365–6381. https://doi.org/10.1007/s00253-022-12150-3

Wiedenbeck, J., & Cohan, F. M. (2011). Origins of bacterial diversity through horizontal genetic transfer and adaptation to new ecological niches. *FEMS Microbiology Reviews, 35*, 957–976. https://doi.org/10.1111/j.1574-6976.2011.00292.x

Wu, D., Wang, W., Yao, Y., et al. (2023). Microbial interactions within beneficial consortia promote soil health. *Science of the Total Environment, 900*. https://doi.org/10.1016/j.scitotenv.2023.165801

Yadav, A. N., Kour, D., Kaur, T., et al. (2021). Biodiversity, and biotechnological contribution of beneficial soil microbiomes for nutrient cycling, plant growth improvement and nutrient uptake. *Biocatalysis and Agricultural Biotechnology, 33*. https://doi.org/10.1016/j.bcab.2021.102009

Yang, Y., Liu, Y., Li, Z., et al. (2020). Significance of soil microbe in microbial-assisted phytoremediation: An effective way to enhance phytoremediation of contaminated soil. *International journal of Environmental Science and Technology, 17*, 2477–2484. https://doi.org/10.1007/s13762-020-02668-2

Yarwood, S. A. (2018). The role of wetland microorganisms in plant-litter decomposition and soil organic matter formation: A critical review. *FEMS Microbiology Ecology, 94*. https://doi.org/10.1093/femsec/fiy175

Zhalnina, K., Zengler, K., Newman, D., & Northen, T. R. (2018). Need for laboratory ecosystems to unravel the structures and functions of soil microbial communities mediated by chemistry. *MBio, 9*. https://doi.org/10.1128/mBio.01175-18

Zhang, Q., Wu, J., Yang, F., et al. (2016). Alterations in soil microbial community composition and biomass following agricultural land use change. *Scientific Reports, 6*. https://doi.org/10.1038/srep36587

Zhang, L., Li, S., Liu, X., et al. (2020). Sensing of autoinducer-2 by functionally distinct receptors in prokaryotes. *Nature Communications, 11*. https://doi.org/10.1038/s41467-020-19243-5

Zhang, L., Chen, F. X., Zeng, Z., et al. (2021a). Advances in metagenomics and its application in environmental microorganisms. *Frontiers in Microbiology, 12*. https://doi.org/10.3389/fmicb.2021.766364

Zhang, N., Nunan, N., Hirsch, P. R., et al. (2021b). Theory of microbial coexistence in promoting soil–plant ecosystem health. *Biology and Fertility of Soils, 57*, 897–911. https://doi.org/10.1007/s00374-021-01586-w

Zhou, J., Wen, Y., Marshall, M. R., et al. (2021). Microplastics as an emerging threat to plant and soil health in agroecosystems. *Science of the Total Environment, 787*. https://doi.org/10.1016/j.scitotenv.2021.147444

The Future of Biocontrol by Plant-Microbe Nexus: Novel Strategies and Applications

17

Gupta Vandana, Patra Sandeep, Nidhi Verma, Salome John, Gupta Aishani, and Balu Ananda Chopade

Abstract

Microbial biocontrol agents (MBCAs) provide a sustainable solution to crop protection. Some of the effective examples of microbial biocontrol agents comprise *Bacillus thuringiensis*, *Trichoderma* species, *and Pseudomonas* species. Despite some successful examples, the development of effective, stable, environmentally, and biologically safe commercial microbial products necessitates wide research and optimization. Standardized protocols for the safety, efficacy, and environmental impact assessments of MBCAs are crucial to warrant their widespread use. Further, awareness among the stakeholders and farmers is critical to the acceptance and use of microbial-based solutions. This chapter discusses the promising MBCAs researched so far, along with their mechanisms of biocontrol, recent advances in genomics-based analysis, and tools to unravel novel biocontrol species, gene families, metabolic pathways, products, and mechanisms of plant-growth-promotion and biocontrol. We have summarized the case studies for successful applications of MCBAs, future prospects, ethical and safety concerns, and regulations pertaining to the commercialization of MBCAs. This

G. Vandana (✉) · P. Sandeep · N. Verma · S. John
Department of Microbiology, Ram Lal Anand College, University of Delhi, New Delhi, India
e-mail: vandanagupta@rla.du.ac.in

G. Aishani
Department of Microbiology, Ram Lal Anand College, University of Delhi, New Delhi, India

Department of Psychology, Indraprastha Women's College, University of Delhi, New Delhi, India

B. A. Chopade
Department of Biotechnology, AKS University, Satna, Madhya Pradesh, India

M. Saraf et al. (eds.), *Climate Change and Soil Microorganisms for Environmental Sustainability*, Microorganisms for Sustainability 28,
https://doi.org/10.1007/978-981-96-3425-5_17

chapter is compiled to provide an overview of the different facets of biocontrol with context to the latest developments in the field.

Keywords

Microbial biocontrol agents · Mechanisms · Genomics · Ethics · Safety · Regulations

17.1 Introduction

The ecosystem is facing the aftermath of excessive usage of chemical fertilizers and pesticides initiated during the Green Revolution for the agriculture of high-yield varieties to meet the food demand of the growing population worldwide in the 1960s. At that time, the negative impact the chemical fertilizers and pesticides was not foreseen. As time passed and slowly these chemicals accumulated in the environment to poisonous levels leading to their biomagnification in aquatic and terrestrial animals and in human beings resulting in diseases and mortalities, to algal blooms severely impacting the usability of the water-bodies and aquatic life, to the creation of ozone hole in the stratosphere resulted in global debates on the use of these approaches versus sustainable approaches. The age-old techniques of organic farming techniques emerged as a solution. The challenge with organic farming is the yield with reduced agricultural land availability and growing population. To keep the yields high while reducing the use of chemicals in agriculture mandates research and development of organic farming techniques.

One of the tried and tested sustainable alternatives to chemical pesticides is the biological control of pest populations in the fields. Public concerns over the hazards of chemical pesticides fueled interest in the biological management of plant pathogens, which may offer superior disease control compared to other methods. Biofertilizers usually promote plant development by phosphate solubilization, nitrogen fixation, phytohormone production, sulfur oxidation, and decomposition of organic compounds, while biocontrol agents promote systemic acquired resistance, express antimicrobial compounds, and certain enzymes, sequester iron using siderophores, and compete with pathogenic microbes in the rhizosphere for space and nutrients (Pirttilä et al., 2021). The coordinated action of microbial communities might enhance plant development and stress tolerance. Numerous unidentified microbial characteristics unique for each strain advantageous for the host plant are likely to be present (Pirttilä et al., 2021). Microorganisms find a crucial place in the ecosystem in controlling various pest populations. They play a crucial role in managing plant pest populations, including insects, nematodes, bacteria, and fungi, thereby protecting crops. Elaborate investigations have been done on the biological control of plant diseases by antagonistic bacteria, focusing often on the suppressive effects of single strains introduced into the soil at high densities. In contrast, crop rotations and organic fertilizers have received not as much attention for managing and manipulating naturally occurring antagonistic microbial communities despite their proven effectiveness in biological control (Hoitink & Boehm, 1999). Several microbes, including bacteria, fungi, viruses, and protozoa, are employed as microbial biocontrol agents (MBCAs). Some of the target pathogenic fungal species include *Fusarium* sp., *Alternaria* sp., *Botryosphaeria* sp., *Phomopsis* sp., *Botrytis* sp., *Phytophthora* sp., *Colletotrichum* sp., *Gaeumannomyces* sp., *Monilia* sp., *Rosellinia* sp., *Rhizoctonia solani*, and *Verticillium* sp. Among bacterial plant pathogens, *Erwinia* sp., *Pseudomonas* sp., and *Xanthomonas* sp. are the important targets, while Lepidoptera (moths & butterflies), Coleoptera (weevils & beetles), Diptera (mosquitoes & flies are some economically important plant infesting insect targets (Bonaterra et al., 2022). MBCAs can directly interact with pathogens through mechanisms such as antibiosis or hyperparasitism (Ghorbanpour et al., 2018). It is possible to select beneficial microbial isolates active against pathogens and cultivate them on artificial media. The practice of using the selected and mass-produced MBCAs, either once or multiple times during a growing season, is known as "augmentative biological control" (Heimpel & Mills, 2017; van Lenteren et al., 2018). With advancements in genomic analysis techniques, it is now possible to analyze microbiomes in suppressive soils (where plants suffer less infection due to suppression of pathogens because of the presence of beneficial microbe) and other natural soils to characterize better biocontrol agents. Genomics can further be exploited to delineate novel inhibitory compounds from MBCAS and their biosynthetic pathways which can be genetically manipulated to achieve better biocontrol. The flip side of biocontrol includes the ethical and safety concerns of releasing MBCAs into the environment. This chapter explores the key facets of biological control including its mechanisms, challenges, ethical and safety concerns, and future prospects.

17.2 Promising Microbial Biocontrol Agents and the Science Behind Biocontrol

The most comprehensively researched and documented microbial biocontrol agents include several genera of bacteria and fungi which employ multiple mechanisms to mitigate various plant pathogenic genera. Several successfully commercialized microbe-based bioinsecticides and biopesticides are in wide use throughout the world (Fig. 17.1). Some effective and well-documented MBCAs along with their mechanisms of action, are detailed in Table 17.1.

MBCA can control pest populations either directly through antibiosis, production of cell-lytic enzymes, and hyperparasitism or indirectly by inducing systemic resistance in plants competing for nutrients and space.

17.2.1 Direct Methods

Nematophagous fungi such as *Arthrobotrys dactyloides* and *Purpureocillium lilacinum* parasitize plant-parasitic nematodes including "*Meloidogyne incognita* and *Meloidogyne javanica*." The fungi produce structures like constrictive rings,

Fig. 17.1 Commercially available microbial biocontrol agent-based biopesticides

adhesive knobs, and adhesive nets on sensing the presence of the nematode in the vicinity, entrapping the nematodes. The nematode-trapping fungi penetrate the body of the nematode through haustoria, suck the nutrients and effectively kill them. Microorganisms secrete a wide variety of volatile low molecular weight organic compounds (VOCs), which are effective in biocontrol due to their antimicrobial properties. Several inhibitory compounds are secreted by MBCAs which offer them a competitive advantage in their natural habitat and are also responsible for direct biocontrol. Some examples of bacterial inhibitory compounds include "iturin,

Table 17.1 Commercial microorganism based biopesticides

S. No.	Biocontrol agent	Target organism	Mechanism of action	Current status	References
1.	*Beauveria bassiana*, *Beauveria brongniartii*	Lepidoptera, Diptera, Coleoptera, Hymenoptera; Fungi and Bacteria; e.g., *Helicoverpa armigera*, *Leptinotarsa decemlineata*, *Cydia pomonella*, *Ostrinia Mubilalis*	Establishment of infection in the target insect through spore attachment on the cuticle; germination of the spore; penetration; multiplication within the host; killing of host insect; and production of new conidia	Mycoinsecticide and Biopesticide, e.g., BotaniGard, Mycotrol, and Naturalis	Sandhu et al. (2012), Singh et al. (2015), and Keswani et al. (2013)
2.	*Pseudomonas* spp.	Oomycets, Nematodes, Aphids, some bacteria, *Globodera rostochiensis*, *Fusarium oxysporum*, *Clavibacter michiganensis*	Antibiosis, Competition for nutrient and space, Induced Systemic Resistance	Commercial formulations available for direct application in the fields	Lahlali et al. (2022) and Lanteigne et al. (2012)
3.	*Bacillus thuringiensis*	Species of the order Lepidoptera (moths & butterflies), Coleoptera (weevils & beetles), Diptera (mosquitoes & flies)	Lysis of Columnar epithelial cells and creation of pores in the midgut of the insects by the action of Bt toxin	Available as commercial bioinsecticide formulations. Transgenic Bt-cotton expressing Cry1Ac has been approved for cultivation in India; Bt Rice, Chickpea, Brinjal, & tomato are at various stages of trials in India	Kumar et al. (2021), Sujayanand et al. (2021), and Jain et al. (2016)
4.	*Trichoderma* spp.	Active against Bacteria, Fungi, Arthropods, & Nematodes, e.g., *Rhizoctonia solani*, *Verticillium dahliae*, *Fusarium* spp., *Botrytis cinerea*, *Ralstonia solanacearum*, *Meloidogyne javanica*, *Leucinodes orbonalis*	Compete for nutrients and space; Fungistatic, causes cell lysis (Antibiosis), or modify the rhizosphere (soil acidification); mycoparasitism	Biofungicide, Biofertilizer, and insecticidal and Nematicidal Sold under trade names such as Binab-T WP, SoilGard, Trichodex, Trichopel, Harzian 20, Bio-Fungus, Root shield, etc.	Tyśkiewicz et al. (2022), Gajera et al. (2013), and Ghazanfar et al. (2018)

(continued)

Table 17.1 (continued)

S. No.	Biocontrol agent	Target organism	Mechanism of action	Current status	References
5.	*Metarhizium brunneun* *Metarhizium anisopliae* *Metarhizium acridum*	Grasshoppers and Locusts, *Spodoptera litura*, Mosquitoes (*Anopheles* spp. & *Aedes* spp.), Termites (*Isoptera*), Ticks (*Rhipicephalus* spp. & *Haemaphysalis* spp.)	Same as *Beauveria*	Mycosecticides and Mycoacaricides; Green Guard®, a mycoacaricides against *Orthoptera* species; BIO 1020, a mycoinsecticide against *Coleoptera* sp.	Aw and Hue (2017)
6.	*Bacillus amyloliquefaciens*	*Ralstonia solanacearum*, *Fusarium oxysporum*, *Alternaria alternata*, *Phytopthora sojae*	Produce antifungals, e.g., Bacillomycin D and Fengycin; Induction of systemic resistance through metabolites such as Surfactin and VOCs; Secrete antibacterials like Bacilysin, Bacteriocins, and Polyketides; compete for root colonization and nutrients	Employed as Biofungicide and Biofertilizer; e.g., RhizoVital 42, AmyloX, Taegro, & Double Nickel 55	Luo et al. (2022), Chowdhury et al. (2015), and Ngalimat et al. (2021)
7.	*Coniothyrium minitans*	*Sclerotinia sclerotiorum*, *Sclerotinia minor*	Production of antifungal metabolites and lytic enzymes including Chitinases and β-1,3 Glucanases, and an antibiotic macrosphelide A	Commercial Biofungicide; Contans WG and KONI	Tomprefa et al. (2009) and Albert et al. (2022)
8.	*Pantoea agglomerans*	*Neofusicoccum parvum*, *Ralstonia solanacearum*, *Erwinia amylovora*, *Alternaria solani*	Competitive Exclusion of pathogens through antibiosis. Produce antibacterials including, herbicolins, pantocins, and Phenazines	Biocontrol of Bacterial wilt, Fire blight, Tomato early blight, Grapevine trunk disease, etc. Commercially registered products are BlossomBless P10c, Pantovital, BlightBan C9–1, Bloomtime E325	Abo-Elyousr and Hassan (2021) and Haidar et al. (2021)

9.	*Lysobacter capsici*	*Sclerotinia minor*, *Pythium ultimum*, *Phytophthora infestans*, *Rhizocotonia solani*	Production of lytic agents active against phytopathogenic bacteria, fungi, and Oomycetes	–	Vlassi et al. (2020) and Afoshin et al. (2020)
10.	*Lactiplantibacillus plantarum*	*Botrytis cinerea*, *Pseudomonas syringae pv. actinidiae*, *Fusarium oxysporum*, *Xanthomonas fragariae*, *Xanthomonas arboricola pv. pruni*, and *Ascosphaera apis*	Compete for key nutrients and secretion of bioactive compounds like Ethanol, Organic acids, Hydrogen peroxide, Diacetyl, Acetoin, and Bacteriocins	Employed as a Biocontrol agent in agricultural sector	Daranas et al. (2019), De Simone et al. (2021), and Iorizzo et al. (2021)
11.	*Coelomomyces*	Mosquito larvae and other aquatic dipterans genera include, *Anopheles*, *Aedes*, *Culex*, & *Opifex*	Infects hemocoel of the insect larvae, production of resting sporangia, and release of meiospores	Used as a Myco-insecticide in the Biological Control of Mosquitoes	Scholte et al. (2004) and Rueda-Páramo et al. (2017)
12.	Baculovirus	Arthropods including *Spodoptera exigua*, *Cydia pomonella*, *Homona magnanima*, *Spodoptera litura*, *Helicoverpa armigera*	Infects and kills target host insects amplify in the insect midgut, followed by a systemic infection resulting in the formation of crystalline occlusion bodies (OBs)	Used as Bioinsecticides Commercially available formulations: Spod-X, Elcar, GemStar LC, Spodopterin, etc.	Sabbahi et al. (2022) and Grzywacz (2017)
13.	Endo-symbiotic *Wolbachia*	Primarily infects mosquitoes including *Aedes* spp., *Culex* spp., *Anopheles* spp., *Culiseta morsitans*	Manipulate host reproduction by Cytoplasmic incompatibility Male killing, Parthenogenesis	Employed in the biocontrol of arboviral disease vectors	Jeffries and Walker (2016) and Manoj et al. (2021)
14.	Nematophagous fungi *Arthrobotrys dactyloides; Purpureocillium lilacinum*	*Meloidogyne incognita*, *Meloidogyne javanica*	Formation of adhesive nets, adhesive knobs, and constrictive rings to immobilize nematodes		

(continued)

Table 17.1 (continued)

S. No.	Biocontrol agent	Target organism	Mechanism of action	Current status	References
15.	*Pythium oligandrum*, strain M1	Mycotoxigenic fungi such as *Fusarium graminearum*, *Fusarium oxysporum*, *Rhizoctonia solani*, and some species of *Verticillium*	Mycoparasitic on pathogenic fungi in rhizosphere; induction of plant systemic resistance; plant-growth promotion	Commercially available as Polyversum®	Pisarčik et al. (2022) and Pellan et al. (2020)
16.	*Aspergillus flavus* NRRL 21882 [Non-toxicogenic]	Toxigenic strains of *Aspergillus flavus*	Competes for nutrients for aflatoxin production; Niche Exclusion; Suppression of AFB_1 by volatile compounds	Commercially sold as Afla-Guard®- GR to reduce Aflatoxin contamination in Corn and Peanuts	Pelles et al. (2021) and Medina et al. (2017)
17.	*Streptomyces griseoviridis* strain K61	*Fusarium*, *Alternaria*, *Botrytis*, *Phomopsis*, *Rhizoctonia*, and *Phytopthora*	Production of Antimicrobial bioactive metabolites and lytic enzymes against pathogens; induction of systemic resistance in plants	Commercially being used as biofungicide under the trade name, Mycostop®	Pellan et al. (2020) and Sharma et al. (2020)
18.	*Gliocladium catenulatum* strain JI446, *Gliocladium virens*	Soil borne fungal pathogens: *Plasmodiophora brassicae*, *Rhizoctonia*, *Pythium*, *Botrytis*, & *Fusarium*	Competes for nutrients with pathogenic fungi in the rhizosphere; induction of systemic resistance	Microbial fungicide under trade names, SoilGard® 12G, SoilGard®, Prima stop soil guard	Lahlali et al. (2022)
19.	*Lecanicillium lecanii* (Formerly known as *Verticillium lecanii*)	It is targeted against scale insects, thrips, whiteflies, mites, mealybugs, peach aphids and other Homopterans	Entomopathogenic mechanism of action similar to *Beauveria* and *Metarhizum* species	Different strains are commercially available with trade names, Biovert Rich, Bio-catch, Sun Agro Verti, Varunastra, Verticim, Vertisoft, & Verti-star	Arakere et al. (2022) and Subramaniam et al. (2021)
20.	*Candida oleophila*, *Candida oleophila* isolate I-182	Post-harvest pathogens such as *Botrytis cinerea*, *Penicillium digitatum*, *Monilinia* spp.	Competes for space and nutrients; hydrolyzes fungal cell walls using enzymes; oxidative stress tolerance; induction of host disease resistance	First yeast-based commercial product for the management and biocontrol of post-harvest diseases; sold under the trade name, Aspire®	Lahlali et al. (2022) and Sui et al. (2020)

fengycin, surfactin, cyanide, chitinases, proteases, pyoluteorin, phenazines, pyrroinitrin, bacilysin, amphisin, voscosinamide, mycosubtilin, beta-glucanases, tensin, and subtilocin". Fungal BCAs are reported to produce ketones, aldehydes, hydrogen cyanide, alcohols, gliotoxin, gliovirin, harzianolide, harzianopyridone, ethylene, heptelidic acid, valinotrocin, viridiol, viridian, 1-hydroxy-3-methylanthraquinone, T39butanolide, and more (Tariq et al., 2020).

17.2.2 Indirect Methods

The release of various metabolites by MBCAs induces systemic resistance in plants. MBCAs compete with the pathogens for nutrients and space, for example, some MBCAs scavenge the available iron by secreting high-affinity siderophores. Also they occupy space by mechanisms such as biofilm formation on the root surface, making it unavailable for colonization by the pathogens. MBCAs also act by promoting plant growth. Different direct and indirect mechanisms of biocontrol are summarized in Fig. 17.2.

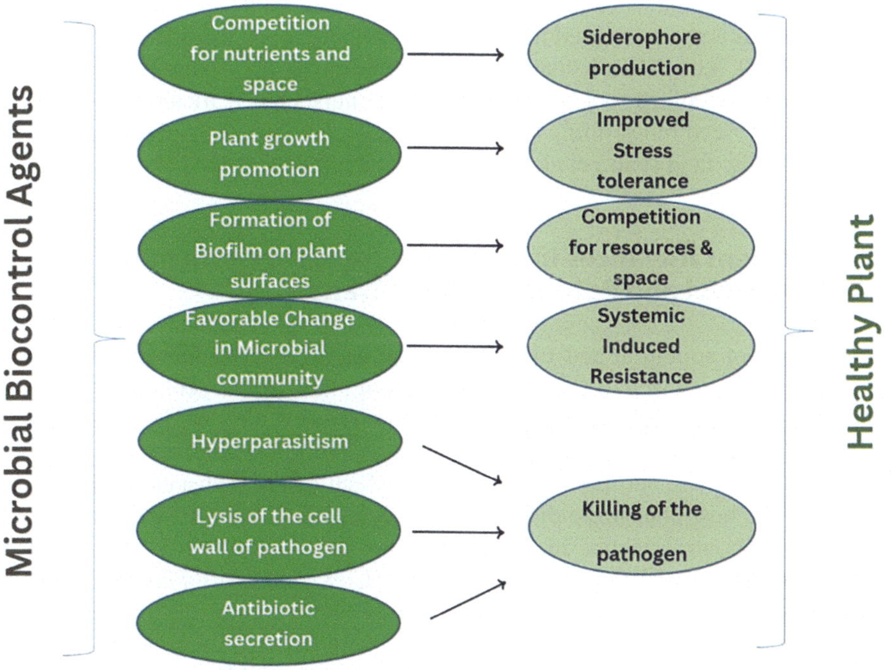

Fig. 17.2 Summary of the different mechanisms of biocontrol by microorganisms

17.3 Field Application of MBCAs

Several formulations of MBCAs are prepared using different carriers and adjuvants. The preparations are commercially available in granulated, powdered, liquid forms and cell immobilization systems. The compounds used in MBCA formulations for application determine the survival and efficacy of the biocontrol. Researchers have explored biopolymers such as chitosan, Agar, and gums from diverse plant sources such as guar bean, α- and κ-carrageenan, xanthan, pectin, lignin, derivatives of cellulose, waxes, alginate, and starch in these formulations to enhance the shelf life of BCAs (reviewed in Patra et al., 2024; Bejarano & Puopolo, 2020; Saberi-Riseh et al., 2023). Adjuvants are often added to the formulations to facilitate the mixing, handling, and dispersal of BCA in the field, enhance the retention and penetration of BCA, decrease evaporation, protect the BCA from pH shifts in the environment, and improve their survival in the field. These additives also enhance the physical properties of formulations (Patra et al., 2024; Bejarano & Puopolo, 2020).

Encapsulating bacteria through immobilization in a polymeric matrix is a feasible strategy as it creates a physical barrier between the encapsulated bacteria and the harsh atmospheric conditions like pH and moisture fluctuations, photo-oxidation, etc. This method offers the advantage of controlled release of the MBCAs in addition to protection against environmental changes. Immobilization of BCA within Alginate beads is the most used method (reviewed in Patra et al., 2024; Bejarano & Puopolo, 2020; Saberi-Riseh et al., 2023). Various methods such as: 1. soil application 2. cutting/seedling root dip application; 3. nursery bed treatment; 4. Foliar application; 5. seed bio-priming 6. seed treatments are used for applying BCA formulations in the field (Patra et al., 2024; Prasad et al., 2023).

17.4 Recent Advances in Genomics-Based Identification of Novel and Potent Microbial Traits

Applications of structural and functional genomics to identify new gene families and their roles can reveal different mechanisms of plant growth promotion and biocontrol in bacteria. The presence of gene clusters pertaining to the biosynthesis of VOCs and antimicrobial compounds, as well as the biosynthesis of surfactin, bacillaene, fengycin, and bacillibactin, all of which exhibit substantial biocontrol activity in the host plant were discovered through genome mining and sequencing of several biocontrol agents, including *B. amyloliquifaciens* GB03, *B. velezensis* FZB42, WRN014, 9912D, M75, RC 218 and CC09, and *B. atrophaeus* GQJK17, respectively (Pirttilä et al., 2021; Wang et al., 2019; Ma et al., 2018; Fan et al., 2018). The presence of genes for non-ribosomal peptide synthetases and polyketide synthases, antibiotic biosynthesis genes like Polymyxin and Fusaricidin, and gene clusters for secondary metabolite biosynthesis have been identified by genome analysis for additional strains of bacteria, including *B. methylotrophicus* B25, *Paenibacillus polymxa* SC2 and E681, biocontrol strains *Streptomyces* sp. JBS5-6 and *B. toyonensis* BAC315, respectively (Pirttilä et al., 2021; Jing et al., 2020;

Lopes et al., 2017; Gerbore et al., 2016). Additionally, it has been observed that the *Rhodopseudomonas palustris* ELI 1980 contains genes for the expression of several antimicrobial compounds, auxin biosynthesis, and nitrogen fixation, all of which aid in plant development and protection against biotic and abiotic stresses (Crovadore et al., 2017). The presence of tricholignan, a biosynthetic gene that facilitates plant iron uptake from the rhizosphere, was identified in the fungal strain *Trichoderma harzianum* t-22 (Pirttilä et al., 2021; Chen et al., 2019).

Biocontrol and host plant homeostasis are greatly influenced by microbial interactions with bacteriophages, endophytic bacteria, and the host plant itself. Gene expression study on *B velezensis* FZB42 revealed the presence of short regulatory RNAs in plant-microbe interactions (Fan et al., 2015). Furthermore, it has been documented that the breakdown of polyhydroxybutyrate in *M. extorquens* DSM13060 plays a protective role in abiotic stress tolerance in bacteria as well as in the host (Müller-Santos et al., 2021). Certain microorganisms can inhibit pathogenic bacteria from colonizing the host plant by interacting with the endophytic populations. For example, citrus variegated chlorosis (CVC) is effectively controlled by *Methylobacterium mesophilicum*, by influencing resident endophytes that suppress *Xylella fastidiosa* colonization in the xylem vessels of sweet orange (Azevedo et al., 2016). Bacteriophages can also be utilized to replace bacterial biocontrol agents since their antimicrobial mechanisms rely on the lysis of bacterial cells. Thus, it is possible to genetically modify these phages to either sensitize the target bacterium to antimicrobials or transfer lethal genes to it (Pirttilä et al., 2021). But this raises apprehensions about the spread of antibiotic resistance to the environment.

17.4.1 Identification of New Promising Microbial Strains with Applications in Biocontrol

Genome alterations in current strains can improve their biocontrol, stress tolerance, and plant growth promotion activities. Biocontrol strains of *Pseudomonas* have previously been genetically engineered and evaluated for effectiveness. For example, genetic manipulation of *P. putida* WCS358 and *P. fluorescens* strains F113 and CHA0 has resulted in increased synthesis of the antibiotic compounds 4-diacetylphloroglucinol (PHL) and pyoluteorin (PLT). *P. fluorescens* CHA0 has also been engineered for the overproduction of Indoleacetic acid (IAA) (Pirttilä et al., 2021). With GMOs being introduced into their respective ecosystems and the potential for harmful trait transmission across microbiomes, there are several ethical concerns with using them as biocontrol or biofertilizers in agricultural practices (Pirttilä et al., 2021).

Systemic bacterial endophytes, which inhabit all internal tissues of a plant, and endosymbionts, which colonize plant cells, are likely to offer the host plant with longer-term advantages than rhizobacteria or apoplastic endophytes (Pirttilä et al., 2021). Given their high degree of adaptation to the intended niche, they would be more resilient to environmental factors such as abiotic stress. Endophytic bacteria share several growth-promoting and biocontrol characteristics with PGPR, including the synthesis of phytohormones and antimicrobial substances, the development of systemic resistance, and nitrogen fixation (Adetunji et al., 2019; Pandey et al., 2019).

Plants that thrive in harsh climatic conditions are ideal targets for identifying novel potent biocontrol agents as they are accustomed to that habitat and are usually dependent on their microbial partners for survival. For instance, Scots pine (*Pinus sylvestris* L.), a conifer species acclimated to hard settings, low soil nutrient levels, drought, and cold temperatures, has been shown to harbor the growth-promoting endosymbiont *M. extorquens* DSM13060 which also has potent biocontrol activity (Matías & Jump, 2012). In the tropical panic grass, *Dichanthelium lanuginosum*, *Curvularia protuberata* an endophytic fungus confers heat tolerance that can be transmitted to crop species including tomato, watermelon, and wheat (Pirttilä et al., 2021). Likewise, *Piriformospora indica*, isolated from a xerophyte, confers resistance to salt and biotic stress in numerous agricultural crops (Pirttilä et al., 2021).

Screening habitats with high microbial variety, such as soil, which constitutes the largest reservoir of microorganisms, makes relevance when searching for novel biofertilizers and biocontrol agents. Forests and other environments rich with distinct plant species likely support a diverse range of microbes that are beneficial to plant growth and survival (Pirttilä et al., 2021). The fungal endophyte communities of grapevine cultivars cultivated on organically managed farms are more varied than those produced using integrated pest-management systems (Pancher et al., 2012). Compared to plants raised in monoculture, the endophytic microbes of bananas are altered when *Erythrina poeppigiana* (Walp.) O. F. Cook or Inga trees are used as green manures which further increases their survival and promotes growth (Pirttilä et al., 2016, 2021). It is more successful in forests than on agricultural land to discover novel microbial strains that could potentially be used as single inoculants in biocontrol and biofertilization, or as members of an intended healthy plant microbiome.

17.5 Case Studies: Successful Applications of MBCAs

The economic impact of infectious diseases in crop and forestry plants has been exacerbated by the globalization of markets and climate change, which facilitate the emergence and rapid spread of new pathogens. Since MBCAs have proven to be effective and environmentally friendly alternatives for controlling plant diseases, promoting plant growth, and enhancing yields, there has been a notable increase in research on the selection, evaluation, characterization, and use of MBCAs over the past few decades (Bonaterra et al., 2022; El-Saadony et al., 2022). Plant growth-promoting rhizobacteria and fungi (PGPR/PGPF), in addition to supporting plant growth, prevent plant diseases by producing inhibitory compounds and stimulating plant immune responses (El-Saadony et al., 2022). The success of PGPR in the biocontrol of phytopathogens under controlled conditions highlights their potential in greenhouse cultivation and commercial horticulture. For example, *Bacillus subtilis*, *Bacillus amyloliquefaciens*, and *Pseudomonas stutzeri* effectively colonize cucumber roots and prevent *Phytophthora capsici* infection (Islam et al., 2016). *B. subtilis* has been demonstrated to protect tomatoes from *Rhizopus stolonifer* and *Penicillium* spp. infections post-harvest.

Nematophagous and endoparasitic fungi successfully reduce root-knot nematode disease. For instance, a talc-based formulation of *Paecilomyces lilacinus* significantly reduced *Meloidogyne incognita* populations in tomato fields (Priya & Kumar, 2006). *Trichoderma harzianum*, *T. viride*, *Pseudomonas fluorescens*, *Pochonia chlamydosporia*, and *P. lilacinus*, effectively reduced nematode numbers in the soil and enhanced plant growth (Kumar et al., 2012).

Bacillus velezensis showed high antibacterial activity against bacterial citrus canker caused by *Xanthomonas citri* subsp. *citri* (Xcc), among several *Bacillus* strains tested. The *B. velezensis* Bv-21 strain reduced citrus canker by 26.30% and decreased the density of the pathogen by 81.68% (Rabbee et al., 2022). *Bacillus ginsengihumi* strain S38 was demonstrated to control the fungus *Botrytis cinerea*, which causes bunch rot of grapes reducing the disease severity by 35–60%, displaying a high consistency and effectiveness similar to the commercial products (Calvo-Garrido et al., 2019).

Phytophthora infestans, responsible for late blight in potatoes and tomatoes, are most vulnerable to biocontrol agents during the pre-infection phase. *Streptomyces*, *Bacillus*, *Pseudomonas*, *Trichoderma* spp., and *Pythium oligandrum*, effectively controlled the pathogen through mechanisms including mycoparasitism, antibiosis, secretion of secondary metabolites, VOCs as well as lytic enzymes (Hashemi et al., 2022). *Curtobacterium flaccumfaciens* pv. *flaccumfaciens* (Cff), which causes bacterial wilt in common beans, was antagonized by rhizobacterial strains. Seed treatment coupled with foliar administration with *Paenibacillus polymyxa*, *Pseudomonas fluorescens*, and *Bacillus cereus* resulted in a considerable reduction in disease severity and frequency (Munene et al., 2023).

Certain newly discovered MBCAs with marketable potential include endophytic *Bacillus* strains from cotton roots, which inhibit *Verticillium dahliae* strain VD-080 and significantly reduce verticillium wilt severity (Hasan et al., 2020). *Bacillus amyloliquefaciens* strains SF14 and SP10 inhibit the brown rot fungus *Monilinia fructigena* and *Monilinia laxa*, respectively in fruits (Lahlali et al., 2022). Additionally, the commercially available *B. amyloliquefaciens* FZB42 significantly reduces lettuce bottom rot caused by *Rhizoctonia solani* (Lahlali et al., 2022). *Pseudomonas chlororaphis* strains-based biopesticides protect plants from several microbial infections, insects, and nematodes through the production of various metabolites (Anderson & Kim, 2018). For instance, *Pseudomonas aeruginosa* from a banana field rhizosphere produces bacteriocin, HCN, and siderophores that suppress phytopathogens including *Fusarium oxysporum*, *Alternaria alternata* and *Aspergillus* sp. (Lahlali et al., 2022). *Pseudomonas* sp. LBUM300 produces compounds such as DAPG and HCN and inhibits the growth of *Clavibacter michiganensis* subsp. *Michiganensis* in vitro and is effective against bacterial canker in tomatoes under natural conditions (Lanteigne et al., 2012).

Fungal species play a role in biocontrol, both as plant endophytes and insect pest entomopathogens (Poveda, 2021). The genus *Trichoderma* is the most extensively studied fungal biocontrol agent and some of its species such as *T. harzianum* are already used as commercial biopesticides and biofertilizers (Alfiky & Weisskopf, 2021; Lahlali et al., 2022; Poveda, 2021). They also indirectly control pathogens in the rhizosphere through the induction of systemic resistance (Nandini et al., 2021). Yeasts such as *Wickerhamomyces anomalus* Wa-32, *Candida lusitaniae* Cl-28, *Candida oleophila* Co-13, *Debaryomyces hansenii* Dh-67, and *Hypopichia pseudoburtonii* Hp-54 were found to be effective against *V. dahlia*, *F. oxysporum* and *P. expansum* (Fernandez-San Millan et al., 2021; reviewed in Patra et al., 2024).

Recent research conducted by Yao et al. showed the presence of 21 gene clusters of secondary metabolites in *Streptomyces albidoflavus* W68, that possess potential biocontrol activity against cucumber damping-off-disease (Yao et al., 2021). It was also shown that the gene cluster involved in the production of candicidin isomers is crucial for antifungal and biocontrol activities of *S. albidoflavus* W68 against *Rhizoctonia solani* infections (Yao et al., 2021).

Lactic acid bacteria (LAB) have recently been shown to possess antifungal and detoxifying properties against mycotoxins released by *Fusarium* species, thereby serving as an effective biocontrol against mycotoxicosis in plants (Krishnan et al., 2024). They have the inherent ability to block mycotoxin production or reduce mycotoxin levels in the soil matrices, adsorb mycotoxins such as deoxynivalenol (DON) on their cell wall constituents, eliminate Zearalenone (ZEA) using enzymes such as zearalenone hydrolases, as well as inhibit mold growth through various metabolites (Krishnan et al., 2024). Numerous bioactive substances, such as phenyllactic acid, reuterin, cyclic dipeptides, antifungal peptides like sakacin, phenolic compounds like 2,4-DTBP, and fatty acids, having potent antifungal action against *Fusarium* mycotoxins, have been reported to be produced by LAB (Krishnan et al., 2024).

Trichoderma guizhouense and *T. koningiopsis* significantly inhibited the pathogen *Plasmodiophora brassicae*, (causes clubroot disease in crucifers) promoting rapeseed growth. *T. guizhouense* exhibited biocontrol efficacy of 44–68% in rapeseed clubroot and in *Arabidopsis thaliana* (Patra et al., 2024; Zhao et al., 2022). *Paenibacillus polymyxa* PJH16, a rhizobacterial strain recovered from the *Cucumis sativus* (cucumber) rhizosphere, showed antifungal activity both in vitro and in vivo against *Fusarium oxysporum* f. sp. *Cucumerinum*, consequently preventing *Fusarium* wilt in cucumbers (Yang et al., 2024). Multiple strategies for biocontrol and plant growth promotion in cucumber have been stated, including colonization at the root along with biofilm production to resist pathogens, production of hydrolases and antimicrobial peptidases like β-1,3-glucanase, cellulases, amylases, and proteases that can act on fungal cell walls, production of organic compounds that regulate the salicylic acid signal for triggering systemic resistance, and production of Indole acetic acid (IAA) and Siderophores to promote plant growth and disease free development (Yang et al., 2024). This strain can be employed in the near future to bioengineer a microbial formulation to act as a biostimulant and biocontrol agent in crop plants.

Lipopeptide biosurfactants produced predominantly from *Bacillus*, *Pseudomonas*, Cyanobacteria, and Actinomycetes species have been shown to exhibit excellent phytopathogenic action, surface activity, and biocontrol activity, rendering them beneficial for the production of biopesticides, fertilizers, and plant wettability agents in the coming years (Sani et al., 2024).

Microbial biocontrol is also gaining attention for post-harvest product quality management (Sellitto et al., 2021). Research on citrus fruits affected by *Penicillium digitatum* and *P. italicum* causing green and blue molds assessed over 180 yeast and bacterial isolates from citrus fruit peels. Five isolates, *Debaryomyces hansenii*, *Candida oleophila*, *Bacillus subtilis*, *B. amyloliquefaciens*, and *B. pumilus*, prevented mold infections and maintained fruit quality and longevity through biofilm formation, antifungal lipopeptides, lytic enzymes, and VOCs (Hammami et al., 2022).

17.6 Ethical and Safety Concerns in Microbial Biocontrol

Biocontrol is considered a safer alternative as it is less likely to harm non-target organisms and does not produce environmentally detrimental substances (Collatz et al., 2021). However, excessive use of microbes can have adverse effects on the environment, as well as human and animal health. Managing MBCAs after their release in the fields is challenging as they can reproduce and spread necessitating policies for risk management to prevent uncontrollable situations. Comprehensive data on survival, and spread of MBCAs and environmental factors, must be evaluated before commercial release (Köhl et al., 2019). Several prerequisites must be met for a commercial MBCA to gain public recognition or registration. Nonetheless, these regulations can hinder the practical implementation of MBCAs. Regulatory bodies often take a long time for the evaluation process due to a lack of expertise with MBCAs (Sundh & Eilenberg, 2021; El-Saadony et al., 2022).

Using MBCAs to control diseases raises ethical and legal concerns as they could threaten local biodiversity. Close collaboration among government institutes, industry, and universities is essential for BCA registration, enabling the evaluation and marketing of novel BCAs and their products (El-Saadony et al., 2022). Before introducing a BCA into the environment, its potential risks, direct or indirect non-target effects, and impacts on biogeochemical cycles must be thoroughly assessed along with safety issues. The extent of direct non-target effects (NTA) caused by BCAs varies widely, with population-level impacts ranging from negligible to harmful (Hinz et al., 2020). For a BCA to be released into the ecosystem, its impacts on non-target species should be minimal or manageable, alongside achieving successful disease suppression (Winding et al., 2004; Hinz et al., 2020).

The release of BCAs into the environment poses the risk of alterations in the density or distribution of native species, competitive displacement of the population of native natural enemies of the pests, loss of biodiversity, potential disruptions to ecological functions such as pollination due to changes in the balance of native insect species, transmission of pathogens to native species, recombination with closely related non-native species leading to gene dilution, or causation of opportunistic infections in humans (De Clercq et al., 2011). Airborne BCAs and their spores can lead to hypersensitive reactions in factory workers routinely exposed to fungal spores, such as those from *Beauveria* or *Metarhizium* spp. (Sandhu et al., 2012). Some microbes produce a broad range of toxic substances that can affect the growth, development, or behavior of non-target species in the environment. These microbial secondary metabolites can have acute toxicity levels that may be significantly higher than those of antibiotics and fungicides and may also be potent carcinogens (Deising et al., 2017).

The Integrated Pest Management (IPM) principle considers chemical control as a last resort when biological control is impracticable and it also identifies the safety concerns associated with biological control (Collatz et al., 2021). Endophytes could be employed as BCAs if they can be successfully transferred from the laboratory to the field to avoid the harmful impacts of pesticides and chemicals (Tewari et al., 2019). Biological formulations derived from endophytic microorganisms would be non-toxic, biodegradable, non-polluting, and release no carbon traces, thus being safe for soil ecology, the environment, and human health (Tewari et al., 2019). The repetitive use of the same BCAs in the natural soils could lead to substantial selection pressure, resulting in the emergence of resistant pathogen strains (Patra et al., 2024). Changes in environmental conditions could render BCAs ineffective (Lahlali et al., 2022). Understanding the degree of environmental variability, biocontrol activity, and survival of the biocontrol agent in different settings, along with their mechanisms of action, is crucial for predicting biocontrol outcomes (Xu & Jeger, 2020).

Competition can be a promising mode of action for effective microbial biocontrol, provided that the ecological niche of MBCA and the target pathogen is similar for the effective colonization by the MBCA. This will avoid repeated applications of the MBCA (Xu & Jeger, 2020). Biotic and abiotic factors in the rhizosphere environment may hinder the disease-controlling functions of MBCAs (Niu et al., 2020; Legein et al., 2020). The survival and efficacy of microbes can be enhanced by Genetic manipulation. However, ethical implications associated with the release of GMOs pose a major issue, as there may be a horizontal gene transfer, such as those of antibiotic-resistant genes, to local microorganisms, leading to mutations. Further research is necessary to develop effective and safe recombinant BCAs (Gosal et al., 2020) (reviewed in Patra et al., 2024).

17.7 Regulation of Commercial Microbial Biocontrol Agents

The commercialization of biofungicides and biopesticides involves multiple steps and is constrained by several regulations (Lahlali et al., 2022). Similar to synthetic pesticides, MBCAs must undergo risk analyses before they can be authorized for commercialization. European Regulation (EC) No. 1107/2009 outlines the marketing regulations for plant protection products based on a comprehensive risk assessment (Lahlali et al., 2022). Additionally, Regulation (EC) No 540/2011 provides a list of approved microorganisms for biocontrol applications across Europe (Villaverde et al., 2014). According to Article 4(3) of Regulation EC 1107/2009 (2011), a plant protection product must meet several criteria: it must be sufficiently effective; it must not pose immediate or delayed risks to human or animal health; it must not have unacceptable negative effects on plants or plant products; it must not cause suffering to vertebrates; and it must not adversely impact the environment and ecosystem. The approval process for commercial biopesticides is extensive, involving numerous evaluations, including toxicological and environmental tests as well as efficacy assessments (Villaverde et al., 2014). The present-day trend in the direction of simplifying regulatory procedures for low-risk substances to reduce chemical pesticide use may facilitate the production and utilization of MBCAs globally.

17.8 Future Prospects

The application of microbial biocontrol agents (MBCAs) in agriculture is anticipated to expand significantly in the coming years as they are increasingly being recognized as safer alternatives or supplements to chemical products (Palmieri et al., 2022). Research focusing on the selection, characterization, and commercialization of MBCAs has surged (Palmieri et al., 2022). Accelerated efforts in research and development in augmentative and classical biocontrol are underway globally. Focused future efforts on identifying novel MBCAs and developing rapid, comprehensive screening methods to evaluate numerous candidates (Bonaterra et al., 2022).

Detailed studies using comparative genomics, transcriptomics, metabolomics, and proteomics on model MBCAs will provide a robust framework for understanding their biological processes and enhancing their beneficial effects (Bonaterra et al., 2022). In the latest research conducted by Sharma et al., genomic characterization of a gram-negative cocci, *Stenotrophomonas maltophilia* BCM showed the existence of genes linked to plant growth promotion, nutrient assimilation, and antimicrobial activity (Sharma et al., 2024). The study also confirmed that *S. maltophila* BCM effectively protects the wheat seedlings from fungal infections caused by *Rhizoctonia solani* and *Fusarium oxysporum* and greatly enhances seed germination percentage, making it a promising candidate for future biofertilizer and biocontrol application in sustainable agriculture (Sharma et al., 2024). Extensive research on the improvement of the strain of *T. harzianum* has been done to enhance its biocontrol activity and more work is underway (Xiao et al., 2023). Future research on other MBCAs is warranted to enhance the desirable characteristics and downregulate the undesirable ones.

Further research is mandated to achieve MBCAs with desirable characteristics such as rapid growth in laboratory conditions, production of diverse bioactive metabolites, plant growth-promoting activity, excellent rhizosphere competence, compatibility with other rhizobacteria/fungi, ecological safety, and resilience to abiotic stresses (Lyu et al., 2020). Effective disease management with plant growth-promoting rhizobacteria-based biocontrol (PGPR) depends on their survival and colonization in the natural soils.

The formulation, shipping, and storage conditions, the procedure of field application, etc., significantly impact PGPR survival and stability (El-Saadony et al., 2022). To achieve high levels of BCA acceptance, advancements in formulation technology, increased product shelf life, maximization of the production of specific microbial types, and ensuring low cost are desirable. Currently the high production costs and regulatory expenses for commercializing new BCAs present significant challenges. Farmers may perceive little economic benefit compared to chemical pesticides, that more consistent and predictable. Awareness programs, including local workshops, training sessions, and free seminars, could expand the use of BCAs in agriculture (El-Saadony et al., 2022).

Commercially available microbes as biofertilizers and BCAs are sometimes less effective in fields than in laboratory or greenhouse experiments and their efficacy against multiple targets can be unsatisfactory, limiting their usage (Gosal et al., 2020; Jaiswal et al., 2022). Introducing non-native microbes may lead to problems if the organisms become invasive, causing outbreaks or ecological imbalances. The use of biocontrol agents and biofertilizers alters the soil microbiome, which needs to be analyzed through large-scale DNA sequencing-based metabarcoding studies. However, most species detected by these analyses have not been functionally defined, making it challenging to understand the implications of microbiome changes (Collatz et al., 2021). Ensuring the safety and effectiveness of biocontrol agents requires considering intricate interactions among microbes, plants, and the environment. A cellular, molecular, and genetic understanding of biocontrol agents can help address associated safety issues (Zehra et al., 2021).

17.9 Conclusion

Biological control, using microorganisms is an effective and sustainable method for managing plant diseases. MBCAs employ several mechanisms to inhibit pests in plants. Numerous sustainable and environmental friendly biocontrol agents have been produced and commercialized. The requirement for MBCAs and their products is expected to rise in the coming years to meet increasing food requirements mandates further research in the area. Various abiotic and biotic factors influence the interactions between plants, pathogens, and MBCAs and determine the efficacy of a particular MBCA in any environment. We discussed relevant well-studied cases of biocontrol and recent advances and challenges in developing these agents. Additionally, future directions for the development of MBCAs are outlined.

Large-scale production and commercialization are hindered by the need for translational research on potent BCAs. Though there are several success stories of microbial biocontrol, research is required for the identification and screening of novel BCAs using reliable and robust high-throughput screening methods. Furthermore, detailed studies of model BCAs through comparative omics and thorough analysis of their mode of action, along with the development of strategies to improve their beneficial effects and their safety profiling could provide a valuable framework for a sustainable future.

References

Abo-Elyousr, K. A. M., & Hassan, S. A. (2021). Biological control of *Ralstonia solanacearum* (Smith), the causal pathogen of bacterial wilt disease by using *Pantoea spp*. *Egyptian Journal of Biological Pest Control, 31*, 113. https://doi.org/10.1186/s41938-021-00460-z

Adetunji, C., Kumar, D., Raina, M., Arogundade, O., & Sarin, N. (2019). *Endophytic microorganisms as biological control agents for plant pathogens: A panacea for sustainable agriculture*. https://doi.org/10.1007/978-3-030-26657-8_1

Afoshin, A. S., Kudryakova, I. V., Borovikova, A. O., et al. (2020). Lytic potential of *Lysobacter capsici* VKM B-2533T: Bacteriolytic enzymes and outer membrane vesicles. *Scientific Reports, 10*, 9944. https://doi.org/10.1038/s41598-020-67122-2

Albert, D., Dumonceaux, T., Carisse, O., Beaulieu, C., & Filion, M. (2022). Combining desirable traits for a good biocontrol strategy against *Sclerotinia sclerotiorum*. *Microorganisms, 10*(6), 1189. https://doi.org/10.3390/microorganisms10061189

Alfiky, A., & Weisskopf, L. (2021). Deciphering trichoderma–plant–pathogen interactions for better development of biocontrol applications. *Journal of Fungi, 7*(1), 61. https://doi.org/10.3390/jof7010061

Anderson, A. J., & Kim, Y. C. (2018). Biopesticides produced by plant-probiotic Pseudomonas chlororaphis isolates. *Crop Protection, 105*, 62–69, ISSN 0261-2194. https://doi.org/10.1016/j.cropro.2017.11.009

Arakere, U. C., Jagannath, S., Krishnamurthy, S., Chowdappa, S., & Konappa, N. (2022). Microbial bio-pesticide as sustainable solution for management of pests. In *Biopesticides* (pp. 183–200). Elsevier. https://doi.org/10.1016/b978-0-12-823355-9.00016-x

Aw, K. M. S., & Hue, S. M. (2017). Mode of infection of *Metarhizium spp*. fungus and their potential as biological control agents. *Journal of Fungi (Basel, Switzerland), 3*(2), 30. https://doi.org/10.3390/jof3020030

Azevedo, J. L., Araújo, W. L., & Lacava, P. T. (2016). The diversity of citrus endophytic bacteria and their interactions with Xylella fastidiosa and host plants. *Genetics and Molecular Biology, 39*(4), 476–491. https://doi.org/10.1590/1678-4685-GMB-2016-0056

Bejarano, A., & Puopolo, G. (2020). Bioformulation of microbial biocontrol agents for a sustainable agriculture. In A. De Cal, P. Melgarejo, & N. Magan (Eds.), *How research can stimulate the development of commercial biological control against plant diseases. Progress in biological control* (Vol. 21). Springer. https://doi.org/10.1007/978-3-030-53238-3_16

Bonaterra, A., Badosa, E., Daranas, N., Francés, J., Roselló, G., & Montesinos, E. (2022). Bacteria as biological control agents of plant diseases. *Microorganisms, 10*(9), 1759. https://doi.org/10.3390/microorganisms10091759

Calvo-Garrido, C., Roudet, J., Aveline, N., Davidou, L., Dupin, S., & Fermaud, M. (2019). Microbial antagonism toward Botrytis bunch rot of grapes in multiple field tests using one Bacillus ginsengihumi strain and formulated biological control products. *Frontiers in Plant Science, 10*, 105.

Chen, M., Liu, Q., Gao, S.-S., Young, A., Jacobsen, S., & Tang, Y. (2019). Genome mining and biosynthesis of a polyketide from a biofertilizer fungus that can facilitate reductive iron assimilation in plant. *Proceedings of the National Academy of Sciences., 116*, 201819998. https://doi.org/10.1073/pnas.1819998116

Chowdhury, S. P., Hartmann, A., Gao, X., & Borriss, R. (2015). Biocontrol mechanism by root-associated Bacillus amyloliquefaciens FZB42 – A review. *Frontiers in Microbiology, 6*, 780. https://doi.org/10.3389/fmicb.2015.00780

Collatz, J., Hinz, H., Kaser, J. M., & Freimoser, F. M. (2021). Benefits and risks of biological control. In *Biological control: Global Impacts, challenges and future directions of pest management* (pp. 142–165). CSIRO Publishing.

Crovadore, J., Xu, S., Chablais, R., Cochard, B., Lukito, D., Calmin, G., & Lefort, F. (2017). Metagenome-assembled genome sequence of Rhodopseudomonas palustris strain ELI 1980, commercialized as a biostimulant. *Genome Announcements, 5*(18), e00221–e00217. https://doi.org/10.1128/genomeA.00221-17

Daranas, N., Roselló, G., Cabrefiga, J., Donati, I., Francés, J., Badosa, E., Spinelli, F., Montesinos, E., & Bonaterra, A. (2019). Biological control of bacterial plant diseases with Lactobacillus plantarum strains selected for their broad-spectrum activity. *Annals of Applied Biology, 174*(1), 92–105. https://doi.org/10.1111/aab.12476

De Clercq, P., Mason, P. G., & Babendreier, D. (2011). Benefits and risks of exotic biological control agents. *BioControl, 56*, 681–698. https://doi.org/10.1007/s10526-011-9372-8

De Simone, N., Capozzi, V., de Chiara, M. L. V., Amodio, M. L., Brahimi, S., Colelli, G., Drider, D., Spano, G., & Russo, P. (2021). Screening of lactic acid bacteria for the bio-control of Botrytis cinerea and the potential of *Lactiplantibacillus plantarum* for eco-friendly preservation of fresh-cut kiwifruit. *Microorganisms, 9*(4), 773. https://doi.org/10.3390/microorganisms9040773

Deising, H. B., Gase, I., & Kubo, Y. (2017). The unpredictable risk imposed by microbial secondary metabolites: How safe is biological control of plant diseases? *Journal of Plant Diseases and Protection, 124*, 413–419. https://doi.org/10.1007/s41348-017-0109-5

El-Saadony, M. T., Saad, A. M., Soliman, S. M., Salem, H. M., Ahmed, A. I., Mahmood, M., El-Tahan, A. M., Ebrahim, A. A. M., Abd El-Mageed, T. A., Negm, S. H., Selim, S., Babalghith, A. O., Elrys, A. S., El-Tarabily, K. A., & AbuQamar, S. F. (2022). Plant growth-promoting microorganisms as biocontrol agents of plant diseases: Mechanisms, challenges and future perspectives. *Frontiers in Plant Science, 13*, 923880. https://doi.org/10.3389/fpls.2022.923880

Fan, B., Li, L., Chao, Y., Förstner, K., Vogel, J., Borriss, R., & Wu, X. Q. (2015). dRNA-Seq reveals genomewide TSSs and noncoding RNAs of plant beneficial rhizobacterium Bacillus amyloliquefaciens FZB42. *PLoS One, 10*(11), e0142002. https://doi.org/10.1371/journal.pone.0142002. PMID: 26540162; PMCID: PMC4634765.

Fan, B., Wang, C., Song, X., Ding, X., Wu, L., Wu, H., Gao, X., & Borriss, R. (2018). *Bacillus velezensis* FZB42 in 2018: The gram-positive model strain for plant growth promotion and biocontrol. *Frontiers in Microbiology, 9*, 2491. https://doi.org/10.3389/fmicb.2018.02491

Fernandez-San Millan, A., Larraya, L., Farran, I., Ancin, M., & Veramendi, J. (2021). Successful biocontrol of major postharvest and soil-borne plant pathogenic fungi by antagonistic yeasts. *Biological Control, 160*, 104683, ISSN 1049-9644. https://doi.org/10.1016/j.biocontrol.2021.104683

Gajera, H., Domadiya, R., Patel, S., Kapopara, M., & Golakiya, B. (2013). Molecular mechanism of Trichoderma as bio-control agents against phytopathogen system–a review. *Current Research in Microbiology and Biotechnology, 1*(4), 133–142.

Gerbore, J., Brutel, A., Lemainque, A., Mairey, B., Médigue, C., Vallenet, D., Lefort, F., & Grizard, D. (2016). Complete genome sequence of Bacillus methylotrophicus strain B25, a potential plant growth-promoting rhizobacterium. *Genome Announcements, 4*(2), e00058-16. https://doi.org/10.1128/genomeA.00058-16

Ghazanfar, M. U., Raza, M., Raza, W., & Qamar, M. I. (2018). Trichoderma as potential biocontrol agent, its exploitation in agriculture: A review. *Plant Protection, 2*(3).

Ghorbanpour, M., Omidvari, M., Abbaszadeh-Dahaji, P., Omidvar, R., & Kariman, K. (2018). Mechanisms underlying the protective effects of beneficial fungi against plant diseases. *Biological Control, 117*, 147–157.

Gosal, S. K., Kaur, J., & Kaur, J. (2020). Microbial biotechnology: A key to sustainable agriculture. In *Phyto-microbiome in stress regulation* (pp. 219–243). Springer.

Grzywacz, D. (2017). Basic and applied research: Baculovirus. In *Microbial control of insect and mite pests* (pp. 27–46). Academic Press.

Haidar, R., Amira, Y., Roudet, J., Marc, F., & Patrice, R. (2021). Application methods and modes of action of Pantoea agglomerans and Paenibacillus sp. to control the grapevine trunk disease-pathogen, Neofusicoccum parvum. *Oeno One, 55*(3), 1–16. https://doi.org/10.20870/oeno-one.2021.55.2.4530

Hammami, R., Oueslati, M., Smiri, M., Nefzi, S., Ruissi, M., Comitini, F., & Sadfi Zouaoui, N. (2022). Epiphytic yeasts and bacteria as candidate biocontrol agents of green and blue molds of citrus fruits. *Journal of Fungi, 8*(8), 818.

Hasan, N., Farzand, A., Heng, Z., Khan, I. U., Moosa, A., Zubair, M., Na, Y., Ying, S., & Canming, T. (2020). Antagonistic potential of novel endophytic Bacillus strains and mediation of plant defense against Verticillium wilt in upland cotton. *Plants, 9*(11), 1438. https://doi.org/10.3390/plants9111438

Hashemi, M., Tabet, D., Sandroni, M., Benavent-Celma, C., Seematti, J., Andersen, C. B., & Grenville-Briggs, L. J. (2022). The hunt for sustainable biocontrol of oomycete plant pathogens, a case study of Phytophthora infestans. *Fungal Biology Reviews, 40*, 53–69.

Heimpel, G. E., & Mills, N. J. (2017). *Biological control*. Cambridge University Press.

Hinz, H. L., Winston, R. L., & Schwarzländer, M. (2020). A global review of target impact and direct nontarget effects of classical weed biological control. *Current Opinion in Insect Science, 38*, 48–54. https://doi.org/10.1016/j.cois.2019.11.006

Hoitink, H. A. J., & Boehm, M. J. (1999). Biocontrol within the context of soil microbial communities: A substrate-dependent phenomenon. *Annual Review of Phytopathology, 37*, 427–446.

Iorizzo, M., Testa, B., Ganassi, S., Lombardi, S. J., Ianiro, M., Letizia, F., Succi, M., Tremonte, P., Vergalito, F., Cozzolino, A., Sorrentino, E., Petrarca, S., De Cristofaro, A., & Coppola, R. (2021). Probiotic properties and potentiality of Lactiplantibacillus plantarum strains for the biological control of chalkbrood disease. *Journal of Fungi (Basel, Switzerland), 7*(5), 379. https://doi.org/10.3390/jof7050379

Islam, S., Akanda, A. M., Prova, A., Islam, M. T., & Hossain, M. M. (2016). Isolation and identification of plant growth promoting rhizobacteria from cucumber rhizosphere and their effect on plant growth promotion and disease suppression. *Frontiers in Microbiology, 6*, 1360. https://doi.org/10.3389/fmicb.2015.01360

Jain, D., Saharan, V., & Pareek, S. (2016). Current status of Bacillus thuringiensis: Insecticidal crystal proteins and transgenic crops. In J. Al-Khayri, S. Jain, & D. Johnson (Eds.), *Advances in plant breeding strategies: Agronomic, abiotic and biotic stress traits*. Springer. https://doi.org/10.1007/978-3-319-22518-0_18

Jaiswal, D. K., Gawande, S. J., Soumia, P. S., Krishna, R., Vaishnav, A., & Ade, A. B. (2022). Biocontrol strategies: An eco-smart tool for integrated pest and diseases management. *BMC Microbiology, 22*(1), 1–5.

Jeffries, C. L., & Walker, T. (2016). Wolbachia biocontrol strategies for arboviral diseases and the potential influence of resident Wolbachia strains in mosquitoes. *Current Tropical Medicine Reports, 3*, 20–25. https://doi.org/10.1007/s40475-016-0066-2

Jing, T., Zhou, D., Zhang, M., Yun, T., Qi, D., Wei, Y., Chen, Y., Zang, X., Wang, W., & Xie, J. (2020). Newly isolated Streptomyces sp. JBS5-6 as a potential biocontrol agent to control Banana Fusarium Wilt: Genome sequencing and secondary metabolite cluster profiles. *Frontiers in Microbiology, 11*, 602591. https://doi.org/10.3389/fmicb.2020.602591

Keswani, C., Singh, S., & Singh, H. (2013). *Beauveria bassiana*: Status, mode of action, applications and safety issues. *Biotech Today, 3*, 16. https://doi.org/10.5958/j.2322-0996.3.1.002

Köhl, J., Booij, K., Kolnaar, R., & Ravensberg, W. J. (2019). Ecological arguments to reconsider data requirements regarding the environmental fate of microbial biocontrol agents in the registration procedure in the European Union. *BioControl, 64*, 469–487.

Krishnan, S. V., Nampoothiri, K. M., Suresh, A., Linh, N. T., Balakumaran, P. A., Pócsi, I., & Pusztahelyi, T. (2024). *Fusarium* biocontrol: Antagonism and mycotoxin elimination by lactic acid bacteria. *Frontiers in Microbiology, 14*, 1260166. https://doi.org/10.3389/fmicb.2023.1260166. PMID: 38235432; PMCID: PMC10791833.

Kumar, V., Singh, A. U., & Jain, R. K. (2012). Comparative efficacy of bioagents as seed treatment for management of Meloidogyne incognita infecting okra. *Nematologia Mediterranea, 40*, 209–211.

Kumar, P., Kamle, M., Borah, R., et al. (2021). Bacillus thuringiensis as microbial biopesticide: Uses and application for sustainable agriculture. *Egyptian Journal of Biological Pest Control, 31*, 95. https://doi.org/10.1186/s41938-021-00440-3

Lahlali, R., Ezrari, S., Radouane, N., Kenfaoui, J., Esmaeel, Q., El Hamss, H., Belabess, Z., et al. (2022). Biological control of plant pathogens: A global perspective. *Microorganisms, 10*(3), 596. MDPI AG. Retrieved from https://doi.org/10.3390/microorganisms10030596

Lanteigne, C., Gadkar, V. J., Wallon, T., Novinscak, A., & Filion, M. (2012). Production of DAPG and HCN by Pseudomonas sp. LBUM300 contributes to the biological control of bacterial canker of tomato. *Phytopathology, 102*(10), 967–973. https://doi.org/10.1094/PHYTO-11-11-0312

Legein, M., Smets, W., Vandenheuvel, D., Eilers, T., Muyshondt, B., Prinsen, E., et al. (2020). Modes of action of microbial biocontrol in the phyllosphere. *Frontiers in Microbiology*, 1619.

Lopes, R., Cerdeira, L., Tavares, G. S., Ruiz, J. C., Blom, J., Horácio, E. C. A., Mantovani, H. C., & Queiroz, M. V. (2017). Genome analysis reveals insights of the endophytic Bacillus toyonensis BAC3151 as a potentially novel agent for biocontrol of plant pathogens. *World Journal of Microbiology & Biotechnology, 33*(10), 185. https://doi.org/10.1007/s11274-017-2347-x

Luo, L., Zhao, C., Wang, E., Raza, A., & Yin, C. (2022). *Bacillus amyloliquefaciens* as an excellent agent for biofertilizer and biocontrol in agriculture: An overview for its mechanisms, 127016, ISSN 0944-5013. *Microbiological Research, 259*. https://doi.org/10.1016/j.micres.2022.127016

Lyu, D., Backer, R., Subramanian, S., & Smith, D. L. (2020). Phytomicrobiome coordination signals hold potential for climate change-resilient agriculture. *Frontiers in Plant Science, 11*, 634. https://doi.org/10.3389/fpls.2020.00634

Ma, J., Wang, C., Wang, H., Liu, K., Zhang, T., Yao, L., Zhao, Z., Du, B., & Ding, Y. (2018). Analysis of the complete genome sequence of *Bacillus atrophaeus* GQJK17 reveals its biocontrol characteristics as a plant growth-promoting rhizobacterium. *BioMed Research International, 2018*, 9473542. https://doi.org/10.1155/2018/9473542

Manoj, R. R. S., Latrofa, M. S., Epis, S., et al. (2021). Wolbachia: Endosymbiont of onchocercid nematodes and their vectors. *Parasites and Vectors, 14*, 245. https://doi.org/10.1186/s13071-021-04742-1

Matías, L., & Jump, A. (2012). Interactions between growth, demography and biotic interactions in determining species range limits in a warming world: The case of Pinus sylvestris. *Forest Ecology and Management., 282*, 10–22. https://doi.org/10.1016/j.foreco.2012.06.053

Medina, A., Mohale, S., Samsudin, N. I. P., Rodriguez-Sixtos, A., Rodriguez, A., & Magan, N. (2017). Biocontrol of mycotoxins: Dynamics and mechanisms of action. In *Current opinion in food science* (Vol. 17, pp. 41–48). Elsevier BV. https://doi.org/10.1016/j.cofs.2017.09.008

Müller-Santos, M., Koskimäki, J. J., Alves, L. P. S., de Souza, E. M., Jendrossek, D., & Pirttilä, A. M. (2021). The protective role of PHB and its degradation products against stress situations in bacteria. *FEMS Microbiology Reviews, 45*(3), fuaa058. https://doi.org/10.1093/femsre/fuaa058

Munene, L., Mugweru, J., & Mwirichia, R. (2023). Management of bacterial wilt caused by Curtobacterium flaccumfaciens pv. flaccumfaciens in common bean (Phaseolus vulgaris) using rhizobacterial biocontrol agents. *Letters in Applied Microbiology, 76*(1), ovac011. https://doi.org/10.1093/lambio/ovac011

Nandini, B., Puttaswamy, H., Saini, R. K., et al. (2021). Trichovariability in rhizosphere soil samples and their biocontrol potential against downy mildew pathogen in pearl millet. *Scientific Reports, 11*, 9517. https://doi.org/10.1038/s41598-021-89061-2

Ngalimat, M. S., Yahaya, R. S. R., Baharudin, M. M. A. A., Yaminudin, S. M., Karim, M., Ahmad, S. A., & Sabri, S. (2021). A review on the biotechnological applications of the operational group Bacillus amyloliquefaciens. *Microorganisms, 9*(3), 614. https://doi.org/10.3390/microorganisms9030614

Niu, B., Wang, W., Yuan, Z., Sederoff, R. R., Sederoff, H., Chiang, V. L., & Borriss, R. (2020). Microbial interactions within multiple-strain biological control agents impact soil-borne plant disease. *Frontiers in Microbiology, 11*, 585404. https://doi.org/10.3389/fmicb.2020.585404

Palmieri, D., Ianiri, G., Del Grosso, C., Barone, G., De Curtis, F., Castoria, R., & Lima, G. (2022). Advances and perspectives in the use of biocontrol agents against fungal plant diseases. *Horticulturae, 8*(7), 577. https://doi.org/10.3390/horticulturae8070577

Pancher, M., Ceol, M., Corneo, P. E., Longa, C. M., Yousaf, S., Pertot, I., & Campisano, A. (2012). Fungal endophytic communities in grapevines (Vitis vinifera L.) respond to crop management. *Applied and Environmental Microbiology, 78*(12), 4308–4317. https://doi.org/10.1128/AEM.07655-11. Epub 2012 Apr 6. PMID: 22492448; PMCID: PMC3370515.

Pandey, P. K., Samanta, R., & Yadav, R. N. S. (2019). Inside the plant: Addressing bacterial endophytes in biotic stress alleviation. *Archives of Microbiology, 201*(4), 415–429. https://doi.org/10.1007/s00203-019-01642-y

Patra, S., Verma, N., Priya, S., Gupta, A., & Gupta, V. (2024). Chapter 10: Microbes as biocontrol agents for sustainable development. In J. Shankar (Ed.), *Microbial approaches for sustainable green technologies* (pp. 196–218). https://doi.org/10.1201/9781003407683-10

Pellan, L., Durand, N., Martinez, V., Fontana, A., Schorr-Galindo, S., & Strub, C. (2020). Commercial biocontrol agents reveal contrasting comportments against two mycotoxigenic fungi in cereals: Fusarium Graminearum and Fusarium Verticillioides. *Toxins (Basel), 12*(3), 152. https://doi.org/10.3390/toxins12030152. PMID: 32121314; PMCID: PMC7150872.

Pelles, F., Sipos, P., Kovács, S., Győri, Z., Pócsi, I., & Pusztahelyi, T. (2021). Biological control and mitigation of aflatoxin contamination in commodities. *Toxins (Basel), 13*(2), 104. https://doi.org/10.3390/toxins13020104. PMID: 33535580; PMCID: PMC7912779.

Pirttilä, A. M., Mohammad Parast Tabas, H., Baruah, N., & Koskimäki, J. J. (2021). Biofertilizers and biocontrol agents for agriculture: How to identify and develop new potent microbial strains and traits. *Microorganisms, 9*(4), 817. https://doi.org/10.3390/microorganisms9040817

Pisarčik, M., Hakl, J., Szabó, O., & Nerušil, P. (2022). Efficacy of Pythium oligandrum on improvement of lucerne yield, root development and disease score under field conditions. *Frontiers in Plant Science, 13*, 1045225. https://doi.org/10.3389/fpls.2022.1045225

Poveda, J. (2021). Trichoderma as biocontrol agent against pests: New uses for a mycoparasite. *Biological Control, 159*, 104634. https://doi.org/10.1016/j.biocontrol.2021.104634

Prasad, D., Singh, R. P., & Singh, R. P. (2023). Biotica research today. *Research Today, 5*(2), 146–148.

Priya, M. S., & Kumar, S. (2006). Dose optimization of *Paecilomyces lilacinus* for the control of Meloidogyne incognita on tomato. *Indian Journal of Nematology, 36*, 27–31.

Rabbee, M. F., Islam, N., & Baek, K. H. (2022). Biocontrol of citrus bacterial canker caused by *Xanthomonas citri* subsp. *citri* by *Bacillus velezensis*. *Saudi Journal of Biological Sciences, 29*(4), 2363–2371. https://doi.org/10.1016/j.sjbs.2021.12.005

Rueda-Páramo, M. E., Montalva, C., Arruda, W., Fernandes, É. K. K., Luz, C., & Humber, R. A. (2017). First report of Coelomomyces santabrancae sp. nov. (Blastocladiomycetes: Blastocladiales) infecting mosquito larvae (Diptera: Culicidae) in central Brazil. *Journal of Invertebrate Pathology, 149*, 114–118. https://doi.org/10.1016/j.jip.2017.08.010

Sabbahi, R., Hock, V., Azzaoui, K., Saoiabi, S., & Hammouti, B. (2022). A global perspective of entomopathogens as microbial biocontrol agents of insect pests. *Journal of Agriculture and Food Research, 10*, 100376.

Saberi Riseh, R., Hassanisaadi, M., Vatankhah, M., Kennedy, J. F. (2023). Encapsulating biocontrol bacteria with starch as a safe and edible biopolymer to alleviate plant diseases: A review. *Carbohydrate Polymers, 302*, 120384. https://doi.org/10.1016/j.carbpol.2022.120384

Sandhu, S., Sharma, A., Beniwal, V., Goel, G., Batra, P., Bhan-Khar, Anil, K., Sundeep, J., & Malhotra, S. (2012). Myco-biocontrol of insect pests: Factors involved, mechanism, and regulation. *Journal of Pathogens, 2012*, 126819. https://doi.org/10.1155/2012/126819

Sani, A., Qin, W. Q., Li, J. Y., Liu, Y. F., Zhou, L., Yang, S. Z., & Mu, B. Z. (2024). Structural diversity and applications of lipopeptide biosurfactants as biocontrol agents against phytopathogens: A review. *Microbiological Research, 278*, 127518. https://doi.org/10.1016/j.micres.2023.127518

Scholte, E. J., Knols, B. G., Samson, R. A., & Takken, W. (2004). Entomopathogenic fungi for mosquito control: A review. *Journal of Insect Science, 4*(1), 19. https://doi.org/10.1673/031.004.1901

Sellitto, V. M., Zara, S., Fracchetti, F., Capozzi, V., & Nardi, T. (2021). Microbial biocontrol as an alternative to synthetic fungicides: Boundaries between pre-and postharvest applications on vegetables and fruits. *Fermentation, 7*(2), 60.

Sharma, V., Sharma, A., Malannavar, A. B., & Salwan, R. (2020). Molecular aspects of biocontrol species of Streptomyces in agricultural crops. In *Molecular aspects of plant beneficial microbes in agriculture* (pp. 89–109). Elsevier. https://doi.org/10.1016/b978-0-12-818469-1.00008-0

Sharma, P., Pandey, R., & Chauhan, N. S. (2024). Biofertilizer and biocontrol properties of *Stenotrophomonas maltophilia* BCM emphasize its potential application for sustainable agriculture. *Frontiers in Plant Science, 15*, 1364807. https://doi.org/10.3389/fpls.2024.1364807

Singh, H. B., Keswani, C., Ray, S., Yadav, S. K., Singh, S. P., Singh, S., & Sarma, B. K. (2015). Beauveria bassiana: Biocontrol beyond lepidopteran pests. *Biocontrol of Lepidopteran Pests, 43*, 219–235. https://doi.org/10.1007/978-3-319-14499-3_10. PMCID: PMC7123603.

Subramaniam, M. S. R., Babu, A., & Deka, B. (2021). Lecanicillium lecanii (Zimmermann) Zare & Gams, as an efficient biocontrol agent of tea thrips, Scirtothrips bispinosus Bagnall (Thysanoptera: Thripidae). *Egyptian Journal of Biological Pest Control, 31*, 38. https://doi.org/10.1186/s41938-021-00380-y

Sui, Y., Wisniewski, M., Droby, S., Piombo, E., Wu, X., & Yue, J. (2020). Genome sequence, assembly, and characterization of the antagonistic yeast *Candida oleophila* used as a biocontrol agent against post-harvest diseases. *Frontiers in Microbiology, 11*, 295. https://doi.org/10.3389/fmicb.2020.00295. PMID: 32158440; PMCID: PMC7052047.

Sujayanand, G. K., Akram, M., Konda, A., et al. (2021). Distribution and toxicity of Bacillus thuringiensis (Berliner) strains from different crop rhizosphere in Indo-Gangetic plains against polyphagous lepidopteran pests. *International Journal of Tropical Insect Science, 41*, 2713–2731. https://doi.org/10.1007/s42690-021-00451-5

Sundh, I., & Eilenberg, J. (2021). Why has the authorization of microbial biological control agents been slower in the EU than in comparable jurisdictions? *Pest Management Science, 77*(5), 2170–2178.

Tariq, M., Khan, A., Asif, M., Khan, F., Ansari, T., Shariq, M., & Siddiqui, M. A. (2020). Biological control: A sustainable and practical approach for plant disease management. *Acta Agriculturae Scandinavica, Section B: Soil & Plant Science, 70*(6), 507–524. https://doi.org/10.1080/09064710.2020.1784262

Tewari, S., Shrivas, V. L., Hariprasad, P., & Sharma, S. (2019). Harnessing endophytes as biocontrol agents. In *Plant health under biotic stress: Volume 2: Microbial interactions* (pp. 189–218). Springer.

Tomprefa, N., McQuilken, M. P., Hill, R. A., & Whipps, J. M. (2009). Antimicrobial activity of Coniothyrium minitans and its macrolide antibiotic macrosphelide A. *Journal of Applied Microbiology, 106*(6), 2048–2056. https://doi.org/10.1111/j.1365-2672.2009.04174.x

Tyśkiewicz, R., Nowak, A., Ozimek, E., & Jaroszuk-Ściseł, J. (2022). Trichoderma: The current status of its application in agriculture for the biocontrol of fungal phytopathogens and stimulation of plant growth. *International Journal of Molecular Sciences, 23*(4), 2329. https://doi.org/10.3390/ijms23042329

van Lenteren, J. C., Bolckmans, K., Köhl, J., Ravensberg, W. J., & Urbaneja, A. (2018). Biological control using invertebrates and microorganisms: Plenty of new opportunities. *BioControl, 63*, 39–59.

Villaverde, J. J., Sevilla-M, B., Sandín-E, P., López-G, C., & Alonso-P, J. L. (2014). Biopesticides in the framework of the European Pesticide Regulation (EC) No. 1107/2009. *Pest Management Science, 70*(1), 2–5. https://doi.org/10.1002/ps.3663. PMID: 24174346.

Vlassi, A., Nesler, A., Perazzolli, M., Lazazzara, V., Büschl, C., Parich, A., Puopolo, G., & Schuhmacher, R. (2020). Volatile organic compounds from Lysobacter capsici AZ78 as potential candidates for biological control of soilborne plant pathogens. *Frontiers in Microbiology, 11*, 1748. https://doi.org/10.3389/fmicb.2020.01748

Wang, J., Xing, J., Lu, J., Sun, Y., Zhao, J., Miao, S., Xiong, Q., Zhang, Y., & Zhang, G. (2019). Complete genome sequencing of Bacillus velezensis WRN014, and comparison with genome sequences of other Bacillus velezensis strains. *Journal of Microbiology and Biotechnology, 29*(5), 794–808. https://doi.org/10.4014/jmb.1901.01040

Winding, A., et al. (2004). Non-target effects of bacterial biological control agents suppressing root pathogenic fungi. *FEMS Microbiology Ecology, 47*(2), 129–141. https://doi.org/10.1016/S0168-6496(03)00261-7

Xiao, Z., Zhao, Q., Li, W., Gao, L., & Liu, G. (2023). Strain improvement of Trichoderma harzianum for enhanced biocontrol capacity: Strategies and prospects. *Frontiers in Microbiology, 14*, 1146210. https://doi.org/10.3389/fmicb.2023.1146210

Xu, X., & Jeger, M. (2020). More ecological research needed for effective biocontrol of plant pathogens. In *How research can stimulate the development of commercial biological control against plant diseases* (pp. 15–30). Springer International Publishing.

Yang, F., Jiang, H., Ma, K., Hegazy, A., Wang, X., Liang, S., Chang, G., Yu, L., Tian, B., & Shi, X. (2024). Genomic and phenotypic analyses reveal *Paenibacillus polymyxa* PJH16 is a potential biocontrol agent against cucumber fusarium wilt. *Frontiers in Microbiology, 15*, 1359263. https://doi.org/10.3389/fmicb.2024.1359263. PMID: 38591040; PMCID: PMC11000672.

Yao, X., Zhang, Z., Huang, J., Wei, S., Sun, X., Chen, Y., Liu, H., & Li, S. (2021). Candicidin isomer production is essential for biocontrol of cucumber *Rhizoctonia* rot by Streptomyces albidoflavus W68. *Applied and Environmental Microbiology, 87*(9), e03078-20. https://doi.org/10.1128/AEM.03078-20

Zehra, A., Raytekar, N. A., Meena, M., & Swapnil, P. (2021). Efficiency of microbial bio-agents as elicitors in plant defense mechanism under biotic stress: A review. *Current Research in Microbial Sciences, 2*, 100054.

Zhao, Y., Chen, X., Cheng, J., Xie, J., Lin, Y., Jiang, D., Fu, Y., & Chen, T. (2022). Application of *Trichoderma* Hz36 and Hk37 as biocontrol agents against clubroot caused by *Plasmodiophora brassicae*. *Journal of Fungi (Basel, Switzerland), 8*(8), 777. https://doi.org/10.3390/jof8080777

Soil Bacteria: An Invisible Army for Climate Resilience and Mitigation

18

Ashwini A. Waoo, Charu Vyas, and Balu Ananda Chopade

Abstract

Soil bacteria, often unseen yet integral to the ecosystem, form an invisible army crucial for climate resilience and mitigation. These microscopic organisms play multifaceted roles, from sequestering carbon and nitrogen fixation to enhancing soil structure and nutrient availability. Their pivotal function extends to mitigating climate change impacts by aiding in carbon storage and reducing greenhouse gas emissions through various metabolic processes. Moreover, certain soil bacteria exhibit resilience to changing environmental conditions. Exploring sustainable agricultural practices and biotechnological interventions offers a promising approach to enhancing climate resilience and mitigating ecological challenges. Research on soil microbes' role in countering climate change effects and understanding their response in diverse ecosystems is crucial for enabling plants to withstand evolving stressors and minimizing adverse impacts. Soil microbes that make them significant in agriculture must be investigated, especially in light of the effects of changing climatic scenarios on crop production and the need to eliminate numerous problems related to the use of pesticides in this sector. Reviewing patents in a particular field is of paramount importance, thus, a thorough description of a patent that was published in the field of microorganisms' role in climate resilience and mitigation was added to the concluding section of this chapter.

Keywords

Bacteria · Climate change · Resilience · Soil microbes · Stress · Sustainable agriculture

A. A. Waoo · C. Vyas · B. A. Chopade (✉)
Department of Biotechnology, Faculty of Life Science and Technology, AKS University, Satna, MP, India

M. Saraf et al. (eds.), *Climate Change and Soil Microorganisms for Environmental Sustainability*, Microorganisms for Sustainability 28,
https://doi.org/10.1007/978-981-96-3425-5_18

18.1 Introduction

Climate resilience entails dealing with the consequences of climate change and proactive planning and adaptation to a changing environment. The light-changing climate entails a combination of individual activities, community efforts, and government policies to create a more resilient and sustainable future. Mitigation might take the form of implementing new technologies and renewable energies, making aging equipment more energy efficient, or altering management procedures or customer behavior.

Reducing the greenhouse gas emissions that are warming the earth is necessary to mitigate climate change. Energy-efficient building retrofitting, the use of renewable energy sources like solar, wind, and small hydro, supporting cities in creating more environmentally friendly transportation options like bus rapid transit, biofuels, and electric vehicles, and encouraging more environmentally friendly land and forest management are some examples of mitigation strategies. Respect for the environment, flora and wildlife, and environmental defense are critical in combating climate change (Doring et al., 2015).

Climate change poses a significant danger to ecosystem sustainability and agricultural crop production. Such conditions necessitate the application of enhanced farming techniques, and ecologically friendly, and climate-resilient technologies. Focusing our research efforts on important issues facing our planet, understanding and predicting how soil microbiomes and the ecosystem services they provide will be impacted by climate change is both a great problem and an immense opportunity.

Microorganisms are the essential players in controlling the Earth's biogeochemical cycles because they are persistent and profusion in the soil ecosystem. The soil microbiomes of the soil regulate the biogeochemical cycling of carbon, nitrogen, and other elements, which is essential for plant and animal growth. By shielding plants from diseases and damaging pests, promoting plant growth, and reducing nutritional and environmental stressors, these microbes may contribute to sustainable crop production by enabling plants to adapt to the different biotic and abiotic stress conditions brought on by the changing climate scenario and reducing its adverse effects (Saini & Sharma, 2019).

Because of rising concerns about climate change, it is critical to efficiently harness beneficial soil bacteria and their interactions to improve agro system resilience to climate change, soil fertility, and health, hence boosting the production of crops. Various qualities of soil microbes that make them significant in agriculture must be investigated, especially in light of the effects of changing climatic scenarios on crop production and the need to eliminate numerous problems related to pesticide use in this sector.

The immense potential of these bacteria is acknowledged, and the scientific community worldwide is engaged in considerable research toward the commercialization and selection of biotechnological and environmentally relevant microbes. Research is needed to determine how soil microbes are affected by climate change in different soil ecosystems that are susceptible to it, and potential strategies for using them to lessen the negative impact of climate change (Jansson & Hofmockel, 2020).

18.2 Climate Change Resilience and Mitigation

Climate resilience is defined as a system, community, or society's ability to predict, prepare for, respond to, recover from, and adapt to climate change and its effects. As the Earth's climate continues to change as a consequence of human activities such as greenhouse gas emissions, it has become critical for individuals, communities, and governments to create resilience to meet the difficulties provided by these changes. The climate crisis is becoming more alarming. Fortunately, there are numerous things we can do to safeguard our future prosperity. These actions are classified as either climate change adaptation or climate change mitigation. These phrases are used interchangeably when discussing climate catastrophe, although they have quite different meanings.

Fossil fuel consumption must be replaced with clean, renewable energy. Until we achieve zero carbon emissions—which suggests that greenhouse gas emissions into the atmosphere are balanced by absorbing and storing those gases in places like tree roots—we must stop deforestation and restore natural habitats. In the long run, we will benefit more by acting sooner to lessen the climate change effects. Though progress has been gradual so far, things are starting to change (Kumar et al., 2012).

The earlier emissions are reduced, the simpler it will be to adapt to the unavoidable changes. The long-term effects of mitigation measures on rising temperatures will be felt shortly, therefore we need to start preparing for this change now that it has arrived.

Countries with high emissions from agriculture, forestry, and fishing are experiencing a decline in natural resource scarcity. Policymakers can promote sustainability, mitigate climate change, and ensure long-term agricultural viability by addressing value addition in these sectors. Strategic interventions, including technology transfer, capacity building, and market-based approaches, can help achieve a more sustainable future while managing resource challenges (Wang & Azam, 2024).

So, mitigation can help reduce climate change so that we do not have increasingly severe storms, adaptation can assist us in preparing for the more severe storms that we are already experiencing, and resilience can help us recover more rapidly.

18.3 Key Aspects of Climate Resilience

Climate resilience involves enhancing adaptive capacities, mitigating risks, and fostering sustainable strategies to withstand and recover from climate change impacts. It encompasses community engagement, infrastructure resilience, and ecosystem preservation to ensure societal and environmental stability in light of climatic challenges (Schimel et al., 2007).

18.3.1 Adaptation

Climate-resilient systems are designed to adapt to changing climates and their consequences. This may entail putting in place safeguards to protect infrastructure, agriculture, and ecosystems against extreme weather events, rising sea levels, and other climate-related risks.

18.3.2 Risk Reduction

Climate change resilience projects aim to lessen the risks associated with climate change by identifying vulnerabilities and implementing mitigation measures. Building flood defenses, adopting drought-resistant farming practices, or engineering buildings to endure stronger storms are examples of such measures (Philippot & Hallin, 2011).

18.3.3 Preparedness and Early Warning Systems

It is critical to be prepared for climate-related calamities. Creating early warning systems can provide early warning of potential risks, allowing for prompt evacuation and response operations.

18.3.4 Infrastructure Resilience

It is critical to ensure that infrastructure is planned and developed to withstand climate-related pressures. Ensuring climate-resilient infrastructure is crucial. This involves building flood-resistant roads and bridges, implementing green infrastructure for stormwater management, and using adaptive materials. Sustainable urban planning, renewable energy investment, and advanced technologies further enhance resilience, ensuring long-term functionality and minimizing environmental impact.

18.3.5 Social and Community Resilience

Social networks and community cohesion can help with climate resilience. Communities that collaborate are better prepared to respond to and recover from climate-related disasters.

18.3.6 Economic Resilience

A resilient economy can resist and recover from shocks caused by climate change. Diversifying industries to lessen reliance on climate-sensitive sectors, for example, or establishing sustainable corporate practices, could be part of this.

18.3.7 Ecosystem Resilience

Healthy, well-managed ecosystems can improve climate resilience by providing services like water filtering, flood prevention, and biodiversity preservation.

18.3.8 Governance and Policy

Climate resilience requires effective policies and governance frameworks. This includes creating and implementing policies governing land use, disaster management, and pollution reduction.

18.4 Role of Microbial Communities

The populations of soil microbes are always changing due to shifting resource availability. While some grow slowly and require more chemically complex substrates, others grow quickly to take advantage of available resources. In general, as environmental conditions change, the resident microorganisms adapt, go dormant, or perish. The degree of disturbance and time required to accumulate mutations, regulate gene transcription and translation, and/or accumulate genes via horizontal gene transfer determine the ability to adapt. Because soil habitats are dynamic systems, most soil microbes have evolved phenotypic methods to adapt to changing environmental conditions. Microorganisms typically reallocate resources from growth strategies to survival strategies in response to the stress of climate change (Schimel et al., 2007).

Nevertheless, very little is known about the fundamental mechanisms at the microbiological scale that control the responses of ecosystems to climate change. Research on how altered physicochemical conditions impact the metabolism of native soil microbial communities is crucial to lowering greenhouse gas emissions in the atmosphere. The two broad ideas that regulate the soil microbiome's response to change are soil habitat connectivity and the combined physiological status of soil microbial community members (i.e., metaphenome), followed by discussions of microbial community and physiological responses to specific climate variations. There is a rising need to find ways to use soil microbial capacity to mitigate the adverse consequences of environmental perturbations as we become more aware of how climate change affects the soil microbiome.

The concept of resilience has become more significant in semi-natural and agricultural settings due to climate change. This includes altering the functions of soil microbial populations directly or indirectly through altering land management practices or by using inoculants as probiotics in the environment.

Resilience is the ability or a system's capacity or person to react (respond) to an outside force (disturbance) while satisfying certain extra requirements after the reaction (outcome). The Latin verb resilience means "to jump back," which is where the word "resilience" originated. The extent to which soil organisms are harmed after stress is known as the soil system's resistance and the speed and scope of recovery are seen as indicators of the system's resilience. At all agricultural levels and domains, it has been characterized as a pertinent and dynamic health criterion (Doring et al., 2015). Soil is an ideal medium for plant and microorganism growth and development. Understanding of the biological underpinnings of soil function resistance and resilience is expanding. Indeed, resistance and resilience have been linked to communities of microbes and the features of indigenous soil microorganisms (Griffiths & Philippot, 2013). Soil microbial communities are an ideal tool to examine resilience since their response to disturbances can be quite fast, with responses occurring within days or weeks (Cregger et al., 2012; Griffiths & Philippot, 2013).

According to Allison and Martiny (2008), a community may be resilient and may as quickly return to its pre-disturbance composition even while the microbial composition is susceptible to disturbance. Numerous traits of microorganisms, such as fast growth rates, horizontal gene exchange, or quick evolution through mutations, could be to blame for this. Microbial communities may therefore be among the elements of an ecosystem that adapt to changing environmental conditions the quickest due to their very short generation times and rapid development under favorable conditions. Because of their wide range of functional and genetic variations, potential for rapid evolution, and enormous dispersal abilities, microbes may be able to mitigate environmental change reactions.

The composition of microbial communities can be resistant to changes in climate as well as variations in plant productivity species composition (Cruz-Martinez et al., 2009). Among fungi, fungal communities and bacteria have shown the ability to adapt to changing conditions dynamically without losing variety (Yuste et al., 2011). Unlike bacterial diversity, fungal diversity was less responsive to seasonal fluctuations in temperature, moisture, and plant activity.

18.5 Benefits of Microbes

A basic requirement for microbial inoculation in agricultural practices is the loss of topsoil, sterility of the soil, low yield index, stunted plant growth, and inadequate diversity of native bacteria. Utilizing plant-microbe interactions would not only aid in climate change mitigation but will also help to develop a green economy and achieve economic stability. It leads to the development of plant cultivars capable of growing under the stress of a warming environment and increased CO_2 levels (Philippot & Hallin, 2011).

Beneficial soil bacteria can support a range of ecosystem activities and processes while reducing dependency on synthetic fertilizers and pesticides, which can have a substantial positive socioeconomic impact on the world economy. Thus, by using beneficial bacteria to modify the soil environment, it is possible to boost nutrient availability for high-yielding, high-quality crops; shield crops from pests, diseases, and weeds; and control other variables affecting output, ecosystem service delivery, and resistance to shocks such as droughts (Lehman et al., 2015).

Various soil microorganisms, like plant growth-promoting rhizobacteria (PGPR), have developed as an environmentally benign method of successfully increasing plant development under biotic and abiotic stress situations.

Additionally, it has been demonstrated that a variety of microbial inoculants can decrease the toxicity of heavy metals, aid in nitrogen fixation, facilitate nutrient transformation, produce siderophores that assist the plant in obtaining sufficient amounts of iron, synthesize indole acetic acid (IAA) and other plant hormones, and give protection against a variety of pathogens (Olanrewaju et al., 2017; Mohanty & Swain, 2018).

A key factor in the effective cycling of nutrients and the growth of plants in farmed soils is the activity of soil microorganisms (Kumar et al., 2012). These PGPRs are being developed as effective biofertilizers for commercial applications. Reducing chemical load while maintaining productivity can be gained by using more natural bacteria as biofertilizers, biopesticides, biofungicides, and other goods. Moreover, adding farm yard manure to the soil improves its resistance to heat stress and enhances its tolerance, while applying only nitrogen fertilizer may lessen the contribution of soil bacteria to plant nutrition.

For plants to grow and develop, bacteria increase the availability of macronutrients like N, P, and K as well as micronutrients like Fe. Plant N consumption and N cycling depend heavily on bioinoculants. According to Gupta et al. (2012), bioinoculants are essential for plant N consumption and N cycling in the soil. The proper quantity and makeup of various macronutrients, micronutrients, and trace elements in the soil are essential for plant growth and health (Hodges, 2010). To influence plant development and productivity, these nutrients must also exist in a physiologically accessible form. For instance, even though the soil contains large amounts of sulfate and phosphate, plants only absorb the soluble ionic form of these nutrients by a variety of processes; leaving the other forms of these minerals unused (Solomon et al., 2003; White & Hammond, 2008).

According to Sahu et al. (2018), microorganisms play a critical role in the mobilization and absorption of nutrients by plants. It has been demonstrated that arbuscular mycorrhizal fungus (AMF) and plant growth-promoting rhizobacteria (PGPR) enhance the nutrition, growth, and health of plants. The overuse of pesticides to increase agricultural productivity has severely disrupted the balance of the microbial community of the soil. As a result, nature's geochemical cycling has been disrupted, resulting in pollution. Their use was frequently advocated to reduce chemical input in agriculture. Microorganisms as bioinoculants are a preferable alternative to chemicals for increasing crop production and quality and restoring soil nutrients for agricultural sustainability and the environment (Alori & Babalola, 2018).

18.6 Role of Bacteria in Climate Resilience and Mitigation Efforts

Soil bacteria help in supporting climate resilience and mitigation efforts. They are a member of the enormous and diverse population of microorganisms found in soil known as the soil microbiome. These microscopic organisms have a considerable impact on soil health and ecosystem functioning, and their actions have the potential to influence global climate patterns in a variety of ways (Griffiths & Philippot, 2013). These microscopic organisms form an unseen army that greatly contributes to a variety of ecological processes such as nutrient cycling, carbon sequestration, and plant health. Soil bacteria create an "invisible army" that is critical to the health and fertility of the Earth's soil. These microorganisms are part of the soil's microbiome, which also contains bacteria, fungi, archaea, viruses, and other microbes. Soil bacteria influence soil structure, nutrient cycling, and plant growth, making them critical for ecosystem health and agricultural output (Cregger et al., 2012).

Here is a quick rundown of how soil bacteria influence climate resilience and mitigation and their exploitation, (Shown in Fig. 18.1).

18.6.1 Carbon Sequestration

Soil microorganisms play a crucial role in the carbon cycle. Photosynthesis occurs when plants absorb CO_2 from the atmosphere and transform it into organic carbon, which is subsequently delivered to the soil via their roots. This organic carbon, once in the soil, can be held for an extended length of time, successfully sequestering

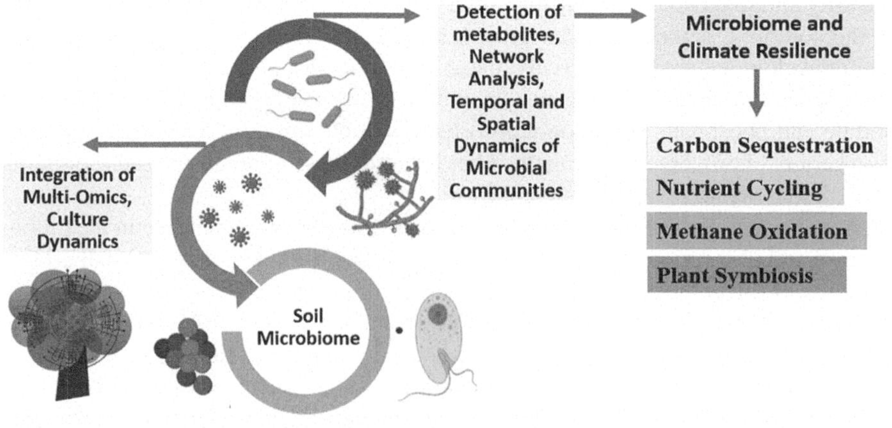

Fig. 18.1 Soil microbiomes for climate resilience studies

carbon and preventing it from being released back into the atmosphere as CO_2, a dangerous greenhouse gas.

Carbon gets trapped into stable, non-gaseous forms via abiotic and/or biotic processes. It is one potential method for lowering atmospheric C stocks. C enters the soil by the assimilation of atmospheric CO_2, primarily by plants and by photoautotrophic bacteria and chemoautotrophic soil microorganisms (Madejón et al., 2007). Up to 20% of photosynthate is directly deposited in the rhizosphere by plants by exudation, and direct root deposition through mycorrhizal fungi (Averill et al., 2014). The symbiotic and free-living organisms that are fostered by carbon inputs spread the carbon throughout the soil matrix. Bioavailable forms of the carbon cycle persist over time as a result of microbial biochemical changes of carbon and the exchange of carbon among communities.

Several approaches are being investigated for utilizing the soil microbiota for carbon storage. To begin with, the soil microbiome possesses a mostly undiscovered and untapped metabolic capacity that might be mined for novel reactions that boost productivity. These reactions could be controlled to direct the carbon breakdown routes toward more resistant and stable end products. Alternatively, indigenous microbial species or consortia with the necessary metabolic pathways could be chosen and then introduced into soil ecosystems to trap C.

Soil C storage can be improved through wise land management (Lal, 2004). Soil management strategies such as no-till farming (Madejón et al., 2007) and diverse crop rotations (King & Hofmockel, 2017) limit soil disturbance and organic matter breakdown, influencing soil C storage. Incorporating hay, fodder, or green manures into cropping systems provides numerous advantages to both the farmer and the environment. These manure or legume-based systems also encourage biologically active soil organic matter (SOM), which can improve soil carbon storage (King & Hofmockel, 2017). Soil microbiomes can also be modified in situ by adding changes that increase their potential to take up and store C in soil. Encourage the generation of stable C from microbial products, which can be stored in deeper soil layers, as one mitigation strategy (Schmidt et al., 2011).

Soil microorganisms can contribute to the creation of soil organic matter by producing table metabolites or by decomposing their own dead biomass (necromass) (Schmidt et al., 2011). Biochar has also been considered as a soil additive for C sequestration (Jansson et al., 2010). Because the extent of microbial activity determines whether biochar is respired or maintained in soil, the soil microbiome is indirectly tied to the promise of biochar.

The optimization of interactions between plants and soil rhizosphere microbiomes is becoming increasingly important to facilitate soil C storage (Jansson et al., 2018; Wallenstein, 2017). Plant genetic manipulation is proposed to produce biomass that is more resistant to microbial breakdown; that is, more lignin maintained in soil. Another approach is to use plant-microbe interactions to improve root exudate deposition by enhancing rhizospheric sink strength and therefore enabling soil C storage (Jansson et al., 2018).

Another possibility is to genetically modify the plant to favor helpful colonization microorganisms. Future solutions may enable us to manage the allocation of photosynthate to optimize the plant-microbe-soil system for optimal plant output as well as soil C deposition (Jansson et al., 2018). Soil microbiologists and plant breeders will need to work cooperatively to establish the greatest combination of specific beneficial microorganisms with specific plant genotypes (Lakshmanan et al., 2017).

Organic stuff, such as dead plants and animals, is decomposed by soil microbes. Carbon dioxide (CO_2) is released into the atmosphere as organic matter decomposes. Some soil bacteria, however, can transform this organic carbon into more stable forms of carbon, such as humus. Humus is a long-lasting organic compound in soil that acts as a carbon sink by storing carbon for lengthy periods. Soil bacteria help decrease CO_2 concentrations in the atmosphere by promoting carbon sequestration, so preventing climate change.

18.6.2 Nutrient Cycling

Soil microorganisms are essential for organic matter decomposition in the soil. They decompose animal remnants and dead plant material, releasing nutrients back into the soil like phosphorus, nitrogen, and sulfur. These nutrients are subsequently made available to plants for usage, enhancing plant growth and increasing ecosystem productivity. Nitrogen-fixing bacteria, for example, transform atmospheric nitrogen into a plant-usable form and production. Proper nutrient cycling improves ecosystem resilience and boosts soil carbon storage capability.

18.6.3 Methane Oxidation

As part of their metabolic operations, some soil bacteria may consume methane, another potent greenhouse gas. These bacteria help to minimize methane emissions from the soil by oxidizing methane into carbon dioxide, which contributes to climate change mitigation.

18.6.4 Plant Symbiosis

Certain soil bacteria, such as nitrogen-fixing bacteria and mycorrhizal fungi create symbiotic associations with plants. Nitrogen-fixing bacteria transform atmospheric nitrogen into a plant's usable form, while mycorrhizal fungus improves the ability of a plant to absorb nutrients from the soil is critical in these interactions to encourage plant growth and health, which helps with carbon sequestration and ecosystem resilience. Soil bacteria help in the formation of soil aggregates, hence improving soil structure (Naylor & Coleman-Derr, 2018).

This improves water retention, reduces erosion risk, and increases ecosystem resilience during droughts and other weather events. Some soil bacteria produce sticky compounds, such as extracellular polymeric substances (EPS), which aid in the bonding of soil particles. This results in more stable soil aggregates, which improves soil structure and reduces erosion. Soil erosion depletes fertile topsoil and releases carbon into the atmosphere, increasing climate change. Soil microorganisms contribute to climate resilience by protecting soil fertility and carbon storage capacity by limiting erosion (Lakshmanan et al., 2017). However, climate change can affect soil microbes. Temperature, precipitation patterns, and land use practices can all have an impact on the composition and functioning of an ecosystem, and soil microbial communities, potentially affecting their contributions to climate resilience and mitigation. It is critical to promote sustainable land management methods that protect soil health, biodiversity, and organic matter content to harness the power of soil bacteria for climate resistance and mitigation. No-till farming, cover cropping, and agroforestry practices can improve soil microbial diversity and activity, resulting in more resilient ecosystems and greater carbon sequestration capacity. We can work toward a more climate-resilient future by knowing and exploiting the potential of this invisible army beneath our feet.

Some soil bacteria operate as natural antagonists to plant pathogens, aiding in the suppression of disease outbreaks that can harm crops and vegetation. This disease control helps to strengthen agricultural systems and natural ecosystems. Soil microorganisms can improve ecosystem resistance to climatic extremes such as droughts and floods. They aid in water retention by increasing soil water-holding capacity during dry periods and aiding drainage following heavy rains (Lakshmanan et al., 2017). Soil microorganisms can help to restore degraded areas. They help in the rehabilitation of ecosystems harmed by human activity or natural disturbances by contributing to, carbon sequestration, nutrient cycling, and soil structure improvement.

Overall, the activity and diversity of soil bacteria are critical for the health and resilience of terrestrial ecosystems, playing an important role in mitigating the climate change impacts. Soil health, particularly bacterial populations, must be protected and promoted to enhance climate resilience and ensure sustainable land management practices.

Certain soil bacteria produce antibiotics or other substances that aid in the suppression of harmful organism growth, thereby protecting plants from disease. Some bacteria synthesize sticky compounds that hold soil particles together, strengthening soil structure and avoiding erosion. This improves water retention and aeration, resulting in a better environment for plant roots. Soil bacteria can break down or degrade a variety of pollutants and toxins, aiding in the detoxification and rehabilitation of polluted places. Soil microbes also contribute to carbon sequestration. They can transform organic carbon from decaying plant and animal matter into more stable forms, assisting in carbon storage.

Soil type, vegetation, climate, and land management practices all have an impact on the total variety and quantity of soil bacteria. Excessive use of chemical fertilizers, and pesticides, as well as intense land cultivation, can alter the equilibrium of soil microorganisms, resulting in decreased soil fertility and overall ecosystem health. Understanding and promoting the roles of soil bacteria is critical for sustainable agriculture environmental protection and maintaining global food security.

Scientists and farmers alike are recognizing the importance of soil microbiomes and implementing strategies that maintain the health of this "invisible army" in our soils.

18.7 Microbes as Beneficial Plant Inoculants

Drought is expected to worsen by climate change, with severe consequences for agricultural productivity. Beneficial plant growth-promoting (PGP) bacteria and fungi found in the rhizosphere may help mitigate the detrimental effects of drought by maximizing plant growth under increasingly harsh conditions (Naylor & Coleman-Derr, 2018). PGP microorganisms can be applied to plants in the field as seed coatings or as liquid or granular supplements. A famous example of a PGP strain is rhizobium inoculants used for biological N fixation in conjunction with legumes. There is currently a rising interest in moving beyond the typical use of inoculants as biofertilizers and biopesticides to harness other beneficial qualities of PGP microorganisms to counteract the negative repercussions.

Several approaches to using beneficial plant-associated soil microorganisms to reduce drought stress in plants are being investigated (Lakshmanan et al., 2017). Soil bacteria, for example, create extracellular polysaccharides (EPS), which result in hydrophobic biofilms that can shield plants from desiccation (Naylor & Coleman-Derr, 2018). There is increased interest in using soil bacteria to alleviate plant stress during drought by producing EPS, which holds water in the soil making it more accessible to plant roots.

Beneficial soil bacteria could boost crop tolerance to drought stress by producing phytohormones that drive plant development, accumulating osmolytes or other protective chemicals, or detoxifying reactive oxygen species (Vurukonda et al., 2016; Lakshmanan et al., 2017). Some Bacillus species, for example, manufacture indole-3-acetic acid (IAA) in the rhizosphere soil, resulting in increased root production (Armada et al., 2015).

Plant root initiation and elongation can be accelerated by adding IAA-producing soil bacteria (Lakshmanan et al., 2017). The longer and more biomass roots can help the plant get water and reduce water stress, and they may also help create soil organic matter. Rhizosphere microorganisms have also been linked to metabolite accumulation in plant cells and relieving osmotic stress caused by water imbalance in plant tissues (Pereyra et al., 2012). Inoculation with *Azospirillum*, for example, helped to protect maize plants subjected to drought stress by raising the Osmo protectant proline (Casanovas et al., 2002), while in another investigation wheat seedlings were protected by boosting water flow through the plant (Pereyra et al., 2012).

Bacterial inoculants can also have an impact on the plant metabolome. Schmidt et al. (2014) investigated the composition of secondary metabolites after inoculating chamomile plants with several bacterial strains. Inoculation with B. subtilis and *Paenibacilluspolymyxa* strains resulted in increased synthesis of apigenein-7-O-glucoside (Schmidt et al., 2014), a bioactive metabolite. Bacteria that produce 1-aminocyclopropane 1-carboxylate (ACC) deaminase can reduce the stress hormone ethylene (Glick, 2014). Some endophytic strains of *Azospirillum* have been proven to influence maize stomatal closure via signal transduction pathways mediated by gibberellins (Cohen et al., 2009) production. Fungal interactions can also improve drought resistance and nutrient acquisition.

Many plants, including the majority of agronomically important ones, have symbiotic associations with beneficial arbuscular mycorrhizal (AM) fungi. These relationships are known to boost plant growth, for example, by enhancing soil phosphorus bioavailability to plants. As the climate changes, these positive interactions may be used to boost plant tolerance to various environmental challenges.AM fungi, for example, have been shown to boost maize plants' ability to cope with drought stress by controlling the synthesis of particular molecules known as aquaporins (Quiroga et al., 2017). Aquaporins have been shown to relieve water stress by controlling plant water loss (Kapilan et al., 2018). AM fungus can also directly get water by extending their mycelia into water-filled soil pores that plant roots cannot otherwise access. These examples show how understanding soil microbiomes opens up a plethora of potential to harness the power of bacteria to adapt to our changing environment and improve ecosystem services.

18.8 Vulnerability to Climate Change

Climate change is a threat to our planet's future, yet there is still time to adapt and limit its impacts. Climate change is a fact that affects a broad range of people worldwide, particularly the most vulnerable, by raising the frequency and ferocity of extreme weather events that inflict material destruction and population relocation. According to the Internal Displacement Monitoring Centre (IDMC), natural catastrophes resulted in the internal displacement (movement within the same nation) of 17.2 million individuals in 2018.

Because of the unparalleled speed with which it has developed, global warming is also the source of the worst environmental hassle in history. According to the Intergovernmental Panel on Climate Change (IPCC), the average world temperature has risen by 0.85 °C since 1880, the Arctic has lost 1.07 million km^2 of ice per decade and the sea levels have risen by 19 cm.

Until recently, the international community had concentrated on reducing CO_2 emissions. However, based on the most recent data, it has broadened its activities, advocating climate change adaptation and mitigation measures to assist in mitigating the repercussions of this phenomenon, the ramifications of which are already being felt around the world. Both tactics complement one another, and while they provide different problems, the result is the same.

The agricultural sector is among the major contributors to greenhouse gas (GHG) emissions. Additionally, fisheries and current livestock systems are greatly impacted by climate change at every stage of the food chain, including processing, production, retailing, transporting, storing, and consumption. The main culprits include increased deforestation, intensive farming practices, animal production, pastoral industries, and overuse (Galanakis, 2023).

Climate change adaptation strategies differ from mitigation strategies in that the former concentrates on reducing potential impacts and addressing the causes of climate change, while the latter aims to lessen its negative effects and take advantage of any opportunities that may present themselves. Climate resilience will be essential to reducing the effects of climate change and laying the groundwork for both our survival and the survival of the rest of the Earth's population if mitigation strategies fall short of emissions reduction targets.

Most of the world's 1.4 billion people get their basic energy from traditional fuels like coal and wood. Not only might this harm the environment, but it also has the potential to prematurely kill millions of people, particularly women and children. By 2035, there will be a more than 50% increase in the world's energy demand, with developing nations experiencing even quicker growth. Clean energy that is safe for the environment and for all of these new clients is needed.

The 2018 Intergovernmental Panel on Climate Change (IPCC) Special Report on Global Warming of 1.5°C highlights the urgent need for immediate climate action. To meet the Paris Agreement's safety targets, global emissions must peak by 2030 and rapidly decrease to net zero by 2050.

18.9 Adaptation and Mitigation Strategies for Climate Change

Addressing the pressing challenges of escalating climate change, this exploration emphasizes the urgent need for effective adaptation and mitigation strategies. It delves into innovative approaches to navigate the impacts of a changing climate, highlighting the importance of proactive measures to foster a sustainable and resilient future. As we stand at a critical juncture, understanding and implementing these strategies are paramount. Although combating climate change is challenging, we can lessen its negative effects by working together and implementing suitable mitigation strategies. Encouraging more travel by rail or shared cars, fewer flights, and an increase in bicycle excursions in towns to advance sustainable mobility and public transit (Saeed et al., 2023).

The 3Rs (reduce, reuse, and recycle) rule, the environmental industry, agriculture, fisheries, animal husbandry, food sustainability, and responsible consumption are all being pushed. By taxing the usage of fossil fuels and creating markets for CO_2 emissions. Along with mitigation steps to slow global warming, adaptation criteria to climate change must be encouraged. Investigating and creating novel solutions to natural disaster prevention and management. Climate change is causing rising worry among governments and policymakers. To address this unavoidable challenge, the SDGs-13 aim emphasizes the significance of creating adaption mechanisms that mitigate its negative consequences and eventually protect both the environment and society. Adaptation preparedness is a critical instrument in societal adaptation capacity building to reduce the consequences of disasters on community living standards (Saeed et al., 2023).

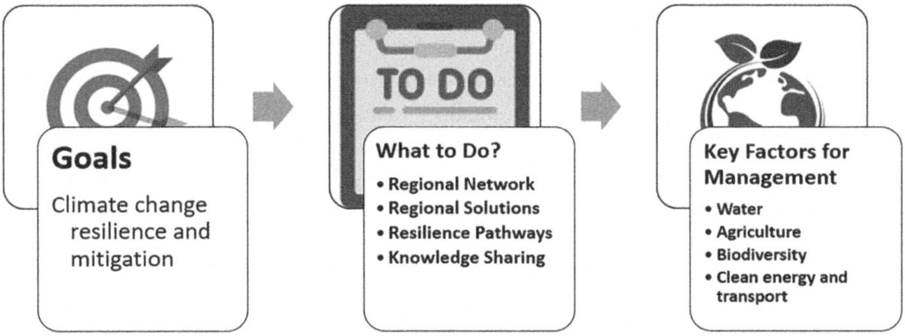

Fig. 18.2 Key factors for climate resilience and mitigation

Wind turbines and biomass power plants have a low environmental impact and should thus be widely used. Renewable energy sources have the potential to decarbonize 90% of the electrical business. By developing the zero-carbon emission decarbonization idea, renewable energy can limit global temperature rise to 1.5 degrees Celsius by 2050 and can replace fossil fuel-derived energy (Osman et al., 2023).

The key factors for climate resilience and mitigation including goals and management strategies are shown in Fig. 18.2.

18.10 Patents on Soil Bacteria for Climate Resilience and Mitigation

Examining patents related to the role of soil microbes in climate resilience and mitigation is essential for advancing sustainable agricultural practices and environmental stewardship. These patents offer insights into innovative techniques and technologies harnessing the power of soil microorganisms to enhance carbon sequestration, improve nutrient cycling, and bolster plant resilience to climate stressors (Kumari et al., 2023). By understanding the intellectual property landscape, researchers and practitioners can leverage patented solutions to optimize soil health and develop climate-smart agricultural strategies. This review not only facilitates the adoption of cutting-edge microbial-based interventions for climate resilience but also encourages collaboration and knowledge-sharing among scientists and stakeholders in the pursuit of effective climate change mitigation strategies.

Crucially, with carbon fixation and organic matter storage, soil bacteria may absorb carbon dioxide (CO_2) from the atmosphere. The potential of microbial-driven carbon storage paves the key to altering climate-smart farming methods and paving the route for long-term production and environmental conservation. The idea behind this patent is carbon sequestration. It is primarily related to a method of raising soil organic carbon (SOC) through acidification of soil-applied fluids using sulfur dioxide adjusted to pH to release soil pores and supply nutrients to carbon dioxide-consuming plants and soil microorganisms. The method for enhancing SOC by opening up soil pores at deeper depths by acidifying land-applied waters with pH-adjusted sulfur dioxide is described in this patent. To sequester a larger net volume increase of SOC, it also explains how to supply sulfur and other nutrients to soil bacteria and plants to boost and raise their overall development and carbon dioxide assimilation (Atkin et al. USOO8092.118B2 US 8,092,118 B2, 2012).

An international patent on a technique for using native ureolytic microorganisms to improve the concentration of metal carbonates in heavy metal-polluted geomaterials was published (international publication number WO 2017/034827 A1). The technique may be applied to bioremediate geomaterials polluted with heavy metals. Therefore, these proteolytic microorganisms can operate as soil bacteria as an unseen army to help with climate resilience and mitigation by sequestering metals as carbonate precipitates.

Developing nations are not the only places where adaptation is required. However, there is a clear need for emerging nations to find various and urgent answers. There has already been a great deal of adaptation. However, when looking through publicly accessible, international channels, answers from developing nations are frequently harder to find. This is not to say that innovation is not happening. In contrast, increased attention is required for adaptation solutions coming from poor nations, not to mention the goal of sharing solutions between emerging nations and the rest of the globe.

Since the early 1990s, a significant increase in biotechnology patents, and sales of genetically modified seeds are now on par with those of herbicides. For crop protection, insecticides, herbicides, and fungicides have long been cost-effective means to protect plants and assure crop stability. However, interest in biological crop protection has increased as a result of agrochemicals' detrimental effects on the environment and soil fertility.

18.10.1 In Africa, Micro-dosing and Targeted Fertilization

One method of applying fertilizer is micro-dosing. It was created by scientists at the CGIAR member International Crops Research Institute for the Semi-Arid Tropics (ICRISAT). Farmers can utilize fertilizers more cost-effectively and efficiently by administering modest dosages of fertilizer at the base of plants a few weeks after planting, or in the planting hole along with the seed during sowing.

Farmers can utilize fertilizers more cost-effectively and efficiently by administering modest dosages of fertilizer at the base of plants a few weeks after planting, or in the planting hole with the seed during sowing. This method has shown to be particularly useful for small-scale farmers with limited resources. The plant's ability to adapt to late-season droughts and other climatic variations is strengthened by the concentration of nutrients at the source, which promotes rapid root growth while preserving nutrients.

18.10.2 Fungi-Based Biofertilizers

Products made of biofertilizers made from fungi are offered by MYCONATIVA. In light of climate change, they are intended to strengthen soils, conserve water, and boost agricultural productivity. Products can be used in bioremediation and ecosystem conservation projects, as well as on fruit trees, crops, and garden plants.

18.10.3 Utilizing Microbes to Improve the Fertility of the Soil

A commercial biological soil enhancement product called Korin Agricultura Fertibokashi was created by the 1994-founded Korin Agropecuária Group. It was developed from the trademarked effective microorganisms (EM) bokashi procedure. This entails taking living things from the natural world, particularly from hostile places like air currents, the seafloor, and volcanoes. Microorganisms are categorized into functional classes based on how quickly they can colonize new areas.

18.10.4 Fixing Nitrogen with Biofertilizer

Kula Bio Kula Bio fixes atmospheric nitrogen into the soil using non-genetically modified organisms (GMO) bacteria. Applying it to conventional and organic crops enables farmers to increase crop productivity while minimizing their impact on the environment. Before being administered to a field, the bacteria in the biofertilizer are strengthened and given a carbon-rich energy source, which allows them to accumulate more nutrients and energy (WIPO, 2022). The structure of microbial communities can be upset, greenhouse gas production can be disturbed, and the prevalence of plant and animal infections can rise as a consequence of climate change. Food production and the health of people may be impacted by these variations. Bacteria can also be employed to lessen the negative consequences of climate change.

Thus, safeguarding the planet's health requires an understanding of the role that microbes play in climate change. Scientists from several fields, such as medicine, microbial engineering, and, ecology are conducting significant research and creating frameworks and methods based on microbes to tackle the issue globally.

Researchers are working hard to understand the rumen microbiome and how to manage the methane generation that goes along with it. A multinational team of scientists discovered a 30% reduction in intestinal methane emissions when cows were fed an inhibitor of a crucial archaeal enzyme involved in methane generation. Building on this finding, the researchers are revealing the impact of the inhibitor on the cow microbiome using omics and models. Bruns, who did not participate in these experiments, pointed out that knowing how the interaction of methanogens with the entire microbiome will be essential for regulating cow feed in a way that lowers methane production during rumination.

Microbes have a wide range of adaptability. When temperatures rise, bacterial respiration rises and bacterial cell size falls. To survive at lower temperatures, viruses that are present in colder waters have evolved to encode proteins linked to the cold shock response. Changes in the makeup of the microbial community are linked to warmer and drier soils and warming oceans. These variations may have an impact on the carbon and nitrogen cycles in those ecosystems.

Changes in community structure can also be facilitated by viruses that infect aquatic and soil microorganisms. For example, viruses known as cyanophases can infect cyanobacteria and subsequently stop fixing carbon dioxide, increasing the amount of greenhouse gases that are available to enter the atmosphere. To fully comprehend the virome's impact on microbial adaptability to climate change, more investigation is necessary. Scientists can forecast the effects of climate change on civilization as a whole by having an understanding of how microbial fluxes of greenhouse gas emissions and carbon will evolve. Microbes are powerful little creatures. They alter the paths of climate.

Microorganisms can lessen the impacts of climate change, even though they also contribute to its harmful effects. Microbes are seen as potential answers by Brajesh Singh, (Ph.D.), According to Singh, "Evidence is mounting that microbial tools can offer practical solutions for both ecosystem adaptation and mitigation of climate change." "Understanding climate change at the microbial level is essential for fruitful outcomes from this line of research and innovation" (asm.org).

18.11 Conclusion

The changing climatic condition poses a significant challenge to agricultural sustainability and necessitates an integrated approach to designing policies that aim to boost agricultural production sustainably. Soil is home to a varied microbial flora that contributes significantly to ecosystem functions. In this shifting climatic scenario, present agricultural management techniques may affect soil fertility and productivity. In the absence of potential climate change management strategies, microbes may lose their natural ability to engage in a variety of biological functions such as suppressing soil-borne plant pathogens, and so on, which are critical for crop plant development, growth, protection, and productivity.

Plant-microbe interactions for agricultural practices like integrated nutrient and soil management, integrated pest and weed management, and organic agriculture involving the utilization of microorganisms as binoculars, biopesticides, biofertilizers, biological weed control, and so on, will help to contribute to climate change. According to the most current national climate assessment for the United States, the climate is expected to continue to change, with weather patterns becoming more irregular and extreme. As a result, it is vital to comprehend and forecast the major and small-scale effects of climate change on the soil microbiome.

Different locations have different soils and soil microbiomes. The biotic and abiotic characteristics of soils differ widely, making it challenging to generalize about how climate change would affect soil microbiomes across a range of soil ecosystems. Local variations in hydrological and biogeochemical variables may affect the kinds of microorganisms found, even within a single soil class. For example, the diversity and richness of soil bacterial populations are significantly impacted by salt and soil pH. Certain soil bacteria that can thrive in such settings and with the resources available are chosen by minerals, nutrients, redox conditions, and organic carbon concentration. Because of this, Earth's numerous soil ecosystems represent a variety of microbial habitats.

References

Allison, S. D., & Martiny, J. B. H. (2008). Resistance, resilience, and redundancy in microbial communities. *Proceedings of the National Academy of Sciences USA, 105*, 11512–11519.

Alori, E. T., & Babalola, O. O. (2018). Microbial inoculants for improving crop quality and human health in Africa. *Frontiers in Microbiology, 9*, 2213.

Armada, E., Azcon, R., Lopez-Castillo, O. M., Calvo-Polanco, M., & Ruiz-Lozano, J. M. (2015). Autochthonous arbuscular mycorrhizal fungi and *Bacillus thuringiensis* from a degraded Mediterranean area can be used to improve physiological traits and performance of a plant of agronomic interest under drought conditions. *Plant Physiology and Biochemistry, 90*, 64–74.

Averill, C., Turner, B. L., & Finzi, A. C. (2014). Mycorrhiza-mediated competition between plants and decomposers drives soil carbon storage. *Nature, 505*, 543–545.

Casanovas, E. M., Barassi, C., & Sueldo, R. J. (2002). Azospirillum inoculation mitigates water stress effects in maize seedlings. *JSTOR, 30*, 343–350.

Cohen, A. C., Travaglia, C. N., Bottini, R., & Piccoli, P. N. (2009). Participation of abscisic acid and gibberellins produced by endophytic Azospirillum in the alleviation of drought effects in maize. *Botany, 87*, 455–462.

Cregger, M. A., Schadt, C. W., McDowell, N. T., Pockman, W. T., & Classena, A. T. (2012). Response of the soil microbial community to changes in precipitation in a semiarid ecosystem. *Applied and Environmental Microbiology, 78*(24), 8587–8594.

Cruz-Martinez, K., Suttle, K. B., Brodie, E. L., Power, M. E., Andersen, G. L., & Banfield, J. F. (2009). Despite strong seasonal responses, soil microbial consortia are more resilient to long-term changes in rainfall than overlying grassland. *The ISME Journal, 3*, 738–744.

Doring, T. F., Vieweger, A., Pautasso, M., Vaarst, M., Finckhe, M. R., & Wolfea, M. S. (2015). Resilience as a universal criterion of health. *Journal of the Science of Food and Agriculture, 95*(3), 455–465.

Galanakis, C. M. (2023). The "vertigo" of the food sector within the triangle of climate change, the post-pandemic world, and the Russian-Ukrainian war. *Food, 12*, 721. https://doi.org/10.3390/foods12040721

Glick, B. R. (2014). Bacteria with ACC deaminase can promote plant growth and help to feed the world. *Microbiological Research, 169*, 30–39.

Griffiths, B. S., & Philippot, L. (2013). Insights into the resistance and resilience of the soil microbial community. *FEMS Microbiology Reviews, 37*, 112–129.

Gupta, R., Bru, D., Bisaria, V. S., Philippot, L., & Sharma, S. (2012). Responses of Cajanus cajan and rhizospheric N-cycling communities to bioinoculants. *Plant and Soil, 358*, 143–154.

Hodges, S. C. (2010). *Soil fertility basics. Soil science extension* (pp. 4927–4932). North Carolina State University.

International publication number WO 2017/034827 A1; (international publication date: 2 March 2017.

Jansson, J. K., & Hofmockel, K. S. (2020). Soil microbiomes and climate change. *Nature Reviews Microbiology, 18*, 35–46.

Jansson, C., Tuskan, G. A., Wullschleger, S. D., & Kalluri, U. C. (2010). Phytosequestration: Carbon biosequestration by plants and the prospects of genetic engineering. *Bioscience, 60*, 685–696.

Jansson, C., Vogel, J., Hazen, S., Brutnell, T., & Mockler, T. (2018). Climate-smart crops with enhanced photosynthesis. *Journal of Experimental Botany, 6*, 3801–3809.

Kapilan, R., Vaziri, M., & Zwiazek, J. J. (2018). Regulation of aquaporins in plants under stress. *Biological Research, 51*, 4.

King, A. E., & Hofmockel, K. S. (2017). Diversified cropping systems support greater microbial cycling and retention of carbon and nitrogen. *Agriculture, Ecosystems & Environment, 240*, 66–76.

Kumar, D., Shivay, Y. S., Dhar, S., Kumar, C., & Prasad, R. (2012). Rhizospheric Flora and the influence of agronomic practices on them: A review. *Proceedings of the National Academy of Sciences, India Section B: Biological Sciences, 83*(1), 1–14.

Kumari, A., Dash, M., Singh, S. K., et al. (2023). Soil microbes: A natural solution for mitigating the impact of climate change. *Environmental Monitoring and Assessment, 195*, 1436. https://doi.org/10.1007/s10661-023-11988-y

Lakshmanan, V., Ray, P., & Craven, K. D. (2017). Toward a resilient, functional microbiome: Drought tolerance-alleviating microbes for sustainable agriculture. *Methods in Molecular Biology, 1631*, 69–84.

Lal, R. (2004). Soil carbon sequestration impacts on global climate change and food security. *Science, 304*, 1623–1627.

Lehman, R. M., Cambardella, C. A., Stott, D. E., Acosta-Martinez, V., Manter, D. K., Buyer, J. S., Maul, J. E., Smith, J. L., Collins, H. P., Halvorson, J. J., Kremer, R. J., Lundgren, J. G., Ducey, T. F., Jin, V. L., & Karlen, D. L. (2015). Understanding and enhancing soil biological health: The solution for reversing soil degradation. *Sustainability, 7*, 988–1027.

Madejón, E., Moreno, F., Murillo, J. M., & Pelegrí, F. (2007). Soil biochemical response to long-term conservation tillage under semi-arid Mediterranean conditions. *Soil and Tillage Research, 94*, 346–352.

Mohanty, S., & Swain, C. K. (2018). Role of microbes in climate smart agriculture. In D. Panpatte, Y. Jhala, H. Shelat, & R. Vyas (Eds.), *Microorganisms for green revolution* (Microorganisms for sustainability) (Vol. 7, pp. 129–140). Springer.

Naylor, D., & Coleman-Derr, D. (2018). Drought stress and root-associated bacterial communities. *Frontiers in Plant Science, 8*. https://doi.org/10.3389/fpls.2017.02223

Olanrewaju, O. S., Glick, B. R., & Babalola, O. O. (2017). Mechanisms of action of plant growth promoting bacteria. *World Journal of Microbiology and Biotechnology, 33*(11), 197.

Osman, A. I., Chen, L., Yang, M., et al. (2023). Cost, environmental impact, and resilience of renewable energy under a changing climate: A review. *Environmental Chemistry Letters, 21*, 741–764. https://doi.org/10.1007/s10311-022-01532-8

Pereyra, M. A., García, P., Colabelli, M. N., Barassi, C. A., & Creus, C. M. (2012). A better water status in wheat seedlings induced by Azospirillum under osmotic stress is related to morphological changes in xylem vessels of the coleoptile. *Applied Soil Ecology, 53*, 94–97.

Philippot, L., & Hallin, S. (2011). Towards food, feed, and energy crops mitigating climate change. *Trends in Plant Science, 16*, 476–480.

Quiroga, G., Erice, G., Aroca, R., Chaumont, F., & Ruiz-Lozano, J. M. (2017). Enhanced drought stress tolerance by the Arbuscular Mycorrhizal Symbiosis in a drought-sensitive maize cultivar is related to a broader and differential regulation of host plant Aquaporins than in a drought-tolerant cultivar. *Frontiers in Plant Science, 8*, 1056.

Saeed, S., Makhdum, M. S. A., Anwar, S., & Yaseen, M. R. (2023). Climate change vulnerability, adaptation, and feedback hypothesis: A comparison of lower-middle, upper-middle, and high-income countries. *Sustainability, 15*, 4145. https://doi.org/10.3390/su15054145

Sahu, A., Bhattacharjya, S., Mandal, A., Thakur, J. K., Atoliya, N., Sahu, N., Manna, M. C., & Patra, A. K. (2018). Microbes: A sustainable approach for enhancing nutrient availability in agricultural soils. In V. Meena (Ed.), *Role of rhizo-spheric microbes in the soil* (pp. 47–75). Springer.

Saini, R., & Sharma, S. (2019) Climate-resilient microbes in sustainable crop production, Chapter 20. *Contaminants in agriculture and environment: Health risks and remediation.*

Schimel, J., Balser, T. C., & Wallenstein, M. (2007). Microbial stress-response physiology and its implications for ecosystem function. *Ecology, 88*, 1386–1394.

Schmidt, M. W. I., et al. (2011). Persistence of soil organic matter as an ecosystem property. *Nature, 478*, 49–56.

Schmidt, R., et al. (2014). Effects of bacterial inoculants on the indigenous microbiome and secondary metabolites of chamomile plants. *Frontiers in Microbiology, 5*, 64.

Solomon, D., Lehmann, J., & Martinez, C. E. (2003). Sulphur K-edge XANES spectroscopy as a tool for understanding Sulphur dynamics in soil organic matter. *Soil Science Society of America Journal, 67*, 1721–1731.

United States Patent, Atkin et al. USOO8092.118B2 US 8,092,118 B2, Jan. 10, 2012 Carbon Sequestration Method.

Vurukonda, S. S. K. P., Vardharajula, S., Shrivastava, M., & Sk, Z. A. (2016). Enhancement of drought stress tolerance in crops by plant growth promoting rhizobacteria. *Microbiological Research, 184*, 13–24.

Wallenstein, M. D. (2017). Managing and manipulating the rhizosphere microbiome for plant health: A systems approach. *Rhizosphere, 3*, 230–232.

Wang, J., & Azam, W. (2024). Natural resource scarcity, fossil fuel energy consumption, and total greenhouse gas emissions in top emitting countries. *Geoscience Frontiers, 15*(2), 101757. https://doi.org/10.1016/j.gsf.2023.101757

White, P. J., & Hammond, J. P. (2008). Phosphorus nutrition of terrestrial plants. In P. J. White & J. P. Hammond (Eds.), *The eco-physiology of plant-phosphorus interactions. Plant ecophysiology* (Vol. 7, pp. 51–81). Springer.

WIPO. (2022). *Green technology book: Solutions for climate change adaptation*. World Intellectual Property Organization. https://doi.org/10.34667/tind.47093

Yuste, J. C., Peñuelas, J., Estiarte, M., Garcia-Mas, J., Mattana, S., Ogaya, R., Pujol, M., & Sardans, J. (2011). Drought-resistant fungi control soil organic matter decomposition and its response to temperature. *Global Change Biology, 17*, 1475–1486.

Part IV

Conclusions

Soil Microbiota as Key Players in Mitigating Climate Change and Achieving a Sustainable Environment: A Conclusion

19

Rajnanda Thaosen, Dona Mazumder, Nandita Das, and Piyush Pandey

Abstract

Soil bacteria are essential to reduce climate change and promote sustainability. These microbes have a significant role in improving soil health, cycling nutrients, and sequestering carbon, all of which lower greenhouse gas emissions. Soil bacteria help in carbon storage and atmospheric CO_2 reduction by encouraging plant development and aiding in the breakdown of organic materials. Additionally, they strengthen the soil's resistance to severe weather, which stabilizes ecosystems. Climate change mitigation can be greatly enhanced by utilizing the potential of soil bacteria through sustainable agricultural methods like biofertilizers and bioremediation. This concluding chapter emphasizes the necessity for more research to maximize the function of soil microbial communities in climate change methods and emphasizes the need to comprehend and utilize these communities to achieve long-term environmental sustainability.

Keywords

Greenhouse gas emissions · Carbon sequestration · Soil microbiome · PGPR

19.1 Introduction

There is a pertinent need for vital discussions in the growing body of literature that addresses the intersection of soil science, microbiology, and environmental sustainability. It underscores the pivotal role that soil microorganisms play in regulating

R. Thaosen · D. Mazumder · N. Das · P. Pandey (✉)
Soil and Environmental Microbiology Laboratory, Department of Microbiology, Assam University, Silchar, India

M. Saraf et al. (eds.), *Climate Change and Soil Microorganisms for Environmental Sustainability*, Microorganisms for Sustainability 28,
https://doi.org/10.1007/978-981-96-3425-5_19

greenhouse gas emissions, enhancing soil health, and fostering plant resilience, all of which are crucial for mitigating the effects of climate change. By exploring the complex interactions between soil microbiota and environmental factors, this book provides valuable insights into how harnessing these natural processes can lead to more sustainable agricultural practices, improved carbon sequestration, and the restoration of degraded ecosystems. As the world faces unprecedented environmental challenges, this comprehensive resource serves as a guide for scientists, policymakers, and environmentalists dedicated to developing innovative strategies for a sustainable future.

19.2 Climate Change, the Challenge

Climate change is defined as the significant shift in climate patterns, such as temperature, humidity, precipitation, droughts, and other extreme weather conditions resulting in adverse consequences in different parts of the world. The world has seen significant changes in climatic conditions over the last few decades. The main cause of climate change has been rising levels of greenhouse gases (GHG) such as carbon dioxide (CO_2), methane (CH_4) and nitrous oxide (N_2O). These gases currently contribute a respective 73%, 19%, and 5% to global total GHG emissions, excluding land use (Olivier et al., 2017). Human activities such as the use of fossil fuels for energy production, industrial activities, automobiles, clearing of forests for land use, and agricultural activities are the major contributing factors to the increase in the level of greenhouse gases. Natural phenomenon such as forest fires, volcanoes, and earthquakes also plays a role in rising greenhouse gas levels (Yue & Gao, 2018). In 2019, global greenhouse gas (GHG) emissions, excluding land-use changes, grew by 1.1% (±1%), reaching 52.4 gigatons of CO_2 equivalent ($GtCO_2$ eq) with a 95% uncertainty range of ±8%. Although this growth rate is half of 2018's, it continues the average annual increase of 1.1% since 2012, significantly lower than the 2.6% average growth in the early 2000s. Including emissions from land-use changes, total GHG emissions for 2019 reached 57.4 $GtCO_2$ eq (±10%), marking a 70% rise compared to 2018, despite a global economic growth of 2.8%. Emissions in 2019 were 59% higher than in 1990 and 44% higher than in 2000 (Olivier et al., 2017).

19.3 Disruption of Biogeochemical Cycles and Ecosystems

Biogeochemical cycles, which are essential for the maintenance of balance of ecosystems and sustenance of life on earth have also been affected due to climate change. Burning of fossil fuels and deforestation have led to higher levels of atmospheric CO_2 altering the carbon cycle. Warming temperatures cause permafrost to thaw, thereby releasing stored CO_2 and methane, further accelerating climate change. Changes in temperature and precipitation can impact nitrogen fixation. There has also been a significant change in the pattern of the oxygen cycle, sulfur cycle, and phosphorous cycle (Gobler, 2020; Oschlies et al., 2018; Jansson & Hofmockel, 2020; Canadell et al., 2021).

Various terrestrial and marine ecosystems have been severely affected by climate change. The coral reef ecosystems of the marine environment, the ecosystems present in polar regions, and the high mountains are the first that have been disrupted (Häder & Barnes, 2019). Rise in temperature, acidification, etc., have resulted in the bleaching of coral reefs as well as a reduction in sea ice, ice formation, etc. (Klein et al., 2024; Zhao et al., 2021; Maksym, 2019). Other ecosystems that are vulnerable include mangrove ecosystems, grassland ecosystems, and aquatic ecosystems (Cinco-Castro & Herrera-Silveira, 2020; Laidre et al., 2020; Du et al., 2022; Prakash, 2021). Adverse weather patterns, drought, and temperature rise have caused severe damage to different ecosystems of the world (Trisos et al., 2020). Apart from these, there have been several cases of climate-induced extreme weather such as tropical cyclones, and frequent wildfires (Barik & Baidya Roy, 2023; Kantamaneni et al., 2022; Jain et al., 2022).

19.4 Addressing Climate Change

Strategies to address climate change include reducing greenhouse gas emissions, particularly CO_2, and adopting renewable energy sources, biofuels, nuclear power, and carbon sequestration, carbon capture, storage, and utilization (CCSU). This also includes reducing dependency on fossil fuels and using renewable energies including solar, wind, hydro, and geothermal power (Hussain et al., 2017; Mathy et al., 2018; Østergaard et al., 2020; Shivakumar et al., 2019). Biofuels offer an alternative energy source. Reduction in waste materials from agriculture, industries, etc., can be achieved by using them as substrates for the production of biofuels (Srivastava et al., 2017; Akalın et al., 2017). Common biofuels include bioethanol, biodiesel, biohydrogen, butanol, and biomethane, which are produced by microorganisms like *Penicillium*, *S. cerevisiae*, *Nannochloropsis*, *Schizochytrium*, *Botryococcus*, *Streptomyces*, *Rhodococcus*, and *Nocardia* (Bardhan et al., 2019).

Carbon sequestration refers to the process of capturing and storing atmospheric CO_2, a critical strategy to mitigate the impacts of global climate change (Joel & Oguanobi, 2024). The USGS is currently evaluating two major types of carbon sequestration: geologic and biological (Moore et al., 2024). Geologic carbon sequestration involves storing CO_2 underground in geological formations, such as porous rock, where it is compressed into a liquid state. This method is often employed during the later stages of an active oil well's life and can be integrated with enhanced oil recovery (Chu et al., 2024). On the other hand, biological carbon sequestration involves capturing and storing CO_2 in natural elements like plants, soil, wood, and water bodies. This approach aims to reduce atmospheric CO_2 levels by promoting the growth of plants, particularly larger ones like trees (Singh et al., 2024). The functional diversity of bacteria in carbon sequestration pathways refers to the various metabolic processes, bacteria use to capture and store carbon (Trivedi et al., 2018). Bacteria such as *Actinobacteria, Chlorobium limicola,* and *Chloroflexus aurantiacus* play a crucial role in the global carbon cycle by participating in pathways that convert inorganic carbon into organic forms, which can then be stored in biomass or sediments (Jiao et al., 2024). Carbon capture, storage, and utilization can be achieved through afforestation, soil carbon sequestration, and direct air carbon capture. Photosynthetic processes capture CO_2 from the atmosphere and store it in biomass, organic matter, and soils, where it can remain for extended periods. Soil carbon sequestration involves enhancing soil carbon content through improved land management practices. Soil carbon levels are determined by the balance between carbon inputs like residues, litter, roots, and manure, and losses through respiration, influenced by soil disturbance (Fuss et al., 2018). The soil microbial community also plays a crucial role in carbon sequestration, influenced by factors such as microbial biomass, community structure, by-products, and soil properties like texture, clay mineralogy, pore size distribution, and aggregate dynamics (Buragohain, 2019; Bhattacharyya et al., 2022; Jiang et al., 2022). Direct air carbon capture and storage (DACCS) is a developing CO_2 removal technology that uses chemical bonding to capture CO_2 directly from the atmosphere for storage or utilization (Pires, 2019; Gambhir & Tavoni, 2019; Liu et al., 2018; Fawzy et al., 2020).

19.5 Role of Soil Microbiota for Environmental Sustainability

Due to the diverse nutrients and vital components in humus, the soil is home to a variety of microorganisms (Hartmann & Six, 2023). The microorganisms provide a number of functions including the biodegradation of contaminants, preservation of soil structure, and circulation of biogenic elements that supply nutrients to plants (Morris & Blackwood, 2024). Moreover, the conservation of biological diversity and ecosystems depends greatly on microbial communities (Philippot et al., 2024). The arrangement of soil particles and their corresponding pore spaces is likely to be one of the most far-reaching soil parameters modified by the soil microbiome and the fertility of the soil is greatly influenced by soil microbiome (Chenu & Cosentino, 2011). There has been evidence of a significant role of soil microorganisms in soil aggregation, where bacteria and fungi have a larger influence than soil fauna (Totsche et al., 2018). Contributions of microorganisms to soil aggregation have recently been examined by Lehmann et al. (2020) using trait-based approaches. It has been found that various microbial groups affect soil aggregation from the smallest scales (mineral modification and interaction) to intermediate scales (soil aggregate stabilization via fungal hypha) and to the largest scale (changes in plant community composition by arbuscular mycorrhizal fungi translating into effects on soil structure). Moreover, bacteria maintain soil structure by producing exopolysaccharides that bind soil particles into stable aggregates, enhancing porosity and water retention. They decompose organic matter into humus, which further stabilizes soil, and interact with fungi and plant roots in the rhizosphere, promoting soil stability and reducing erosion (Daunoras et al., 2024).

19.5.1 Soil Microbiota in Cycling of Nutrients

The global nutrient cycling regulates around 90% of the energy flux in soil by microbes (Saccá et al., 2017). In the nitrogen cycle, bacteria fix atmospheric nitrogen into forms that plants can use, decompose organic matter, and convert nitrogen compounds to maintain soil fertility and ecosystem health (Kuypers et al., 2018). Biological nitrogen fixation is carried out by plant symbionts like *Rhizobium* and *Bradyrhizobium* or free-living diazotrophs like *Azospirillum* and *Azotobacter* (Clark et al., 2021). Additionally, in the phosphorus cycle, bacteria break down organic matter to release phosphorus, making it available to plants, and helping recycle phosphorus in soils, preventing nutrient depletion and supporting sustainable agriculture. Phosphate-solubilizing bacteria and fungi such as *Bacillus, Pseudomonas, Enterobacter, Burkholderia,* and Arbuscular mycorrhizal fungi, *Rhizophagus,* also absorb and translocate solubilized phosphorus to the host plant (Li et al., 2021).

19.5.2 Soil Microbiota in Plant Growth Promotion and Rhizoremediation

Plant growth-promoting rhizobacteria (PGPR) includes bacteria, which are free-living in nature and obtained from the rhizosphere having the capability to produce and secrete metabolites, which promote plant growth after colonizing their roots (Beneduzi et al., 2012). It is estimated that ~2–5% of rhizobacteria act as PGPR (Glick, 2015). They are a useful group of rhizosphere microorganisms that can assist plant development through a variety of mechanisms which include synthesis of indole acetic acid (IAA) production, phosphate solubilization atmospheric nitrogen fixation, siderophore synthesis, ACC (1-aminocyclopropane-1-carboxylate) deaminase (Singha et al., 2018a, Singha et al., 2018b). In addition, it was also demonstrated (Debnath et al., 2023a, Debnath et al., 2023b) that bacteria encourage root growth and regulate the generation of stress hormones. Besides, rhizoremediation can effectively remove petroleum hydrocarbons, from the environment through total mineralization (Ruley et al., 2022). As per the published research article by Kotoky et al. (2018), obligate hydrocarbonoblastic bacteria such as *Nesiotobacter, Nitratireductor, Acinetobacter, Marinobacter, Pseudoalteromonas,* and *Sphingomonas* could degrade polyaromatic hydrocarbon (PAH) from the soil. Similarly, microorganisms can degrade a wide range of hydrocarbons, ranging from aliphatic, monoaromatic, and polyaromatic to resins. Some species of the bacterial genera *Acinetobacter, Brevibacterium, Cellulomonas, Marinobacter, Ochrobactrum, Pseudomonas, Arthrobacter, Stenotrophomaonas, Corynebacterium,* and *Bacillus* have been reported as potential hydrocarbon degraders (Singha & Pandey, 2020). Recent reports suggest that combining a plant-bacterial consortium with NPK biostimulation significantly enhances the degradation of petroleum hydrocarbons (up to 67% with *A. indica* and 55% with *D. regia*) in crude oil-contaminated soils (Bhuyan et al., 2023).

19.5.3 Soil Microbiota in Maintaining Economic Stability and Disease Suppression

One of the main problems of agriculture is the loss of soil fertility, this generates a reduction in crop productivity, mainly due to the loss of ecosystem functions of soil microorganisms (Gouda et al., 2018). The knowledge of soil microbial communities contributes to the development of sustainable and environmentally friendly agriculture, and it is important to promote their care and proliferation, also in connection with the type of fertilization (Dincă et al., 2022). According to Ma et al. (2020), compost is one of the most used and ancient fertilizers. Compost promotes soil biodiversity and is a beneficial practice for soil restoration as it provides an important source of nutrients, improves soil structure, enhances fertility, increases the productivity of agricultural soils, and stimulates the rate of vegetation establishment and root development (Toundou et al., 2021). The dominant bacterial phyla in the composting process are Firmicutes, Actinobacteria, Proteobacteria, Bacteroidetes, and Chloroflexi (Siedt et al., 2021). In addition, soil microbiota exerts a pivotal influence on augmenting the suppressive properties of natural soil, thus enhancing its ability to inhibit soil-borne phytopathogens (De Corato, 2020). These pathogens cause diseases including root rot, damping-off, and wilts that directly affect plant growth and survival (Janvier et al., 2007). The suppressive attributes can be elucidated by the collective antimicrobial effects exerted by molecules and microorganisms, or through mechanisms involving antagonistic interactions between pathogens and microbes. The bacteria, including *Paenibacillus, Bacillus, and Pseudomonas;* actinomycetes such as *Streptomyces,* and filamentous fungi like *Aspergillus, Trichoderma, Penicillium,* and *Gliocladium,* constitute the biocontrol-based microbiome that can activate all pathways involved in disease suppression (De Corato, 2020). Figure 19.1 illustrates the multifaceted roles of microorganisms in influencing climate change, facilitating nutrient cycling, promoting plant growth, degrading environmental contaminants, and controlling plant diseases.

19.6 Contributions and Considerations

A detailed discussion on "Climate Change and the Role of Soil Microbiota for Environmental Sustainability" is not only highly important but also urgently needed in today's context. As climate change continues to pose significant threats to ecosystems, food security, and human livelihoods, understanding the role of soil microbiota becomes crucial. These microorganisms are integral to soil health, nutrient cycling, and carbon sequestration, all of which are vital processes that influence climate dynamics. Deliberations on this topic are essential to uncover how soil microbiota can be leveraged to mitigate climate change effects, enhance agricultural productivity, and restore degraded lands. Such discussions also help in identifying innovative, nature-based solutions for sustainable development, making it a critical area of focus for researchers, policymakers, and environmentalists.

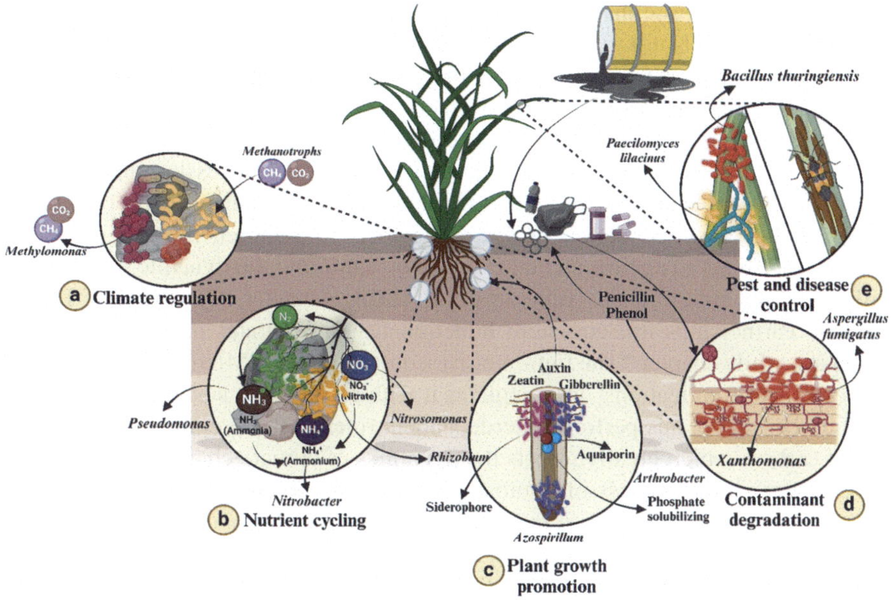

Fig. 19.1 Role of microbes in (**a**) climate change, (**b**) nutrient cycling, (**c**) plant growth, (**d**) contaminant degradation, and (**e**) disease control. The details have been described in the text

Quorum sensing is a key mechanism by which microbes communicate and coordinate their behavior based on population density (Moreno-Gámez et al., 2023). This process involves the production and detection of signaling molecules called autoinducers, which regulate essential functions such as biofilm formation, virulence, and antibiotic resistance (Wu & Luo, 2021; Zhou et al., 2020). In soil, such communications can have a significant impact on regulating functions of the soil microbial community (Labouyrie et al., 2023; Hemkemeyer et al., 2021). Hence, it is significant to discuss this aspect in organisms like *Pseudomonas* spp. and other organisms (Qin et al., 2022). Through this process, bacteria can optimize their behavior, enhance survival, and increase their competitive advantage in various environments (Abisado et al., 2018). This intra-community communication is not only critical for the survival of individual species but also has profound implications for microbial ecology and the development of strategies to control microbial functions in stress-induced conditions (Garcia-Contreras et al., 2016; Zhao et al., 2019; Wang et al., 2020). This has been discussed in chapters "Soil Actinomycetes with a Perspective of Cellulolytic Activity" and "Dialogue in the Dirt: Plant–Microbe Interaction in Soil".

Advanced techniques such as metagenomics and meta-transcriptomics provide a full understanding of microbial diversity and function in complicated settings such as soil, thereby complementing established methods (Kotoky et al., 2018; Singha & Pandey 2021). By sequencing the whole genome of a soil sample, a process known as metagenomics, scientists may identify and describe every species of microbe present, including those that are unculturable and are frequently overlooked by traditional methods (Kotoky et al., 2017; Rajkumari & Pandey, 2019; Kotoky & Pandey, 2021). By using this method, the genetic potential of the community is exposed, including genes that support plant growth, bioremediation, and nutrient cycling. By examining the RNA of the community, meta-transcriptomics goes one step further and offers insights into the patterns of gene expression and ongoing metabolic processes under particular environmental circumstances (Das et al., 2021, 2022, 2023). These techniques enable researchers to comprehend how microbial communities react to environmental changes by capturing the dynamic interactions within the soil microbiome (Bhuyan et al., 2023). Chapter "The Role of Plant–Microbe Partnerships in Ecosystem Functionality" deliberates in detail on omics strategies, while chapter "The Future of Biocontrol by Plant-Microbe Nexus: Novel Strategies and Applications" also explains it with reference to sustainability.

Microbes and plants perform interdependent roles in the cycling and sequestration of carbon, which greatly affects the dynamics of climate change (Nayak et al., 2022). Through the process of photosynthesis, plants absorb carbon dioxide from the atmosphere and transform it into organic matter, which serves as the foundation for terrestrial carbon stores (Sha et al., 2022; Chen et al., 2024). Then, through plant leftovers, litterfall, and root exudates, this carbon is incorporated into the soil (Wang et al., 2021). This organic material is broken down by bacteria and fungus in the soil, which either releases carbon dioxide back into the atmosphere through respiration or keeps it stable in the soil as humus and other organic compounds (Das et al., 2024a, b). The balance of these processes, influenced by environmental factors such as temperature, moisture, and nutrient availability, can either mitigate or exacerbate climate change (Bolan et al., 2024). Understanding the intricate plant-microbe interactions and their role in carbon dynamics is vital for developing strategies to enhance carbon sequestration and reduce greenhouse gas emissions (Das et al., 2024a, b; Singha & Pandey, 2023). Chapter "Metabolic Mysteries: New Perspectives on Soil Bacterial Metabolism" has a detailed discussion on carbon cycling, while chapter "The Role of Plant–Microbe Partnerships in Ecosystem Functionality" also has interesting overviews. Chapters "Microbes to the Rescue—The Future of Plant Protection" and "Plants, Microbes, and Carbon: The Interplay of Carbon Cycling and Sequestration" bring insights into the role of PGPR, while 12 and 14 would be interesting with such perspectives.

Plant growth-promoting rhizobacteria (PGPR) are becoming more well-acknowledged for their role in reducing the effects of climate change and for their substantial environmental advantages (Kshetri et al., 2024). By fixing atmospheric nitrogen, solubilizing phosphates, and generating phytohormones, these advantageous bacteria promote plant development and lessen the need for chemical fertilizers, which increase greenhouse gas emissions (Maheshwari et al., 2023; Debnath et al., 2023a, Debnath et al., 2023b). Furthermore, PGPR can strengthen plants' resistance to abiotic stressors like salinity and drought, which are made worse by climate change, promoting sustainable agriculture (Pandey & Dheeman, 2022; Debnath et al., 2023a, Debnath et al., 2023b). Through encouraging root development and improving soil carbon sequestration, PGPR increases soil organic matter, which functions as a sink for carbon (Bisht et al., 2015). Their potential in environmental management techniques aimed at mitigating the effects of climate change is further highlighted by their participation in bioremediation, which involves the degradation of contaminants and the improvement of soil health (Rajkumari et al., 2018; Kotoky et al., 2019; Kotoky & Pandey, 2020a, Kotoky & Pandey, 2020b). Chapters "Leveraging Plant Growth Promoting Microbes with the Goal of Minimising the Impact of Climate Change on Agriculture" and "Soil Bacteria: An Invisible Army for Climate Resilience and Mitigation" have discussed these topics in detail.

The plant-soil-microbe interaction is a critical determinant of soil ecosystem health, influencing nutrient cycling, plant productivity, and soil structure (Kotoky & Pandey, 2020a, Kotoky & Pandey, 2020b; Singha et al., 2018a, Singha et al., 2018b). Plants exude a variety of organic compounds through their roots, which serve as nutrients for soil microbes, including bacteria, fungi, and archaea (Singha et al., 2018a, Singha et al., 2018b). These microbes, in turn, perform essential functions such as nitrogen fixation, phosphorus solubilization, and organic matter decomposition, which enhance soil fertility and promote plant growth (Nevita et al., 2018). The study of these interactions, often through advanced techniques like metagenomics and meta-transcriptomics, reveals how microbial communities respond to plant signals and environmental changes, facilitating the development of sustainable agricultural practices (Diwan et al., 2022; Jain et al., 2024). For instance, understanding how beneficial microbes suppress soil-borne pathogens or improve plant stress tolerance can lead to the creation of bioinoculants and other biological inputs that reduce reliance on chemical fertilizers and pesticides (Devi et al., 2017; Iqbal et al., 2023). This holistic approach not only enhances crop yield and soil health but also contributes to long-term agricultural sustainability by maintaining ecological balance and improving soil resilience against climate change (Pandey et al., 2012). Chapter "Metabolic Mysteries: New Perspectives on Soil Bacterial Metabolism" discusses this issue in detail.

In order to manage plant diseases, biocontrol rhizobacteria are essential since they provide a sustainable substitute for chemical pesticides that have negative effects on the environment (Gahatraj et al., 2024). Through a variety of strategies, including the synthesis of antimicrobial chemicals, rivalry for nutrients and available space, and the development of plant systemic resistance, these advantageous microorganisms inhibit soil-borne diseases (El-Saadony et al., 2022; Ghosh et al., 2024). Biocontrol rhizobacteria improve agricultural output and quality while lowering the need for chemical inputs, which are frequently linked to greenhouse gas emissions and environmental degradation, by decreasing the frequency of plant illnesses (Kshetri et al., 2019; Tripathi et al., 2024). Furthermore, these bacteria help to promote soil carbon sequestration in agriculture by encouraging healthier plant growth and lowering crop losses (Ahmed et al., 2019). This double advantage of mitigating climate change and managing diseases emphasizes how crucial it is to incorporate biocontrol rhizobacteria into sustainable agriculture operations (Zuma et al., 2023). Biocontrol agents such as *Bacillus thuringiensis, Trichoderma species*, and *Pseudomonas* species offer sustainable alternatives to chemical pesticides, playing a key role in reducing agriculture's carbon footprint (Santoyo et al., 2019; Khatoon et al., 2022). Despite their success, the development of effective, stable, and environmentally safe commercial products necessitates extensive research and optimization (Bataineh et al., 2024). Standardized protocols for evaluating the safety, efficacy, and environmental impact of these agents are crucial not only for their widespread adoption but also for their potential to mitigate climate change by lowering greenhouse gas emissions associated with conventional agrochemicals (Datta et al., 2022; Gamage et al., 2023; Brunelle et al., 2024). This chapter explains the processes of biocontrol and their significance in climate change.

Overall, in conclusion, the role of soil microbiota in addressing climate change and promoting environmental sustainability cannot be overstated. These microscopic organisms are key players in maintaining soil health, supporting plant growth, and mitigating the impacts of climate change through natural processes like carbon sequestration and nutrient cycling. As the world grapples with the pressing challenges of environmental degradation and climate change, it is imperative to deepen our understanding of soil microbiota and integrate this knowledge into sustainable land management practices. By harnessing the potential of these organisms, we can develop innovative strategies that not only protect our environment but also ensure a sustainable future for generations to come.

References

Abisado, R. G., Benomar, S., Klaus, J. R., Dandekar, A. A., & Chandler, J. R. (2018). Bacterial quorum sensing and microbial community interactions. *MBio, 9*(3). https://doi.org/10.1128/mBio.02331-17

Ahmed, A. A. Q., Odelade, K. A., & Babalola, O. O. (2019). Microbial inoculants for improving carbon sequestration in agroecosystems to mitigate climate change. In W. Leal Filho (Ed.), *Handbook of climate change resilience*. Springer. https://doi.org/10.1007/978-3-319-71025-9_119-1

Akalın, M. K., Tekin, K., & Karagöz, S. (2017). Supercritical fluid extraction of biofuels from biomass. *Environmental Chemistry Letters, 15*, 29–41. https://doi.org/10.1007/s10311-016-0593-z

Bardhan, P., Gupta, K., & Mandal, M. (2019). Microbes as bio-resource for sustainable production of biofuels and other bioenergy products. In *New and future developments in microbial biotechnology and bioengineering* (pp. 205–222). Elsevier. https://doi.org/10.1016/B978-0-444-64191-5.00015-8

Barik, A., & Baidya Roy, S. (2023). Climate change strongly affects future fire weather danger in Indian forests. *Communications Earth & Environment, 4*(1), 452. https://doi.org/10.1038/s43247-023-01112-w

Bataineh, M. J., Sánchez-Sellero, P., & Ayad, F. (2024). Green is the new black: How research and development and green innovation provide businesses a competitive edge. *Business Strategy and the Environment, 33*(2), 1004–1023. https://doi.org/10.1002/bse.3533

Beneduzi, A., Ambrosini, A., & Passaglia, L. M. (2012). Plant growth-promoting rhizobacteria (PGPR): Their potential as antagonists and biocontrol agents. *Genetics and Molecular Biology, 35*, 1044–1051. https://doi.org/10.1590/S1415-47572012000600020

Bhattacharyya, S. S., Ros, G. H., Furtak, K., Iqbal, H. M., & Parra-Saldívar, R. (2022). Soil carbon sequestration–An interplay between soil microbial community and soil organic matter dynamics. *Science of the Total Environment, 815*, 152928. https://doi.org/10.1016/j.scitotenv.2022.152928

Bhuyan, B., Kotoky, R., & Pandey, P. (2023). Impacts of rhizoremediation and biostimulation on soil microbial community, for enhanced degradation of petroleum hydrocarbons in crude oil-contaminated agricultural soils. *Environmental Science and Pollution Research, 30*(41), 94649–94668. https://doi.org/10.1007/s11356-023-29033-3

Bisht, S., Pandey, P., Bhargava, B., Sharma, S., Kumar, V., & Krishan, D. (2015). Bioremediation of polyaromatic hydrocarbons (PAHs) using rhizosphere technology. *Brazilian Journal of Microbiology, 46*(1), 7–21. https://doi.org/10.1590/S1517-838246120131354

Bolan, S., Padhye, L. P., Jasemizad, T., Govarthanan, M., Karmegam, N., Wijesekara, H., Amarasiri, D., Hou, D., Zhou, P., Biswal, B. K., Balasubramanian, R., Wang, H., Siddique, K. H. M., Rinklebe, J., Kirkham, M. B., & Bolan, N. (2024). Impacts of climate change on the fate of contaminants through extreme weather events. *Science of the Total Environment, 909*(2023, November), 168388. https://doi.org/10.1016/j.scitotenv.2023.168388

Brunelle, T., Chakir, R., Carpentier, A., Dorin, B., Goll, D., Guilpart, N., Maggi, F., Makowski, D., Nesme, T., Roosen, J., & Tang, F. H. M. (2024). Reducing chemical inputs in agriculture requires a system change. *Communications Earth and Environment, 5*(1), 1–9. https://doi.org/10.1038/s43247-024-01533-1

Buragohain, P. (2019). Role of microbes on carbon sequestration. *International Journal of Microbiology Research, 11*(1), 1464–1468. ISSN: 0975-5276.

Canadell, J. G., Monteiro, P. M. S., Costa, M. H., Da Cunha, L. C., Cox, P. M., et al. (2021). Chapter 5: Global carbon and other biogeochemical cycles and feedbacks.

Chen, Z., Wang, W., Forzieri, G., & Cescatti, A. (2024). Transition from positive to negative indirect CO2 effects on the vegetation carbon uptake. Nature. *Communications, 15*(1). https://doi.org/10.1038/s41467-024-45957-x

Chenu, C., & Cosentino, D. (2011). *Microbial regulation of soil structural dynamics*. Biol. Soils Life Inn. Space. https://doi.org/10.1079/9781845935320.0037

Chu, H., Huang, Z., Zhang, Z., Yan, X., Qiu, B., & Xu, N. (2024). Integration of carbon emission reduction policies and technologies: Research progress on carbon capture, utilization and storage technologies. *Separation and Purification Technology*, 127153. https://doi.org/10.1016/j.seppur.2024.127153

Cinco-Castro, S., & Herrera-Silveira, J. (2020). Vulnerability of mangrove ecosystems to climate change effects: The case of the Yucatan Peninsula. *Ocean & Coastal Management, 192*, 105196. https://doi.org/10.1016/j.ocecoaman.2020.105196

Clark, I. M., Hughes, D. J., Fu, Q., Abadie, M., & Hirsch, P. R. (2021). Metagenomic approaches reveal differences in genetic diversity and relative abundance of nitrifying bacteria and archaea in contrasting soils. *Scientific Reports, 11*(1), 15905. https://doi.org/10.1038/s41598-021-95100-9

Das, N., Kotoky, R., Maurya, A. P., Bhuyan, B., & Pandey, P. (2021). Paradigm shift in antibiotic-resistome of petroleum hydrocarbon contaminated soil. *Science of the Total Environment, 757*(xxxx), 143777. https://doi.org/10.1016/j.scitotenv.2020.143777

Das, N., Bhuyan, B., & Pandey, P. (2022). Correlation of soil microbiome with crude oil contamination drives detection of hydrocarbon degrading genes which are independent to quantity and type of contaminants. *Environmental Research, 215*(P1), 114185. https://doi.org/10.1016/j.envres.2022.114185

Das, N., Das, A., Das, S., Bhatawadekar, V., Pandey, P., Choure, K., Damare, S., & Pandey, P. (2023). Petroleum hydrocarbon catabolic pathways as targets for metabolic engineering strategies for enhanced bioremediation of crude-oil-contaminated environments. *Fermentation, 9*(2), 1–31. https://doi.org/10.3390/fermentation9020196

Das, S., Das, A., Das, N., Nath, T., Langthasa, M., Pandey, P., Kumar, V., Choure, K., Kumar, S., & Pandey, P. (2024a). Harnessing the potential of microbial keratinases for bioconversion of keratin waste. *Environmental Science and Pollution Research, 31*, 57478. https://doi.org/10.1007/s11356-024-34233-6

Das, N., Maheshwari, D. K., & Pandey, P. (2024b). Energy crop-based rhizoremediation and lignocellulosic biomass production as sustainable bioeconomy-driven solution for biofuel generation and waste mitigation. *Progress in Energy and Combustion Science, 103*, 101161. https://doi.org/10.1016/j.pecs.2024.101161

Datta, P., Behera, B., & Rahut, D. B. (2022). Climate change and Indian agriculture: A systematic review of farmers' perception, adaptation, and transformation. *Environmental Challenges, 8*(January), 100543. https://doi.org/10.1016/j.envc.2022.100543

Daunoras, J., Kačergius, A., & Gudiukaitė, R. (2024). Role of soil microbiota enzymes in soil health and activity changes depending on climate change and the type of soil ecosystem. *Biology, 13*(2), 85. https://doi.org/10.3390/biology13020085

De Corato, U. (2020). Soil microbiota manipulation and its role in suppressing soil-borne plant pathogens in organic farming systems under the light of microbiome-assisted strategies. *Chemical and Biological Technologies in Agriculture, 7*, 1–26. https://doi.org/10.1186/s40538-020-00183-7

Debnath, S., Chakraborty, S., Langthasa, M., Choure, K., Agnihotri, V., Srivastava, A., Rai, P. K., Tilwari, A., Maheshwari, D. K., & Pandey, P. (2023a). Non-rhizobial nodule endophytes improve nodulation, change root exudation pattern and promote the growth of lentil, for prospective application in fallow soil. *Frontiers in Plant Science, 14*(April), 1–19. https://doi.org/10.3389/fpls.2023.1152875

Debnath, S., Das, A., Maheshwari, D. K., & Pandey, P. (2023b). Treatment with atypical rhizobia, Pararhizobium giardinii and Ochrobactrum sp. modulate the rhizospheric bacterial community, and enhances Lens culinaris growth in fallow-soil. *Microbiological Research, 267*, 127255. https://doi.org/10.1016/j.micres.2022.127255

Devi, K. A., Pandey, G., Rawat, A. K. S., Sharma, G. D., & Pandey, P. (2017). The endophytic symbiont-Pseudomonas aeruginosa stimulates the antioxidant activity and growth of Achyranthes aspera L. *Frontiers in Microbiology, 8*(Sep), 1–14. https://doi.org/10.3389/fmicb.2017.01897

Dincă, L. C., Grenni, P., Onet, C., & Onet, A. (2022). Fertilization and soil microbial community: A review. *Applied Sciences, 12*(3), 1198. https://doi.org/10.3390/app12031198

Diwan, D., Rashid, M. M., & Vaishnav, A. (2022). Current understanding of plant-microbe interaction through the lenses of multi-omics approaches and their benefits in sustainable agriculture. *Microbiological Research, 265*(September), 127180. https://doi.org/10.1016/j.micres.2022.127180

Du, Q., Sun, Y., Guan, Q., Pan, N., Wang, Q., Ma, Y., Li, H., & Liang, L. (2022). Vulnerability of grassland ecosystems to climate change in the Qilian Mountains, northwest China. *Journal of Hydrology, 612*, 128305. https://doi.org/10.1016/j.jhydrol.2022.128305

El-Saadony, M. T., Saad, A. M., Soliman, S. M., Salem, H. M., Ahmed, A. I., Mahmood, M., El-Tahan, A. M., Ebrahim, A. A. M., Abd El-Mageed, T. A., Negm, S. H., Selim, S., Babalghith, A. O., Elrys, A. S., El-Tarabily, K. A., & AbuQamar, S. F. (2022). Plant growth-promoting microorganisms as biocontrol agents of plant diseases: Mechanisms, challenges and future perspectives. *Frontiers in Plant Science, 13*(October), 1–19. https://doi.org/10.3389/fpls.2022.923880

Fawzy, S., Osman, A. I., Doran, J., & Rooney, D. W. (2020). Strategies for mitigation of climate change: A review. *Environmental Chemistry Letters, 18*, 2069–2094. https://doi.org/10.1007/s10311-020-01059-w

Fuss, S., Lamb, W. F., Callaghan, M. W., Hilaire, J., Creutzig, F., Amann, T., et al. (2018). Negative emissions—Part 2: Costs, potentials and side effects. *Environmental Research Letters, 13*(6), 063002.

Gahatraj, I., Roy, R., Sharma, A., et al. (2024). Identification of molecular interactions of pesticides with keratinase for their potential to inhibit keratin biodegradation. *In Silico Pharmacology, 12*, 54. https://doi.org/10.1007/s40203-024-00229-w

Gamage, A., Gangahagedara, R., Gamage, J., Jayasinghe, N., Kodikara, N., Suraweera, P., & Merah, O. (2023). Role of organic farming for achieving sustainability in agriculture. *Farming System, 1*(1), 100005. https://doi.org/10.1016/j.farsys.2023.100005

Gambhir, A., & Tavoni, M. (2019). Direct air carbon capture and sequestration: how it works and how it could contribute to climate-change mitigation. *One Earth, 1*(4), 405–409.

Garcia-Contreras, R., Maeda, T., & Wood, T. K. (2016). Can resistance against quorum-sensing interference be selected? *ISME Journal, 10*(1), 4–10. https://doi.org/10.1038/ismej.2015.84

Ghosh, K., Das, S., Sorongpong, S., Das, N., & Pandey, P. (2024). Emergence of Lasiodiplodia theobromae induced leaf necrosis in tea (Camellia sinensis [L.] O. Kuntze) from India. *Archives of Microbiology, 206*(6), 2–4. https://doi.org/10.1007/s00203-024-04018-z

Glick, B. R. (2015). *Beneficial plant-bacterial interactions* (Vol. 243). Springer.

Gobler, C. J. (2020). Climate change and harmful algal blooms: insights and perspective. *Harmful Algae, 91*, 101731. https://doi.org/10.1016/j.hal.2019.101731

Gouda, S., Kerry, R. G., Das, G., Paramithiotis, S., Shin, H. S., & Patra, J. K. (2018). Revitalization of plant growth promoting rhizobacteria for sustainable development in agriculture. *Microbiological Research, 206*, 131–140. https://doi.org/10.1016/j.micres.2017.08.016

Häder, D. P., & Barnes, P. W. (2019). Comparing the impacts of climate change on the responses and linkages between terrestrial and aquatic ecosystems. *Science of the Total Environment, 682*, 239–246. https://doi.org/10.1016/j.scitotenv.2019.05.024

Hartmann, M., & Six, J. (2023). Soil structure and microbiome functions in agroecosystems. *Nature Reviews Earth & Environment, 4*(1), 4–18. https://doi.org/10.1038/s43017-022-00366-w

Hemkemeyer, M., Schwalb, S. A., Heinze, S., Joergensen, R. G., & Wichern, F. (2021). Functions of elements in soil microorganisms. *Microbiological Research, 252*(March), 126832. https://doi.org/10.1016/j.micres.2021.126832

Hussain, A., Arif, S. M., & Aslam, M. (2017). Emerging renewable and sustainable energy technologies: State of the art. *Renewable and Sustainable Energy Reviews, 71*, 12–28. https://doi.org/10.1016/j.rser.2016.12.033

Iqbal, B., Li, G., Alabbosh, K. F., Hussain, H., Khan, I., Tariq, M., Javed, Q., Naeem, M., & Ahmad, N. (2023). Advancing environmental sustainability through microbial reprogramming in growth improvement, stress alleviation, and phytoremediation. *Plant Stress, 10*(November), 100283. https://doi.org/10.1016/j.stress.2023.100283

Jain, P., Castellanos-Acuna, D., Coogan, S. C., Abatzoglou, J. T., & Flannigan, M. D. (2022). Observed increases in extreme fire weather driven by atmospheric humidity and temperature. *Nature Climate Change, 12*(1), 63–70.

Jain, A., Sarsaiya, S., Singh, R., Gong, Q., Wu, Q., & Shi, J. (2024). Omics approaches in understanding the benefits of plant-microbe interactions. *Frontiers in Microbiology, 15*. https://doi.org/10.3389/fmicb.2024.1391059

Jansson, J. K., & Hofmockel, K. S. (2020). Soil microbiomes and climate change. *Nature Reviews Microbiology, 18*(1), 35–46. https://doi.org/10.1038/s41579-019-0265-7

Janvier, C., Villeneuve, F., Alabouvette, C., Edel-Hermann, V., Mateille, T., & Steinberg, C. (2007). Soil health through soil disease suppression: which strategy from descriptors to indicators? *Soil Biology and Biochemistry, 39*(1), 1–23. https://doi.org/10.1016/j.soilbio.2006.07.001

Jiang, P., Xiao, L. Q., Wan, X., Yu, T., Liu, Y. F., & Liu, M. X. (2022). Research progress on microbial carbon sequestration in soil: A review. *Eurasian Soil Science, 55*(10), 1395–1404. https://doi.org/10.1134/S1064229322100064

Jiao, N., Luo, T., Chen, Q., Zhao, Z., Xiao, X., Liu, J., Jian, Z., Xie, S., Thomas, H., Herndl, G. J., Benner, R., Gonsior, M., Chen, F., Cai, W. F., & Robinson, C. (2024). The microbial carbon pump and climate change. *Nature Reviews Microbiology*, 408. https://doi.org/10.1038/s41579-024-01018-0

Joel, O. T., & Oguanobi, V. U. (2024). Geotechnical assessments for renewable energy infrastructure: ensuring stability in wind and solar projects. *Engineering Science & Technology Journal, 5*(5), 1588–1605. https://doi.org/10.51594/estj.v5i5.1110

Kantamaneni, K., Panneer, S., Krishnan, A., Shekhar, S., Bhat, L., & Rice, L. (2022). Appraisal of climate change and cyclone trends in Indian coastal states: A systematic approach towards climate action. *Arabian Journal of Geosciences, 15*(9), 814. https://doi.org/10.1007/s12517-022-10076-8

Khatoon, Z., del Carmen Orozco-Mosqueda, M., Huang, S., Nascimento, F. X., & Santoyo, G. (2022). Peptide antibiotics produced by Bacillus Species: First line of attack in the biocontrol of plant diseases. In M. T. Islam, M. Rahman, & P. Pandey (Eds.), *Bacilli in agrobio-*

technology. Bacilli in climate resilient agriculture and bioprospecting. Springer. https://doi.org/10.1007/978-3-030-85465-2_2

Klein, S. G., Roch, C., & Duarte, C. M. (2024). Systematic review of the uncertainty of coral reef futures under climate change. *Nature Communications, 15*(1), 2224. https://doi.org/10.1038/s41467-024-46255-2

Kotoky, R., & Pandey, P. (2020a). Difference in the rhizosphere microbiome of Melia azedarach during removal of benzo(A)pyrene from cadmium co-contaminated soil. *Chemosphere, 258*, 127175. https://doi.org/10.1016/j.chemosphere.2020.127175

Kotoky, R., & Pandey, P. (2020b). Rhizosphere mediated biodegradation of benzo(A)pyrene by surfactin producing soil bacilli applied through Melia azadirachta rhizosphere. *International Journal of Phytoremediation, 22*(4), 363–372. https://doi.org/10.1080/15226514.2019.1663486

Kotoky, R., & Pandey, P. (2021). The genomic attributes of Cd-resistant, hydrocarbonoclastic Bacillus subtilis SR1 for rhizodegradation of benzo(A)pyrene under co-contaminated conditions. *Genomics, 113*(1P2), 613–623. https://doi.org/10.1016/j.ygeno.2020.09.057

Kotoky, R., Singha, L. P., & Pandey, P. (2017). Draft genome sequence of heavy metal-resistant soil bacterium *Serratia marcescens* S2I7, which has the ability to degrade polyaromatic hydrocarbons. *Genome Announcements, 5*(48), 1–2.

Kotoky, R., Rajkumari, J., & Pandey, P. (2018). The rhizosphere microbiome: Significance in rhizoremediation of polyaromatic hydrocarbon contaminated soil. *Journal of Environmental Management, 217*, 858–870. https://doi.org/10.1016/j.jenvman.2018.04.022

Kotoky, R., Nath, S., Kumar Maheshwari, D., & Pandey, P. (2019). Cadmium resistant plant growth promoting rhizobacteria Serratia marcescens S2I7 associated with the growth promotion of rice plant. *Environmental Sustainability, 2*(2), 135–144. https://doi.org/10.1007/s42398-019-00055-3

Kshetri, L., Naseem, F., & Pandey, P. (2019). Role of Serratia sp. as biocontrol agent and plant growth stimulator, with prospects of biotic stress management in plant. In R. Sayyed (Ed.), *Plant growth promoting Rhizobacteria for sustainable stress management. Microorganisms for sustainability* (Vol. 13). Springer. https://doi.org/10.1007/978-981-13-6986-5_6

Kshetri, L., Kotoky, R., Debnath, S., Maheshwari, D. K., & Pandey, P. (2024). Shift in the soil rhizobacterial community for enhanced solubilization and bioavailability of phosphorus in the rhizosphere of Allium hookeri Thwaites, through bioaugmentation of phosphate-solubilizing bacteria. *3 Biotech, 14*(8), 1–16. https://doi.org/10.1007/s13205-024-04026-2

Kuypers, M. M., Marchant, H. K., & Kartal, B. (2018). The microbial nitrogen-cycling network. *Nature Reviews Microbiology, 16*(5), 263–276.

Labouyrie, M., Ballabio, C., Romero, F., Panagos, P., Jones, A., Schmid, M. W., Mikryukov, V., Dulya, O., Tedersoo, L., Bahram, M., Lugato, E., van der Heijden, M. G. A., & Orgiazzi, A. (2023). Patterns in soil microbial diversity across Europe. *Nature Communications, 14*(1). https://doi.org/10.1038/s41467-023-37937-4

Laidre, K. L., Atkinson, S., Regehr, E. V., Stern, H. L., Born, E. W., Wiig, Ø., et al. (2020). Interrelated ecological impacts of climate change on an apex predator. *Ecological Applications, 30*(4), e02071. https://doi.org/10.1002/eap.2071

Lehmann, A., Zheng, W., Ryo, M., Soutschek, K., Roy, J., Rongstock, R., Maaß, S., & Rillig, M. C. (2020). Fungal traits important for soil aggregation. *Frontiers in Microbiology, 10*, 2904. https://doi.org/10.3389/fmicb.2019.02904

Li, J. T., Lu, J. L., Wang, H. Y., Fang, Z., Wang, X. J., Feng, S. W., Wang, Z., Yuan, T., Zhang, S. C., Ou, S. N., & Liang, J. L. (2021). A comprehensive synthesis unveils the mysteries of phosphate-solubilizing microbes. *Biological Reviews, 96*(6), 2771–2793. https://doi.org/10.1111/brv.12779

Liu, S., Zhang, Y., Jiang, H., Wang, X., Zhang, T., & Yao, Y. (2018). High CO2 adsorption by amino-modified bio-spherical cellulose nanofibres aerogels. *Environmental Chemistry Letters, 16*, 605–614. https://doi.org/10.1007/s10311-017-0701-8

Ma, S., Xiong, J., Cui, R., Sun, X., Han, L., Xu, Y., Kan, Z., Gong, X., & Huang, G. (2020). Effects of intermittent aeration on greenhouse gas emissions and bacterial community succession during large-scale membrane-covered aerobic composting. *Journal of Cleaner Production, 266*, 121551. https://doi.org/10.1016/j.jclepro.2020.121551

Maheshwari, D. K., Das, A., Dheeman, S., & Pandey, P. (2023). An overall insight into the attributes, interactions, and future applications of "microbial consortium" for plant growth promotion with contemporary approaches. In D. K. Maheshwari & S. Dheeman (Eds.), *Sustainable agrobiology. Microorganisms for sustainability* (Vol. 43). Springer. https://doi.org/10.1007/978-981-19-9570-5_1

Maksym, T. (2019). Arctic and Antarctic sea ice change: Contrasts, commonalities, and causes. *Annual Review of Marine Science, 11*(1), 187–213.

Mathy, S., Menanteau, P., & Criqui, P. (2018). After the Paris Agreement: Measuring the global decarbonization wedges from national energy scenarios. *Ecological Economics, 150*, 273–289. https://doi.org/10.1016/j.ecolecon.2018.04.012

Moore, E. J., Karplus, V. J., & Morgan, M. G. (2024). Expert elicitation of the timing and uncertainty to establish a geologic sequestration well for CO2 in the United States. *Proceedings of the National Academy of Sciences, 121*(1), e2307984120. https://doi.org/10.1073/pnas.2307984120

Moreno-Gámez, S., Hochberg, M. E., & van Doorn, G. S. (2023). Quorum sensing as a mechanism to harness the wisdom of the crowds. *Nature Communications, 14*(1), 1–10. https://doi.org/10.1038/s41467-023-37950-7

Morris, S. J., & Blackwood, C. B. (2024). The ecology of soil biota and their function. In *Soil microbiology, ecology and biochemistry* (pp. 275–302). Elsevier. https://doi.org/10.1016/B978-0-12-822941-5.00010-7

Nayak, N., Mehrotra, R., & Mehrotra, S. (2022). Carbon biosequestration strategies: A review. *Carbon Capture Science and Technology, 4*(July), 100065. https://doi.org/10.1016/j.ccst.2022.100065

Nevita, T., Sharma, G. D., & Pandey, P. (2018). Differences in rice rhizosphere bacterial community structure by application of lignocellulolytic plant-probiotic bacteria with rapid composting traits. *Ecological Engineering, 120*, 209–221. https://doi.org/10.1016/j.ecoleng.2018.06.007

Olivier, J. G., Schure, K. M., & Peters, J. A. H. W. (2017). Trends in global CO2 and total greenhouse gas emissions. *PBL Netherlands Environmental Assessment Agency, 5*, 1–11.

Oschlies, A., Brandt, P., Stramma, L., & Schmidtko, S. (2018). Drivers and mechanisms of ocean deoxygenation. *Nature Geoscience, 11*(7), 467–473. https://doi.org/10.1038/s41561-018-0152-2

Østergaard, P. A., Duic, N., Noorollahi, Y., Mikulcic, H., & Kalogirou, S. (2020). Sustainable development using renewable energy technology. *Renewable Energy, 146*, 2430–2437. https://doi.org/10.1016/j.renene.2019.08.094

Pandey, P., & Dheeman, S. (2022). Conclusions: The rhizobial eminence for biological nitrogen fixation—Revisited and refined. In D. K. Maheshwari, R. Dobhal, & S. Dheeman (Eds.), *Nitrogen fixing bacteria: Sustainable growth of non-legumes. Microorganisms for sustainability* (Vol. 36). Springer. https://doi.org/10.1007/978-981-19-4906-7_17

Pandey, P., Bisht, S., Sood, A., Aeron, A., Sharma, G. D., & Maheshwari, D. K. (2012). Consortium of plant-growth-promoting bacteria: Future perspective in agriculture. In D. Maheshwari (Ed.), *Bacteria in agrobiology: Plant probiotics*. Springer. https://doi.org/10.1007/978-3-642-27515-9_10

Philippot, L., Chenu, C., Kappler, A., Rillig, M. C., & Fierer, N. (2024). The interplay between microbial communities and soil properties. *Nature Reviews Microbiology, 22*(4), 226–239. https://doi.org/10.1038/s41579-023-00980-5

Pires, J. C. M. (2019). Negative emissions technologies: A complementary solution for climate change mitigation. *Science of the Total Environment, 672*, 502–514. https://doi.org/10.1016/j.scitotenv.2019.04.004

Prakash, S. (2021). Impact of climate change on aquatic ecosystem and its biodiversity: An overview. *International Journal of Biological Innovations, 3*(2). https://doi.org/10.46505/IJBI.2021.3210

Qin, S., Xiao, W., Zhou, C., Pu, Q., Deng, X., Lan, L., Liang, H., Song, X., & Wu, M. (2022). Pseudomonas aeruginosa: Pathogenesis, virulence factors, antibiotic resistance, interaction with host, technology advances and emerging therapeutics. *Signal Transduction and Targeted Therapy, 7*(1), 1–27. https://doi.org/10.1038/s41392-022-01056-1

Rajkumari, J., & Pandey, P. (2019). Genomic insights and comparative genomics of Bacillus species having diverse mechanisms of biocontrol against fungal phytopathogens. In M. Islam, M. Rahman, P. Pandey, M. Boehme, & G. Haesaert (Eds.), *Bacilli and agrobiotechnology: Phytostimulation and biocontrol. Bacilli in climate resilient agriculture and bioprospecting*. Springer. https://doi.org/10.1007/978-3-030-15175-1_12

Rajkumari, J., Paikhomba Singha, L., & Pandey, P. (2018). Genomic insights of aromatic hydrocarbon degrading Klebsiella pneumoniae AWD5 with plant growth promoting attributes: A paradigm of soil isolate with elements of biodegradation. *3 Biotech, 8*(2), 1–22. https://doi.org/10.1007/s13205-018-1134-1

Ruley, J. A., Amoding, A., Tumuhairwe, J. B., & Basamba, T. A. (2022). Rhizoremediation of petroleum hydrocarbon–contaminated soils: A systematic review of mutualism between phytoremediation species and soil living microorganisms. *Phytoremediation*, 263–296. https://doi.org/10.1016/B978-0-323-89874-4.00008-X

Saccá, M. L., Barra Caracciolo, A., Di Lenola, M., & Grenni, P. (2017). Ecosystem services provided by soil microorganisms. In *Soil biological communities and ecosystem resilience* (pp. 9–24). Springer International Publishing. https://doi.org/10.1007/978-3-319-63336-7_2

Santoyo, G., et al. (2019). Plant growth promotion by ACC deaminase-producing Bacilli under salt stress conditions. In M. Islam, M. Rahman, P. Pandey, M. Boehme, & G. Haesaert (Eds.), *Bacilli and agrobiotechnology: Phytostimulation and biocontrol. Bacilli in climate resilient agriculture and bioprospecting*. Springer. https://doi.org/10.1007/978-3-030-15175-1_5

Sha, Z., Bai, Y., Li, R., Lan, H., Zhang, X., Li, J., Liu, X., Chang, S., & Xie, Y. (2022). The global carbon sink potential of terrestrial vegetation can be increased substantially by optimal land management. *Communications Earth and Environment, 3*(1), 1–10. https://doi.org/10.1038/s43247-021-00333-1

Shivakumar, A., Dobbins, A., Fahl, U., & Singh, A. (2019). Drivers of renewable energy deployment in the EU: An analysis of past trends and projections. *Energy Strategy Reviews, 26*, 100402. https://doi.org/10.1016/j.esr.2019.100402

Siedt, M., Schäffer, A., Smith, K. E., Nabel, M., Roß-Nickoll, M., & Van Dongen, J. T. (2021). Comparing straw, compost, and biochar regarding their suitability as agricultural soil amendments to affect soil structure, nutrient leaching, microbial communities, and the fate of pesticides. *Science of the Total Environment, 751*, 141607. https://doi.org/10.1016/j.scitotenv.2020.141607

Singh, S., Kiran, B. R., & Mohan, S. V. (2024). Carbon farming: A circular framework to augment CO2 sinks and to combat climate change. *Environmental Science: Advances, 3*(4), 522–542. https://doi.org/10.1039/D3VA00296A

Singha, L. P., & Pandey, P. (2020). Rhizobacterial community of Jatropha curcas associated with pyrene biodegradation by consortium of PAH-degrading bacteria. *Applied Soil Ecology, 155*, 103685. https://doi.org/10.1016/j.apsoil.2020.103685

Singha, L. P., & Pandey, P. (2021). Rhizosphere assisted bioengineering approaches for the mitigation of petroleum hydrocarbons contamination in soil. *Critical Reviews in Biotechnology, 41*(5), 749–766. https://doi.org/10.1080/07388551.2021.1888066

Singha, K. M., & Pandey, P. (2023). Endophytic bacteria with host-supportive genetic determinants in their genomes induce growth and antioxidant activity related gene functions in transcriptome of black rice (Oryza sativa L.). *Environmental and Experimental Botany, 213*, 105396. https://doi.org/10.1016/j.envexpbot.2023.105396

Singha, B., Mazumder, P. B., & Pandey, P. (2018a). Characterization of plant growth promoting rhizobia from root nodule of two legume species cultivated in Assam, India. *Proceedings of the National Academy of Sciences India Section B – Biological Sciences, 88*(3), 1007–1016. https://doi.org/10.1007/s40011-016-0836-6

Singha, L. P., Sinha, N., & Pandey, P. (2018b). Rhizoremediation prospects of Polyaromatic hydrocarbon degrading rhizobacteria, that facilitate glutathione and glutathione-S-transferase mediated stress response, and enhance growth of rice plants in pyrene contaminated soil. *Ecotoxicology and Environmental Safety, 164*(August), 579–588. https://doi.org/10.1016/j.ecoenv.2018.08.069

Srivastava, N., Srivastava, M., Manikanta, A., Singh, P., Ramteke, P. W., & Mishra, P. K. (2017). Nanomaterials for biofuel production using lignocellulosic waste. *Environmental Chemistry Letters, 15*, 179–184. https://doi.org/10.1007/s10311-017-0622-6

Totsche, K. U., Amelung, W., Gerzabek, M. H., Guggenberger, G., Klumpp, E., Knief, C., Lehndorff, E., Mikutta, R., Peth, S., Prechtel, A., Ray, N., & Kögel-Knabner, I. (2018). Microaggregates in soils. *Journal of Plant Nutrition and Soil Science, 181*(1), 104–136. https://doi.org/10.1002/jpln.201600451

Toundou, O., Pallier, V., Feuillade-Cathalifaud, G., & Tozo, K. (2021). Impact of agronomic and organic characteristics of waste composts from Togo on Zea mays L. nutrients contents under water stress. *Journal of Environmental Management, 285*, 112158. https://doi.org/10.1016/j.jenvman.2021.112158

Tripathi, A., Pandey, V. K., Jain, D., Singh, G., Brar, N. S., Taufeeq, A., Pandey, I., Dash, K. K., Samrot, A. V., & Rustagi, S. (2024). An updated review on significance of PGPR-induced plant signalling and stress management in advancing sustainable agriculture. *Journal of Agriculture and Food Research, 16*(April), 101169. https://doi.org/10.1016/j.jafr.2024.101169

Trisos, C. H., Merow, C., & Pigot, A. L. (2020). The projected timing of abrupt ecological disruption from climate change. *Nature, 580*(7804), 496–501. https://doi.org/10.1038/s41586-020-2189-9

Trivedi, P., Wallenstein, M. D., Delgado-Baquerizo, M., & Singh, B. K. (2018). Microbial modulators and mechanisms of soil carbon storage. In *Soil carbon storage* (pp. 73–115). Academic Press. https://doi.org/10.1016/B978-0-12-812766-7.00003-2

Wang, Y., Rattray, J. B., Thomas, S. A., Gurney, J., & Brown, S. P. (2020). In silico bacteria evolve robust cooperation via complex quorum-sensing strategies. *Scientific Reports, 10*(1), 1–10. https://doi.org/10.1038/s41598-020-65076-z

Wang, W., Hu, K., Huang, K., & Tao, J. (2021). Mechanical fragmentation of leaf litter by fine root growth contributes greatly to the early decomposition of leaf litter. *Global Ecology and Conservation, 26*, e01456. https://doi.org/10.1016/j.gecco.2021.e01456

Wu, L., & Luo, Y. (2021). Bacterial Quorum-Sensing Systems And Their Role In Intestinal Bacteria-Host Crosstalk. *Frontiers in Microbiology, 12*(January), 1–11. https://doi.org/10.3389/fmicb.2021.611413

Yue, X. L., & Gao, Q. X. (2018). Contributions of natural systems and human activity to greenhouse gas emissions. *Advances in Climate Change Research, 9*(4), 243–252. https://doi.org/10.1016/j.accre.2018.12.003

Zhao, K., Liu, L., Chen, X., Huang, T., Du, L., Lin, J., Yuan, Y., Zhou, Y., Yue, B., Wei, K., & Chu, Y. (2019). Behavioral heterogeneity in quorum sensing can stabilize social cooperation in microbial populations. *BMC Biology, 17*(1), 1–15. https://doi.org/10.1186/s12915-019-0639-3

Zhao, H., Yuan, M., Strokal, M., Wu, H. C., Liu, X., Murk, A., Kroeze, C., & Osinga, R. (2021). Impacts of nitrogen pollution on corals in the context of global climate change and potential strategies to conserve coral reefs. *Science of the Total Environment, 774*, 145017. https://doi.org/10.1016/j.scitotenv.2021.145017

Zhou, L., Zhang, Y., Ge, Y., Zhu, X., & Pan, J. (2020). Regulatory mechanisms and promising applications of quorum sensing-inhibiting agents in control of bacterial biofilm formation. *Frontiers in Microbiology, 11*(October), 1–11. https://doi.org/10.3389/fmicb.2020.589640

Zuma, M., Arthur, G., Coopoosamy, R., & Naidoo, K. (2023). Incorporating cropping systems with eco-friendly strategies and solutions to mitigate the effects of climate change on crop production. *Journal of Agriculture and Food Research, 14*(July), 100722. https://doi.org/10.1016/j.jafr.2023.100722

The manufacturer's authorised representative in the EU is Springer Nature Customer Service Centre GmbH, Europaplatz 3, 69115 Heidelberg, Germany. If you have any concerns regarding our products, please contact ProductSafety@springernature.com

Printed and bound by CPI Group (UK) Ltd, Croydon, CR0 4YY
01/10/2026
02225929-0015